Satellite
Meteorology
AN INTRODUCTION

This book is dedicated to the memory of Verner E. Suomi.

Stanley Q. Kidder
The University of Alabama in Huntsville

Thomas H. Vonder Haar
Colorado State University

Academic Press

San Diego New York Boston London Sydney Tokyo Toronto

This book is printed on acid-free paper. ♾

Academic Press
A Harcourt Science and Technology Company
525 B Street, Suite 1900, San Diego, California 92101-4495, USA
http://www.academicpress.com

Academic Press
Harcourt Place, 32 Jamestown Road, London NW1 7BY, UK
http://www.academicpress.com

Library of Congress Cataloging-in-Publication Data

Kidder, Stanley Q.
Satellite meteorology : an introduction / by Stanley Q. Kidder,
Thomas H. Vonder Haar.
p. cm.
Includes index.
ISBN 0-12-406430-2
1. Satellite meteorology. I. Vonder Haar, Thomas H. II. Title.
QC879.5.K53 1995
551.6'354—dc20 94-42859
CIP

PRINTED IN THE UNITED STATES OF AMERICA
01 02 03 04 05 06 MM 10 9 8 7 6 5 4

Contents

3 Radiative Transfer

4 Meteorological Satellite Instrumentation

5 Image Interpretation

6 Temperature and Trace Gases

7 Winds

8 Clouds and Aerosols

9 Precipitation

10 Earth Radiation Budget

11 The Future

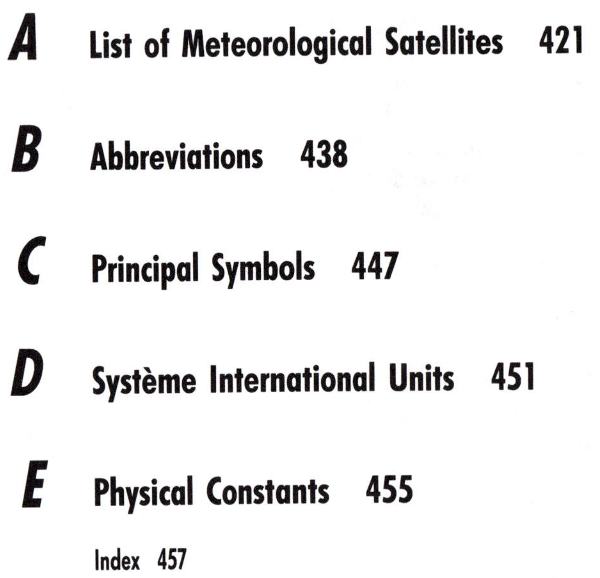

APPENDIX

Preface

During more than three decades since the launch of Vanguard 2, Explorer 7, and TIROS 1, no textbook has appeared to guide students in the use of what has become an indispensable tool for meteorology: the weather satellite. This book is an attempt to remedy this problem at the introductory level. The book is intended for upper level undergraduates or beginning graduate students who have a background in calculus and physics and have been introduced to atmospheric science (meteorology). It can be useful to students in any field which deals with the atmosphere, in particular with atmospheric measurements.

This is not an "advanced" book. We have included the basics—what we think every atmospheric scientist should know about satellite meteorology. Measurement of basic meteorological quantities (temperature, wind, etc.) is stressed over ways to analyze or forecast various features. Our philosophy is that with a thorough understanding of how the basic quantities are measured and how accurate they are, the reader can proceed to use the data in many types of investigations and will be able to understand more advanced techniques. Unfortunately this means that many useful and interesting applications of weather satellite data have not been included in this book. Some of these advanced applications can be found in the bibliography at the end of each chapter.

We place special emphasis on the physical understanding of measurements from space because it is this understanding which will allow both the useful application of current techniques and the development of future techniques. We also tend to emphasize operational techniques over experimental techniques. We do this in the belief that most readers will not do their own information or parameter retrievals from raw satellite data; they will use parameters retrieved by others. Most often these parameters will be retrieved using operational, near real-time methods. A thorough exploration of the operational techniques is therefore important.

Extensive lists of original references are included at the ends of chapters. There is no substitute for reading the technical literature, not even this book. One of our major goals is to enable students to read and understand the literature: articles on which this book is based, articles which we did not cover, and future articles which will be written in this rapidly evolving area of atmospheric science.

We would like to acknowledge the research support of NOAA and Colorado State University, especially the staff at the Cooperative Institute for Research in the Atmosphere, throughout the period involved in writing this book. In addition, other research sponsors have provided support necessary to explore the field and to complete this book. Many people contributed generously to this effort. We spoke to a substantial fraction of the authors of papers cited and were constantly amazed and pleased at the time and effort which they were willing to donate. Many errors and misconceptions were detected and eliminated by these colleagues, for which we are grateful. In particular, we would like to acknowledge the assistance of the men and women of the National Aeronautics and Space Administration (NASA), and of the National Oceanic and Atmospheric Administration (NOAA), without whose help this book would have been impossible. For many of these scientists and engineers, the daily operation and use of weather satellites is their profession. We gratefully acknowledge the assistance of Judy Sorbie Dunn, who expertly drafted the figures; Fredi Boston, our librarian, who found all those references for us; and Loretta Wilson, our special assistant. We extend special thanks to our colleague, Donald L. Reinke, for his substantial contributions to the layout and content of Chapter 5, Image Interpretation. We thank Dr. William W. Vaughan who carefully read and made valuable comments on the draft.

Finally, we thank and dedicate this book to our wives, Bobbie Poole and Dee Vonder Haar—and to our children of all ages—whose support and encouragement was matched only by their patience throughout the extended period of authorship.

Stanley Q. Kidder
Thomas H. Vonder Haar

1

Introduction

IN THE MORE than 30 years since the first meteorological satellites were launched, they have become indispensable for study of the Earth's atmosphere. Indeed, together with their land- and ocean-sensing cousins, meteorological satellites view the Earth from a global perspective which is unmatched and unmatchable by any other observing system. In this book we explore what has become a very broad field: satellite meteorology. As we explore, we attempt to reveal the excitement that new observing capabilities bring to science.

To begin, we present an overview of the history and important milestones of satellite meteorology. This, we hope, will give readers the background necessary to understand today's satellites and techniques. In the second section of this chapter, we offer a preview of the remainder of the book to give readers some perspective on where we are taking them.

1.1 HISTORY OF SATELLITE METEOROLOGY

Satellite meteorology predates the launch of the first meteorological satellite (*metsat*). As early as 1860, Jules Verne wrote about "Lunanauts" observing cloud systems (Vaughan, 1982). By the late 1940s, rockets carrying cameras were being launched into suborbital flights. The photographs that they returned gave rise by the early 1950s to serious scientific discussion of the possibility of observing the weather from space (e.g., Wexler, 1954). Several groups, notably the U.S. Army Evans Signal Laboratory and the University of Wisconsin, pursued the idea of launching a weather satellite. These efforts were intensified after the launch by the Soviet Union on 4 October 1957 of the first successful Earth satellite, Sputnik 1. The first successful U.S. satellite, Explorer 1, was launched 123 days later on 31 January 1958. These early days are chronicled in proceedings volumes edited by Vaughan (1982) and Vonder Haar *et al.* (1982).

Of fundamental importance to space flight, in general, and satellite meteorology, in particular, was the formation of the National Aeronautics and Space Administration (NASA) on 1 October 1958. For more than 30 years, NASA has lead the development of all types of scientific satellites used for civilian purposes. Appendix A of this book attempts to list all satellites that have made atmospheric measurements. A large fraction were developed by NASA. Involved from the first in satellite meteorology were agencies that now are components of the U.S. National Oceanic and Atmospheric Administration (NOAA), particularly the U.S. Weather Bureau (now the National Weather Service). Today operational U.S. meteorological satellites are controlled by NOAA and the U.S. Air Force.

The first satellite with a meteorological instrument was Vanguard 2, launched 17 February 1959. Developed by the U.S. Army's Evans Signal Laboratory, Vanguard 2 had a pair of photocells behind lenses that, much like today's scanning radiometers, were supposed to sweep out a visible Earth image as the satellite orbited and spun. Unfortunately, the satellite wobbled on its axis, causing the scan lines to crisscross, which rendered the data unusable.

Explorer 6, launched 7 August 1959, was the second satellite with meteorological instruments. It carried an imaging system and a Suomi radiometer (see below). It went into a highly elliptical orbit, however, and was essentially unusable, although it did return the first Earth photo.

The first successful meteorological instrument on an orbiting satellite was the Suomi radiometer, which flew on Explorer 7, launched 13 October 1959. Developed by Verner Suomi and colleagues at the University of Wisconsin, it consisted of hemispheres, painted either black or white, backed by aluminum mirrors, and mounted on the equator of a spinning satellite. The mirrors reflected the scene back to the hemispheres, such that the hemispheres acted like spheres isolated in space. Since the satellite spun, the spheres sampled solar radiation and terrestrial radiation independent of the orientation of the satellite's spin axis. The temperature of each hemisphere was monitored, and its time rate of change was related to the net gain or loss of radiative energy at the sensor. The black hemispheres absorbed all radiation; the white hemisphere reflected solar radiation but absorbed infrared radiation. The difference between the radiation balance of the hemispheres indicated solar radiation. With these data, coarse maps of the solar radiation reflected by the Earth and the infrared radiation emitted by the Earth were made for the first time.

The first satellite completely dedicated to satellite meteorology was launched on 1 April 1960. TIROS 1 (Television and Infrared Observational Satellite), the 22nd successfully launched satellite, was hatbox-shaped, about 57 cm in height and 107 cm in diameter. Its mass was 120 kg. Figure 1.1 shows a sketch of TIROS 1 along with Vanguard 2, Explorer 7, and Nimbus 1.

The image-making instrument on TIROS 1 was a vidicon camera, which was an adaptation of a standard television camera. Essentially, a lens focused the image on the light-sensitive face of a cathode ray tube (CRT) about 12.7 mm square. The bright and dark areas of the image resulted in a pattern of electrical charge on the CRT. An electron beam scanned the CRT face to measure the charge. The scanning was similar to normal television, it had 500 lines, each with 500 elements. Scanning an entire image took 2 s. The voltages measured by the vidicon camera were telemetered to the ground to be assembled into an image.

Figure 1.2 shows the first image returned by TIROS 1. Although crude by today's standards, TIROS 1 images generated immense excitement. For the first time we could view the Earth and its weather systems as a whole. Not only have satellite observations become essential for meteorology; we believe that they have fundamentally changed our perception of the Earth from a set of distant, isolated continents to an integrated, inseparable system of land, ocean, atmosphere, and living things. Nearly 23,000 images were returned in the 79-day lifetime of TIROS 1.

Nine additional satellites were launched in the TIROS series; the last, TIROS 10, was launched on 2 July 1965. Several technological improvements were introduced in the TIROS series. TIROS 2 introduced a scanning radiometer, the Medium Resolution Infrared Radiometer (MRIR), which was similar to today's imaging instruments (see Chapter 4). TIROS 3, 4, and 7 also carried improved versions of the Suomi radiometer.

The first four TIROS were launched into 48°-inclination orbits (see Section 2.2.4). Beginning with TIROS 5, they were launched into 58°-inclination orbits to expand coverage 10° toward the poles.

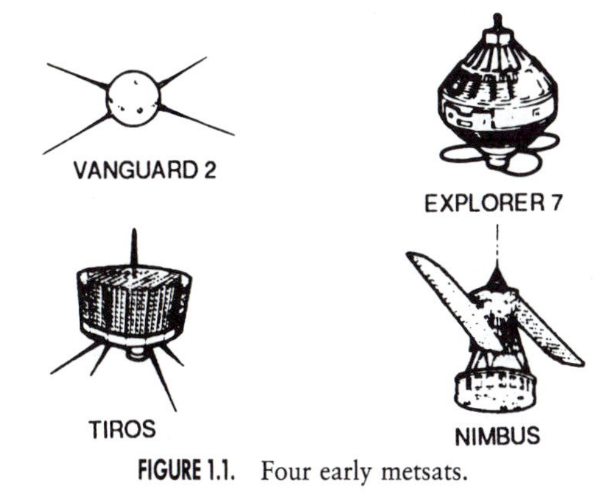

FIGURE 1.1. Four early metsats.

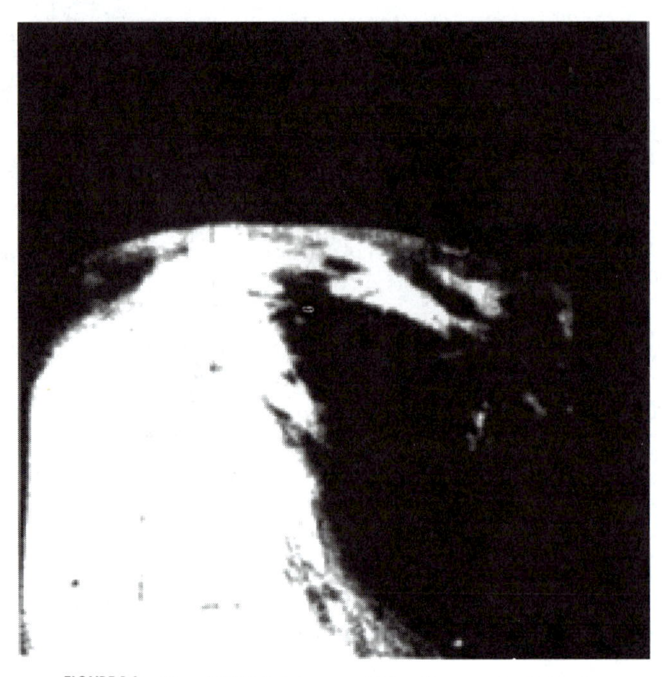

FIGURE 1.2. First TIROS 1 image. [After Rao *et al.* (1990).]

TIROS 8, launched 21 December 1963, introduced Automatic Picture Transmission (APT). A new vidicon camera with 800-line resolution was scanned at the slow rate of 4 lines per second, and the data were immediately broadcast to the Earth at VHF (very high frequency) frequencies. The slow transmission rate meant that inexpensive equipment could be used to receive and display the images. Thus anyone with the proper equipment could directly receive weather satellite images as the satellite passed by twice each day. APT is still an important function on today's polar-orbiting weather satellites.

To maintain their orientation in space, many satellites spin. In the absence of external torques, angular momentum is conserved, and the spin axis points in a constant direction in space as the satellite orbits the Earth. TIROS spun at about 12 revolutions per minute (rpm). This caused a viewing problem with the first eight TIROS, however. The vidicon cameras on these satellites were on the "bottom" of the craft; that is, they pointed parallel to the spin axis and, therefore, in a constant direction in space. The Earth was in their field of view only about 25% of the time. During the remaining 75% of each orbit, they viewed space. TIROS 9, launched 22 January 1965, introduced a new configuration, the "cartwheel" configuration. The satellite's spin axis was tilted to be perpendicular to the orbital plane, and the cameras were reoriented to point out the side of the spacecraft. Thus, the satellite "rolled" like the wheel of a cart in its orbit about the Earth. With each rotation of the satellite, the cameras would point toward Earth, and a picture could be taken. This allowed global composite images to be made; an example is shown in Fig. 1.3.

FIGURE 1.3. TIROS 9 Earth mosaic. [Courtesy of NASA and NOAA.]

In 1964 an extremely important series of experimental meteorological satellites was initiated, the Nimbus series. Nimbus 1, launched 28 August 1964 (see Fig. 1.1), had two notable firsts. It was the first *three-axis stabilized* metsat; that is, with the use of momentum wheels (flywheels inside the spacecraft) controlled by horizon sensors, it rotated once per orbit (by placing torque on the appropriate momentum wheels) so that its instruments constantly pointed toward Earth. Its APT camera was therefore much more useful than that of TIROS 8, which only viewed the Earth 25% of the time. Figure 1.4 shows a Nimbus 1 APT photo.

Nimbus 1 also was the first sunsynchronous satellite, which means that it passed over any point on Earth at approximately the same time each day (see

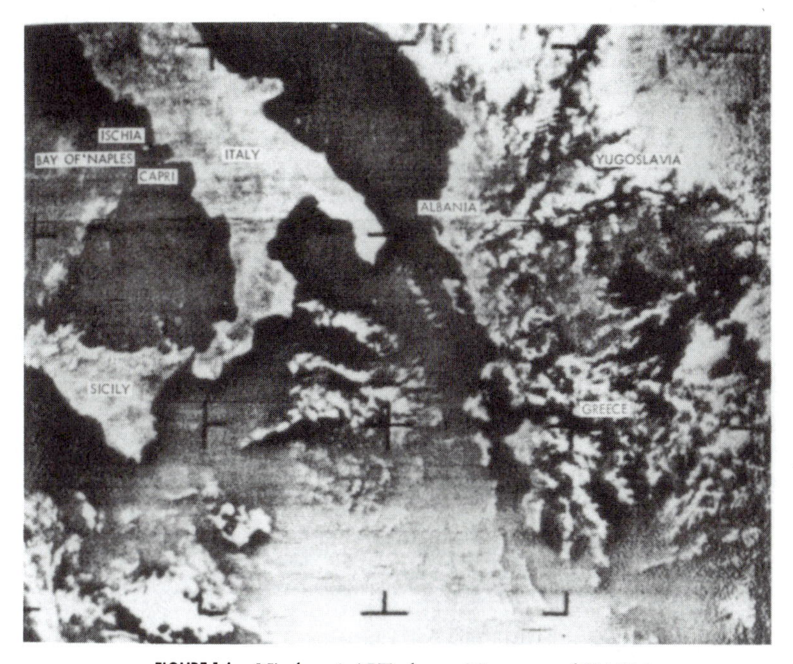

FIGURE 1.4. Nimbus 1 APT photo. [Courtesy of NASA.]

Section 2.4.1). This regularity increased its utility in operational forecasting. The sunsynchronous orbit has been used ever since for U.S. operational metsats in near-polar orbits. Nimbus 1's High Resolution Infrared Radiometer (HRIR), a scanning radiometer quite similar to those in operational use today, provided night and day coverage. Figure 1.5 shows an HRIR image of Hurricane Gladys.

An important accomplishment of satellite meteorology is that since sometime in the mid-1960s when metsat coverage became continuous, there have been no undetected tropical cyclones anywhere on Earth. These ocean-born storms, which for centuries menaced seafarers and coastal and island dwellers, can no longer surprise potential victims. Lives are still lost to tropical cyclones, but many are now saved because of the warnings that metsats make possible.

In total, seven Nimbus satellites were launched. Some experiments on the last one, Nimbus 7, launched 24 October 1978, were still operational as this book was being written! The Nimbus series tested many new concepts that have lead to the operational instruments in use today. These instruments will be discussed elsewhere in this book.

The first 5 years of satellite meteorology are also documented by Hubert and Lehr (1967), to which the reader is referred for interesting details to augment the references noted at the beginning of this section.

By 1966 the United States was ready to initiate an operational (as opposed to experimental) series of metsats. The Environmental Science Service Administration (NOAA's predecessor) commissioned nine satellites, ESSA 1 through 9, which were launched between 3 February 1966 and 26 February 1969. Each was essentially like TIROS 9; each flew in the cartwheel configuration, but in sunsyn-

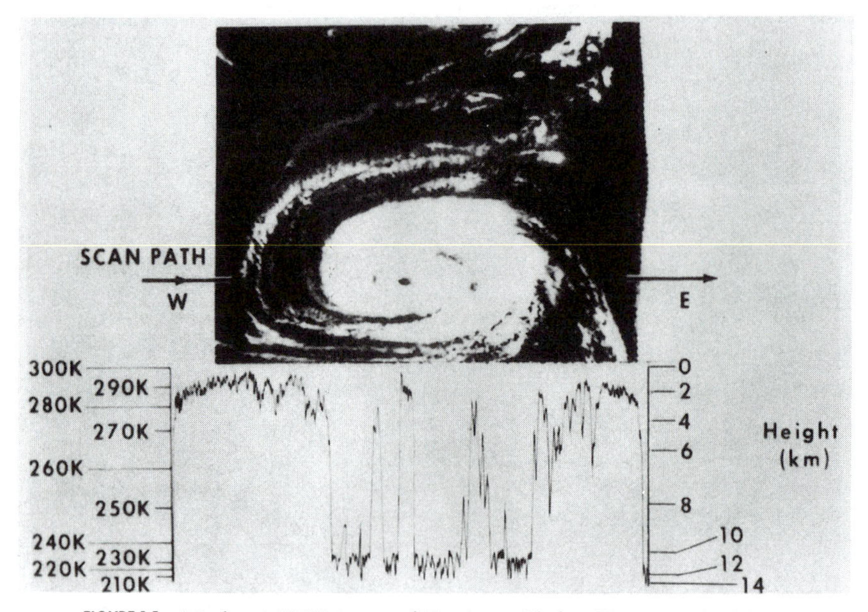

FIGURE 1.5. Nimbus 1 HRIR image of Hurricane Gladys. [Courtesy of NASA.]

chronous orbit. The odd-numbered satellites had Advanced Vidicon Camera Systems (AVCSs), which were proven on the Nimbus satellites, and which could record images for later playback to Earth receiving stations. The even-numbered satellites had APT cameras for immediate broadcast to Earth. Once again, Suomi-type radiometers measured the Earth's energy budget from some of these satellites.

On 16 September 1966 the U.S. Air Force (USAF) launched the first in a series of metsats, the Defense Meteorological Satellite Program (DMSP). This series was called DMSP Block 4. It included seven satellites; the last was launched on 23 July 1969.

On 7 December 1966 satellite meteorology took another leap forward with the launch of the first Applications Technology Satellite (ATS 1). Carrying a Spin Scan Cloud Camera, developed by Verner Suomi and Robert Parent at the University of Wisconsin, ATS 1 was placed into geostationary orbit (see Section 2.4.2) to view the Western Hemisphere in visible light. For the first time, rapid-imaging of nearly an entire hemisphere was possible. We could watch, fascinated, as storm systems developed and moved and were captured in a time series of images. Today such images are an indispensable part of weather analysis and forecasting. ATS 3, launched 5 November 1967, carried a Multicolor Spin Scan Cloud Camera, which employed a filter wheel to make the first color images of Earth (Plate 1).

Virtually all parts of the electromagnetic spectrum are useful for satellite meteorology. The microwave portion of the spectrum was first explored by the Soviet satellite Kosmos 243, launched 23 September 1968.

In 1969 the Soviets began a lengthy series of operational metsats, which, in its third generation, continues today. The first Meteor-1 was launched on 26 March 1969. The 31st member of the Meteor-1 series was launched on 10 July 1981.

Another very important event occurred on 14 April 1969 with the launch of Nimbus 3. It carried two instruments designed to provide atmospheric soundings from space. For the first time, satellite data were used quantitatively in numerical weather-prediction models. The Satellite Infrared Spectrometer (SIRS), made measurements in the 15-μm portion of the spectrum. It was the forerunner of today's operational sounding instruments. A second instrument, the Infrared Interferometer Spectrometer (IRIS), measured spectra in the infrared from about 6 to 25 μm. (See Section 4.3.1.) IRIS also flew on the Voyager spacecraft to Jupiter, Saturn, Uranus, and Neptune, and a similar instrument is under consideration for future geostationary spacecraft.

The second series of U.S. operational metsats began on 23 January 1970 with the launch of TIROS M, also called the Improved TIROS Operational System (ITOS). The NOAA 1 through 5 satellites completed the series. These satellites were three-axis stabilized and flew in sunsynchronous orbits.

An extremely long-lasting instrument, and the first ultraviolet instrument in space, was the Backscatter Ultraviolet (BUV) launched on Nimbus 4 on 8 April 1970. BUV measured ozone (see Section 6.5) for nearly 10 years.

Similar in appearance to the Nimbus satellites, the Landsat series was designed for land remote sensing. Its sensors have extremely high resolution, 80 m in the first satellite and up to 30 m in the latest satellite (Landsat 5). Landsat 1, also called the Earth Resources Technology Satellite (ERTS), was launched on 23 July 1972. Landsat data are used in meteorology primarily to study small clouds and surface features that may influence weather.

The first generation of semioperational geostationary metsats began with the launch of the Synchronous Meteorological Satellite 1 (SMS 1) on 17 May 1974. SMS 2 was launched on 6 February 1975. These satellites carried the first Data Collection Platform (DCP) repeater. Data from meteorological or other platforms on the surface (Fig. 1.6) could be relayed by the satellite to a central receiving site. Thus data from remote ground sites could be easily obtained for the first time. The cloud cameras on the ATS satellites made images in the visible portion of the spectrum only. SMS and the succeeding GOES have an infrared radiometer as well. Since 27 June 1974, when SMS 1 became operational, we have had continuous, uninterrupted, 24-hour-per-day monitoring of most of the Western Hemisphere from space.

On 11 September 1976 the DMSP Block 5D series of the USAF began. The primary Block 5D instrument is the Operational Linescan System (OLS). It is still the highest-resolution (600 m) meteorological instrument in space. OLS is interesting in another way, also. Its shortwave sensor has a broad passband, which means that it can collect enough radiation to make some images by moonlight, and it can sense city lights at night. (See Section 5.5.2.4.)

The second series of operational Soviet metsats began on 11 July 1975 with the launch of Meteor-2 1. Eighteen satellites have been launched in the series.

The first truly operational geostationary metsat, the Geostationary Operational Environmental Satellite 1 (GOES 1), was launched on 16 October 1975. GOES

FIGURE 1.6. Data Collection Platform (DCP).

2 and 3 were similar. Since the launch of SMS 2, the United States has generally maintained two geostationary satellites in orbit, one at 75° west longitude, and one at 135° west longitude.

In 1977 and 1978 two more geostationary metsats were launched: Japan's Geostationary Meteorological Satellite 1 (GMS 1) was stationed it at 140° east longitude, and the European Space Agency's Meteosat 1 was stationed at the prime meridian. Meteosat was the first geostationary satellite to make images of mid- to upper-troposphere water vapor at 6.7 μm (Fig. 1.7) in addition to visible and 10–12-μm infrared.

The third generation of U.S. polar-orbiting metsats began on 13 October 1978 with the launch of TIROS N. This series, which continues today, is discussed in Chapter 4.

India has been quite active in satellite meteorology. Two Indian polar orbiters have been launched: Bhaskara 1 on 7 June 1979 and Bhaskara 2 on 20 November 1981. On 31 August 1983, the geostationary Insat 1B was launched from the Space Shuttle. Stationed at 74° east longitude, Insat 1B completed geostationary coverage of the tropics and midlatitudes around the Earth. (Although in 1978 and 1979, GOES 1 was temporarily stationed at 55° east longitude to provide global coverage for the First GARP Global Experiment.) Insat was also the first three-axis stabilized geostationary metsat.

On 9 September 1980 GOES 4, the first in the second generation of GOES satellites, was launched. This series of satellites is discussed in detail in Chapter 4.

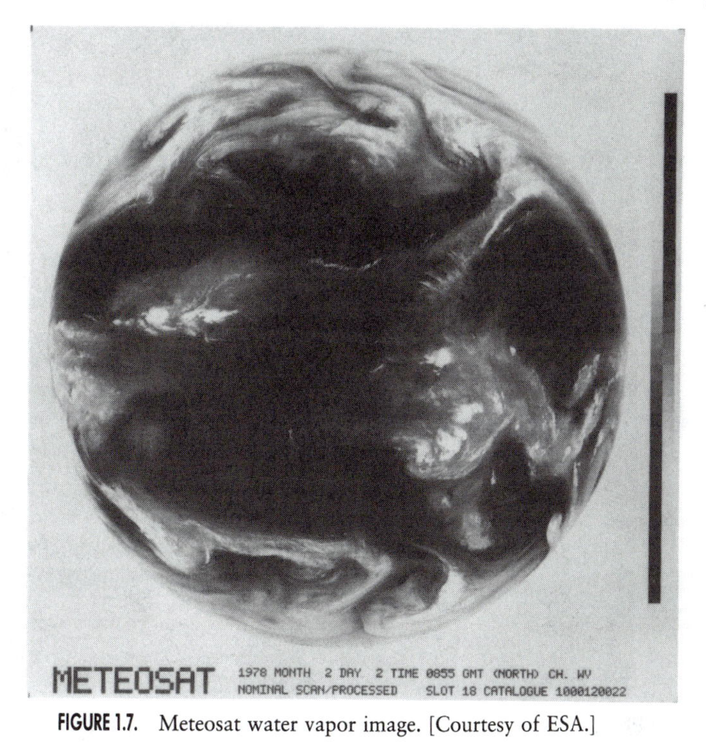

FIGURE 1.7. Meteosat water vapor image. [Courtesy of ESA.]

The first satellite dedicated to climate research was launched on 5 October 1984 from the Space Shuttle Challenger (Fig 1.8). Called the Earth Radiation Budget Satellite (ERBS), it carried two instruments: the Earth Radiation Budget Experiment (ERBE), which is discussed in Chapter 10, and the Stratospheric Aerosol and Gas Experiment II, which is discussed in Chapter 8. ERBS flies in a nonsunsynchronous orbit so that its measurements will sample all local times. It was teamed with two NOAA satellites carrying identical ERBE instruments. This sampling strategy is discussed in Section 2.6.

The final satellite series, which completes the suite of operational satellites in use as this book was being completed, began with the launch by the Soviet Union of Meteor-3 1 on 26 July 1988.

The history of satellite meteorology has many facets in addition to the hardware that has been launched into orbit. In particular, the explosion in computer and communications technology during the space age has literally made weather satellites possible. This technology has also made possible the dissemination of the satellite data and products to the operational forecasting and research sites where they are needed. Figure 1.9 shows what today seems unremarkable, a satellite image transmitted in 1980 from the Colorado State University Direct Readout Ground Station for GOES to the National Weather Service Forecast Office in Denver. Without such communications capabilities, and without the computing

FIGURE 1.8. ERBS being launched from the Space Shuttle. [Courtesy of NASA.]

FIGURE 1.9. GOES image transmitted in near-realtime to a remote site.

power necessary to turn raw counts received from the satellite into useful measurements, satellite meteorology would still be in its infancy. Although the reader will see in this book only glimpses of the communications and computer technology that underlies all of satellite meteorology, it should not be forgotten.

In the remainder of this chapter, we offer the reader a preview of the contents and philosophy of this book.

1.2 SCOPE OF THE BOOK

Within the context of related areas of study, Fig. 1.10 denotes the scope of this basic, introductory book. As shown in the figure, the technical subareas that constitute the core of satellite meteorology are (1) time and space sampling of weather and climate features, (2) algorithms and interpretation methods, (3) satellite senors, and (4) weather and climate products and applications.

Chapters 2 through 4 present basic material necessary to understand satellites and how they can be used in meteorology. Chapter 2 discusses satellite orbits. Satellites are not free to travel any path in space, they must follow those dictated by the laws of physics. Knowledge of these laws and of possible orbits is essential to understanding satellite meteorology. Chapter 3 discusses electromagnetic radia-

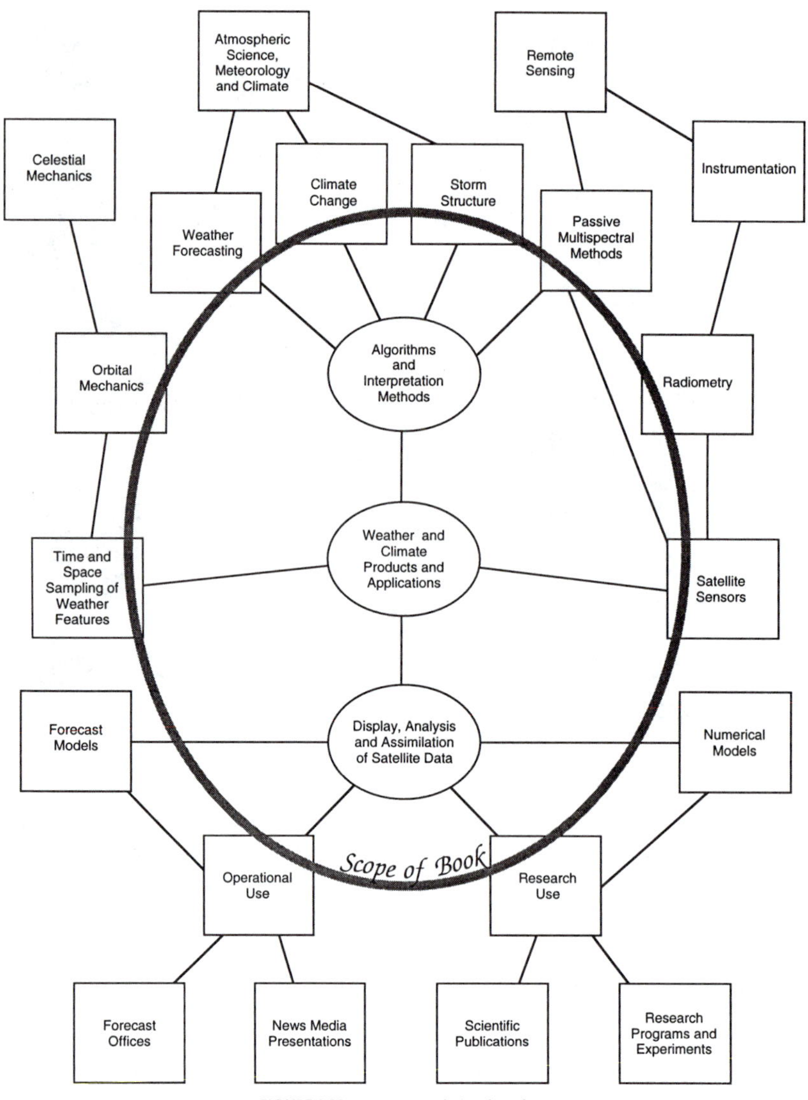

FIGURE 1.10. Scope of the book.

tion, which is the only quantity that meteorological satellites directly measure. The sources of this radiation and its interaction with the Earth's surface and atmosphere are explored. Chapter 4 discusses the instruments that make the measurements. Using instruments on current satellites as examples, the basic operation and capabilities of meteorological instruments are detailed.

Chapter 5 presents the basics of weather satellite image analysis, that is, interpreting the pictures returned by metsats. This was the first, and for many is still the only, application of meteorological satellite data. Because this subject can (and does) fill volumes, we present only what we think everyone should know about satellite images. References at the end of the chapter direct the reader to more detailed discussions.

Chapters 6 through 10 discuss the fundamental parameters that can be retrieved from meteorological satellite data. We concentrate on three issues: what meteorological parameters can be retrieved, how the parameters are retrieved from the electromagnetic measurements, and how accurate the retrieved parameters are. The parameters covered in Chapters 6 through 10, respectively, are temperature and trace-gas concentration, winds, clouds and aerosols, precipitation, and radiation budget. Many specialized applications of satellite data are not covered in this book. We have confined ourselves to the basic parameters in the interest of brevity, but also in the belief that the fundamental parameters are better able to be combined with other measurements such as rawinsonde data or radar data to further our knowledge of the atmosphere.

Finally, in Chapter 11 we look into the future as best we can to indicate what the next decade may bring to satellite meteorology. The reader is warned that the technological, political, and economic rates of change are such that making forecasts of the future of satellite meteorology is a difficult task, yet pleasant. In planning one's career, or indeed one's next project, however, it is necessary have some idea of what the future holds.

Now, dear reader, welcome to the world of satellite meteorology. We hope that you enjoy the following tour.

Bibliography

Hubert, L. F., and P. E. Lehr (1967). *Weather Satellites.* Blaisdell Publishing Co., Waltham, MA., 120 pp.

Rao, P. K., S. J. Holmes, R. K. Anderson, J. S. Winston, and P. E. Lehr (eds.) (1990). *Weather Satellites: Systems, Data, and Environmental Applications.* American Meteorological Society, Boston.

Vaughan, W. W. (ed.) (1982). *Meteorological Satellites—Past, Present, and Future.* NASA Conference Publication 2227, Washington, DC, 60 pp.

Vonder Haar, T. H., W. W. Vaughan, M. H. Davis, and M. A. Cook (eds.) (1982). *The Conception, Growth, Accomplishments, and Future of Meteorological Satellites.* NASA Conference Publication 2257, Washington, DC, 101 pp.

Wexler, H. (1954). Observing the weather from a satellite vehicle. *J. Br. Interplanetary Sci.*, 13, 269–276.

2

Orbits and Navigation

TO FULLY UNDERSTAND and use satellite data it is necessary to understand the orbits in which satellites are constrained to move and the geometry with which they view the Earth. This chapter begins with a review of basic physical principles which reveal the shape of a satellite orbit and how to orient the orbital plane in space. This knowledge allows us to calculate the position of a satellite at any time. Orbit perturbations and their effects on meteorological satellite orbits are then discussed. Next the geometry of satellite tracking and Earth location of the measurements made from the satellites are explored. This leads to a discussion of space–time sampling. The chapter concludes with a brief overview of satellite launch vehicles and orbit insertion options.

2.1 NEWTON'S LAWS

Isaac Newton[1] discovered the basic principles that govern the motions of satellites and other heavenly bodies.

[1] English physicist and mathematician, 1642–1727.

Newton's Laws of Motion

1. Every body will continue in its state of rest or of uniform motion in a straight line except insofar as it is compelled to change that state by an impressed force.
2. The rate of change of momentum is proportional to the impressed force and takes place in the line in which the force acts.
3. Action and reaction are equal and opposite.

Since momentum is the product of the mass of a body and its velocity, Newton's Second Law is the familiar

$$F = ma = m\frac{dv}{dt}, \tag{2.1}$$

where F is force, m is mass, a is acceleration, v is velocity, and t is time. In addition, Newton gave us the functional form of the force that determines satellite motion:

Newton's Law of Universal Gravitation

The force of attraction between two point masses m_1 and m_2 separated by a distance r is

$$F = \frac{Gm_1m_2}{r^2}, \tag{2.2}$$

where G is the Newtonian (or universal) gravitation constant (see Appendix E).

Consider the simple circular orbit shown in Fig. 2.1. Assuming that the Earth is a sphere, we can treat it as a point mass. The centripital force required to keep the satellite in a circular orbit is mv^2/r, where v is the orbital velocity of the satellite. The force of gravity that supplies this centripital force is Gm_em/r^2, where m_e is the mass of the Earth (Appendix E) and m is the mass of the satellite. Equating the two forces gives

$$\frac{mv^2}{r} = \frac{Gm_em}{r^2}. \tag{2.3}$$

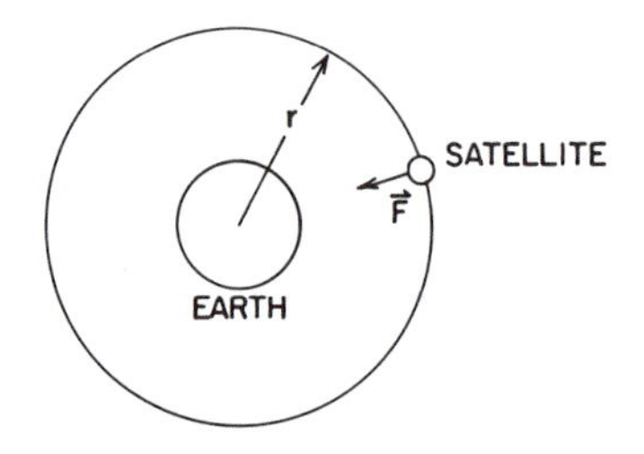

FIGURE 2.1. A circular satellite orbit.

Division by m eliminates the mass of the satellite from the equation, which means that the orbit of a satellite is independent of its mass. The *period* of the satellite is the orbit circumference divided by the velocity: $T = 2\pi r/v$. Substituting in Eq. 2.3 gives

$$T^2 = \frac{4\pi^2}{Gm_e} r^3. \tag{2.4}$$

The current NOAA satellites orbit at approximately 850 km above the Earth's surface.[2] Since the equatorial radius of the Earth is about 6378 km, the orbit radius is about 7228 km. Substituting in Eq. 2.4 shows that the NOAA satellites have a period of about 102 min.

As a second example, we calculate the radius required for a satellite in *geosynchronous* orbit, that is an orbit in which the satellite has the same angular velocity as the Earth. The angular velocity of a satellite is

$$\xi = \frac{2\pi}{T}. \tag{2.5}$$

Substituting Eq. 2.5 in Eq. 2.4 gives

$$r^3 = \frac{Gm_e}{\xi^2}. \tag{2.6}$$

Inserting the angular velocity of the Earth (Appendix E), the required radius for a geosynchronous orbit is 42,164 km, or about 35,786 km above the Earth's surface.

2.2 KEPLERIAN ORBITS

Satellites, however, do not travel in perfect circles, although a circular orbit is the goal for most meteorological satellites. It is possible to derive the exact form of a satellite's orbit from Newton's laws of motion and the law of universal gravitation.[3] The results of this derivation are neatly summarized in Kepler's laws and in Kepler's equation.

2.2.1 Kepler's Laws

Johannes Kepler[4] died 12 years before Newton was born and, therefore, did not have the advantage of Newton's work. Kepler formulated his laws by analyzing a mass of data on the position of the planets. This task was complicated by the rotation of the Earth and the motion of the Earth about the sun, which make

[2] Specifications call for them to orbit at either 833 or 870 km; 850 km is a representative value.

[3] The reader is referred to Escobal (1965) and Goldstein (1950) for two quite different, but equally lengthy, derivations.

[4] German astronomer, 1571–1630.

planetary motions seem very complex. In modern form, Kepler's laws may be stated as follows:

Kepler's Laws

1. All planets travel in elliptical paths with the sun at one focus.
2. The radius vector from the sun to a planet sweeps out equal areas in equal times.
3. The ratio of the square of the period of revolution of a planet to the cube of its semimajor axis is the same for all planets revolving around the sun.

The same laws apply if we substitute *satellite* for *planet* and *Earth* for *sun*. Equation 2.4 is a statement of Kepler's third law for the special case of a circular orbit.

2.2.2 Ellipse Geometry

The parameters that are used to specify satellite orbits are based in part on geometric terminology. Figure 2.2 illustrates the geometry of an elliptical orbit. The point where the satellite most closely approaches the Earth is the *perigee*, or more generally, the *perifocus*. The point where the satellite is furthest from the Earth is called the *apogee* or *apofocus*. The distance from the center of the ellipse to the perigee (or apogee) is the *semimajor axis* and will be denoted by the symbol a. The distance from the center of the ellipse to one focus (to the center of the Earth) divided by the semimajor axis is the *eccentricity* and will be denoted by the symbol ε. For an ellipse, the eccentricity is a number between zero and one ($0 \leq \varepsilon < 1$). A circle is an ellipse with zero eccentricity. The equation for the ellipse, that is, the path that the satellite follows, is given in polar coordinates

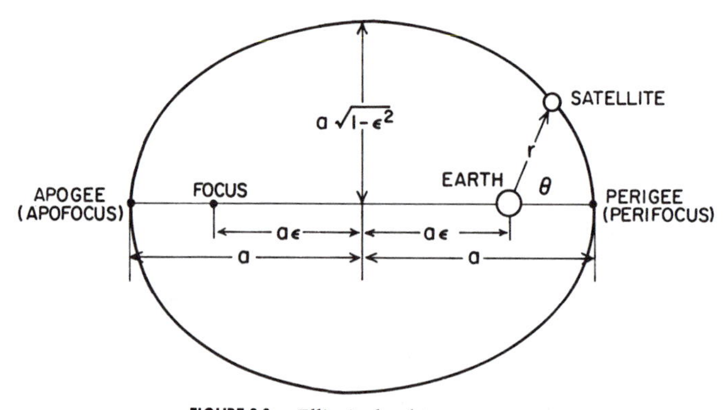

FIGURE 2.2. Elliptical orbit geometry.

with the Earth as origin as

$$r = \frac{a(1 - \varepsilon^2)}{1 + \varepsilon \cos\theta}. \tag{2.7}$$

The angle θ is the *true anomaly* and is always measured counterclockwise (the direction of satellite motion) from the perigee.

2.2.3 Kepler's Equation

A satellite in a circular orbit undergoes uniform angular velocity. By Kepler's Second Law, however, a satellite in an elliptical orbit cannot have uniform angular velocity; it must travel faster when it is closer to Earth. The position of the satellite as a function of time can be found by applying Kepler's equation:

$$M = n(t - t_\mathrm{p}) = e - \varepsilon \sin e, \tag{2.8}$$

where M is the *mean anomaly*; M increases linearly in time at the rate n, called the *mean motion constant*, given by

$$n = \frac{2\pi}{T} = \sqrt{\frac{Gm_\mathrm{e}}{a^3}}. \tag{2.9}$$

By definition M is zero when the satellite is at perigee; therefore, t_p is the time of perigeal passage. The angle e is the *eccentric anomaly*. It is geometrically related to the *true anomaly* (Fig. 2.3):

$$\cos\theta = \frac{\cos e - \varepsilon}{1 - \varepsilon \cos e}, \tag{2.10a}$$

$$\cos e = \frac{\cos\theta + \varepsilon}{1 + \varepsilon \cos\theta}. \tag{2.10b}$$

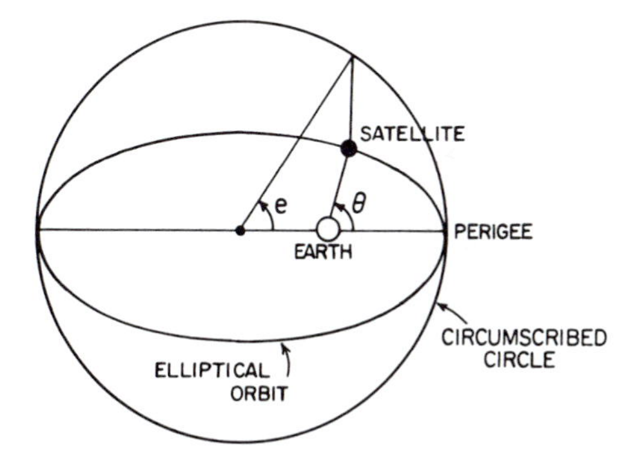

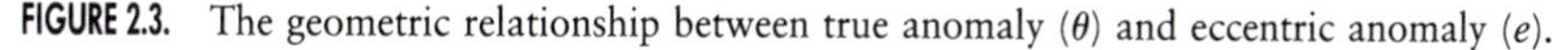
FIGURE 2.3. The geometric relationship between true anomaly (θ) and eccentric anomaly (e).

2.2.4 Orientation in Space

By calculating r and θ at time t, we have positioned the satellite in the plane of its orbit; now we must position the orbital plane in space. To do so requires the definition of a coordinate system. This coordinate system must be an *inertial coordinate system*, that is, a nonaccelerating system in which Newton's Laws of Motion are valid. A coordinate system fixed to the rotating Earth is not such a system. We will adopt an astronomical coordinate system called the *right ascension–declination coordinate system.*[5] In this system (Fig. 2.4) the z axis is aligned with the Earth's spin axis. The x axis is chosen such that it points from the center of the Earth to the sun at the moment of the *vernal equinox*, when the sun is crossing the equatorial plane from the Southern Hemisphere to the Northern Hemisphere.[6] The y axis is chosen to make it a right-handed coordinate system. In this system, the *declination* of a point in space is its angular displacement measured northward from the equatorial plane, and the *right ascension* is the angular displacement, measured counterclockwise from the x axis, of the projection of the point in the equatorial plane (Fig. 2.5).

Three angles are used to position an elliptical orbit in the right ascension–declination coordinate system (Fig. 2.6): the inclination angle, the right ascension of ascending node, and the argument of perigee.

The *inclination angle* (i) is the angle between the equatorial plane and the orbital plane. By convention, the inclination angle is zero if the orbital plane coincides with the equatorial plane *and* if the satellite rotates in the same direction as the Earth. If the two planes coincide but the satellite rotates opposite to the Earth, the inclination angle is 180°. *Prograde* orbits are those with inclination angles less than 90°; *retrograde* orbits are those with i greater than 90°.

The *ascending node* is the point where the satellite crosses the equatorial plane going north (*ascends*). The right ascension of this point is the *right ascension of ascending node* (Ω). It is measured in the equatorial plane from the x axis (vernal equinox) to the ascending node. In practice, the right ascension of ascending node has a more general meaning. It is the right ascension of the intersection of the orbital plane with the equatorial plane; thus it is always defined, not just when the satellite is at an ascending node.

Finally, the *argument of perigee* (ω) is the angle measured in the orbital plane between the ascending node (equatorial plane) and the perigee.

[5] Because the origin of this coordinate system moves about the sun with the Earth, it is not truly inertial. However, the sun's gravity causes the satellite to rotate around the sun as does the Earth. Therefore, the satellite acts as if the right ascension–declination coordinate system were inertial.

[6] This x axis is also referred to as the *First Point of Aries* because it used to point at the constellation Aries. Because of the influence of the sun and moon on the nonspherical Earth, the Earth's spin axis precesses like a top with a period of 25,781 years. This causes the vernal equinox to change. Today, the x axis points to the constellation Pisces, but it is still referred to as the First Point of Aries.

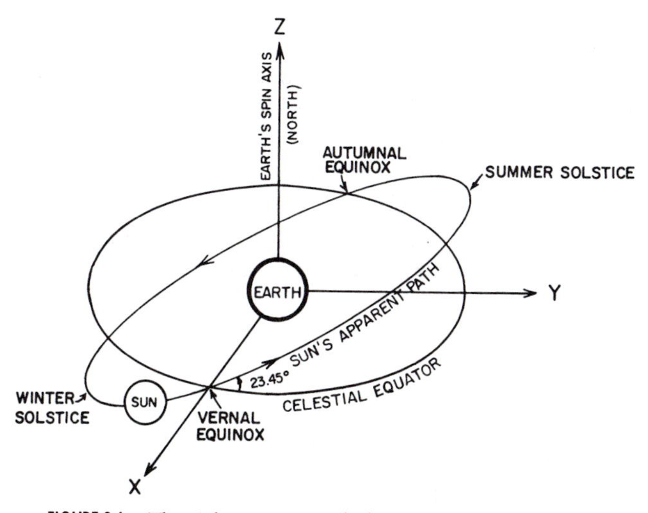

FIGURE 2.4. The right ascension–declination coordinate system.

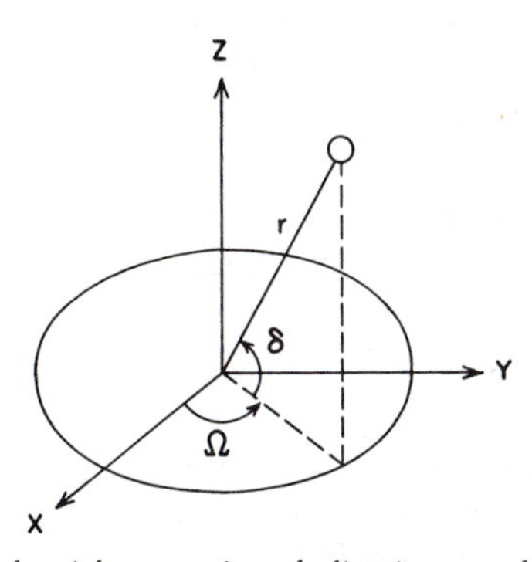

FIGURE 2.5. Coordinates used in the right ascension–declination coordinate system: right ascension (Ω), declination (δ), and radius (r).

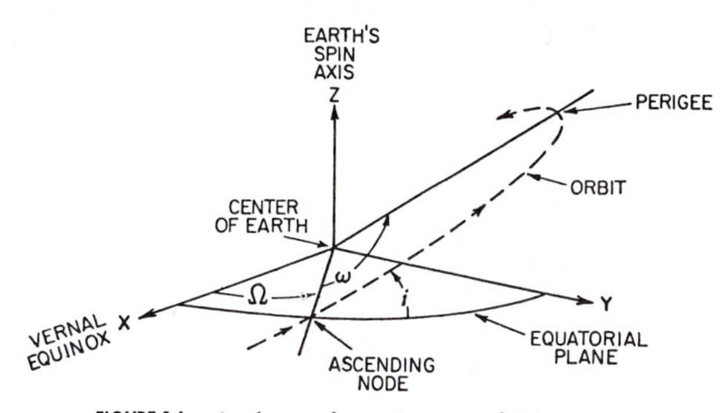

FIGURE 2.6. Angles used to orient an orbit in space.

2.2.5 Orbital Elements

The just-discussed parameters for location of a satellite in space are collectively known as the *classical orbital elements* (Table 2.1) or as *Brouwer mean orbital elements* (Brouwer and Clemence, 1961). These parameters may be determined by optical, radar, or radio ranging observations or by matching features of known locations on the Earth's surface (*landmarks*) with observations made by instruments on the satellite (Dubyago, 1961; Escobal, 1965). The orbital elements for particular satellites are available from the agencies that operate them: NOAA or NASA in the United States,[7] the European Space Agency (ESA) in Europe, etc. A final parameter, included in Table 2.1, is the time when these elements are observed or are "valid." This time is called the *epoch time* (t_0). Some of the orbital elements change with time, as we shall see below. A subscript "o" on an orbital element indicates a value at the epoch time.

There is some variation in how the orbital elements are specified. ESA, for example, substitutes true anomaly for mean anomaly. Also, in less formal descriptions of satellite orbits, one frequently sees the height of the satellite above the Earth's surface substituted for the semimajor axis. Since the Earth is not round, the height of a satellite will vary according to its position in the orbit. Specifying the semimajor axis is a much better way to describe a satellite orbit.

Orbits in which the classical orbital elements (except M) are constant are called *Keplerian orbits*. Viewed from space, Keplerian orbits are simple. The satellite moves in an elliptical path with the center of the Earth at one focus. The ellipse maintains a constant size, shape, and orientation with respect to the stars (Fig 2.7a). Perhaps surprisingly, the only effect of the sun's gravity on the satellite is to move the focus of the ellipse (the Earth) in an elliptical path around the sun (the Earth's orbit).

Viewed from the Earth, Keplerian orbits appear complicated because the Earth rotates on its axis as the satellite orbits the Earth (Fig. 2.8). The rotation of the Earth beneath a fixed orbit results in two daily passes of the satellite near a point on the Earth (assuming that the period is substantially less than a day and that the inclination angle is greater than the latitude of the point). One pass occurs during the ascending portion of the orbit; the other occurs during the descending portion of the orbit. This usually means that one pass occurs during daylight and one during darkness.

2.3 ORBIT PERTURBATIONS

Although satellites travel in nearly Keplerian orbits, these orbits are perturbed by a variety of forces (Table 2.2). Forces arising from the last five processes are small and can be viewed as causing essentially random perturbations in the orbital elements. Operationally they are dealt with simply by periodically (1) observing

[7] NOAA orbital elements for the polar-orbiting satellites are broadcast in the form of "TBUS bulletins." Barnes and Smallwood (1982) explain how to interpret these bulletins.

TABLE 2.1. Classical Orbital Elements

Element	*Symbol*
Semimajor axis	a
Eccentricity	ε
Inclination	i
Argument of perigee	ω_0
Right ascension of ascending node	Ω_0
Mean anomaly	M_0
Epoch time	t_0

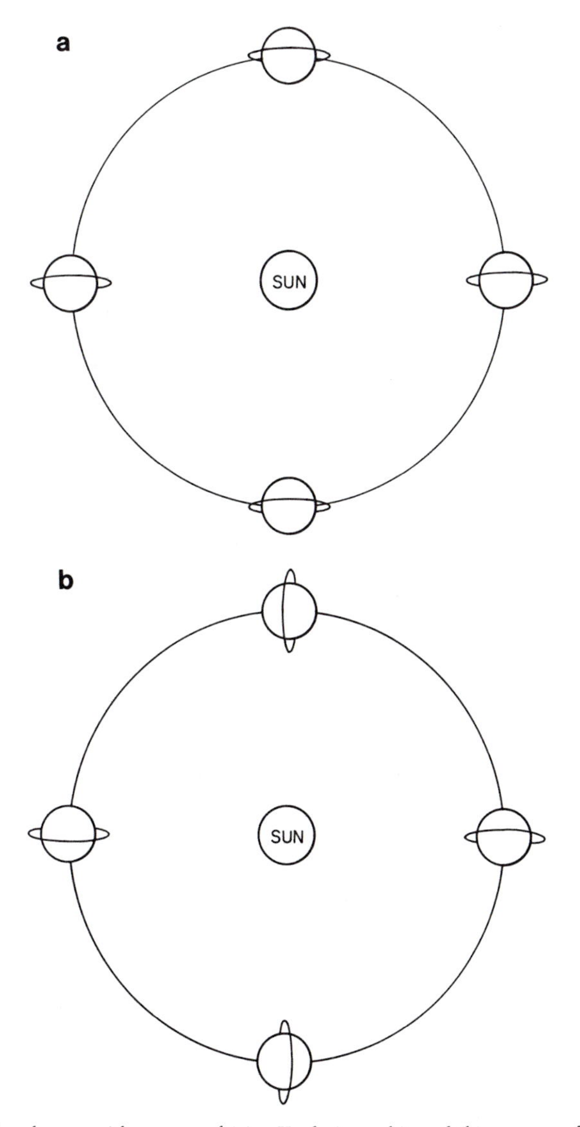

FIGURE 2.7. The change with season of (a) a Keplerian orbit and (b) a sunsynchronous orbit.

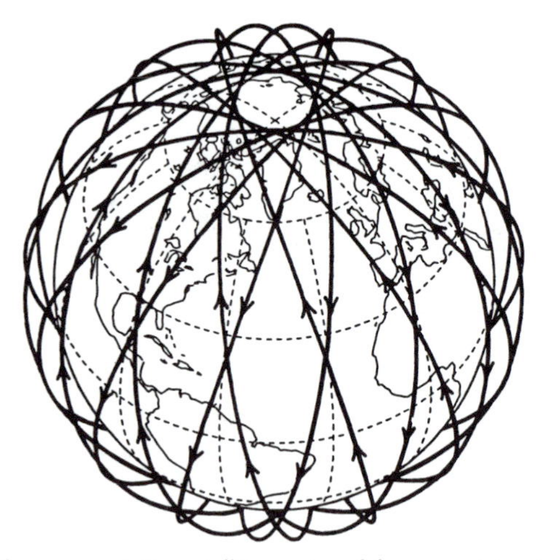

FIGURE 2.8. The orbit of a representative satellite as viewed from a point rotating with the Earth.

the orbital elements and (2) adjusting the orbit with on-board thrusters. Forces due to the nonspherical Earth cause secular (linear with time) changes in the orbital elements. These forces can be predicted theoretically and indeed are useful.

The gravitational potential of Earth is a complicated function of the Earth's shape, the distribution of land and ocean, and even the density of crustal material. As a first-order correction to a spherical shape, we may treat the Earth as an oblate spheroid of revolution. In cross section the Earth is approximately elliptical. The distance from the center of the Earth to the equator is, on average, 6378.140 km, whereas the distance to the poles is 6356.755 km. One can think of the Earth as a sphere with a 21-km-thick "belt" around the equator. The gravitational potential of the Earth is approximately given by

$$U = -\frac{Gm_e}{r}\left[1 + \frac{1}{2}J_2\left(\frac{r_{ee}}{r}\right)^2(1 - 3\sin^2\delta) + \cdots\right], \tag{2.11}$$

where r_{ee} is the equatorial radius of the Earth, δ is the declination angle, and J_2 is the coefficient of the quadrupole term (Appendix E). The higher-order

TABLE 2.2. Orbit Perturbing Forces

Force	*Source*
Nonspherical gravitational field	Nonspherical, nonhomogeneous Earth
Gravitational attraction of auxiliary bodies	Moon, planets
Radiation pressure	Sun's radiation
Particle flux	Solar wind
Lift and drag	Residual atmosphere
Electromagnetic forces	Interaction of electrical currents in the satellite with the Earth's magnetic field

terms are more than two orders of magnitude smaller than the quadrupole term and will not be considered here, although they are necessary for very accurate calculations.

How does this belt of extra mass affect a satellite's orbit? One might expect it to cause the satellite to orbit at a different speed, and indeed it does. The time rate of change of the mean anomaly (dM/dt) is given by the mean motion constant n in the unperturbed orbit and by the *anomalistic mean motion constant*, $\bar{n}$, in a perturbed orbit. Considering only the quadrupole term, Escobal (1965) shows that

$$\frac{dM}{dt} = \bar{n} = n\left[1 + \frac{3}{2}J_2\left(\frac{r_{ee}}{a}\right)^2(1-\varepsilon^2)^{-3/2}\left(1 - \frac{3}{2}\sin^2 i\right)\right]. \tag{2.12}$$

When the inclination angle is less than 54.7°, $\bar{n}$ is greater than n; the satellite orbits faster than it would in an unperturbed orbit. However, for larger inclinations, the satellite orbits more slowly than it otherwise would.

Because the belt exerts an equatorward force, one might also expect that it would have an effect on the inclination angle. This force, however, affects the right ascension of the ascending node rather than the inclination angle. Just as the force of gravity causes a top to precess rather than to fall over, so the attraction of the belt causes the orbit to precess about the z axis rather than to change its inclination angle. Escobal (1965) gives the rate of change of the right ascension of ascending node as

$$\frac{d\Omega}{dt} = -\bar{n}\left[\frac{3}{2}J_2\left(\frac{r_{ee}}{a}\right)^2(1-\varepsilon^2)^{-2}\cos i\right]. \tag{2.13}$$

The final effect of the belt is to cause the argument of perigee to rotate or precess. Escobal (1965) gives

$$\frac{d\omega}{dt} = \bar{n}\left[\frac{3}{2}J_2\left(\frac{r_{ee}}{a}\right)^2(1-\varepsilon^2)^{-2}\left(2 - \frac{5}{2}\sin^2 i\right)\right]. \tag{2.14}$$

The other three orbital elements, a, ε, and i, undergo small, oscillatory changes that may be neglected.

If SI Units are used, Eqs. 2.9, 2.12, 2.13, and 2.14 result respectively in values of n, $\bar{n}$, $d\Omega/dt$, and $d\omega/dt$ whose units are radians per second.

The *anomalistic period* of a perturbed orbit is simply

$$\bar{T} = \frac{2\pi}{\bar{n}}. \tag{2.15}$$

However, because M is measured from perigee, the anomalistic period is the time for the satellite to travel from perigee to moving perigee. Of more use is the *synodic* or *nodal period*, $\tilde{T}$, which is the time for the satellite to travel from one ascending node to the next ascending node. An exact value of $\tilde{T}$ must be calculated numerically; however, to very good approximation

$$\tilde{T} = \frac{2\pi}{\left(\bar{n} + \frac{d\omega}{dt}\right)}. \tag{2.16}$$

In summary, then, the first-order effects of the nonspherical gravitational potential of the Earth consist of a slow, linear change in two of the classical orbital elements, the right ascension of ascending node and the argument of perigee, and a small change in the mean motion constant.

2.4 METEOROLOGICAL SATELLITE ORBITS

Nearly all present meteorological satellites are in one of two orbits, sunsynchronous or geostationary, but other orbits are also useful.

2.4.1 Sunsynchronous Orbits

The nonspherical gravitational perturbation of Earth, far from being a problem, has a very useful application. As shown in Fig. 2.7a, the angle between the lines that join the sun and the ascending node to the center of the Earth changes in a Keplerian orbit because the orbital plane is fixed while the Earth rotates around the sun. This causes the satellite to pass over an area at different times of the day. For example, if the satellite passes over near noon and midnight in the spring, it will pass over near 6:00 am and 6:00 pm in the winter. Several problems result; among them are (1) the data do not fit conveniently into operational schedules, (2) orientation of solar cell panels is difficult, and (3) dawn or dusk visible images may not be as useful as images made at other times. Fortunately, the perturbations caused by the nonspherical Earth can be employed to keep the sun–Earth–satellite angle constant.

The Earth makes one complete revolution about the sun (2π radians) in one tropical year (31,556,925.9747 s). Thus the right ascension of the sun changes at the average rate of 1.991064×10^{-7} rad s^{-1} (0.9856473° day^{-1}). If the inclination of the satellite is correctly chosen, the right ascension of its ascending node can be made to precess at this same rate. An orbit that is so synchronized with the sun is called a *sunsynchronous orbit*. For a satellite with a semimajor axis of 7228 km and zero eccentricity, Eq. 2.13 requires an inclination of 98.8° to be sunsynchronous. Figure 2.7b shows the change with season of a sunsynchronous orbit.

Because the sun–Earth–ascending node angle is constant[8] in a sunsynchronous orbit, the satellite is often said to cross the equator at the same local time every

[8] Apart from small changes due to the elliptical orbit of the Earth.

day. Unfortunately, *local time* is an ambiguous term. We will use it to mean

$$LT \equiv t_U + \frac{\Psi}{15^\circ}, \tag{2.17}$$

where t_U is coordinated universal time in hours and Ψ is the longitude in degrees of a particular point.[9] *Equator crossing time* (ECT) is the local time when a satellite crosses the equator:

$$ECT \equiv t_U + \frac{\Psi_N}{15^\circ}, \tag{2.18}$$

where Ψ_N is the longitude of ascending or descending node. If

$$\Psi_{sun} = -15^\circ(t_U - 12) \tag{2.19}$$

is the longitude of the sun, and if $\Delta\Psi \equiv \Psi_N - \Psi_{sun}$, it is easy to show that

$$ECT = 12 + \frac{\Delta\Psi}{15^\circ}. \tag{2.20}$$

If $\Delta\Psi$ is constant, as it is for a sunsynchronous satellite, then ECT is constant.

Sunsynchronous satellites are classified by their ECTs. *Noon satellites* (or *noon–midnight satellites*) ascend (or descend) near noon LT (local time). They must, therefore, descend (or ascend) near local midnight. *Morning satellites* ascend (or descend) between 06 and 12 h LT, and descend (or ascend) between 18 and 24 h LT. Afternoon satellites ascend (or descend) between 12 and 18 h LT, and descend (or ascend) between 00 and 06 h LT.

The highest latitude reached by the subsatellite point (in any orbit) is equal to the inclination angle (or the supplement of i, in the case of retrograde orbits). Since sunsynchronous orbits reach high latitudes, they are referred to as *near-polar orbits*. This is frequently shortened to *polar orbits*, although they do not cross directly over the poles. These orbits are also called *low Earth orbits* (*LEOs*) to distinguish them from geostationary orbits (*GEOs*). Note, however, that *polar orbiter* is a general term for a satellite that passes near the poles, and *low Earth orbiter* is a general term for a satellite that orbits not far above the Earth's surface. While sunsynchronous satellites are of necessity polar orbiters and LEOs, the converse is not necessarily true.

The *ground track* of a satellite is the path of the point on the Earth's surface that is directly between the satellite and the center of the Earth (the *subsatellite point*). Figure 2.9 shows the ground track for three orbits of the sunsynchronous NOAA 11 satellite.

[9] The other use of *local time* refers to the time on one's watch, that is, the time in a particular time zone. Time zones are defined as areas where time is agreed to be the local time (in our sense) on a particular meridian. Eastern Standard Time, for example, is the local time on the 75° west meridian ($\Psi = -75^\circ$).

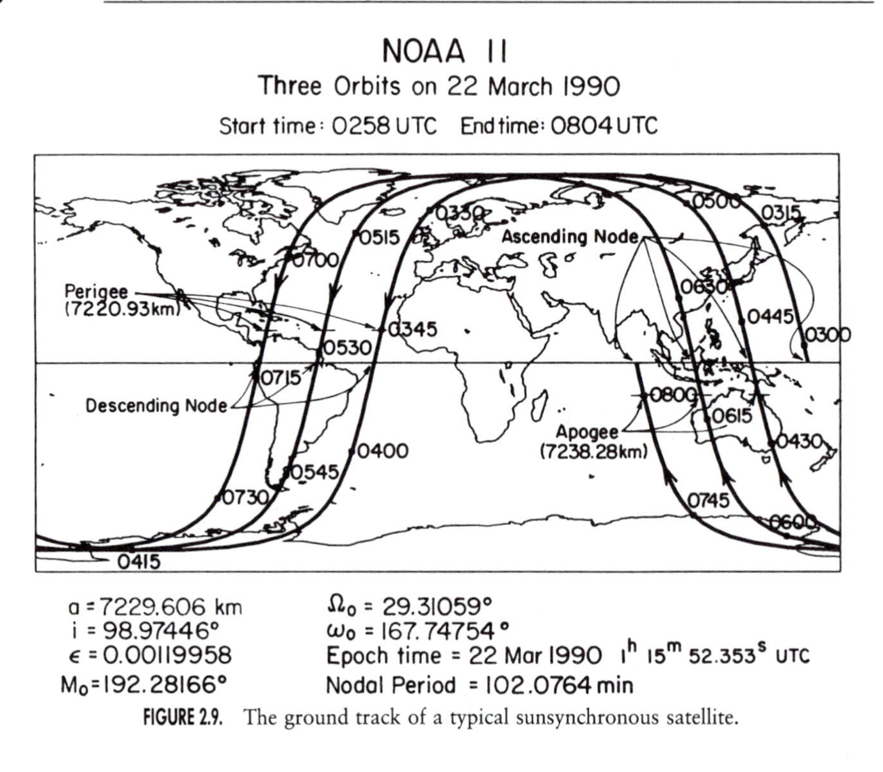

FIGURE 2.9. The ground track of a typical sunsynchronous satellite.

2.4.2 Geostationary Orbits

In Section 2.1 we calculated the radius of a geosynchronous orbit to be 42,164 km. Perturbations due to the nonspherical Earth, however, require a slight adjustment in this figure. The adjustment is small because geosynchronous orbit is about 6.6 Earth radii, and the correction terms are inversely proportional to the square of this ratio. For an orbit with zero eccentricity and zero inclination, Eqs. 2.12, 2.14, and 2.16 require a semimajor axis of 42,168 km to be geosynchronous.

The terms *geosynchronous* and *geostationary* are often used interchangeably. In fact, they are not the same. *Geosynchronous* means that the satellite orbits with the same angular velocity as the Earth. *Geostationary* orbit is geosynchronous, but it is also required to have zero inclination angle and zero eccentricity. Geostationary satellites, therefore, remain essentially motionless above a point on the equator. They are classified by the longitude of their subsatellite point.

Second-order perturbations cause a geostationary satellite to drift from the desired orbit. Periodic maneuvers, performed as frequently as once a week, are required to correct the orbit. These maneuvers keep operational geostationary satellites very close to the desired orbit. For example, on 11 March 1990, the GOES 7 satellite had an inclination angle of 0.05°; therefore, it did not venture more than 0.05° latitude from the equator. Figure 2.10 shows the ground track for a geostationary satellite that is no longer used for imaging and therefore whose orbit is not so carefully maintained.

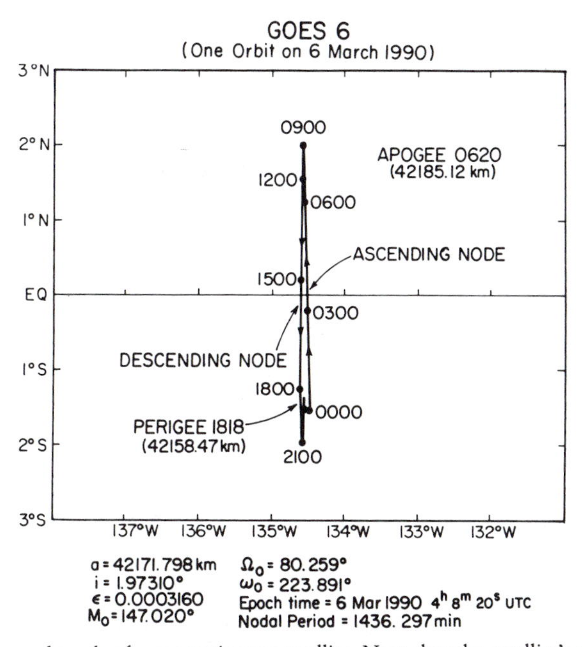

FIGURE 2.10. The ground track of a geostationary satellite. Note that the satellite's orbit is not quite geostationary; it drifts west slightly each day.

2.4.3 Other Orbits

Geostationary and sunsynchronous are only two of infinite possible orbits. Others have been and will become useful for meteorological satellites.

The Earth Radiation Budget Satellite (ERBS) was launched from the Space Shuttle and orbits at an altitude of 600 km with an inclination angle of 57°. It was placed in this orbit so that it would precess with respect to the sun and sample all local times (see Section 2.6) over the course of a month.

The former Soviet Union placed its Meteor satellites in low Earth orbit with inclination angles of about 82° (see Appendix A). The former Soviet Union also used a highly elliptical orbit for Molniya communications satellites. It has been suggested that this orbit would be useful for meteorological observations of the high latitudes (Kidder and Vonder Haar, 1990). The *Molniya orbit* has an inclination angle of 63.4°, at which the argument of perigee is motionless (Eq. 2.14); thus the apogee, from which measurements are made, stays at a given latitude. The semimajor axis is chosen such that the satellite makes two orbits while the Earth turns once with respect to the plane of the orbit. The eccentricity is made as large as possible so that the satellite will stay near apogee longer. However, the eccentricity must not be so large that the satellite encounters significant atmospheric drag at perigee. A semimajor axis of 26,554 km and an eccentricity of 0.72 result in a perigee of 7378 km (1000 km above the equator), an apogee of 45,730 km (39,352 km above the equator), and a period of 717.8 min. The attractiveness of this orbit is that it functions as a high-latitude, part-time, nearly geostationary satellite. For about 8 h centered on apogee, the satellite is synchronized with the Earth so that it is nearly stationary in the sky. For a meteorological satellite in a Molniya orbit, the rapid imaging capability, which is so useful from geostationary orbit, would be available in the high latitudes.

As meteorological satellite instruments become more specialized, more custom orbits are likely to be used.

2.5 SATELLITE POSITIONING, TRACKING, AND NAVIGATION

It is important to be able to calculate the position of a satellite in space, to track it from Earth, and to know where its instruments are pointing. These topics are discussed in turn in this section.

2.5.1 Positioning in Space

To locate a satellite in a perturbed orbit at time t, one needs current values of the orbital elements. The three constant elements, a, ε, and i, are taken directly from a recent bulletin.[10] The other three, M, Ω, and ω, are calculated:

$$M = M_o + \frac{dM}{dt}(t - t_o), \tag{2.21a}$$

$$\Omega = \Omega_o + \frac{d\Omega}{dt}(t - t_o), \tag{2.21b}$$

$$\omega = \omega_o + \frac{d\omega}{dt}(t - t_o). \tag{2.21c}$$

Then the satellite is positioned by one of several methods. We find two methods useful: the vector rotation method and the spherical geometry method.

2.5.1.1 The Vector Rotation Method

Figure 2.11 illustrates what we call the *vector rotation method*. It is discussed in a somewhat different form by Escobal (1965) and others. In the first step, the satellite is located in the plane of its orbit; that is, the true anomaly θ and the radius r are calculated. This is done by (1) solving for e using Eq. 2.8, (2) calculating θ using Eq. 2.10a, and (3) calculating r using Eq. 2.7. (For a circular orbit, this step is simplified because the mean anomaly, the eccentric anomaly, and the true anomaly are identical, and r is constant.)

In the second step, a vector is formed that points from the center of the Earth to the satellite in the right ascension–declination coordinate system. The Cartesian

[10] Such bulletins are available from a variety of sources. Because these sources change rapidly, we suggest that the interested reader contact the agencies listed in Section 4.4 to find a convenient source of satellite bulletins.

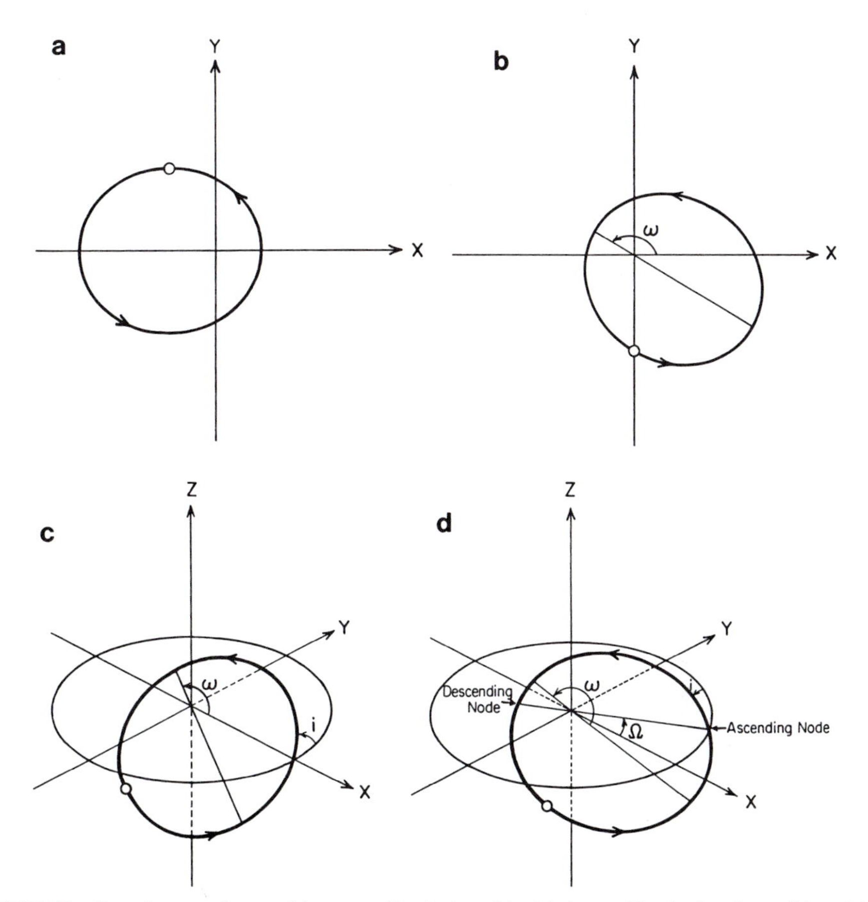

FIGURE 2.11. Rotations used to position a satellite in its orbit: (a) the satellite in the plane of its orbit, (b) rotation about the z axis through the argument of perigee (ω), (c) rotation about the x axis through the inclination angle (i), and (d) rotation about the z axis through the right ascension of ascending node (Ω).

coordinates of this vector are

$$\begin{pmatrix} x \\ y \\ z \end{pmatrix} = \begin{pmatrix} r\cos\theta \\ r\sin\theta \\ 0 \end{pmatrix}. \tag{2.22}$$

At this point, the orbital ellipse is assumed to lie in the x–y plane with the perigee on the positive x axis (Fig 2.11a).

In the final three steps, the vector is rotated so that the orbital plane is properly oriented in space.

In the third step, the vector is rotated about the z axis through the argument of perigee (Fig. 2.11b). This rotation is conveniently accomplished by multiplying the vector by a rotation matrix, in this case

$$\begin{pmatrix} x' \\ y' \\ z' \end{pmatrix} = \begin{pmatrix} \cos\omega & -\sin\omega & 0 \\ \sin\omega & \cos\omega & 0 \\ 0 & 0 & 1 \end{pmatrix} \begin{pmatrix} x \\ y \\ z \end{pmatrix} = \begin{pmatrix} x\cos\omega - y\sin\omega \\ x\sin\omega + y\cos\omega \\ z \end{pmatrix}. \tag{2.23}$$

In the fourth step, the vector is rotated about the x axis through the inclination angle (Fig. 2.11c).

$$\begin{pmatrix} x'' \\ y'' \\ z'' \end{pmatrix} = \begin{pmatrix} 1 & 0 & 0 \\ 0 & \cos i & -\sin i \\ 0 & \sin i & \cos i \end{pmatrix} \begin{pmatrix} x' \\ y' \\ z' \end{pmatrix} = \begin{pmatrix} x' \\ y'\cos i - z'\sin i \\ y'\sin i + z'\cos i \end{pmatrix}. \tag{2.24}$$

In the fifth and final step, the vector is rotated about the z axis through the right ascension of the ascending node (Fig. 2.11d).

$$\begin{pmatrix} x''' \\ y''' \\ z''' \end{pmatrix} = \begin{pmatrix} \cos\Omega & -\sin\Omega & 0 \\ \sin\Omega & \cos\Omega & 0 \\ 0 & 0 & 1 \end{pmatrix} \begin{pmatrix} x'' \\ y'' \\ z'' \end{pmatrix} = \begin{pmatrix} x''\cos\Omega - y''\sin\Omega \\ x''\sin\Omega + y''\cos\Omega \\ z'' \end{pmatrix}. \tag{2.25}$$

The vector (x''', y''', z''') is the location of the satellite in the right ascension–declination coordinate system at time t. This vector may be converted into the radius, declination, and right ascension of the satellite by

$$r_s = \sqrt{x'''^2 + y'''^2 + z'''^2} = r, \tag{2.26a}$$

$$\delta_s = \sin^{-1}\left(\frac{z'''}{r_s}\right), \tag{2.26b}$$

$$\Omega_s = \tan^{-1}\left(\frac{y'''}{x'''}\right). \tag{2.26c}$$

After one has calculated the right ascension, declination, and radius of the satellite, it is useful to calculate the latitude and longitude of the subsatellite point. Assuming that the Earth is a sphere, the latitude (known as the *geocentric latitude*) is simply equal to the declination. The longitude of the subsatellite point is the difference between the right ascension of the satellite and the right ascension of the prime meridian (0° longitude) which passes through Greenwich, England (Fig. 2.12). The right ascension of Greenwich can be calculated knowing its right ascension at a given time and the rotation rate of the Earth.[11] Since the rotation rate changes very slightly, due to the actions of the wind and ocean currents, very accurate knowledge of the right ascension of Greenwich requires observations. Some satellite bulletins give the right ascension of Greenwich in addition to the satellite orbital elements.

[11] If nothing else is available, one can use the following: at 0000 UTC on 1 January 1990 the right ascension of Greenwich was 100.38641°, and the rotation rate was $7.292115922 \times 10^{-5}$ radians per second or 360.9856507° per day.

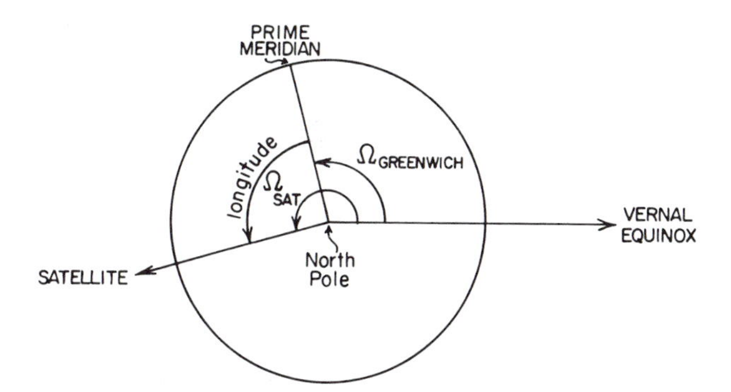

FIGURE 2.12. The relationship between Earth longitude and right ascension.

The inverse problem of finding when a satellite passes over (or close to) a particular point is solved iteratively by (1) estimating the time, (2) calculating the position of the satellite, and (3) correcting the time estimate. Steps 2 and 3 are repeated until a satisfactory solution is found.

2.5.1.2 The Spherical Geometry Method

The *spherical geometry method* can be derived using spherical geometry (Madden and Parsons, 1973), but it is also a distillation of the vector rotation method. Let Γ, the *argument of latitude*, be the angle, measured in the orbital plane, from the ascending node to the satellite. Numerically,

$$\Gamma \equiv \theta + \omega, \tag{2.27}$$

where θ is the true anomaly and ω is the argument of perigee. Working through the mathematics of the vector rotation method results in[12]

$$r_s = r, \tag{2.28a}$$

$$\Theta_s = \delta_s = \sin^{-1}(\sin\Gamma \sin i), \tag{2.28b}$$

$$\Psi_s = \tan^{-1}\left(\frac{\sin\Gamma \cos i}{\cos\Gamma}\right) + \Omega_o - \Omega_e(t_o) - \left(\frac{d\Omega_e}{dt} - \frac{d\Omega}{dt}\right)(t - t_o). \tag{2.28c}$$

Here r is the distance of the satellite calculated with Eq. 2.7; Θ_s and Ψ_s are its latitude and longitude, respectively. $\Omega_e(t_o)$ is the right ascension of Greenwich at the epoch time, and therefore, $\Omega_o - \Omega_e(t_o)$ is the longitude of ascending node at the epoch time. The quantity $(d\Omega_e/dt - d\Omega/dt)$ is the *relative Earth rotation rate*, that is, the rotation rate of the Earth relative to the orbital plane. For a sunsynchronous satellite, it must be exactly 2π radians per day.

[12] Normally the arctangent term would be written $\tan^{-1}(\tan\Gamma \cos i)$. The form in Eq. 2.28c is used because it allows the quadrant of the angle to be determined unambiguously. In Fortran, for example, ATAN2(sinΓcos*i*,cosΓ) will result in the correct angle.

For a circular orbit, or one which is so nearly circular that no significant error occurs from neglecting its elliptical nature,

$$\Gamma(t) = \Gamma_o + \left(\bar{n} + \frac{d\omega}{dt}\right)(t - t_o), \tag{2.29}$$

and the spherical geometry method is particularly easy to apply. Polar orbiters can often be treated with this approximation. When they are so treated, the orbital parameters may come in a different form. The supplied parameters may be (1) the longitude of ascending node, (2) the nodal period, (3) the radius (or semimajor axis), (4) the inclination, (5) the time of ascending node, and (6) the nodal longitude increment (ΔLON), which is the difference in longitude[13] between successive ascending nodes:

$$\Delta\text{LON} = \left(\frac{d\Omega_e}{dt} - \frac{d\Omega}{dt}\right)\tilde{T}, \tag{2.30}$$

where $\tilde{T}$ is the nodal period. The above equations still apply, but one must remember that $\Gamma_o = 0$, $(\bar{n} + d\omega/dt) = 2\pi/\tilde{T}$, and $(d\Omega_e/dt - d\Omega/dt) = \Delta\text{LON}/\tilde{T}$.

2.5.2 Tracking

A list of time versus position of a celestial body is called an *ephemeris* (plural: *ephemerides*). To *track* a satellite, one must be able to point one's antenna at it. The *elevation angle*, measured from the local horizontal, and the *azimuth angle*, measured clockwise from the north, can be calculated from the ephemeris data as follows.

Suppose the subsatellite point is at latitude Θ_s and longitude Ψ_s, and that the satellite is at radius r_s from the center of the Earth. Suppose also that the antenna is located at latitude Θ_e, longitude Ψ_e, and radius r_e (the radius of the Earth). The Cartesian coordinates of the satellite, then, are

$$\vec{r}_s = \begin{pmatrix} x_s \\ y_s \\ z_s \end{pmatrix} = \begin{pmatrix} r_s \cos\Theta_s \cos\Psi_s \\ r_s \cos\Theta_s \sin\Psi_s \\ r_s \sin\Theta_s \end{pmatrix}, \tag{2.31}$$

whereas the Cartesian coordinates of the antenna are

$$\vec{r}_e = \begin{pmatrix} x_e \\ y_e \\ z_e \end{pmatrix} = \begin{pmatrix} r_e \cos\Theta_e \cos\Psi_e \\ r_e \cos\Theta_e \sin\Psi_e \\ r_e \sin\Theta_e \end{pmatrix}. \tag{2.32}$$

The difference vector $(\vec{r}_D \equiv \vec{r}_s - \vec{r}_e)$ points from the antenna to the satellite (Fig. 2.13). Assuming a spherical Earth, the vector $\vec{r}_e$ points to the local vertical (Fig.

[13] That is, the next ascending node occurs ΔLON *west* of the current ascending node.

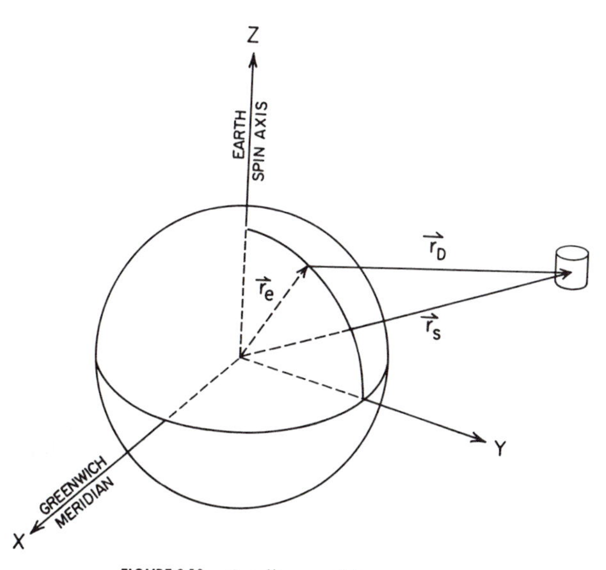

FIGURE 2.13. Satellite tracking geometry.

2.14). The cosine of the satellite's *zenith angle* ζ (the complement of the elevation angle) is given by

$$\cos\zeta = \frac{\vec{r}_e \cdot \vec{r}_D}{|\vec{r}_e||\vec{r}_D|}. \tag{2.33}$$

Finding the azimuth angle is a little more difficult. First, we need to find two vectors in the tangent plane at the antenna. The first points north:

$$\vec{r}_N = \begin{pmatrix} x_N \\ y_N \\ z_N \end{pmatrix} = \begin{pmatrix} -\sin\Theta_e \cos\Psi_e \\ -\sin\Theta_e \sin\Psi_e \\ \cos\Theta_e \end{pmatrix}. \tag{2.34}$$

The second is the horizontal projection of $\vec{r}_D$. If we define unit vectors in the

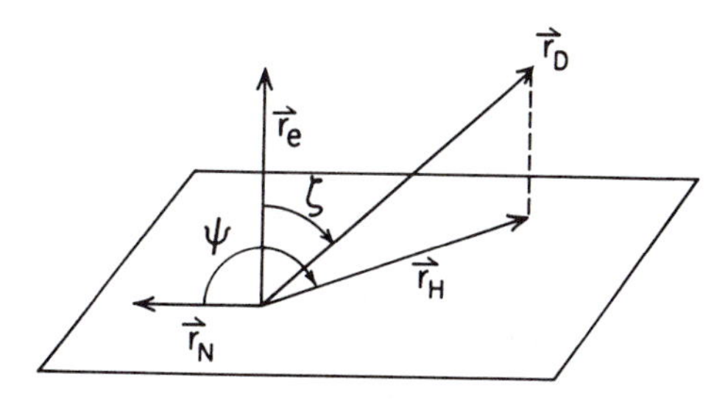

FIGURE 2.14. Definition of zenith angle (ζ) and azimuth angle (ψ).

directions of $\vec{r}_e$ and $\vec{r}_D$ as

$$\hat{r}_e \equiv \frac{\vec{r}_e}{|\vec{r}_e|}, \tag{2.35a}$$

$$\hat{r}_D \equiv \frac{\vec{r}_D}{|\vec{r}_D|}, \tag{2.35b}$$

the required horizontal vector is

$$\vec{r}_H = \vec{r}_D - (\hat{r}_e \cdot \vec{r}_D)\hat{r}_e = \vec{r}_D - |\vec{r}_D|\cos\zeta\,\hat{r}_e = |\vec{r}_D|(\hat{r}_D - \cos\zeta\,\hat{r}_e). \tag{2.36}$$

The azimuth angle is then given by

$$\cos\psi = \frac{\vec{r}_N \cdot \vec{r}_H}{|\vec{r}_N||\vec{r}_H|}. \tag{2.37}$$

One must be careful when taking the inverse cosine. If the satellite is west of the antenna, ψ will be greater than 180°. It also must be noted that these equations assume a spherical Earth. Fortunately, most receiving antennas are insensitive to the slight errors this assumption causes.

2.5.3 Navigation

In addition to knowing where a satellite is in its orbit, it is necessary to know the Earth coordinates (latitude, longitude) of the particular scene it is viewing. The problem of calculating the Earth coordinates is known as the *navigation problem*; fundamentally, it is a complex geometry problem. It requires an accurate knowledge of where the satellite is in its orbit, the orientation of the satellite (its *attitude*), and the scanning geometry of the instrument involved.

In simplified form, we can proceed as follows. Suppose that at a particular time a satellite is at position (x_s, y_s, z_s) with respect to the center of the Earth in the right ascension–declination coordinate system. Suppose further, that the telescope is pointing in a direction specified by declination δ_T and right ascension Ω_T. A unit vector in the direction that the telescope is pointing is given by

$$\begin{pmatrix} x_T \\ y_T \\ z_T \end{pmatrix} = \begin{pmatrix} \cos\delta_T \cos\Omega_T \\ \cos\delta_T \sin\Omega_T \\ \sin\delta_T \end{pmatrix}. \tag{2.38}$$

Figure 2.15 shows that the ray from which the telescope receives radiation (that is, the line in space through the satellite and in the direction of the telescope) is given by

$$\begin{pmatrix} x \\ y \\ z \end{pmatrix} = \begin{pmatrix} x_s + s\,x_T \\ y_s + s\,y_T \\ z_s + s\,z_T \end{pmatrix}, \tag{2.39}$$

where s is the distance from the satellite.

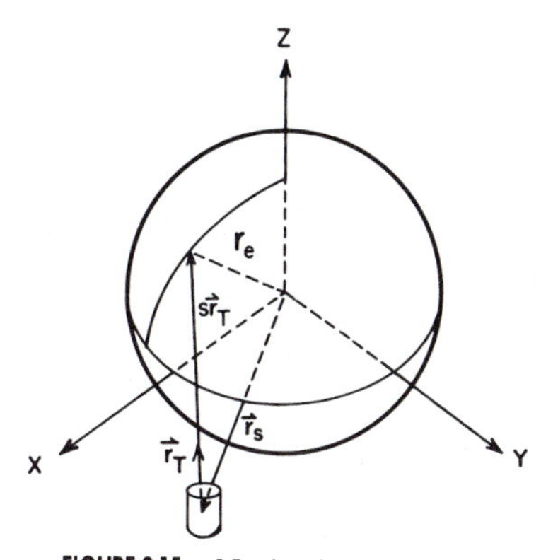

FIGURE 2.15. Navigation geometry.

The location at which this ray strikes the spherical Earth is the solution of the equation

$$(x_s + s\,x_T)^2 + (y_s + s\,y_T)^2 + (z_s + s\,z_T)^2 = r_e^2. \tag{2.40}$$

This is a quadratic equation in s that has no real roots, if the ray does not intersect the Earth; or two real roots, if it does. The smaller root is to be chosen; the larger root represents the location from which the ray reemerges from the opposite side of the Earth. When the ray is just tangent to the Earth, the two roots are equal.[14]

After a solution for s has been found, Eq. 2.39 gives the Cartesian coordinates in the right ascension–declination coordinate system of the point on the Earth's surface being viewed. The latitude and longitude can then be found as in Section 2.5.1.

Satellite images are usually the result of a scanning instrument. The data come in the form of scan lines, each divided into elements or samples known as *pixels* or *scan spots*. Because scanning is very carefully timed, each pixel has a unique time associated with it. Therefore, calculating the latitude and longitude of a pixel is accomplished using the equations of Sections 2.5.1 and 2.5.3 in the forward direction; time yields satellite position and telescope pointing angles, which then yield latitude and longitude. The opposite problem, finding the pixel which observed a particular point on Earth (latitude and longitude), must be solved in an iterative manner because the exact time when the point was observed is unknown. In brief, the time of observation is estimated, the actual point being observed at that time is calculated, and a correction is made in the time which moves the point of observation closer to the desired point. This procedure is iterated until satisfactory convergence is achieved.

The scheme outlined here for finding latitude and longitude is simple and very general. It is applicable to a wide variety of satellite orbits and instruments. For

[14] For a geostationary satellite, this occurs about 81° from the satellite subpoint and explains why geostationary satellites never observe the poles.

each instrument, the difficult part is to determine the telescope pointing angles δ_T and Ω_T.

2.5.3.1 Geostationary Geometry

Until recently, all geostationary meteorological satellites were spin stabilized. They spin on an axis which is maintained nearly parallel to the Earth's spin axis. The rotation of the satellite changes the right ascension of the telescope and provides scanning across the Earth. Scanning in the north–south direction is accomplished by a tilting mirror (see Chapter 4), which changes the telescope declination. Thus δ_T and Ω_T are natural coordinates for spin-stabilized geostationary satellites.

Unfortunately, the satellite's spin axis is not exactly parallel to the Earth's spin axis. Furthermore, although the radiometer's telescope is rigidly oriented with respect to the principal axis of the satellite, the spin axis deviates slightly from the principal axis, which causes deviations similar to pitch, roll, and yaw in a low Earth orbiter (see next section). Corrections for these effects and for the non-spherical Earth can be made. The interested reader is referred to Hambrick and Phillips (1980).

The parameters that describe the satellite orbit and attitude must be accurately known to perform accurate navigation. These parameters can be determined by the use of landmarks. Normally the orbit and attitude parameters are accurate, as is navigation performed with them. However, for up to 18 h after the thrusters are fired in an orbit- or attitude-correcting maneuver, navigation parameters are poorly known, and pixels can be significantly misplaced. These errors can be partially corrected by displaying the data as an image and shifting the image up or down and right or left until a landmark is properly positioned. Rotation of the image is sometimes necessary to achieve good navigation, especially if a large (continent-size) area is being studied.

It is interesting to note that the GOES satellites can detect a few stars at the edges of the image frame. These stars can be used to very accurately determine the attitude of the satellite. Then landmarks can be used to determine the orbital elements (Hambrick and Phillips, 1980).

2.5.3.2 Low Earth Orbit Geometry

Low Earth orbit satellite instruments have many scanning patterns. Navigation of these data can be achieved using different approachs. We outline an approach, based on the discussion above, which is general enough for use with many scanning patterns. The basis of the technique is that if we can determine where a scan spot is in relation to the satellite, then we can use nearly the same rotation matrices with which we position the satellite to position the scan spot. First we must define the angles and a coordinate system used to specify satellite attitude.

The instruments on many low Earth orbit satellites are mounted on the underside of the satellite and scan perpendicular to the velocity vector through the subsatellite point (see Chapter 4). A convenient coordinate system (Fig. 2.16) is one in which the z axis points from the satellite toward the center of the Earth,

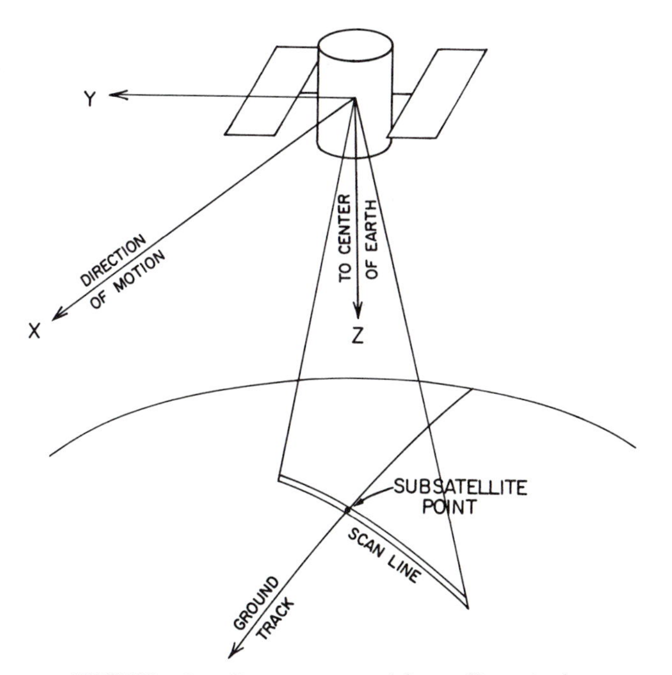

FIGURE 2.16. Coordinate system used for satellite attitude.

the x axis points in the direction of satellite motion, and the y axis is chosen to complete a right-handed coordinate system. Three angles specify the orientation of the satellite in this coordinate system. Rotation about the y axis is called *pitch*, rotation about the x axis is called *roll*, and rotation about the z axis is called *yaw*.

A combination of these angles can be used to specify nearly any scan geometry. Instruments that scan through nadir perpendicular to the satellite motion vector are described by changing the roll angle. Instruments that scan in a cone can be described by a constant pitch plus a variable yaw. Instruments that scan through the subpoint but at an oblique angle with respect to the satellite motion vector can be described with a roll plus a constant yaw.

To calculate the position of a scan spot with respect to the satellite, we proceed as follows. First, position the satellite at radius r_s, declination zero, and right ascension zero, and let its velocity vector point east. Assume that the telescope is pointing straight down,[15] or $\delta_T = 0$ and $\Omega_T = \pi$. That is, the telescope pointing vector is $x_T = -1$, $y_T = 0$, $z_T = 0$. Next, rotate the telescope vector through the pitch, roll, and yaw angles that describe the position of the telescope at time t,

[15] If the satellite is not pointing straight down, its deviation is described by pitch, roll, and yaw *bias errors*, which are usually small. For example, horizon sensors and a sun sensor on the current NOAA satellites maintain pitch, roll and yaw bias errors to within ±0.2° of zero. If the bias errors are known, and if the desired precision of the calculation requires it, the initial telescope pointing vector can be corrected for bias errors at this point.

or for the scan position desired. *At the assumed position and orientation of the satellite, in the right ascension–declination coordinate system*, the pitch rotation matrix is

$$\begin{pmatrix} \cos\alpha_P & \sin\alpha_P & 0 \\ -\sin\alpha_P & \cos\alpha_P & 0 \\ 0 & 0 & 1 \end{pmatrix},$$

where α_P is the pitch angle; the roll rotation matrix is

$$\begin{pmatrix} \cos\alpha_R & 0 & -\sin\alpha_R \\ 0 & 1 & 0 \\ \sin\alpha_R & 0 & \cos\alpha_R \end{pmatrix},$$

where α_R is the roll angle; and the yaw rotation matrix is

$$\begin{pmatrix} 1 & 0 & 0 \\ 0 & \cos\alpha_Y & -\sin\alpha_Y \\ 0 & \sin\alpha_Y & \cos\alpha_Y \end{pmatrix},$$

where α_Y is the yaw angle.[16]

After the telescope pointing vector has been determined, the distance s from the satellite to the scan spot is calculated using Eq. 2.40, and the position of the spot relative to the satellite is calculated using Eq. 2.39.

Finally, the scan spot is moved along with the satellite to its actual position by (1) rotating about the z axis through the argument of latitude, (2) rotating about the x axis through the inclination angle, and (3) rotating about the z axis through the right ascension of the ascending node minus the right ascension of Greenwich.

An advantage to this method is that if the orbit is sufficiently circular, the vectors to the scan spots can be calculated in advance and simply rotated into position at successive times.

Note that limb scanners, which scan the atmosphere above the Earth's horizon, can be treated with this procedure except that Eq. 2.40 is not applicable because the ray does not strike the Earth. Instead, the distance to the tangent point, that is, the point where the ray most closely approaches the Earth, can be used for s. If α is the angle between the initial telescope pointing vector (straight down) and the final vector, then

$$s = r_s \cos\alpha. \tag{2.41}$$

Finally, we would like to outline a simple calculation that is frequently useful in satellite meteorology: how to find the distance of a scan spot from the subsatellite

[16] If the satellite is thought of as an airplane, a positive pitch angle is defined here as the nose pointing up, a positive roll as the right wing pointing up, and a positive yaw as a counterclockwise rotation of the plane as viewed from above.

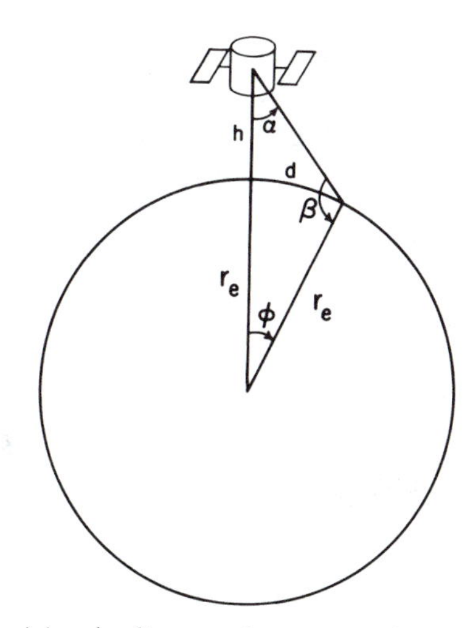

FIGURE 2.17. Determining the distance of a scan spot from the subsatellite point.

point. Figure 2.17 shows the geometry of this calculation. If α is the scan angle, then the law of sines gives the angle β as

$$\sin\beta = \left(\frac{r_s}{r_e}\right)\sin\alpha. \tag{2.42}$$

The angle measured from the center of the Earth is

$$\phi = \pi - \beta - \alpha, \tag{2.43}$$

and the distance from the subsatellite point to the scan spot is ϕr_e.

The *swath width* is the width of the entire scan of the satellite instrument. If the instrument scans equally on each side of the ground track, then the swath width is $2\phi r_e$, where α is the maximum scan angle.[17]

2.6 SPACE–TIME SAMPLING

To select an orbit for a satellite or a scan pattern for a particular instrument, several questions must be answered: What areas will the orbit and scan pattern allow the instrument to observe? How often will an area be observed? At what local times will the observations be made? At what viewing zenith and azimuth angles will the observations be made? These questions are all aspects of what is called *space–time sampling*.

Geostationary satellites are designed to be nearly stationary over a point on the equator. They therefore view a fixed area (about 42% of the globe). Any

[17] With this definition, the swath width is the distance between the centers of the extreme scan spots (see Fig. 4.10). Sometimes, the halfwidth of the radiometer field of view is added to each end of the swath, so that the swath width describes all that the radiometer senses.

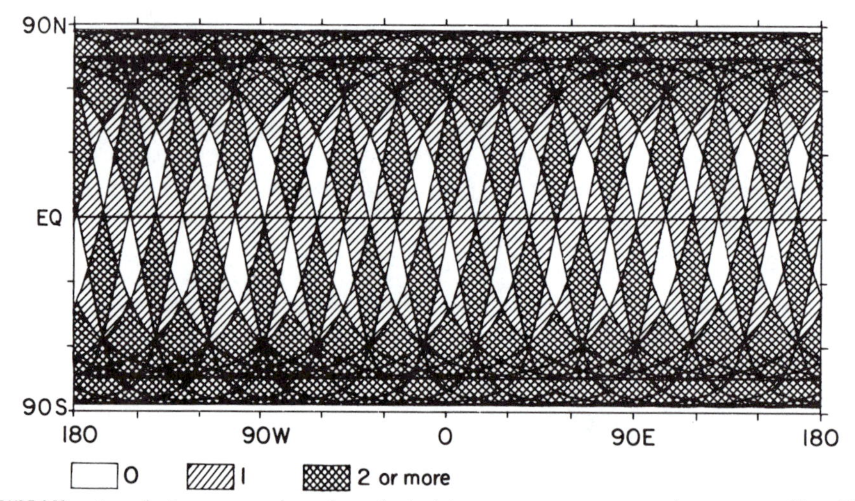

FIGURE 2.18. One day's coverage by a hypothetical instrument on a sunsynchronous satellite. The coverage at the equator is 50%. The orbit is circular with a semimajor axis of 7228 km and an inclination angle of 98.8°. Note that slightly different orbit parameters can result in quite different patterns.

point in this area can be observed as frequently as their instruments will allow; that is, it can be observed at any local time. However, since each point has a fixed geometric relationship to the satellite, it is viewed at only one zenith and one azimuth angle.

For a satellite in low Earth orbit, these questions depend on the satellite's orbit and the scanning geometry of its instruments. Most meteorological satellite instruments are designed such that the area viewed on one orbit touches or overlaps the area viewed on previous and successive orbits. If the satellite's inclination angle is large enough, the instrument views every point on Earth twice per day, at least. The poles are observed on every orbit. Usually each point is viewed at a wide range of zenith and azimuth angles. Many meteorological satellites are in sunsynchronous orbits, which have constant equator crossing times. These satellites view each point (except near the poles) only in a small range of local times ($\pm\frac{1}{2}\tilde{T}$) centered on the two equator crossing times.

For instruments whose scans on successive orbits do not overlap, it is often best to plot the coverage for a day and to determine visually which areas are observed and which are not. Figure 2.18, for example, shows the one-day coverage of a hypothetical instrument in a sunsynchronous orbit which has 50% coverage at the equator.[18] Some areas are not observed, some are observed once, and some are observed twice or more. This pattern will be the same on succeeding days, *except that it will drift in longitude.* The drift rate can be calculated as follows. Divide the length of day by the nodal period and round to the nearest integer,

[18] That is, the swath width divided by the sine of the inclination angle is 50% of ΔLON.

N. The westward longitude increment in N complete orbits is $N\Delta$LON. If we express the nodal longitude increment ΔLON (Eq. 2.30) in degrees, then the change in the pattern per day is

$$Drift = -N\Delta\text{LON} + 360°. \tag{2.44}$$

Sunsynchronous satellites have a relative Earth rotation rate of exactly 360° per day. They have the interesting property that if they make an integral number of orbits in an integral number of days, then they must arrive exactly at the longitude where they started and repeat the ground track. If such a satellite makes $N + k/m$ orbits per day, where k and m are integers, then the orbit track repeats every m days after making $mN + k$ orbits. Furthermore, if k and m have no common factors, all m of the ground tracks, spaced ΔLON/m, will be traversed. Earth remote sensing satellites have utilized this property. Landsats 1, 2, and 3, for example, were designed to make $13\frac{17}{18}$ orbits per day. Thus they had a nodal period of 103.27 min, which means that ΔLON was 25.82°. The distance between ascending nodes at the equator was about 2874 km. The Multispectral Scanner (MSS) scanned across the satellite track with a ground swath width of only 185 km; only a small fraction of the equator was observed on any one day. The daily longitude drift was −1.43° (−ΔLON/18), or about 160 km west of an ascending node on the previous day. Since the swath width was greater than the westward movement, the swaths on successive days overlapped. In 18 days the satellites observed every point on the equator and began the cycle anew.

The French SPOT satellites (see *SPOT User's Handbook*) utilize the same type of repeat cycle, except that they orbit $14\frac{5}{26}$ times per day. The swaths on successive days do not overlap, but in 26 days the entire Earth is imaged. Landsats 4 and 5 have similar orbits.

Note that this repeat cycle is very sensitive to the semimajor axis. If the orbital altitude of Landsats 1, 2, and 3 had been decreased by only 19 km, they would have made exactly 14 orbits in one day. There would have been no westward progression of the swaths. Some parts of the Earth would be observed every day; the rest would never be observed.

For studies of diurnal variation, a point must be observed at local times throughout the day. Since sunsynchronous satellites view a point at nearly the same two local times every day, they are not useful for diurnal variation studies. A satellite designed specifically to measure diurnal variation is the Earth Radiation Budget Satellite (ERBS; see Chapter 10). It is in a 57°-inclination orbit at an altitude of 600 km. The right ascension of ascending node moves west by 3.955° per day, while the mean sun moves east 0.986° per day (360° in one year). Thus the angle between the sun and the ascending node changes 4.94° per day. Because the satellite makes observations both as it ascends and as it descends, all local times will be sampled when the sun–Earth–ascending node angle has changed by 180°. The ERBS, then, samples all local times in about 36 days.

Many space–time sampling strategies are possible. The reader is encouraged to use the equations presented above to investigate some of the possibilities.

2.7 LAUNCH VEHICLES AND PROFILES

A discussion of satellite orbits would not be complete without mention of the launch vehicles used and the strategies available for achieving orbit.

2.7.1 Launch Vehicles

U.S. meteorological satellites have been launched by a variety of vehicles, but recently two have predominated (Fig. 2.19). The Delta 3914 was used to place the 345-kg GOES 4–7 satellites into 35,790-km geostationary orbits. The Atlas E/F is used to boost the 1421-kg NOAA satellites into 850-km sunsynchronous orbits. Other meteorological satellites have been placed into orbit by the French Ariane rocket, the Soviet F-2 rocket, and the Japanese N-2 rocket, and the U.S. Space Shuttle (see Appendix A). As satellite launches become more commercial and more competitive, many different rockets are likely to be used for meteorological satellites.

It is interesting that a far larger rocket is used to launch the low Earth orbit satellites than the geostationary satellites. In part this is because the energy required to achieve orbit is proportional to the mass of the satellite. However, Fig. 2.20 shows that the Earth is at the bottom of a deep gravitational potential well. The first step into space is the energy-consuming step. It takes approximately 35 MJ kg^{-1} to lift a satellite into an 850 km orbit; it takes only about 65% (23 MJ kg^{-1}) additional energy to increase that orbit by a factor of 42 to geostationary altitude.

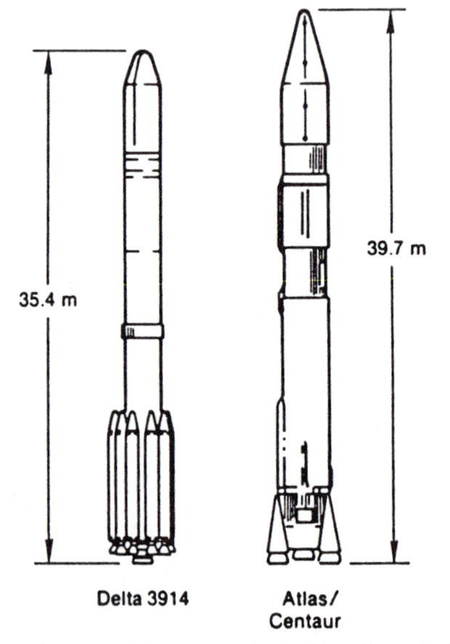

FIGURE 2.19. Rockets used to launch recent U.S. meteorological satellites. [After Chen (1985). Reprinted by permission of Academic Press.]

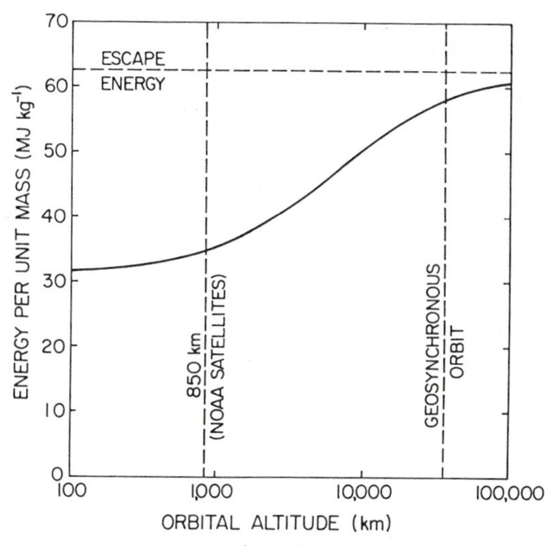

FIGURE 2.20. The energy per unit mass required to place a satellite in orbit as a function of orbital altitude. Note that the considerable energy required to lift the rocket itself is not included in this figure.

2.7.2 Launch Profiles

Three basic strategies are available for orbit insertion. In *power-all-the-way ascents* the rocket burns steadily until the desired orbit is achieved. This launch profile is more costly, but less risky, than the others because rockets do not have to be restarted in space. This profile is used for manned space flights.

The second type of launch profile is called *ballistic ascent* because of its similarity with artillery. A large first-stage rocket is used in the early part of the flight to propel the payload to high velocity. It then coasts to the location of the desired orbit, where a second-stage rocket is fired to adjust the trajectory to the desired orbit.

The third type of launch profile is called *elliptical ascent.* Orbit insertion is achieved in three steps. First the payload is placed in a low Earth orbit by either of the above means. This first orbit is referred to as a *parking orbit.* In the next phase, a rocket is fired to move the payload into an elliptical *transfer orbit* whose perigee is the parking orbit and whose apogee is the desired orbit. When the payload reaches apogee, a rocket (sometimes called an "apogee kick motor") modifies the orbit to the desired (usually circular) shape. Elliptical ascent is used for geostationary satellites.

Bibliography

Barnes, J. C., and M. D. Smallwood (1982). *TIROS-N Series Direct Readout Services Users Guide.* NOAA, Washington, DC.

Brouwer, D., and G. M. Clemence (1961). *Methods of Celestial Mechanics.* Academic Press, New York.

Chen, H. S. (1985). *Space Remote Sensing Systems: An Introduction.* Academic Press, Orlando.

Dubyago, A. D. (1961). *The Determination of Orbits.* The Macmillan Company, New York.
Escobal, P. R. (1965). *Methods of Orbit Determination.* John Wiley and Sons, New York.
Goldstein, H. (1950). *Classical Mechanics.* Addison–Wesley, Reading, MA.
Hambrick, L. N., and Phillips, D. R. (1980). *Earth Locating Image Data of Spin-Stabilized Geosynchronous Satellites.* NOAA Tech. Memo. NESS 111, Washington, DC.
Kidder, S. Q., and T. H. Vonder Haar (1990). On the use of satellites in Molniya orbits for meteorological observation of middle and high latitudes. *J. Atmos. Ocean. Tech.*, 7, 517–522.
Madden, R., and Parsons, C. (1973). A technique for real-time, quantitative display of APT Scanning Radiometer data. *J. Appl. Meteor.*, **12**, 381–385.
Short, N. M., P. D. Lowman, Jr., S. C. Freden, and W. A. Finch, Jr. (1976). *Mission to Earth: Landsat Views the World.* NASA, Washington, DC.
Smith, E. A. (1980). *Orbital Mechanics and Analytic Modeling of Meteorological Satellite Orbits.* Colo. State Univ. Atmos. Sci. Pap. 321, Fort Collins, CO.
SPOT User's Handbook. SPOT Image Corp., Reston, VA.

3

Radiative Transfer

ALL OF THE information received by a satellite about the Earth and its atmosphere comes in the form of electromagnetic radiation. It is necessary, therefore, to understand the mechanisms by which this radiation is generated and how it interacts with the atmosphere. Several texts listed in the Bibliography explore atmospheric radiation in detail. Here we concentrate on those aspects that are essential for satellite meteorology.

3.1 BASIC QUANTITIES

Electromagnetic radiation consists of alternating electric and magnetic fields (Fig. 3.1). The electric field vector is perpendicular to the magnetic field vector, and the direction of propagation is perpendicular to both. Radiation is often specified by its *wavelength,* which is the distance between crests of the electric or magnetic field. Figure 3.2 shows the electromagnetic spectrum. A broad range of wavelengths from the ultraviolet to the microwave region is useful in satellite meteorology.

An alternate way to describe radiation is to give its *frequency,* which is the rate at which the electric or magnetic field oscillates when observed at a point.

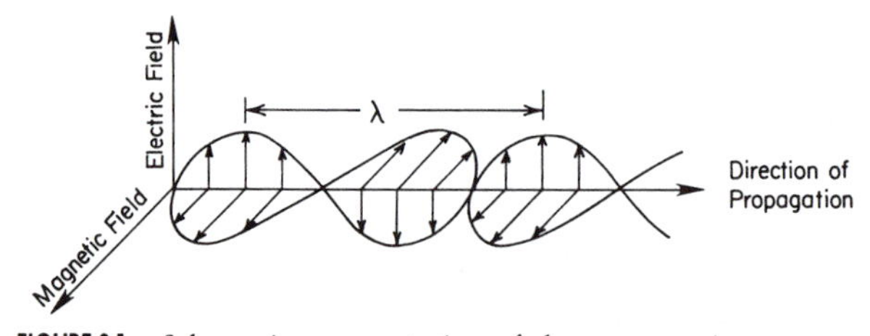

FIGURE 3.1. Schematic representation of electromagnetic waves.

The fundamental unit of frequency is the hertz (Hz), or one cycle per second. The frequency ν is related to the wavelength λ by

$$\nu = \frac{c}{\lambda}, \tag{3.1}$$

where c is the speed with which electromagnetic radiation travels and is known as the *speed of light*. In a vacuum the speed of light is 2.99792458×10^8 m s^{-1}. In the atmosphere, it travels slightly more slowly, due to interaction with air molecules.

The *index of refraction*, n, of a substance is the ratio of the speed of light in vacuum to the speed with which electromagnetic radiation travels in that substance. At sea level, the index of refraction of air is approximately 1.0003. For

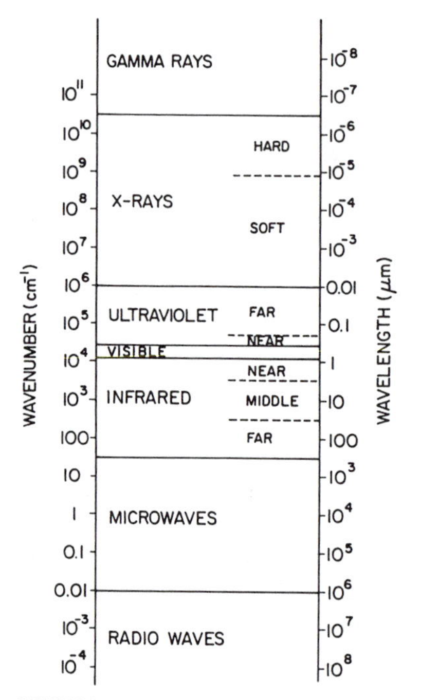

FIGURE 3.2. The electromagnetic spectrum.

TABLE 3.1. Radiation Symbols[a] and Units

Quantity	*Recommended symbol*	*SI Unit*
Frequency	ν	Hz
Wavelength	λ	m
Wavenumber	κ	m^{-1}
Radiant energy	Q	J
Radiant exposure	H	$J\ m^{-2}$
Radiant flux	Φ	W
Radiant flux density	M, E	$W\ m^{-2}$
Radiant exitance	M	$W\ m^{-2}$
Irradiance	E	$W\ m^{-2}$
Radiance	L	$W\ m^{-2}\ sr^{-1}$
Emittance	ε	Unitless
Absorptance	α	Unitless
Reflectance	ρ	Unitless
Transmittance	τ	Unitless
Absorption coefficient	σ_a	m^{-1}
Scattering coefficient	σ_s	m^{-1}
Extinction coefficient	σ_e	m^{-1}
Single-scatter albedo	$\tilde{\omega}$	Unitless
Absorption number	$\tilde{\alpha}$	Unitless
Vertical optical depth	δ	Unitless
Slant-path optical depth	δ_{sl}	Unitless
Scattering angle	ψ_s	rad
Scattering phase function	$p(\psi_s)$	sr^{-1}
Bidirectional reflectance	γ_r	sr^{-1}
Anisotropic reflectance factor	ξ_r	Unitless
Albedo	A	Unitless

[a] We use primarily the radiation symbols recommended by the Radiation Commission of the International Association of Meteorology and Atmospheric Sciences (IAMAS) as described in Raschke (1978).

most purposes, therefore, the speed of light in a vacuum can be safely used even in the atmosphere. However, strong vertical gradients of atmospheric density and humidity result in strong vertical gradients of n (see Section 3.5.3). These cause bending of electromagnetic rays and can cause slight mislocation of satellite scan spots.

One also sees radiation specified by *wavenumber*, κ, which is the reciprocal of the wavelength. Traditionally, wavenumber is expressed in inverse centimeters. Radiation with a 15-μm wavelength has a 667-cm^{-1} wavenumber, for example. Since wavenumber is inversely proportional to wavelength, it is directly proportional to frequency.

A fundamental property of electromagnetic radiation is that it can transport energy. Many of the units used to quantify electromagnetic radiation are based on energy.[1] These units are summarized in Table 3.1. The basic unit of *radiant*

[1] Some units, such as the lumen and the candela, are based on how bright an object appears to the human eye. These units are no longer used in satellite meteorology.

energy is the joule. *Radiant flux* is radiant energy per unit time, measured in watts [W; joules per second (J s^{-1})]. Radiant flux depends on area, which is often inconvenient; it is usually normalized by surface area. *Radiant flux density* is radiant flux crossing a unit area. It is measured, of course, in watts per square meter (W m^{-2}). Radiant flux density is so frequently used that it is subdivided to indicate which way the energy is traveling. *Radiant exitance* (M) is radiant flux density emerging from an area, and *irradiance* (E) is radiant flux density incident on an area.

In nature, radiation is a function of direction. The directional dependence is taken into account by employing the *solid angle*. If one draws lines from the center of the unit sphere to every point on the surface of an object, the area of the projection on the unit sphere is the solid angle (Fig. 3.3). The solid angle of an object that completely surrounds a point is 4π steradians (sr), the area of the unit sphere. The solid angle subtended by an infinite plane is 2π sr. For an object with cross-sectional area A_c at a distance r from a point ($A_c \ll r^2$), the solid angle is A_c/r^2. Solid angle is traditionally represented by the symbol Ω. If θ represents the zenith angle (the angle measured from the vertical or from the normal to a surface), and ϕ represents the azimuth angle (Fig. 3.4), then a differential element of solid angle is mathematically given by

$$d\Omega = \sin\theta d\theta d\phi = -d\mu d\phi, \tag{3.2}$$

where $\mu \equiv \cos\theta$.

Radiant flux density per unit solid angle is known as *radiance* and is preferably assigned the symbol L. Suppose that a small element of surface is emitting radiation with radiance L. A question that arises is: What is the radiant exitance, that is, what is the total amount of radiation leaving the surface? This question is answered by integrating the radiance over the 2π sr above the surface. However, radiance represents the radiation leaving (or incident on) an area *perpendicular* to the beam. For other directions, we must weight the radiance by $\cos\theta$. Therefore, the

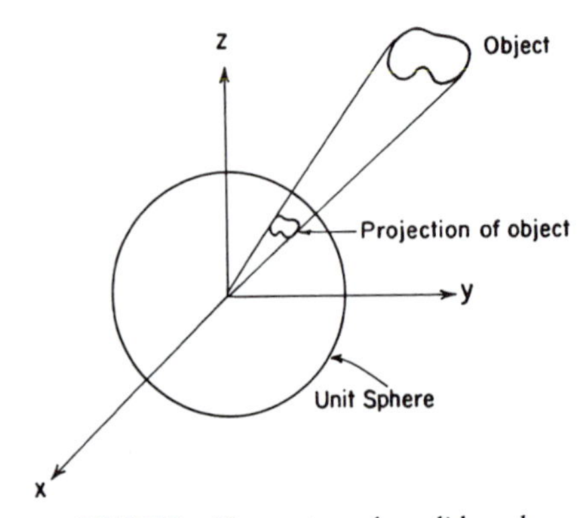

FIGURE 3.3. Illustration of a solid angle.

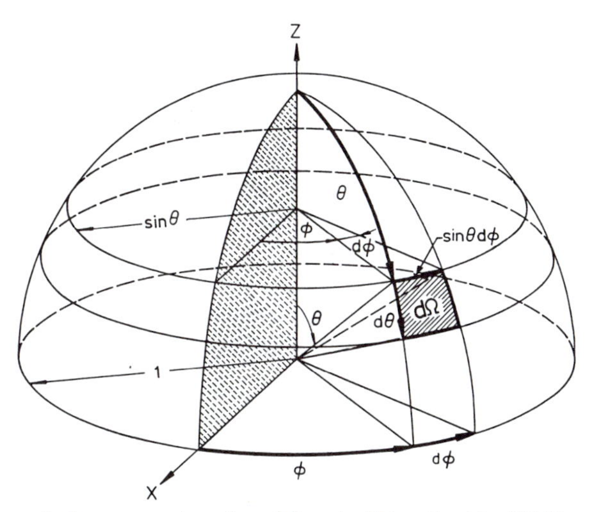

FIGURE 3.4. Mathematical representation of a solid angle. [After Raschke (1978), with permission of the International Association of Meteorology and Atmospheric Sciences.]

radiant exitance is

$$M = \int_0^{2\pi} L(\theta,\phi) \cos\theta \, d\Omega = \int_0^{2\pi} \int_0^{\pi/2} L(\theta,\phi) \cos\theta \sin\theta \, d\theta \, d\phi \tag{3.3}$$

$$= \int_0^{2\pi} \int_0^1 L(\mu,\phi) \, \mu \, d\mu \, d\phi$$

If the radiance is independent of direction (*isotropic*), then

$$M = L \int_0^{2\pi} \int_0^{\pi/2} \cos\theta \sin\theta \, d\theta \, d\phi = \pi L. \tag{3.4}$$

The above energy-based quantities may all be prefixed with the word *monochromatic* or *spectral* to indicate the wavelength dependence of the radiation. A subscript (λ, ν, or κ) is used to indicate whether wavelength, frequency, or wavenumber is being considered. Because radiance is simply the integral over all wavelengths (frequencies, wavenumbers) of monochromatic radiance, we must have

$$L = \int_0^{\infty} L_\lambda d\lambda = \int_0^{\infty} L_\nu d\nu = \int_0^{\infty} L_\kappa d\kappa \tag{3.5}$$

or

$$L_\lambda = -\frac{d\nu}{d\lambda} L_\nu = \frac{\nu^2}{c} L_\nu = \kappa^2 L_\kappa. \tag{3.6}$$

In other words, radiation per unit wavelength interval is the same as radiation per unit frequency or wavenumber interval, if we take into account the different size of the intervals.

The most fundamental radiation unit for satellite meteorology is monochromatic radiance, which is the energy per unit time per unit wavelength (frequency, wavenumber) per unit solid angle crossing a unit area perpendicular to the beam. The reason radiance is most fundamental is that the basic satellite instrument, the radiometer (see Chapter 4), has a detector of a certain area whose output is proportional to the energy per unit time striking it. Further, the sensor is usually at the focus of a telescope which collects radiation from a certain solid angle and which has filters that pass radiation of only a certain narrow range of wavelengths. Normalizing the sensor output by area, solid angle, and wavelength range results in a quantity that is most closely related to monochromatic radiance.

Radiance also has the useful property that it is independent of distance from an object as long as the viewing angle and the amount of intervening matter are not changed. Consider a satellite viewing a small object. The irradiance reaching the satellite from the object will decrease inversely as the square of the distance of the satellite. However, the solid angle of the object subtended at the satellite will also decrease inversely as the square of the distance of the satellite. The radiance of the object as viewed by the satellite, which is simply the irradiance divided by the solid angle, is, therefore, independent of distance. Of course, if the object were at the Earth's surface, its radiance measured at the Earth's surface would be different from that measured at the satellite, due to the intervening atmosphere.

3.2 BLACKBODY RADIATION

All material above absolute zero in temperature emits radiation. Explaining the nature of this radiation was one of the chief problems facing physicists in the nineteenth century. As usual, though, nature guards her secrets. If one looks at two different kinds of material, each at the same temperature, one finds that the radiation being emitted by them is different. This led physicists to invent the perfect emitter, known as a *blackbody*, which emits the maximum amount of radiation at each wavelength. Although some materials come very close to being perfect emitters in some wavelength ranges, no real material is a perfect blackbody. Fortunately, the radiation inside a cavity whose walls are thick enough to prevent any radiation from passing directly through them can be shown to be the radiation that would be emitted by a blackbody. By observing the radiation inside cavities (through small holes) physicists knew, by the late nineteenth century, the empirical relationship between blackbody radiation and the two variables on which it depends: temperature and wavelength.

3.2.1 The Planck Function

In searching for a theoretical derivation of blackbody radiation, Planck[2] made the revolutionary assumption that an oscillating atom in the wall of a cavity can exchange energy with the radiation field inside a cavity only in discrete bundles called *quanta* given by $\Delta E = h\nu$, where h is known as *Planck's constant* (see Appendix E). With this assumption, he showed that the radiance being emitted by a blackbody is given by

$$B_\lambda(T) = \frac{2hc^2\lambda^{-5}}{\exp\left(\dfrac{hc}{\lambda kT}\right) - 1}, \tag{3.7}$$

where k is Boltzmann's[3] constant (see Appendix E), and T is the absolute temperature. Equation 3.7 is known as the *Planck function*; it earned him the Nobel Prize in 1918. The Planck function is more conveniently written as

$$B_\lambda(T) = \frac{c_1\lambda^{-5}}{\exp\left(\dfrac{c_2}{\lambda T}\right) - 1}, \tag{3.8}$$

where c_1 and c_2 are the first and second radiation constants (see Appendix E). Since the radiance from a blackbody is independent of direction, the radiant exitance from a blackbody is simply πB_λ.

Figure 3.5 shows $B_\lambda(T)$ plotted versus wavelength and temperature. It is important to note that $B_\lambda(T)$ is a monotonically increasing function of T. For a particular wavelength $\lambda = \Lambda$, if T_1 is less than T_2, then $B_\Lambda(T_1)$ is less than $B_\Lambda(T_2)$. $B_\lambda(T)$ is not monotonic in λ. For any temperature T, $B_\lambda(T)$ has a single maximum at a wavelength that may be determined by setting the partial derivative of $B_\lambda(T)$ with respect to λ equal to zero. The result is known as *Wein's*[4] *displacement law:*

$$\lambda_m T = 2897.9\ \mu\text{m K}, \tag{3.9}$$

where λ_m is the wavelength (expressed in micrometers) of maximum emission for a blackbody at temperature T (expressed in kelvins). Wein's displacement law was discovered empirically; he was awarded the Nobel Prize for it in 1911. The dashed line in Fig. 3.5 shows this relationship.

Another important aspect of the Planck function is its integral over wavelength. The total radiant exitance from a blackbody is

$$M_{BB} = \int_0^\infty \pi B_\lambda(T)\,d\lambda = \frac{\pi^5}{15}c_1c_2^{-4}T^4 = \sigma T^4, \tag{3.10}$$

[2] Max Karl Ernst Ludwig Planck, German physicist, 1858–1947.
[3] Ludwig Eduard Boltzmann, Austrian physicist, 1844–1906.
[4] Wilhelm Wein, German physicist, 1864–1928.

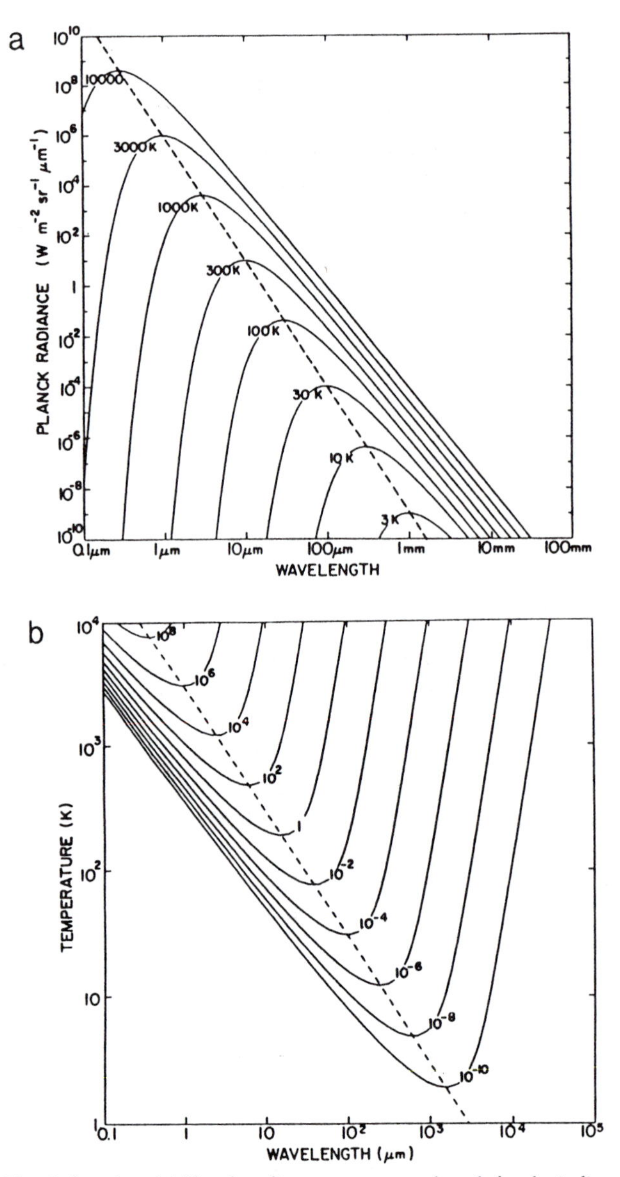

FIGURE 3.5. The Planck function. (a) Planck radiance versus wavelength for the indicated temperature, (b) temperature versus wavelength for the indicated Planck radiances (W $m^{-2}sr^{-1}\mu m^{-1}$), (c) Planck radiance versus temperature for the indicated wavelengths. The dashed line is Wein's displacement law, which gives the wavelength of maximum emission as a function of temperature.

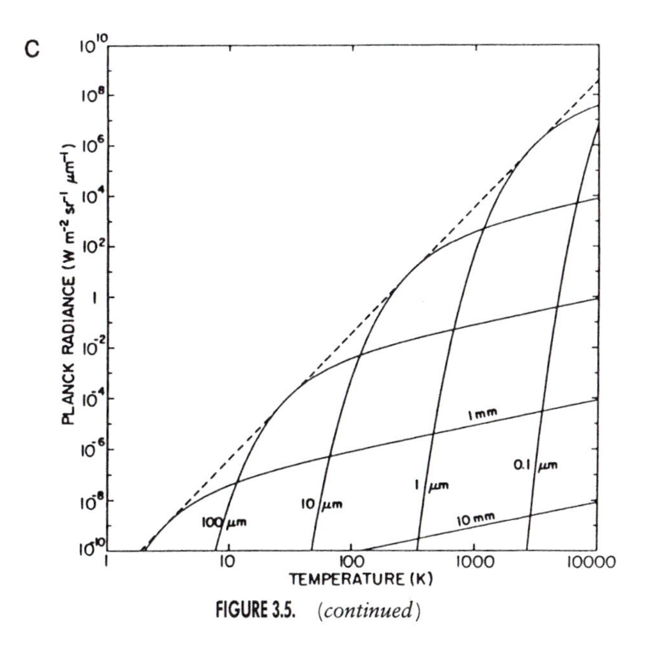

FIGURE 3.5. *(continued)*

where σ is called the *Stefan–Boltzmann constant* (see Appendix E), and Eq. 3.10 is called the *Stefan–Boltzmann law.*[5]

Finally, we would like to discuss a useful approximation to the Planck function. At millimeter and centimeter wavelengths, for temperatures encountered on the Earth and in its atmosphere, $c_2/\lambda T \ll 1$. Thus $\exp(c_2/\lambda T)$ can be approximated by $1 + c_2/\lambda T$. The Planck function then becomes

$$B_\lambda(T) = \frac{c_1}{c_2}\lambda^{-4}T. \tag{3.11}$$

This is known as the *Rayleigh–Jeans*[6] *approximation.* It says that in the microwave portion of the spectrum, radiance is simply proportional to temperature. In fact, in the microwave region, it is customary to divide radiance values by $(c_1/c_2)\lambda^{-4}$ and to refer to the quotient as *brightness temperature.* Brightness temperature is also used in the infrared portion of the spectrum, where it is known as *equivalent blackbody temperature.* However, equivalent blackbody temperature must be found by inverting the Planck function rather than by simple division.

[5] Named after Josef Stefan, Austrian physicist, 1835–1893, who discovered it by observing the cooling rates of hot bodies; and Boltzmann, who demonstrated it using thermodynamics.

[6] After John William Strutt, 3rd Baron Rayleigh, English physicist, 1842–1919; and Sir James Hopwood Jeans, English mathematician and astronomer, 1877–1946.

3.2.2 Nonblackbodies

Since real material is not perfectly black, a way must be devised to quantify how closely it approximates a blackbody. The *emittance* of a body is defined as

$$\varepsilon_\lambda \equiv \frac{\text{emitted radiation at } \lambda}{B_\lambda(T)}. \tag{3.12}$$

Emittance can be a function of temperature and viewing geometry as well as wavelength. For a blackbody, ε_λ is identically one. Three related quantities describe the fate of radiation incident on a body:

$$\alpha_\lambda = \text{absorptance} \equiv \frac{\text{absorbed radiation at } \lambda}{\text{incident radiation at } \lambda}, \tag{3.13a}$$

$$\rho_\lambda = \text{reflectance} \equiv \frac{\text{reflected radiation at } \lambda}{\text{incident radiation at } \lambda}, \tag{3.13b}$$

$$\tau_\lambda = \text{transmittance} \equiv \frac{\text{transmitted radiation at } \lambda}{\text{incident radiation at } \lambda}. \tag{3.13c}$$

Because these three processes are the only possibilities for the incident radiation,[7] by energy conservation, each quantity must be between zero and one, and

$$\alpha_\lambda + \rho_\lambda + \tau_\lambda \equiv 1. \tag{3.14}$$

Kirchhoff[8] discovered that a body is exactly as good an absorber as it is an emitter.[9] This is summarized in *Kirchhoff's law:*

$$\alpha_\lambda \equiv \varepsilon_\lambda. \tag{3.15}$$

This law applies only to material that is in *local thermodynamic equilibrium*, which means that it can be characterized by a single thermodynamic temperature. This is a good assumption below about 100 km in the Earth's atmosphere. Above 100 km, collisions between molecules are rare enough that different chemical species can have different thermodynamic temperatures. For most satellite meteorology applications, however, the Earth's atmosphere can be considered to be in local thermodynamic equilibrium.

Since the emittance of a blackbody is by definition one, its absorptance must also be one; that is, a blackbody, in addition to being a perfect emitter, is also a perfect absorber. It therefore appears black, thus the name *blackbody.*

The emittance of real materials is enormously variable. Shown in Fig. 3.6 is the emittance of two materials used in the Suomi radiometer discussed in Chapter 1. The black paint is supposed to approximate a blackbody by absorbing all radiation incident on it. Anodized aluminum looks like flat white paint. It is supposed to reflect solar radiation and absorb infrared radiation emitted by the

[7] We will not consider Raman scattering or fluorescence, in which radiation is absorbed at one wavelength and reemitted at another.

[8] Gustav Robert Kirchhoff, German physicist, 1824–1887.

[9] Assuming the contrary can be shown to violate the second law of thermodynamics.

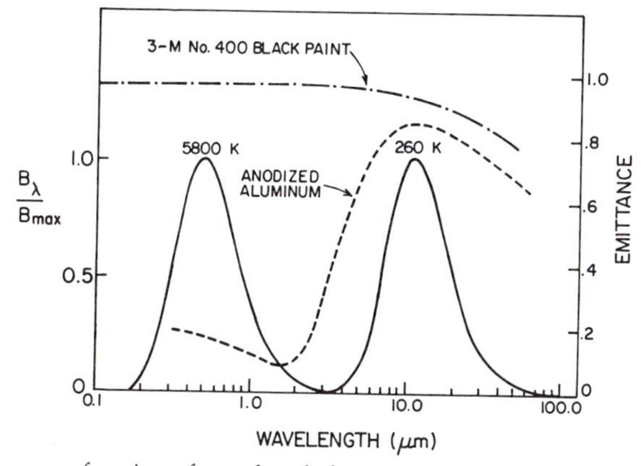

FIGURE 3.6. Emittance as a function of wavelength for two materials used in a satellite radiometer. Normalized Planck curves representing solar radiation (5800 K) and terrestrial radiation (260 K) are shown. (Do not be deceived by these normalized curves. The sun's radiance is larger than the Earth's at every wavelength.) [Adapted from Smith, W. L. (1985). Satellites. In D. D. Houghton (ed.), *Handbook of Applied Meteorology*. Copyright © 1985, John Wiley & Sons, Inc. Reprinted by permission of John Wiley & Sons, Inc.]

Earth and its atmosphere (*terrestrial radiation*). The difference between measurements made with instruments coated with these materials is related to the amount of incident solar radiation.

3.3 THE RADIATIVE TRANSFER EQUATION

At last we are ready to discuss the transfer of electromagnetic radiation through the atmosphere. Consider radiation incident on a differential volume of atmosphere (or, more generally, material) shown in Fig. 3.7. The equation that will be developed deals with the change in radiance as the radiation passes through the volume. Radiance is the appropriate variable because, as explained above, if there were no material in the volume, the radiance would not change.

If we do not consider polarization effects, four processes can change the radiance as it passes through the volume:

A. Radiation from the beam can be absorbed by the material.
B. Radiation can be emitted by the material.
C. Radiation can be scattered out of the beam into other directions.
D. Radiation from other directions can be scattered into the beam.

The rate of change of radiance with distance, dL_λ/ds, then consists of the above four terms:

$$\frac{dL_\lambda}{ds} = A + B + C + D. \tag{3.16}$$

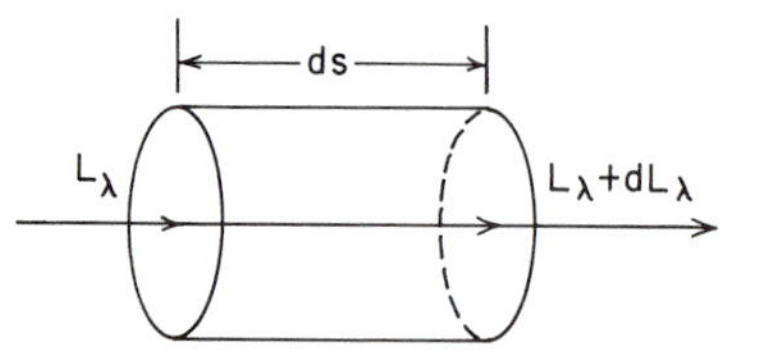

FIGURE 3.7. Differential volume element containing material which alters a beam of radiation passing through it.

Terms A and C remove radiation from the beam and are known as *depletion terms*. Terms B and D add radiation to the beam and are known as *source terms*.

Beer's law[10] states that the rate of decrease in the intensity of radiation as it passes through a medium is proportional to the intensity of the radiation. Term A, therefore, takes the form $-\sigma_a(\lambda)L_\lambda$, where $\sigma_a(\lambda)$ is the *volume absorption coefficient* and is equal to $\rho\beta_a(\lambda)$, where ρ is the density of absorber and $\beta_a(\lambda)$ is the *mass absorption coefficient*.[11] If Beer's law is integrated over a finite depth of absorber from a to b, then

$$L_\lambda = L_o \exp\left(-\int_a^b \sigma_a(\lambda)\,ds\right), \tag{3.17}$$

where s is distance along the path and L_o is the radiance incident on the absorber. If there were no scattering, the transmittance (Eq. 3.13c) would be

$$\tau_\lambda(a, b) = \exp\left(-\int_a^b \sigma_a(\lambda)\,ds\right), \tag{3.18}$$

and the absorptance (Eq. 3.13a) would be $1 - \tau_\lambda$.

The emission with which we will be concerned is Planckian emission. Since by Kirchhoff's law a material is as good an emitter as an absorber, term B takes the form $+\sigma_a(\lambda)B_\lambda(T)$.

Scattering of radiation out of the beam follows Beer's law. Term C takes the form $-\sigma_s(\lambda)L_\lambda$, where $\sigma_s(\lambda)$ is the *volume scattering coefficient*.

Finally, Term D describes the amount of radiation from other directions that is scattered into the beam. This term is complicated by the fact that all directions must be considered. If we are concerned with radiation traveling in a direction specified by the angles θ and ϕ, then Term D is given by

$$\text{Term } D = +\frac{\sigma_s(\lambda)}{4\pi}\int_0^{2\pi}\int_0^{\pi} L_\lambda(\theta',\phi')\,p(\psi_s)\,\sin\theta'\,d\theta'\,d\phi', \tag{3.19}$$

[10] After Wilhelm Beer, German astronomer, 1779–1850. It is also known as *Bouguer's law*, after the French mathematician Pierre Bouguer (1698–1758), who first published it in 1729; and as *Lambert's law*, after the German mathematician Johann Heinrich Lambert (1728–1777).

[11] σ_a has SI units of m^{-1} (reciprocal meters); ρ, of course, has SI units of kg m^{-3}; and β_a has SI units of m^2 kg^{-1}.

where (θ',ϕ') represents the direction of incoming radiation, and ψ_s is the *scattering angle* (the angle between θ, ϕ and θ', ϕ'):

$$\cos\psi_s = \cos\theta\cos\theta' + \sin\theta\sin\theta'\cos(\phi - \phi'). \tag{3.20}$$

$p(\psi_s)$ is called the *scattering phase function*. Basically, it tells what portion of the radiation from direction (θ',ϕ') is scattered into direction (θ,ϕ). $p(\psi_s)$ has the property that

$$\frac{1}{4\pi}\int_0^{2\pi}\int_0^{\pi} p(\psi_s)\sin\theta'\,d\theta'\,d\phi' \equiv 1. \tag{3.21}$$

(For an isotropic scatterer, $p(\psi_s) \equiv 1$.) Term D can be thought of as the product of $\sigma_s(\lambda)$ and a directionally weighted average $[p(\psi_s)$ is the weight$]$ of the radiance from all directions:

$$\text{Term } D = \sigma_s(\lambda)\langle L'_\lambda\rangle, \tag{3.22}$$

where

$$\langle L'_\lambda\rangle \equiv \frac{1}{4\pi}\int_0^{2\pi}\int_0^{\pi} L_\lambda(\theta',\phi')\,p(\psi_s)\sin\theta'\,d\theta'\,d\phi'. \tag{3.23}$$

Combining all terms, the *radiative transfer equation* for nonpolarized radiation[12] becomes

$$\frac{dL_\lambda}{ds} = -\sigma_a(\lambda)L_\lambda(\theta,\phi) - \sigma_s(\lambda)L_\lambda(\theta,\phi) + \sigma_a(\lambda)B_\lambda(T) + \frac{\sigma_s(\lambda)}{4\pi}\int_0^{2\pi}\int_0^{\pi} L_\lambda(\theta',\phi')\,p(\psi_s)\sin\theta'\,d\theta'\,d\phi'. \tag{3.24}$$

This is a very complex equation. It is useful to examine its physical meaning before attempting to solve it.

Substituting Eq. 3.23 and rearranging slightly, the radiative transfer equation becomes

$$\frac{dL_\lambda}{ds} = \sigma_a(\lambda)[B_\lambda(T) - L_\lambda(\theta,\phi)] + \sigma_s(\lambda)[\langle L'_\lambda\rangle - L_\lambda(\theta,\phi)]. \tag{3.25}$$

Consider a beam of radiation upwelling through a thin atmospheric layer on its way to a satellite. The first term on the right-hand side of Eq. 3.25 represents the effects of absorption and emission. If $\sigma_a(\lambda)$ is zero, the layer is transparent absorptionally, and the beam passes through it unchanged. If $\sigma_a(\lambda)$ is not zero, the temperature of the layer and the radiance itself determine the change in the beam. The beam is augmented if $B_\lambda(T)$ is greater than $L_\lambda(\theta,\phi)$; it is diminished if $B_\lambda(T)$ is less than $L_\Lambda(\theta,\phi)$. The second term on the right represents the effects

[12] The radiative transfer equation for polarized radiation is not much more complicated. Polarized radiation can be described by a vector whose elements are the four Stokes parameters. The scattering phase function becomes a 4 × 4 matrix. The reader is referred to Liou (1980) for further details.

of scattering. If scattering particles are absent, $\sigma_s(\lambda)$ is zero, and scattering has no effect on the beam. If $\sigma_s(\lambda)$ is nonzero, the beam is augmented if the directionally weighted average radiance $\langle L'_\lambda \rangle$ is greater than the beam radiance $L_\lambda(\theta,\phi)$; it is diminished if $\langle L'_\lambda \rangle$ is less than $L_\lambda(\theta,\phi)$.

The radiative transfer equation can be formally solved. Since we will not use the formal solution, we refer readers to Goody and Yung (1989) for further discussion of it. Instead, we will concentrate on several simplifications that are used in satellite meteorology.

The first two terms on the right-hand side of Eq. 3.24 can be combined to form $-\sigma_e(\lambda) L_\lambda(\theta,\phi)$, where

$$\sigma_e(\lambda) \equiv \sigma_a(\lambda) + \sigma_s(\lambda) \tag{3.26}$$

is called the *volume extinction coefficient.*

It is convenient to divide the radiative transfer equation by the extinction coefficient and to introduce a new variable, δ_{sl}, the *slant path optical depth*:

$$\delta_{sl}(s_1,s_2) \equiv \int_{s_1}^{s_2} \sigma_e(\lambda,s)\,ds. \tag{3.27}$$

Because most meteorological variables are known as a function of height z rather than of slant path s (Fig. 3.8), we will use the *vertical optical depth*

$$\delta_\lambda(z_1,z_2) \equiv \int_{z_1}^{z_2} \sigma_e(\lambda,z)\,dz. \tag{3.28}$$

Since the Earth's atmosphere is thin in comparison with the radius of the Earth, the two optical depths are related by

$$\delta_{sl}(s_1,s_2) = \frac{\delta_\lambda(z_1,z_2)}{\mu}, \tag{3.29}$$

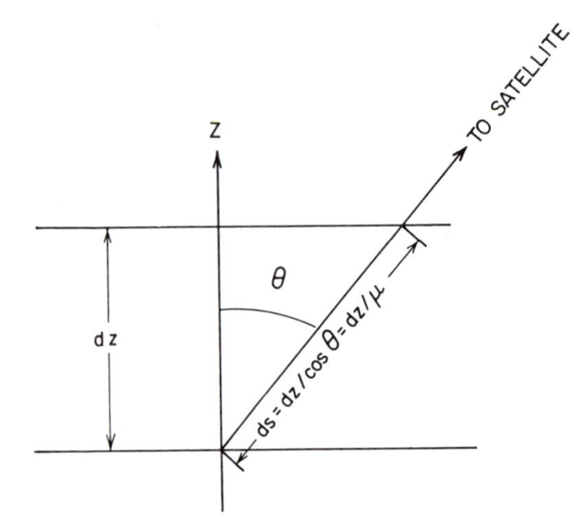

FIGURE 3.8. Relationship of depth to slant path through an infinitesimal atmospheric slab.

where μ is the cosine of the zenith angle θ, z_1 corresponds to s_1, and z_2 corresponds to s_2. We will use the symbol δ_λ without arguments to mean the vertical optical depth between the surface and level z, that is, $\delta_\lambda = \delta_\lambda(0,z)$, and therefore, $d\delta_\lambda = \sigma_e dz = \mu\sigma_e ds$.

Two additional definitions are useful. The *absorption number*, $\tilde{\alpha}_\lambda$, is defined as

$$\tilde{\alpha}_\lambda \equiv \frac{\sigma_a(\lambda)}{\sigma_e(\lambda)}, \tag{3.30}$$

and the *single-scatter albedo*, $\tilde{\omega}_\lambda$, is defined as

$$\tilde{\omega}_\lambda \equiv \frac{\sigma_s(\lambda)}{\sigma_e(\lambda)}. \tag{3.31}$$

With these definitions, the radiative transfer equation becomes

$$\mu \frac{dL_\lambda}{d\delta_\lambda} = -L_\lambda(\theta,\phi) + \tilde{\alpha}_\lambda B_\lambda(T) + \frac{\tilde{\omega}_\lambda}{4\pi}\int_0^{2\pi}\int_{-1}^{1} L_\lambda(\mu',\phi')\,p(\psi_s)\,d\mu'\,d\phi'. \tag{3.32}$$

Note that we have substituted μ for $\cos\theta$ in this equation, as is commonly done.

Now we are ready to simplify and solve the radiative transfer equation. We will explore two common simplifications: the no-scattering case and the no-emission case.

3.3.1 No-Scattering Equations

A common assumption is that scattering is negligible. This is a good assumption in the infrared portion of the spectrum in the absence of clouds. If there is no scattering, the single-scatter albedo is zero, and the absorption number is one. The radiative transfer equation then becomes

$$\mu \frac{dL_\lambda}{d\delta_\lambda} = -L_\lambda(\theta,\phi) + B_\lambda(T), \tag{3.33}$$

which is known as *Schwarzchild's equation.*[13]

The infrared radiance observed by a satellite (in the absence of scattering) can be calculated by formally integrating Schwarzchild's equation. Since Schwarzchild's equation is a first-order, linear, ordinary differential equation, its integration is straightforward. A detailed derivation is justified, however, because the result is arguably the most important equation in satellite meteorology. To start, we rearrange Eq. 3.33 and multiply by the integrating factor $\exp(\delta_\lambda/\mu)/\mu$:

$$\exp\left(\frac{\delta_\lambda}{\mu}\right)\frac{dL_\lambda}{d\delta_\lambda} + \frac{1}{\mu}\exp\left(\frac{\delta_\lambda}{\mu}\right)L_\lambda(\theta,\phi) = \frac{1}{\mu}\exp\left(\frac{\delta_\lambda}{\mu}\right)B_\lambda(T). \tag{3.34}$$

The left-hand side is now an exact differential:

$$\frac{d}{d\delta_\lambda}\left[\exp\left(\frac{\delta_\lambda}{\mu}\right)L_\lambda(\theta,\phi)\right] = \frac{1}{\mu}\exp\left(\frac{\delta_\lambda}{\mu}\right)B_\lambda(T). \tag{3.35}$$

[13] After Karl Schwarzchild, German astronomer, 1873–1916.

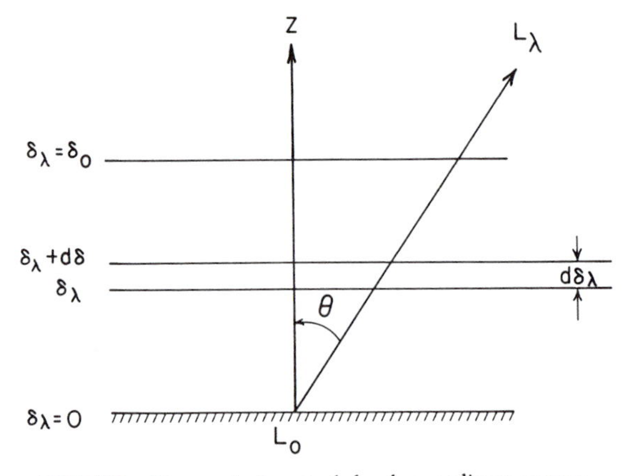

FIGURE 3.9. The vertical optical depth coordinate system.

Using δ_λ as the vertical coordinate (Fig. 3.9), Eq. 3.35 can be integrated from the Earth's surface ($\delta_\lambda = 0$) to the satellite ($\delta_\lambda = \delta_o$):[14]

$$\int_0^{\delta_o} \frac{d}{d\delta_\lambda}\left[\exp\left(\frac{\delta_\lambda}{\mu}\right) L_\lambda(\theta,\phi)\right] d\delta_\lambda = \int_0^{\delta_o} \exp\left(\frac{\delta_\lambda}{\mu}\right) B_\lambda(T)\,\frac{d\delta_\lambda}{\mu}. \tag{3.36}$$

Because the left-hand side of Eq. 3.36 is the integral of an exact differential, the integral is simply

$$\left[\exp\left(\frac{\delta_\lambda}{\mu}\right) L_\lambda\right]_0^{\delta_o} = \exp\left(\frac{\delta_o}{\mu}\right) L_\lambda - L_o = \int_0^{\delta_o} \exp\left(\frac{\delta_\lambda}{\mu}\right) B_\lambda(T)\,\frac{d\delta_\lambda}{\mu}, \tag{3.37}$$

where L_λ is the radiance reaching the satellite, and L_o is the radiance leaving the surface. Solving for L_λ:

$$L_\lambda = L_o \exp\left(-\frac{\delta_o}{\mu}\right) + \int_0^{\delta_o} \exp\left(-\frac{(\delta_o - \delta_\lambda)}{\mu}\right) B_\lambda(T)\,\frac{d\delta_\lambda}{\mu}. \tag{3.38}$$

This equation forms the basis for sounding the atmosphere and for corrections necessary for surface parameter estimation (Chapter 6).

The vertical transmittance (see Section 3.2.2) of the atmospheric layer between optical depths δ_1 and δ_2 is

$$\tau_\lambda(\delta_1,\delta_2) \equiv \exp(-|\delta_2 - \delta_1|). \tag{3.39}$$

We will use the symbol τ_λ without arguments to mean the vertical transmittance of the layer between level δ_λ and the satellite (δ_o),[15] that is, $\tau_\lambda \equiv \tau_\lambda(\delta_\lambda,\delta_o)$. The

[14] Meteorological satellites orbit well above the effective top of the atmosphere; therefore, δ_o is the vertical optical depth of the entire atmosphere.

[15] This definition is a little unusual. The customary definition is $\tau_\lambda \equiv \exp(-\delta_\lambda)$. However, this transmittance is that between the surface and level δ_λ. Since the transmittance between δ_λ and δ_o is a much more useful quantity, we adopt $\tau_\lambda \equiv \exp[-(\delta_o - \delta_\lambda)]$.

transmittance between the surface and the satellite is $\tau_o \equiv \tau_\lambda(0,\delta_o)$. With the aid of these definitions, the integrated form of Schwarzchild's equation has a simple physical explanation.

Equation 3.38 says that the radiance reaching a satellite comes from two sources. The first term on the right-hand side is the surface term. L_o is the radiance leaving the surface; $\exp(-\delta_o/\mu) = \tau_o^{1/\mu}$ is the transmittance of the entire atmosphere along the slant path to the satellite. The product is that portion of the surface radiance which reaches the satellite. The second term is the contribution of the atmosphere. Since we are not considering scattering,

$$\frac{d\delta_\lambda}{\mu} = \sigma_a \frac{dz}{\mu}, \tag{3.40}$$

that is, $d\delta_\lambda/\mu$ is the emittance of the atmospheric layer between δ_λ and $\delta_\lambda + d\delta_\lambda$. Therefore, $B_\lambda(T)d\delta_\lambda/\mu$ is the radiance emitted in the direction of the satellite by the layer. The factor $\exp[-(\delta_o - \delta_\lambda)/\mu] = \tau_\lambda^{1/\mu}$ is the slant path transmittance from δ_λ to the satellite. Therefore, the product $\exp[-(\delta_o - \delta_\lambda)/\mu]B_\lambda(T)d\delta_\lambda/\mu$ is that portion of the radiance emitted by the atmospheric layer between δ_λ and $\delta_\lambda + d\delta_\lambda$ which reaches the satellite. The integral indicates that the contributions of all atmospheric layers are to be summed.

A useful rule of thumb is that a satellite receives the maximum radiation from the layer which is one optical depth into the atmosphere, that is, where $(\delta_o - \delta_\lambda)/\mu = 1$. This rule is exact for the uninteresting case of an isothermal atmosphere and is approximately correct for other atmospheres.

Note that it is sometimes convenient to use transmittance as the vertical coordinate:

$$L_\lambda = L_o\tau_o^{1/\mu} + \int_{\tau_o}^{1} B_\lambda(T)\, \tau_\lambda^{(1/\mu-1)} \frac{d\tau_\lambda}{\mu}. \tag{3.41}$$

For overhead viewing ($\mu = 1$),

$$L_\lambda = L_o\tau_o + \int_{\tau_o}^{1} B_\lambda(T)\, d\tau_\lambda. \tag{3.42}$$

3.3.2 No-Emission Equations

Neither the Earth nor its atmosphere emits significant radiation at visible or near-infrared wavelengths. For these wavelengths, Planck emission $[B_\lambda(T)]$ may be neglected, and the radiative transfer equation becomes

$$\mu \frac{dL_\lambda}{d\delta_\lambda} = -L_\lambda(\mu,\phi) + \frac{\tilde{\omega}_\lambda}{4\pi} \int_0^{2\pi} \int_{-1}^{1} L_\lambda(\mu',\phi')\, p(\psi_s)\, d\mu'\, d\phi'. \tag{3.43}$$

Neither liquid water nor water vapor absorbs much radiation in the visible portion of the spectrum. At visible wavelengths, it is often acceptable to assume that absorption in clouds is zero, or equivalently that $\tilde{\omega}_\lambda = 1$. This approximation is called *conservative scattering* because radiative energy is conserved: all of the energy that is incident on a cloud is assumed to emerge from it, although in various directions.

The source of ultraviolet, visible, and near-infrared radiation is, of course, the sun. It is convenient and traditional to divide this radiation into *direct-beam radiation*, which comes directly from the sun without any interaction with the atmosphere, and *diffuse radiation*, which has undergone at least one scattering event in the atmosphere or reflection from the surface. Direct-beam radiation follows Beer's law:

$$L_\lambda^{\text{direct}} = L_\lambda^{\text{sun}} \exp\left(-\frac{\delta_o - \delta_\lambda}{\mu_{\text{sun}}}\right), \tag{3.44}$$

where L_λ^{sun} is the sun's radiance at the top of the atmosphere, and μ_{sun} is the cosine of the solar zenith angle. The diffuse radiation is described by Eq. 3.43, but one correction must be made. When the second term of Eq. 3.43 is calculated, both direct and diffuse radiation must be considered. Since the sun's radiation comes from essentially only one direction $(\theta_{\text{sun}}, \phi_{\text{sun}})$, the direct-beam term can be taken out of the integral, leaving only the diffuse radiance:

$$\mu \frac{dL_\lambda^{\text{diffuse}}}{d\delta_\lambda} = -L_\lambda^{\text{diffuse}}(\mu,\phi) + \frac{\tilde{\omega}_\lambda}{4\pi} \int_0^{2\pi} \int_{-1}^{1} L_\lambda^{\text{diffuse}}(\mu',\phi')\, p(\psi_s)\, d\mu'\, d\phi'$$

$$+ \frac{\tilde{\omega}_\lambda}{4\pi} L_\lambda^{\text{sun}} \Omega_{\text{sun}} \exp\left(-\frac{\delta_o - \delta_\lambda}{\mu_{\text{sun}}}\right) p(\psi_{\text{sun}}), \tag{3.45}$$

where Ω_{sun} is the solid angle subtended by the sun, and ψ_{sun} is the scattering angle between the sun and the observed direction (θ,ϕ). The product $L_\lambda^{\text{sun}}\Omega_{\text{sun}}$ is the solar spectral irradiance, E_λ^{sun}. Equation 3.45 is used to study clouds' effects on radiation. It includes the multiple scattering of radiation that occurs in thick clouds, and it has challenged atmospheric scientists for many years. The reader is referred to Liou (1980) for a discussion of the ways it has been approached.

A further approximation, which is useful for the study of aerosols (see Chapter 8) and thin clouds, is that only the first scattering need be considered. This is called the *single-scattering approximation*, and it is equivalent to neglecting the second term in Eq. 3.45, which deals with the scattering of diffuse (already scattered) radiation.

In order to apply the radiative transfer equation, we must have a knowledge of the absorption and scattering properties of the Earth's atmosphere, the reflection properties of the Earth's surface, and the characteristics of the sun's radiation. These will be dealt with in the next four sections.

3.4 GASEOUS ABSORPTION

The study of absorption and emission by gases is the field of spectroscopy, an old and complex science which was largely explained by quantum mechanics in the first half of the twentieth century (Herzberg, 1950; McCartney, 1983). Radiation can interact with atmospheric gases in five ways: ionization–dissociation interactions, electronic transitions, vibrational transitions, rotational transitions, and forbidden transitions.

3.4.1 Radiative Interactions

In *ionization–dissociation interactions*, an electron is stripped from an atom or molecule, or a molecule is torn apart. These interactions occur primarily at ultraviolet and shorter wavelengths. Because any energy greater than the threshold energy will ionize or dissociate a molecule, these interactions produce relatively smooth *spectra* (plots of absorption coefficient versus wavelength or wavenumber), unlike the spectra produced by the processes described below.

If one divides the absorption coefficient by the number density (number of molecules per unit volume), one arrives at a quantity with the units of area that spectroscopists call the *absorption cross section*. It is a molecule's effective area, that is, a measure of how effectively it absorbs radiation. Figure 3.10 shows a semischematic representation of the absorption cross sections of O_3, O_2, and N_2 in the ultraviolet and visible portions of the spectrum.

All solar radiation shorter than about 0.1 μm in wavelength is absorbed in the upper atmosphere by ionizing atmospheric gases, particularly atomic oxygen; this ionization produces the ionosphere. Wavelengths between 0.1 and 0.2 μm are absorbed by dissociation of molecular oxygen (O_2) into atomic oxygen. Short of about 0.2423 μm, O_2 dissociates in the Herzberg continuum. Short of about

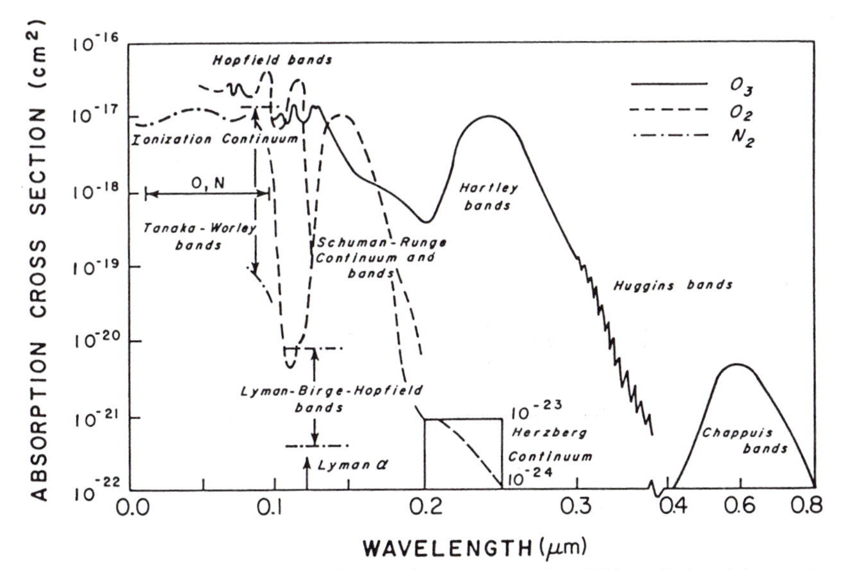

FIGURE 3.10. Absorption cross sections of atmospheric gases in the visible and ultraviolet portions of the spectrum. Note that these curves are intended to indicate the relative significance of various absorbers and should not be taken as the source of quantitative data. [After Liou (1980). Reprinted by permission of Academic Press.]

0.1750 μm, O_2 dissociates in the much stronger Schuman–Runge continuum. Virtually all radiation between 0.2 and 0.3 μm is absorbed by dissociation of ozone (O_3) in the Hartley bands. Ozone also dissociates in the weak Chappuis bands in the visible portion of the spectrum. Ionization–dissociation interactions are important to satellite meteorology primarily for measuring ozone concentration profiles.

In *electronic transitions*, an orbital electron jumps between quantized energy levels. These transitions occur mostly in the ultraviolet and visible portions of the spectrum. Of importance to satellite meteorology are the ultraviolet Huggins bands of ozone, which are used to measure integrated (total) ozone, and a weak 0.77-μm band (the A band) of molecular oxygen, which may someday be used to estimate surface pressure by measuring the total amount of O_2 in an atmospheric column. Figure 3.11 shows the vertical transmittance of the Earth's atmosphere in the visible, ultraviolet, and near-infrared portions of the spectrum.

In *vibrational transitions*, a molecule changes vibrational energy states. These transitions occur mostly in the infrared portion of the spectrum and are extremely important for satellite meteorology. They are discussed in the next section. At temperatures found in the Earth's atmosphere, most molecules are in the ground vibrational state. The spectrum of vibrational transitions, therefore, is caused primarily by transitions between the ground state and the first vibrational excited state.

In *rotational transitions*, a molecule changes rotational energy states. These occur in the far-infrared and microwave portion of the spectrum. However, rota-

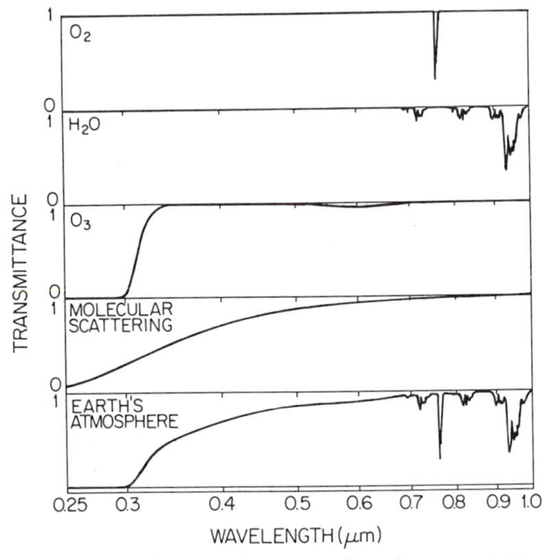

FIGURE 3.11. Vertical transmittance of the Earth's atmosphere between 0.25 and 1.0 μm. Rayleigh scattering by air molecules is the chief limitation to the transfer of visible radiation through the clear atmosphere. The effects of aerosols have not been included in these curves. [Calculated using LOWTRAN 6 (Kneizys *et al.*, 1983).]

tional transitions can occur at the same time as a vibrational transition, which complicates the spectrum. Figure 3.12 shows the infrared spectrum of the radiatively most important atmospheric gases. The far-infrared spectrum is dominated by rotational transitions of water vapor. Measurement of water vapor in the microwave region is an important use of rotational transitions.

Finally, *forbidden transitions* are those transitions which are not caused by the interaction between the electric field of the radiation and the electric dipole moment of a molecule. Some forbidden transitions do, in fact, occur. A meteorologically important forbidden transition is caused by the reorientation of unpaired electron spins in the O_2 molecule. This results in an absorption band in the 5-mm region which is used for temperature sounding. Figure 3.13 shows the microwave spectrum of the Earth's atmosphere.

Figure 3.14 shows the complete spectrum of the Earth's atmosphere under very low spectral resolution. Although much of the spectrum is opaque because of absorption by atmospheric gases, there are several important areas, called *windows*, where the atmosphere is relatively (but *not* absolutely) transparent. The most important of these are the visible window, the 3.7-μm window, the microwave windows (2–4 mm and >6 mm), and the 8.5–12.5 μm window. This

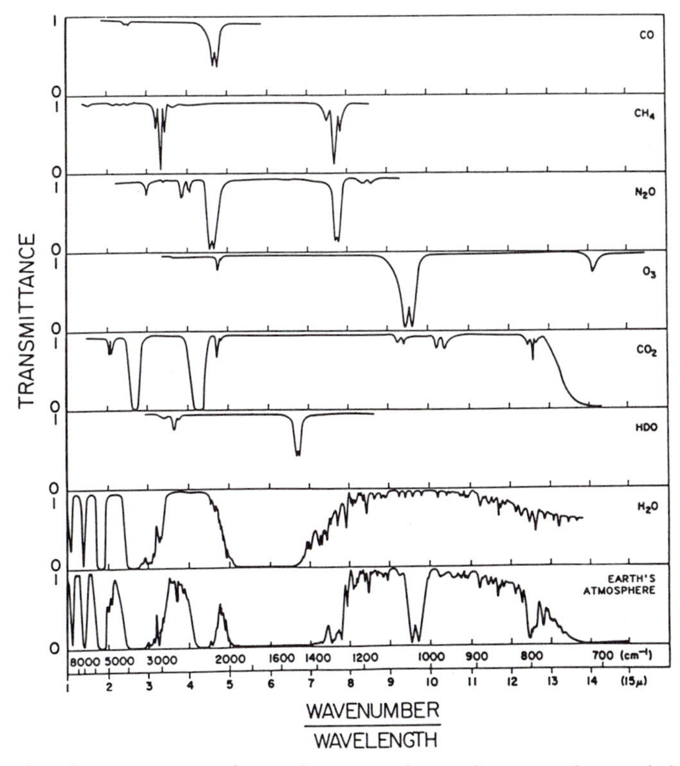

FIGURE 3.12. Infrared transmittance of several gases in the Earth's atmosphere and the combined atmospheric transmittance. [After Valley (1965).]

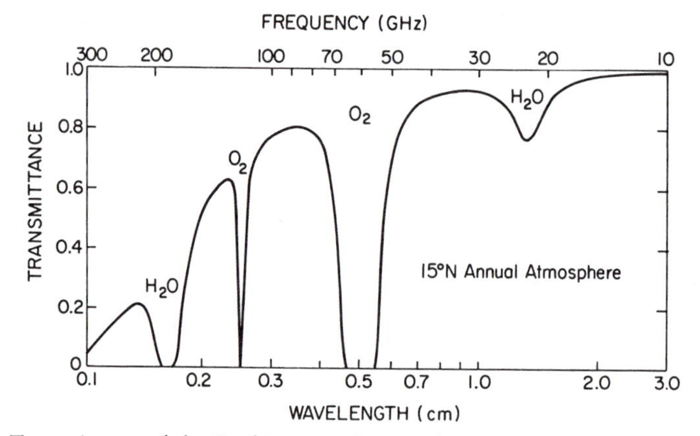

FIGURE 3.13. Transmittance of the Earth's atmosphere in the microwave portion of the spectrum. [Calculated using the model of Liebe and Gimmestad (1978).]

last window is punctuated by the 9.6-μm ozone absorption band (vibrational), and it is affected by water vapor absorption.

3.4.2 Vibrational Transitions

Because vibrational transitions are so important in satellite meteorology, it is necessary to consider them in detail. The two chief absorbers in the infrared region of the spectrum are carbon dioxide and water vapor. Shown in Fig. 3.15 are the three ways that each molecule can vibrate and the wavelengths (and wavenumbers) of these vibrations. Not all of these vibration modes are radiatively active, however. The symmetric stretching mode of the linear CO_2 molecule, for example, has neither a static nor a dynamic electric dipole moment because the symmetry of the molecule is maintained. If a molecule has no electric dipole moment, the electric field of incident radiation cannot interact with the molecule. The infrared spectrum of CO_2 (Fig. 3.12) shows that there is no absorption at 7.46 μm. For the same reason, CO_2 cannot have a pure rotation spectrum. The lack of an electric dipole moment also explains why the two most abundant gases in the Earth's atmosphere, N_2 and O_2, are transparent in the infrared.

Quantum mechanics tells us that vibrational transitions occur only at discrete frequencies. These results, however, are for isolated, motionless molecules. In the

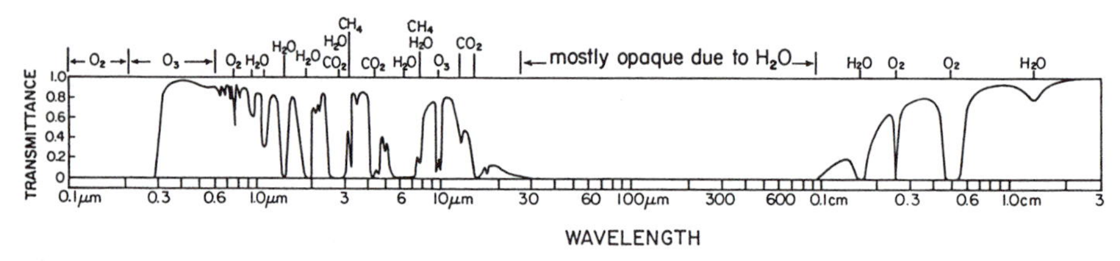

FIGURE 3.14. Spectrum of the Earth's atmosphere. [Adapted from Goody and Yung (1989), with permission from Oxford University Press.]

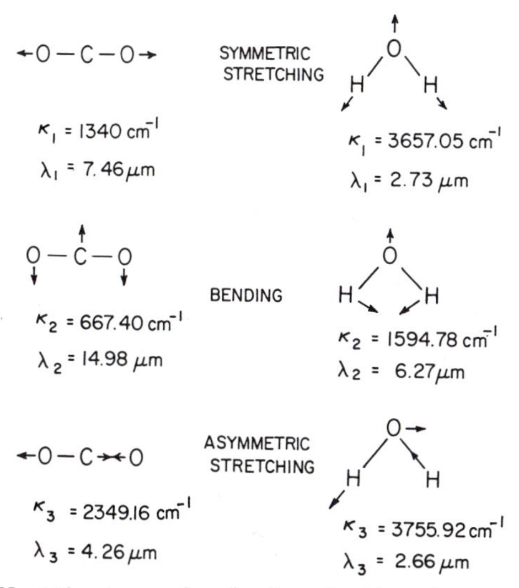

FIGURE 3.15. Vibration modes of carbon dioxide and water vapor.

Earth's atmosphere, absorption lines are broadened chiefly by collisions, which distort the molecule and cause it to absorb (or emit) at slightly different wavelengths. Infrared lines are usually modeled as having the *Lorentz lineshape* (Fig. 3.16):

$$\beta_a(\kappa) = \frac{(\Delta\kappa)S}{\pi[(\kappa - \kappa_o)^2 + (\Delta\kappa)^2]}, \tag{3.46}$$

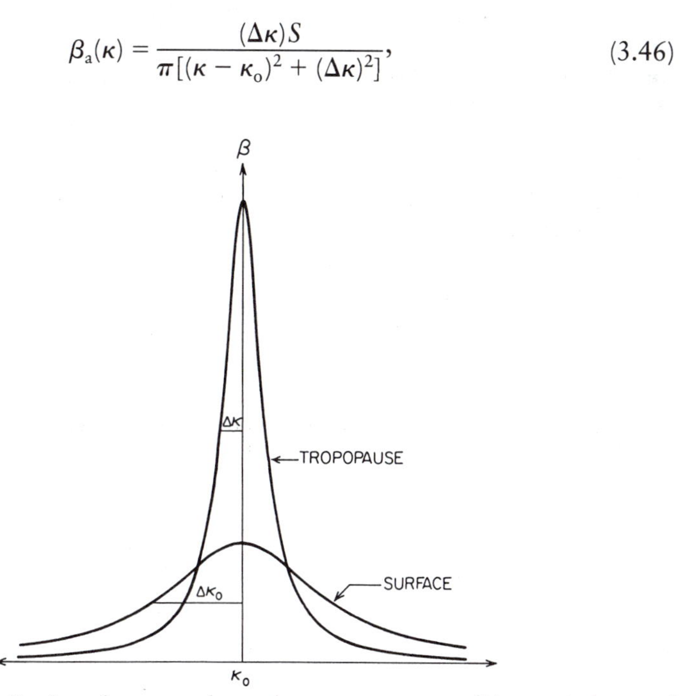

FIGURE 3.16. The Lorentz lineshape for near-surface and near-tropopause conditions; κ_o is the central wavenumber, and $\Delta\kappa$ is the halfwidth, which is a function of temperature and pressure.

where $\beta_a(\kappa)$ is the mass absorption coefficient, κ is wavenumber, κ_o is the central wavenumber, S is the line strength (independent of $\Delta\kappa$), and $\Delta\kappa$ is the halfwidth. The halfwidth is proportional to the collision frequency of a molecule and can be shown to depend on temperature and pressure as

$$\Delta\kappa(T,P) = \Delta\kappa_o \left(\frac{P}{P_o}\right)\left(\frac{T_o}{T}\right)^{1/2}, \tag{3.47}$$

where T_o is a reference temperature, and P_o is a reference pressure. The line strength S is a function of temperature and the energy of the lower level (E'') of the transition. Four numbers, then, specify each line: κ_o, S_o, $\Delta\kappa_o$, and E'' (McClatchey *et al.*, 1973). If two or more absorption lines overlap, which is usually the case, the *volume* absorption coefficients are summed. Calculation of absorption coefficients by this method (the *line-by-line method*) would not be too difficult except that there are more than 100,000 absorption lines between 1 and 25 μm!

How can there be so many lines with only a few absorbing gases? Close examination of the 15-μm band of CO_2 gives the answer. Figure 3.17 shows what serves as a high-resolution spectrogram of CO_2 in the wavenumber range 600–740 cm^{-1} (16.67–13.51 μm). The bending mode vibrational transition occurs at 667.40 cm^{-1} (14.98 μm). Quantum-mechanical selection rules, however, allow rotational transitions to accompany vibrational transitions. The rotational quantum number J may change by $+1$, forming the higher frequency *R branch*; 0, forming the central *Q branch*; or -1, forming the lower frequency *P branch*. In addition, the energy of the Jth rotational state is proportional to $J(J+1)$. The difference between the J and $J+1$ energy levels is therefore linearly related to J. Since many excited rotational states are populated, the P and R branches consist of a series of equally spaced lines, the strengths of which are related to the number of molecules in the Jth state.

The spectrum is further complicated by the presence of isotopes that vibrate at slightly different frequencies due to differing molecular masses. The dominant form of carbon dioxide in the Earth's atmosphere is $C^{12}O_2^{16}$. At 649 and 662 cm^{-1}, the Q branches of isotopes $C^{13}O_2^{16}$ and $C^{12}O^{16}O^{18}$, respectively, can be seen. The less intense P and R branches of these isotopes are masked by the $C^{12}O_2^{16}$ spectrum. Other, less abundant isotopes are also present.

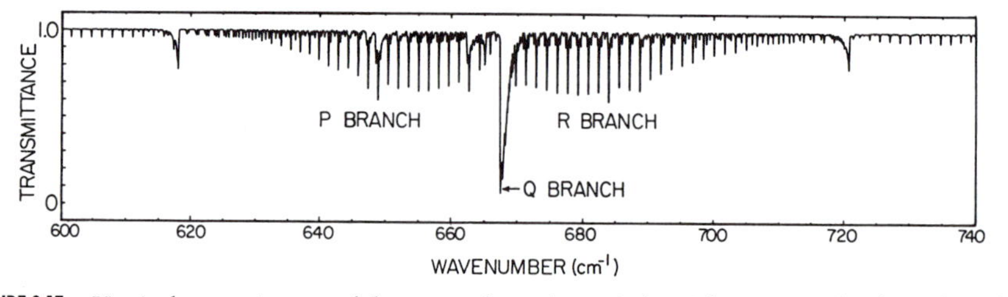

FIGURE 3.17. Vertical transmittance of the atmosphere above 40 km. The P, Q, and R branches of the bending vibrational mode of CO_2 can be seen along with absorption due to isotopes of CO_2 (see text). [Adapted from Kyle and Goldman (1975).]

The spectrum is still further complicated by combination and overtone bands. A *combination band* is one in which two transitions take place simultaneously. At 618 and 721 cm^{-1}, for example, one can see bands which arise from the transition of the symmetric stretching mode from the ground state to the first excited state simultaneous with the transition of the bending mode from the first excited state to the ground state. An *overtone band* is one in which a molecule absorbs (or emits) a photon of sufficient energy to cause it to jump two or more levels. Water vapor, for example, absorbs in a band centered at 3151 cm^{-1}, which is the result of double excitation of its bending mode. When all of these complications are taken into account, 100,000 is not an unreasonable number of absorption lines, even though only a handful of atmospheric gases contribute to the infrared spectrum.

Comparing Fig. 3.17 with Fig. 3.12, one might ask: Why are the rotational transitions not evident in Fig. 3.12? The answer has two parts. First, Fig. 3.12 shows the transmittance of the entire atmosphere. Near the surface, the individual rotation lines are considerably broadened (Fig. 3.16) so that they overlap each other. Second, Fig. 3.12 is what is called a *low-resolution spectrum*. The spectrum is averaged over a passband and therefore appears smooth. The result of these two effects is that in Fig. 3.12, only the envelopes of the P, Q, and R branches of absorption bands are discernable. Note also that not all molecules have a Q branch. Ozone, for example, has a bimodal absorption band centered at 9.6 μm, which consists of the envelope of its P and R branches.

A complete compilation of vibration–rotation lines important for satellite meteorology has been made by the U.S. Air Force Geophysics Laboratory (McClatchey *et al.*, 1973) and an optimized computer codes to calculate atmospheric transmittances have been written, for example, FASCOD (Clough *et al.*, 1982). Readers who need to do line-by-line calculations are advised to obtain one of these codes.

Vibrational transitions are the basis of temperature sounding and measurement of a wide variety of trace gases in the atmosphere. The 15-μm and 4.3-μm CO_2 bands are used for temperature sounding. The 6.3-μm H_2O band is used for water vapor measurements. The 9.6-μm band of ozone is used to make total ozone measurements. Several other important gases have been measured in the stratosphere using vibrational transitions. Among them are NO, N_2O, NO_2, HNO_3, HCl, HF, CH_4, CO, and CO_2.

3.4.3 Practical Problems

Two problems arise in the application of absorption line data in satellite meteorology. First, the parameters that specify absorption lines are not perfectly known. Calculated atmospheric transmittances agree with laboratory measurements only to within about 10%. This uncertainty results in errors in temperature retrievals, for example.

The second problem is that the radiative transfer equation is strictly only valid at a single wavelength. A typical satellite sounder channel, however, is about 20

cm^{-1} wide (see Chapter 4), which includes several lines. Calculation of the radiance that a satellite radiometer would measure requires the calculation of radiance at many closely spaced wavelengths followed by integration over the passband of the radiometer channel:

$$L_{\text{measured}} = \frac{\int_{\lambda_1}^{\lambda_2} L_\lambda f(\lambda)\, d\lambda}{\int_{\lambda_1}^{\lambda_2} f(\lambda)\, d\lambda}, \tag{3.48}$$

where $f(\lambda)$ is the response function of the radiometer channel. Even with fast line-by-line routines such as FASCOD, these calculations are too time consuming to be of practical use in many applications. Instead, one turns either to band models or to polynomial expansions.

In a *band model* the transmittance of an atmospheric layer in an entire band is calculated at one time using an appropriate parameterization, which is usually based on fitting line-by-line calculations to a function. Goody and Yung (1989, Chapter 4) explain several parameterizations. The LOWTRAN computer code (McClatchey *et al.*, 1972; Kneizys *et al.*, 1983), in which transmittance averaged over a 20-cm^{-1} band is calculated, is an example of this type of model.

In polynomial expansions, the atmosphere is divided into layers. For each layer and for each channel of the radiometer, transmittances are calculated line-by-line for a variety of temperatures and humidities. Polynomial functions of temperature and humidity are then fitted to the transmittances. To calculate the transmittance of the atmosphere from a particular level to a satellite, one uses an estimate of atmospheric temperature and humidity and calculates the transmittance of each layer. The total atmospheric transmittance is the product of the layer transmittances. Polynomial expansions are used extensively in satellite sounding (see McMillin and Fleming, 1976).

3.5 SCATTERING

Radiation scattered from a particle is a function of several things: particle shape, particle size, particle index of refraction, wavelength of radiation, and, of course, viewing geometry.[16] In 1908 Mie[17] applied Maxwell's[18] equations, which describe electromagnetic radiation, to the case of a plane electromagnetic wave incident on a sphere. The far-field radiation (that observed at many radii from the sphere) is the scattered radiation. [See Liou (1980) for this lengthy derivation.] Mie showed that for a spherical scatterer, the scattered radiation is a function of only viewing angle, index of refraction, and the *size parameter* defined as

$$\chi \equiv \frac{2\pi r}{\lambda}, \tag{3.49}$$

[16] See van de Hulst (1957) for a thorough discussion of scattering by particles.

[17] Gustav Mie, German physicist, 1868–1957.

[18] James Clerk Maxwell, Scottish mathematician and physicist, 1831–1879.

where r is the radius of the sphere. The size parameter can be used to divide scattering into three regimes (Fig. 3.18).

3.5.1 Mie Scattering

For size parameters in the range 0.1–50, the wavelength of the radiation and the circumference of the particle are comparable. Radiation strongly interacts with the particle, and, therefore, the full Mie equations must be used. These equations have been applied extensively to the detection of raindrops by radar. The study of aerosols (smoke, dust, haze) using visible radiation falls in the Mie regime. Also in the Mie regime is the interaction of cloud droplets with infrared radiation.

A complete discussion of Mie scattering is outside the scope of this book because the scattering equations are very complicated. Insight into the results, however, can be gained as follows. If the volume absorption coefficient is divided by the number of scattering particles per unit volume and by the cross-sectional area of each scatterer, the result is the *scattering efficiency* (Q_s) for a single scatterer. Q_s is the ratio of the total scattered radiation (regardless of direction) to the incident radiation. Q_s is a function of the size parameter and of the index of refraction of the particle.

Many substances absorb radiation as well as scatter it. This can be conveniently taken into account by letting the index of refraction become a complex number

$$m \equiv n - in', \tag{3.50}$$

where n, the real part of the index of refraction, is as defined above, and n', the imaginary part, accounts for absorption inside the scatterers. Figure 3.19 shows

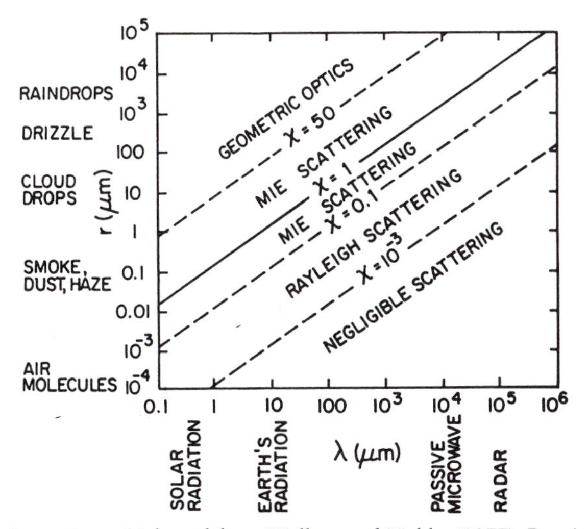

FIGURE 3.18. Scattering regimes. [Adapted from Wallace and Hobbs (1977). Reprinted by permission of Academic Press.]

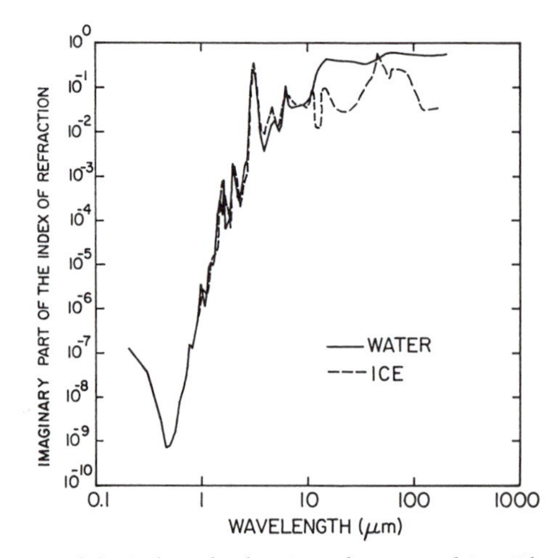

FIGURE 3.19. Imaginary part of the index of refraction of water and ice. [Plotted from data in Irvine and Pollack (1968).]

n' as a function of wavelength for water and ice. In the visible portion of the spectrum, n' is negligibly small, but in the infrared it becomes significant.

Figure 3.20 shows the scattering efficiency for water drops ($n = 1.33$) as a function of size parameter for several values of n'. Scattering efficiency in the Mie regime is quite clearly a complicated function. In clouds, there is usually a distribution of drop sizes. Suppose that $N(r)dr$ is the number of drops per unit volume in the radius range r to $r + dr$. If the scatterers are sufficiently far apart (many wavelengths) that they act independently, the scattering coefficient is given by

$$\sigma_s(\lambda) = \int_0^\infty \pi r^2 Q_s N(r) dr \tag{3.51}$$

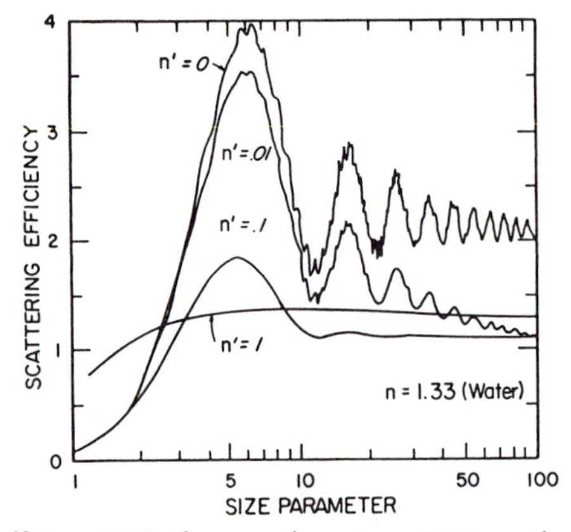

FIGURE 3.20. Scattering efficiency (Q_s) of water spheres ($n = 1.33$) as a function of size parameter (χ) for several values of n'. [Adapted from Liou (1980) and Hansen and Travis (1974). Reprinted by permission of Academic Press, Inc. and Kluwer Academic Publishers.]

Integration over the size distribution smooths the scattering efficiency. In any case, it must be noted that scattering is a much smoother function of wavelength than is gaseous absorption.

Also of interest is the scattering phase function, which determines in which direction the radiation is scattered. Figure 3.21 shows the scattering phase function for water drops for several size parameters. As the size parameter increases, the phase function becomes strongly peaked in the forward direction; relatively little radiation is backscattered toward the source of the radiation. Finally, we note that in general scattering polarizes radiation; in some applications polarization must be taken into account.

3.5.2 Geometric Optics

For χ greater than about 50, the sphere is large in comparison with the wavelength of radiation. This is the realm of *geometric optics*, where rays, which are reflected and refracted at the surface of a scatterer, can be traced. Ray tracing can be used with scatterers that are nonspherical. As shown in Fig. 3.18, the interaction of solar radiation with virtually all types of hydrometeors falls in this regime. A wide variety of optical phenomena such as rainbows and halos can be explained with geometric optics. The interaction of infrared radiation with precipitation-size particles also falls within this regime.

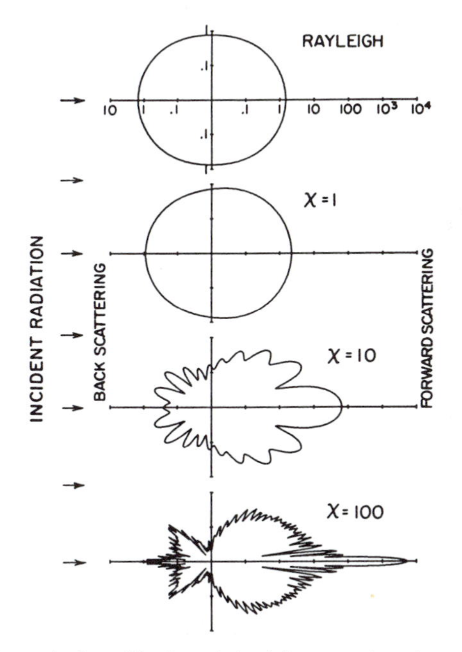

FIGURE 3.21. Polar plots (note the logarithmic scales) of the scattering phase function of water drops for several size parameters. [Plotted from data supplied by Steven A. Ackerman, Cooperative Institute for Meteorological Satellite Studies, University of Wisconsin–Madison.]

For nonabsorbing spheres, Fig. 3.20 shows that the scattering efficiency asymptotically approaches 2 as the size parameter increases. Not only is the radiation that directly strikes the drop scattered, but an equal amount of radiation that comes near the drop is *refracted* around it. Since the refracted radiation changes direction, it has been scattered.

3.5.3 Rayleigh Scattering

For χ less than about 0.1, only the first term in the Mie equations must be considered. The scattering efficiency becomes

$$Q_s = \frac{8\chi^4}{3}\left|\frac{m^2-1}{m^2+2}\right|^2, \tag{3.52}$$

where m is the complex index of refraction. Because the particle is small in comparison with the wavelength of the radiation, Rayleigh scattering is insensitive to particle shape; Eq. 3.52 works for nonspherical as well as spherical particles. Because of the χ^4 dependence, scattering becomes negligible for χ less than 10^{-3}.

Air molecules act as Rayleigh scatterers for visible and ultraviolet radiation. The scattering coefficient for air molecules should be simply

$$\sigma_s = \pi r^2 Q_s N, \tag{3.53}$$

where N is the number of air molecules per unit volume. However, this formula can be made more useful and more accurate as follows. First, we multiply by a factor $f = 1.061$ to correct for the anisotropic behavior of air molecules. Second, we note that the index of refraction of air is very close to one, and if only scattering is to be considered, only the real part of the index of refraction need be used. Therefore, $m^2 + 2 \approx 3$, and $m^2 - 1 = (n+1)(n-1) \approx 2(n-1)$. Substituting these and $\chi = 2\pi r/\lambda$ into Eqs. 3.52 and 3.53 gives

$$\sigma_s = f\frac{32\pi^3}{3\lambda^4}(n-1)^2 N\left(\frac{4\pi r^3}{3}\right)^2. \tag{3.54}$$

Third, the volume of each scatterer is $4\pi r^3/3$. Since air fills any volume that it occupies, this volume must be equal to the reciprocal of the number of scatterers per unit volume (N). Therefore,

$$\sigma_s = f\frac{32\pi^3}{3\lambda^4 N}(n-1)^2. \tag{3.55}$$

Finally, both N and $n-1$ are proportional to atmospheric density ρ. For an ideal gas, N is given by

$$N = \left(\frac{N_A}{W_m}\right)\rho, \tag{3.56}$$

where N_A is Avogadro's[19] number (Appendix E), and W_m is the molecular weight of the gas (28.966 kg $kmol^{-1}$ for dry air). The index of refraction of air is given by

$$n - 1 = \left(\frac{\rho}{\rho_o}\right)(n_o - 1), \tag{3.57}$$

where ρ_o and n_o are the sea-level values. The Rayleigh scattering coefficient for air then becomes

$$\sigma_s = f\frac{32\pi^3 W_m \rho}{3\rho_o^2 N_A \lambda^4}(n_o - 1)^2. \tag{3.58}$$

Liou (1980) states that $n_o - 1$ can be approximated as

$$(n_o - 1) \times 10^6 = 64.328 + \frac{29498.1}{146 - \lambda^{-2}} + \frac{255.4}{41 - \lambda^{-2}}, \tag{3.59}$$

where λ is in micrometers. Equation 3.58 states that for Rayleigh scattering, $\sigma_s(\lambda)$ is proportional to ρ, as it must be, and to the familiar λ^{-4}.

Figure 3.11 shows the vertical transmittance of the Earth's atmosphere due to Rayleigh (molecular) scattering. At visible and ultraviolet wavelengths, Rayleigh scattering by air molecules must be taken into account. Aerosol optical depth can be measured over the ocean *if* the observed radiance is corrected for Rayleigh scattering by air molecules (see Section 8.5). Rayleigh scattering of air molecules must also be taken into account when making ultraviolet measurements of ozone (see Section 6.5).

The Rayleigh scattering phase function for unpolarized incident radiation is given by

$$p_{\text{Rayleigh}}(\psi_s) = \tfrac{3}{4}(1 + \cos^2\psi_s), \tag{3.60}$$

where ψ_s is the angle between the incoming and the scattered radiation (Fig. 3.21). It is important to note that unlike larger particles, Rayleigh particles scatter equally well in the forward direction ($\psi_s = 0°$) and the backward direction ($\psi_s = 180°$).

Rayleigh particles, for example, cloud droplets, can also absorb radiation. The Rayleigh absorption efficiency is given by

$$Q_a = 4\chi I_m\left(\frac{m^2 - 1}{m^2 + 2}\right), \tag{3.61}$$

where I_m indicates the imaginary part. Because Q_a varies with χ while Q_s varies with χ^4, absorption can quickly dominate scattering for Rayleigh particles that have even the smallest amount of absorption. The absorption coefficient due to a number of particles is (similar to Eq. 3.51)

$$\sigma_a(\lambda) = \int_0^\infty \pi r^2 Q_a N(r)\, dr. \tag{3.62}$$

[19] Amedeo Avogadro, count of Quaregna, Italian physicist, 1776–1856.

In the Rayleigh regime, Q_a is proportional to r; thus $\sigma_a(\lambda)$ is proportional to the total volume of scatterer per unit volume of atmosphere. In a cloud, for example, absorption due to Rayleigh-size cloud droplets is proportional to the liquid water content. This is important in the microwave region where cloud droplets absorb, but do not scatter. Using microwave measurements, the vertically integrated liquid water content of clouds can be estimated (see Section 8.3).

3.5.4 Clouds

Because clouds play a large role in satellite meteorology, it is important to understand in a general way the interaction of clouds with radiation. Clouds consist of water drops or ice crystals with radii on the order of 10 μm. Drops with radii of 100 μm have significant fall speed and constitute drizzle. Drops with radii of 1000 μm (1 mm) are raindrops. Clouds have drop concentrations on the order of 10^8 m^{-3}; that is, the drops are on the order of 1 mm apart.

In the visible portion of the spectrum ($\lambda \sim 0.5$ μm), cloud drops are geometric scatterers; therefore, the scattering efficiency is approximately 2. The scattering coefficient, then, is ~0.1 m^{-1}. A photon's *mean free path* (the average distance between scattering events) is the reciprocal of the volume scattering coefficient, or ~10 m. Therefore, a cloud only a few tens of meters thick is sufficient to scatter all of the visible radiation incident on it. Since liquid water does not absorb visible radiation well, very little of the radiation is absorbed; most emerges from the cloud somewhere after being scattered many times. Welch *et al.* (1980), for example, have made calculations for a 2-km-thick stratus cloud with the sun directly over head (Table 3.2). At 0.55 μm, the cloud absorbs only 0.2% of incident radiation; 79.8% is scattered out the top of the cloud, and 20.0% is scattered out the bottom of the cloud. Since clouds have a distribution of drop sizes and the size parameter is large, all visible wavelengths are scattered nearly

TABLE 3.2. Calculated Radiative Properties of a 2-km-thick Stratus Cloud[a]

Wavelength (μm)	*Absorbed* (%)	*Scattered* (%)	
		Out top	*Out bottom*
0.55	0.2	79.8	20.0
0.765	0.5	80.6	18.9
0.95	8.1	76.3	15.5
1.15	17.9	70.4	11.7
1.4	47.4	49.9	2.7
1.8	61.9	37.6	0.5
2.8	99.6	0.4	0.0
3.35	99.4	0.6	0.0
6.6	99.05	0.95	0.0
Total	10.0	73.8	16.6

[a] From Table 2.12a of Welch *et al.* (1980).

equally well. Clouds therefore appear white. Solar radiation extends into the near infrared region of the spectrum. In the near infrared, absorption due to water vapor increases as does the absorption of liquid water. Averaged over the solar spectrum, Welch *et al.* (1980) calculate that the cloud absorbs 10.0%, scatters 73.8% out the top of the cloud, and scatters 16.2% out the bottom of the 2-km-thick cloud.

In the 8.5–12.5-μm window, cloud droplets are Mie scatterers. The scattering efficiency is roughly in the range 1–3, but in contrast to visible wavelengths, the absorption efficiency is on the order of 1. Therefore, clouds absorb nearly all of the infrared radiation incident on them. They act essentially as blackbodies.

In the microwave portion of the spectrum ($\chi \sim 0.01$), absorption due to cloud drops is very small. The transmittance of a typical nonraining cloud is greater than 90%. Scattering is negligibly small. Raindrop-size particles, however, interact strongly with microwave radiation. Therefore, clouds are nearly transparent in the microwave region, but raining clouds are not. This forms one basis of microwave detection of precipitation (see Chapter 9).

Cirrus clouds have a higher transmittance than water clouds because ice clouds have far fewer particles per unit volume than water clouds and because water is a better absorber than ice. Cirrus clouds can also be vertically thinner than water clouds. In general thin cirrus clouds are difficult to detect with satellite radiometers, yet their effects are not negligible. They can cause problems in the retrieval of atmospheric soundings (see Chapters 6 and 8). The detection of cirrus clouds with satellite instruments is an area of active research.

3.6 SURFACE REFLECTION

Reflected radiation, particularly reflected solar radiation and reflected microwave radiation, is very important to satellite meteorology. Several quantities are used to describe reflected radiation. The most basic is the *bidirectional reflectance*, γ_r, which is related to the fraction of radiation from incident direction (θ_i, ϕ_i) that is reflected in the direction (θ_r, ϕ_r). Perhaps the best way to define it is to write the formula for the radiance reflected from a small element of surface:

$$L_r(\theta_r, \phi_r) = \int_0^{2\pi} \int_0^{\pi/2} L_i(\theta_i, \phi_i)\, \gamma_r(\theta_r, \phi_r; \theta_i, \phi_i) \cos\theta_i \sin\theta_i \, d\theta_i \, d\phi_i . \quad (3.63)$$

Basically, radiation from incident direction (θ_i, ϕ_i) illuminates a small element of surface. Taking into account the effect of incident angle, $L_i \cos\theta_i$ is available to be reflected. A fraction $\gamma_r(\theta_i, \phi_i; \theta_r, \phi_r)$ is reflected into direction (θ_r, ϕ_r). Integrating over all incident solid angles gives the reflected radiance in direction (θ_r, ϕ_r).

An important property of the bidirectional reflectance is known as the *Helmholtz reciprocity principle.* It states that the bidirectional reflectance is invariant if the directions of incoming and outgoing radiation are interchanged:

$$\gamma_r(\theta_r, \phi_r; \theta_i, \phi_i) = \gamma_r(\theta_i, \phi_i; \theta_r, \phi_r). \quad (3.64)$$

Equation 3.64 is used in the construction of tables of bidirectional reflectance from satellite observations because not all incident radiation angles can be observed (see Chapter 10).

Probably the most frequently used reflection quantity is the *albedo* (A), which is the ratio of radiant exitance (M, due to reflection) to irradiance (E):

$$E = \int_0^{2\pi} \int_0^{\pi/2} L_i(\theta_i, \phi_i) \cos\theta_i \sin\theta_i \, d\theta_i \, d\phi_i, \tag{3.65a}$$

$$M = \int_0^{2\pi} \int_0^{\pi/2} L_r(\theta_r, \phi_r) \cos\theta_r \sin\theta_r \, d\theta_r \, d\phi_r, \tag{3.65b}$$

$$A \equiv \frac{M}{E}. \tag{3.65c}$$

Albedo is a unitless ratio between zero and one. As defined in Eqs. 3.65, it is a function of neither the incoming nor the outgoing angles; however, this does not mean that the albedo is constant. If the incoming radiation changes, the albedo will change. This is most easily understood by restricting the incoming radiation to direct-beam solar radiation, which comes from a very narrow range of angles. In this case,

$$E = L_{sun} \Omega_{sun} \cos\theta_{sun}, \tag{3.66a}$$

$$L_r(\theta_r, \phi_r) = L_{sun} \Omega_{sun} \cos\theta_{sun} \, \gamma_r(\theta_r, \phi_r; \theta_{sun}, \phi_{sun}), \tag{3.66b}$$

$$M = L_{sun} \Omega_{sun} \cos\theta_{sun} \int_0^{2\pi} \int_0^{\pi/2} \gamma_r(\theta_r, \phi_r; \theta_{sun}, \phi_{sun}) \cos\theta_r \sin\theta_r \, d\theta_r \, d\phi_r, \tag{3.66c}$$

$$A(\theta_{sun}, \phi_{sun}) = \frac{M}{E} = \int_0^{2\pi} \int_0^{\pi/2} \gamma_r(\theta_r, \phi_r; \theta_{sun}, \phi_{sun}) \cos\theta_r \sin\theta_r \, d\theta_r \, d\phi_r, \tag{3.66d}$$

where Ω_{sun} is the solid angle of the sun subtended at the Earth. Since the direction of the sun remains in the equation, the albedo is a function of solar direction.

As an example of reflecting surfaces, two limiting cases are useful. A *Lambertian* or *isotropic reflector* reflects radiation uniformly in all directions. If its albedo is A, then its bidirectional reflectance is a constant A/π. Flat white paint approximates a perfect ($A = 1$, independent of wavelength) Lambertian reflector. A *specular reflector* is like a mirror; its bidirectional reflectance is strongly peaked. Solar radiation from a perfect specular reflector would be observed only at the zenith angle equal to the solar zenith angle and at the azimuth angle equal to the solar azimuth angle plus 180°. Water surfaces are similar to specular reflectors, except that real water surfaces are always somewhat rough, so the solar reflection is blurred and larger than the sun. This is called *sun glint* or *sun glitter* (see Chapter 5).

Although we have not used the subscript λ with the reflectance quantities discussed here, they are functions of wavelength. They can be integrated over the passband of a satellite radiometer or over the solar spectrum. To so integrate albedo, integration is performed separately for E and M over the passband, and

TABLE 3.3. Albedo (%) of Various Surfaces Integrated over Solar Wavelengths[a]

Bare soil	10–25
Sand, desert	25–40
Grass	15–25
Forest	10–20
Snow (clean, dry)	75–95
Snow (wet and/or dirty)	25–75
Sea surface (sun > 25° above horizon)	<10
Sea surface (low sun angle)	10–70

[a] Adapted from Kondratyev (1969) by Wallace and Hobbs (1977).

then the ratio is taken. Readers of the literature should be careful to determine whether "albedo," in particular, refers to a monochromatic quantity or to one integrated over some passband. The albedo of various surfaces, integrated over solar wavelengths, is given in Table 3.3.

A function closely related to albedo and bidirectional reflectance is the *anisotropic reflectance factor*

$$\xi_r(\theta_r, \phi_r; \theta_i, \phi_i) \equiv \frac{\pi}{A} \gamma_r(\theta_r, \phi_r; \theta_i, \phi_i). \tag{3.67}$$

Suppose a non-Lambertian surface has albedo A. ξ_r compares the radiance from the surface to that from a Lambertian surface also with albedo A; it is greater than one where the surface reflects more than a Lambertian surface; it is less than one where the surface reflects less than a Lambertian surface. Since bidirectional reflectance is usually applied to solar radiation, it is convenient to use the azimuth angle of the sun as the reference azimuth. ξ_r is therefore usually written as $\xi_r(\theta_r, \theta_{sun}, \phi_r - \phi_{sun})$. Figure 3.22 shows ξ_r for four surfaces: snow, cloud, land, and ocean. Note that although isotropic reflection ($\xi_r \approx 1$) is not a bad assumption in some cases, sun glint is evident in the water reflectance. Most surfaces deviate significantly from Lambertian surfaces at low solar elevation angles (high solar zenith angles). Finally, note that Eqs. 3.66 and 3.67 require that for any incident direction $(\theta_{sun}, \phi_{sun})$

$$\int_0^{2\pi} \int_0^{\pi/2} \xi_r(\theta_r, \phi_r; \theta_{sun}, \phi_{sun}) \cos\theta_r \sin\theta_r \, d\theta_r \, d\phi_r \equiv \pi. \tag{3.68}$$

This equation is useful for checking experimentally determined values of ξ_r.

3.7 SOLAR RADIATION

The solar radiation reaching the Earth originates (for our purposes) from a layer of the sun called the *photosphere*, which coincides with the visible disk of

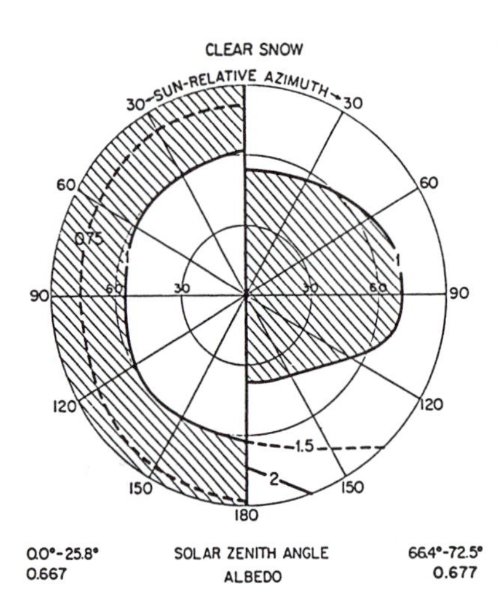

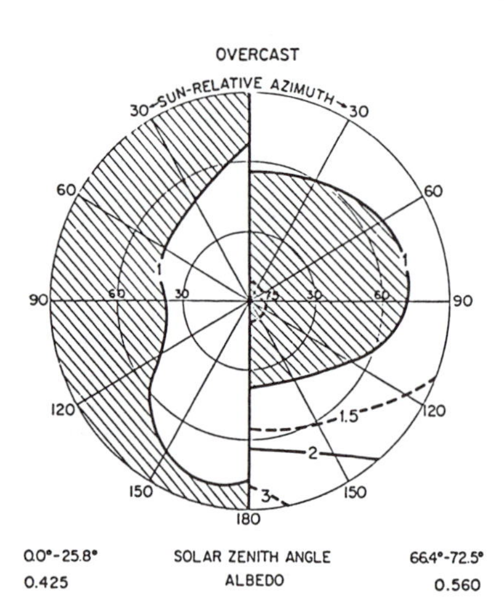

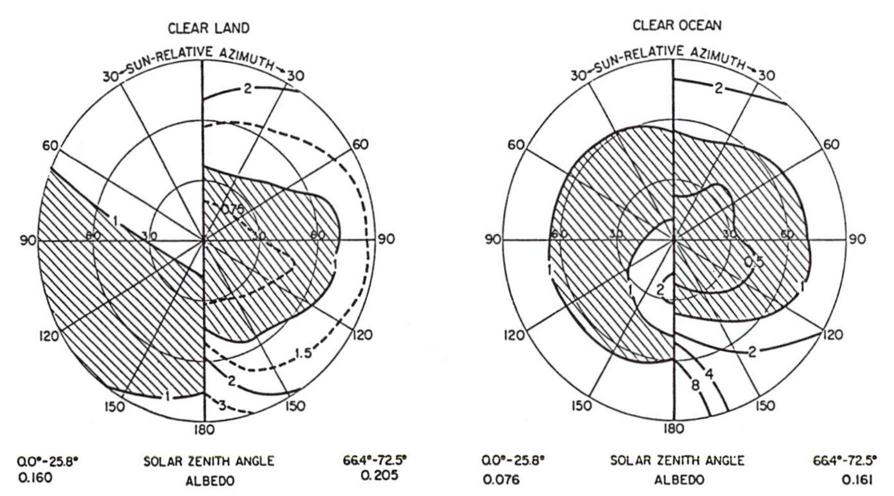

FIGURE 3.22. Anisotropic reflectance factor for snow, middle cloud (overcast), clear land, and clear ocean from the Nimbus 7 Earth Radiation Budget Experiment. The radial lines represent the viewing azimuth angle relative to the sun: At the top of each plot (0° relative azimuth), reflection is back toward the sun; at the bottom of each plot (180° relative azimuth), reflection is away from the sun. The circles represent the viewing zenith angle: The center (0° zenith angle) is looking straight down on the scene; the outer circle (90° zenith angle) represents viewing parallel to the surface. Shaded areas are those in which the scene reflects less radiation (appears darker) than an isotropic (Lambertian) reflector; unshaded areas reflect more radiation (appear brighter) than an isotropic reflector. [Plotted from data in Suttles *et al.* (1988).]

the sun and has a radius of 6.96×10^5 km. The radiation leaving the photosphere is very nearly that of a 6000-K blackbody. However, before reaching Earth, solar radiation must traverse the solar atmospheric layers of the *chromosphere* and the *corona*. The gases in these layers are both cooler and warmer than 6000 K; therefore, they both absorb and emit radiation at their characteristic wavelengths. After leaving the solar atmosphere, the radiation travels (on average) the 1.49598×10^8 km to Earth. Because the solid angle subtended by the sun at the Earth is so small (6.8×10^{-5} sr) solar radiation all comes from essentially the same direction. It is customary, therefore, to use solar irradiance rather than solar radiance. The irradiance reaching the top of the Earth's atmosphere is the radiant exitance leaving the top of the sun's atmosphere times the square of the ratio of the radius of the photosphere to the Earth–sun distance ($L_{sun}\Omega_{sun}$). The radiation reaching the Earth's surface is further modified by scattering and gaseous absorption in the atmosphere. Figure 3.23 shows the solar spectral irradiance reaching the top of the Earth's atmosphere and that reaching the surface. Shown for comparison is the spectral irradiance which would reach the Earth if the sun were a 6000-K blackbody.

Solar irradiance reaching the Earth peaks in the visible portion of the spectrum near 0.48 μm, whereas infrared radiation emitted by the Earth peaks near 10 μm. The Earth emits essentially no visible radiation; likewise, the Earth receives negligible amounts of 10 μm solar radiation, due to a combination of the sun's

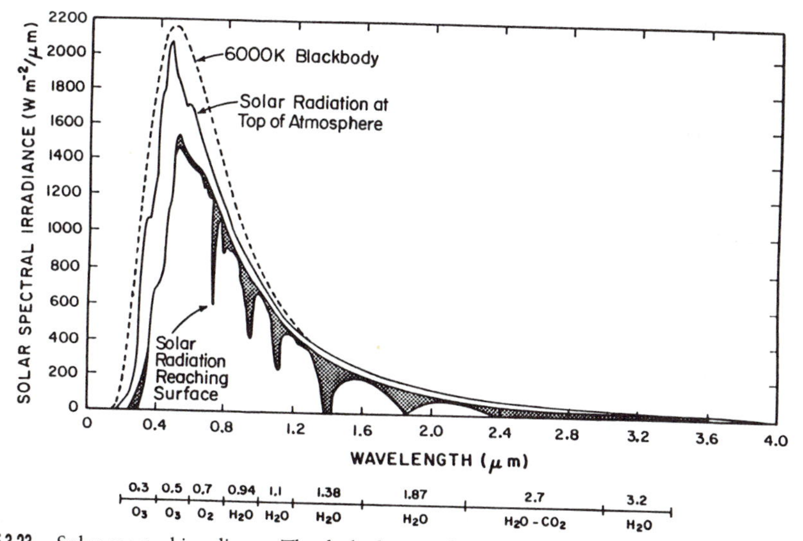

FIGURE 3.23. Solar spectral irradiance. The dashed curve shows the approximate irradiance that would be received at the Earth if the sun were a 6000-K blackbody. The top solid curve shows the spectral irradiance at the top of the atmosphere. (The integral under this curve is the solar constant.) The bottom solid curve represents the approximate solar irradiance reaching the Earth's surface after absorption and scattering in the atmosphere. The shaded area represents absorption by atmospheric gases, and the difference between the top solid curve and the envelope of the shaded area represents scattering. [Adapted from Liou (1980). Reprinted by permission of Academic Press.]

temperature and its very great distance from Earth. As a result of this separation in wavelength, solar radiation is often called *shortwave radiation*, and terrestrial radiation is called *longwave radiation*. The separation between solar and terrestrial radiation is not quite complete, however. Suppose that we represent the Earth as a 250-K blackbody and that we represent the sun as a black, 5774-K sphere whose radius is that of the sun and whose distance is the mean Earth–sun distance. Then solar radiation and terrestrial radiation are equal in magnitude at about 5.7 μm. Solar radiation is one tenth terrestrial radiation at 7.7 μm, and terrestrial radiation is one tenth solar radiation at 4.5 μm. During daylight hours, satellite data must be carefully interpreted near these wavelengths.

The annual average total irradiance reaching the top of the Earth's atmosphere is known as the *solar constant* (S_{sun}). Accurate determination of the solar constant is important for solar physics and astronomy as well as meteorology. Several satellites are currently making measurements of the solar constant. A "consensus" value of S_{sun}, based on satellite radiometer measurements,[20] is about 1368 W m^{-2} (equivalent to a 5774-K blackbody). Unfortunately the solar "constant" is not constant. It appears to follow the 11-year sunspot cycle, varying about $\pm$0.6 W m^{-2} (see Section 10.1). Of course the solar irradiance varies $\pm$3.4% from S_{sun} during the year due to the eccentricity of the Earth's orbit about the sun.

Bibliography

Clough, S. F., F. Kneizys, W. Gallery, L. Rothman, L. Abreu, and J. Chetwynd (1982). *FASCOD*. U.S. Air Force Geophys. Lab. AFGL-TR-78-0081, Hanscom AFB, MA.

Goody, R. M., and Y. L. Yung (1989). *Atmospheric Radiation: Theoretical Basis*, 2nd ed. Oxford Univ. Press, Oxford and New York.

Hansen, J. E., and L. D. Travis (1974). Light scattering in planetary atmospheres. *Space Sci. Rev.*, 16, 527–610.

Herzberg, G. (1950). *Atomic Spectra and Atomic Structure*. Dover, New York.

Irvine, W. M., and J. B. Pollack (1968). Infrared optical properties of water and ice spheres. *Icarus*, 8, 324–360.

Kneizys, F., E. Shuttle, W. Gallery, J. Chetwynd, L. Abreu, J. Selby, S. Clough, and R. Fenn (1983). *Atmospheric Transmittance/Radiance Computer Code LOWTRAN 6*. U.S. Air Force Geophys. Lab. AFGL-83-0187, Hanscom AFB, MA.

Kondratyev, K. Ya. (1969). *Radiation in the Atmosphere*. Academic Press, New York.

Kyle, T. G., and A. Goldman (1975). *Atlas of Computed Infrared Atmospheric Absorption Spectra*. National Center for Atmospheric Research, NCAR-TN/STR-112, Boulder, CO.

Liebe, H. J., and G. G. Gimmestad (1978). Calculation of clear air EHF refractivity. *Radio Science*, 13, 245–251.

Liou, K.-N. (1980). *An Introduction to Atmospheric Radiation*. Academic Press, New York.

McCartney, E. J. (1983). *Absorption and Emission by Atmospheric Gases*. Wiley, New York.

McClatchey, R. A., R. W. Fenn, J. E. A. Selby, F. E. Volz, and J. S. Garing (1972). *Optical Properties*

[20] Based on measurements from the Nimbus 7 Earth Radiation Budget (ERB) experiment, the Active Cavity Radiometer Irradiance Monitor (ACRIM) on board the Solar Maximum Mission satellite, and the Earth Radiation Budget Experiment (ERBE) on board the Earth Radiation Budget Satellite, NOAA 9, and NOAA 10. A systematic difference between the instruments of approximately 7 W m^{-2} has yet to be resolved (see Section 10.1).

of the Atmosphere (3rd ed.). U.S. Air Force Cambridge Res. Lab. AFCRL-TR-72-0497, Hanscom AFB, MA.

McClatchey, R. A., W. S. Benedict, S. A. Clough, D. E. Burch, R. F. Calfee, K. Fox, L. S. Rothman, and J. S. Garing, (1973). *AFCRL Atmospheric Absorption Line Parameters Compilation.* U.S. Air Force Cambridge Res. Lab. AFCRL-TR-73-0096, Hanscom AFB, MA.

McMillin, L. M., and H. E. Fleming (1976). Atmospheric transmittance of an absorbing gas: A computationally fast and accurate transmittance model for absorbing gases with constant mixing ratios in inhomogeneous atmospheres. *Appl. Optics*, **15**, 358–363.

Raschke, E. (ed.), (1978): *Terminology and Units of Radiation Quantities and Measurements.* International Association of Meteorology and Atmospheric Sciences (IAMAS, formerly the International Association of Meteorology and Atmospheric Physics) Radiation Commission, Boulder, CO. (Printed by the National Center for Atmospheric Research, Boulder, CO 80307.)

Smith, W. L. (1985). Satellites. In D. D. Houghton (ed.), *Handbook of Applied Meteorology.* John Wiley & Sons, Inc., New York.

Suttles, J. T., R. N. Green, P. Minnis, G. L. Smith, W. F. Staylor, B. A. Wielicki, I. J. Walker, D. F. Young, V. R. Taylor, and L. L. Stowe (1988). *Angular Radiation Models for Earth-Atmosphere System, Volume I: Shortwave Radiation.* NASA Ref. Pub. RP 1184, vol. I, Langley Research Center, Hampton, VA.

Valley, S. L. (ed.) (1965). *Handbook of Geophysics and Space Environments.* Air Force Cambridge Research Laboratory, Hanscom AFB, MA.

van de Hulst, H. C. (1957). *Light Scattering by Small Particles.* John Wiley & Sons, New York.

Wallace, J. M., and P. V. Hobbs (1977). *Atmospheric Science: An Introductory Survey.* Academic Press, New York.

Welch, R. M., S. K. Cox, and J. M. Davis (1980). Solar radiation and clouds. *Meteorol. Monogr.*, **17**, 96 pp.

4

Meteorological Satellite Instrumentation

CENTRAL TO SATELLITE meteorology is the instrumentation with which the basic radiometric measurements are made. A book devoted entirely to satellite instruments is Chen (1985). We approach this subject by discussing in some detail the instruments on currently operational meteorological satellites[1] because these are the sources of the data that readers will most likely use. Next we discuss important but not yet operational instrumentation on other satellites. In the final section of the chapter, we list data archives, the locations where satellite data are stored.

4.1 OPERATIONAL POLAR-ORBITING SATELLITES

Two countries maintain systems of operational polar-orbiting satellites: the former Soviet Union and the United States. The Soviet satellites are called *Meteor* satellites; we defer discussion of them to Section 4.1.7.

[1] *Operational satellites* are designed for and used in day-to-day operations of weather services. The opposite is *experimental satellites*, which may find use in weather service operations, but are designed to test new instruments.

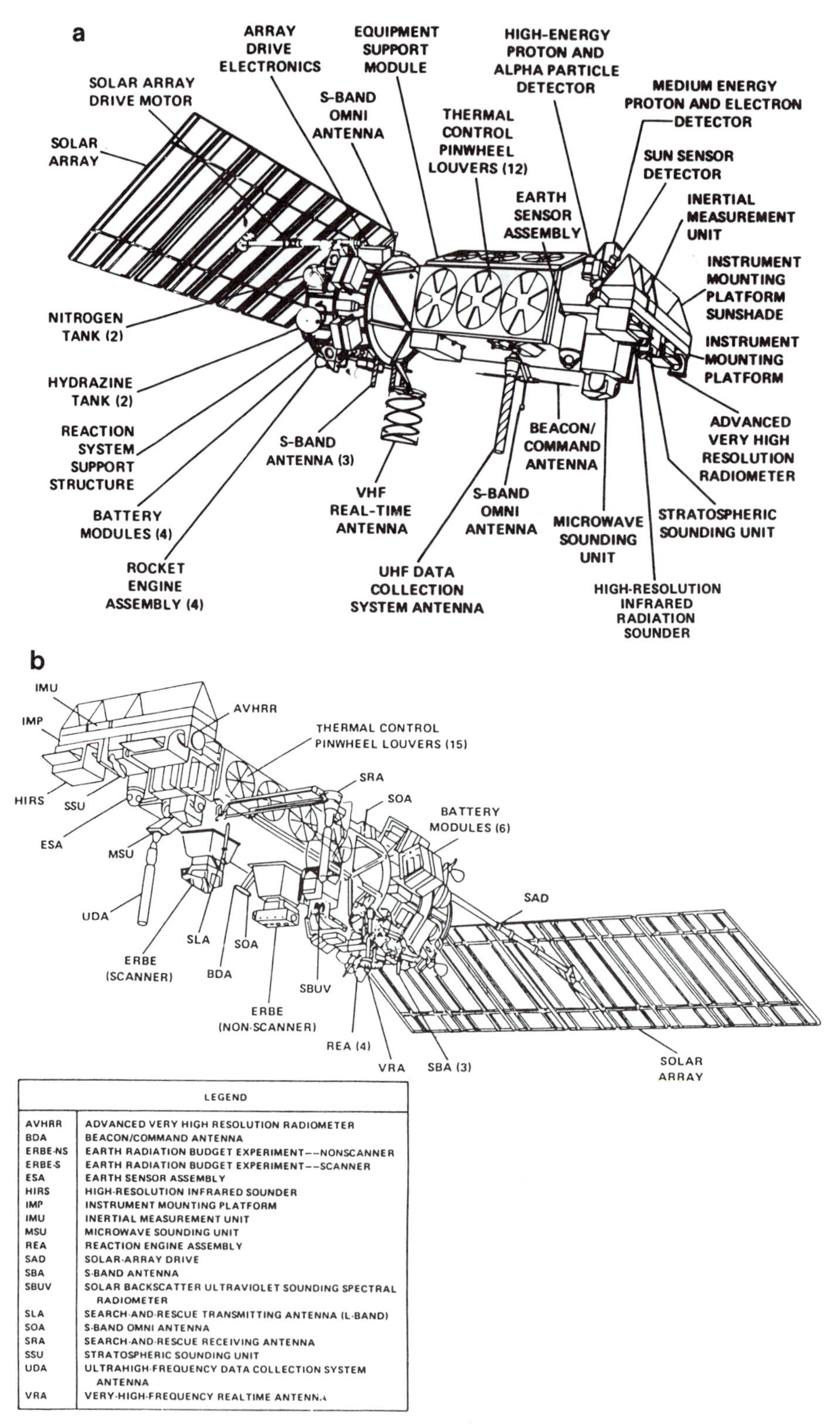

FIGURE 4.1. The TIROS N-series satellite (a) TIROS N ; (b) Advanced TIROS N. [After Schwalb (1978, 1982).]

The United States polar-orbiting meteorological satellites began with TIROS 1, launched on 1 April 1960; 10 satellites were launched in the series ending in 1965. The first operational series of meteorological satellites was the ESSA series; 9 satellites were launched between 1966 and 1969. The second series of operational satellites was the ITOS series; 6 satellites (ITOS and NOAA 1–5) were launched between 1970 and 1976. The current polar-orbiting series began with TIROS N, which was launched 13 October 1978. Figure 4.1a shows a diagram of the TIROS N satellite; the NOAA 6, 7, and 12 satellites are similar. The NOAA 8, 9, 10, 11, 13, and 14 satellites are modified versions of TIROS N and are called *Advanced TIROS N* (ATN; Fig. 4.1b). Table 4.1 summarizes the physical characteristics of these satellites.

The *satellite bus* (the part of the satellite excluding instruments) for the TIROS N series satellites is built by RCA Corporation (now Martin Marietta Astro Space). The duties of the bus are many. It contains radio transmitters, which send

TABLE 4.1. NOAA Satellite Summary

Parameter	*TIROS N*	*Advanced TIROS N*
Satellite mass	1421 kg (including expendables)	1009 kg (excluding expendables)
Payload mass (including tape recorders)	194 kg	386 kg
Satellite size		
Length	3.71 m	3.71 m
Diameter	1.88 m	1.88 m
Solar array		
Length	4.91 m	4.91 m
Width	2.37 m	2.37 m
Power output (worst case)	420 W	515 W
Power required (full operation)	330 W	475 W
Attitude control system		
Absolute (all axes)	0.2°	0.2°
Determination (all axes)	0.14°	0.14°
Communications frequencies (MHz)		
Command link	148.56	148.56
Beacon	136.77	137.77
	137.77	137.77
S-band	1698.0	1698.0
	1702.5	1702.5
	1707.0	1707.0
APT	137.50	137.50
	137.62	137.62
DCS (uplink)	401.65	401.65
S&R	—	1544.5
S&R (uplink)	—	121.5
	—	243.0
	—	406.0
Launch vehicle	Atlas E/F	Atlas E/F
Planned lifetime	2 years	2 years

TABLE 4.2. NOAA Satellite Instruments

Satellite	*Orbit*	*AVHRR*	*HIRS/2*	*MSU*	*SSU*	*ERBE*	*SBUV/2*	*SEM*	*DCS*	*S&R*
TIROS N	PM	1	X	X	X			X	X	
NOAA 6	AM	1	X	X	X			X	X	
NOAA 7	PM	2	X	X	X			X	X	
NOAA 8	AM	1	X	X	X			X	X	X
NOAA 9	PM	2	X	X	X	X	X	X	X	X
NOAA 10	AM	1	X	X	X	X		X	X	X
NOAA 11	PM	2	X	X	X		X	X	X	X
NOAA 12	AM	1	X	X					X	
NOAA 13	PM	2	X	X	X		X	X	X	X
NOAA 14	AM	2	X	X			X		X	X

data collected by the instruments to Earth stations, and radio receivers, which acquire operational commands from the ground. The bus contains the computers that process the data from the instruments and the tape recorders on which data are recorded for later transmission to Earth. The bus must supply electrical power to itself and the instruments, control the environment (particularly temperature) for the instruments and electronics, maintain proper orientation, and be capable of adjusting its orbit. These are difficult tasks in the harsh environment of space. Satellite bus technology is a subject unto itself, and we will not pursue it further. We concentrate on the instruments which make the measurements.

Table 4.2 lists the instruments carried on the TIROS N series satellites. Several of these instruments are not directly related to meteorology. For example, the Search and Rescue (S&R[2]) system detects signals from downed aircraft and other vessels and gives a precise estimate of their location to aid in rescue operations. The Space Environment Monitor (SEM) measures energetic particles (protons, electrons, alpha particles) for solar and ionospheric studies. Finally, the Data Collection System (DCS) relays meteorological and other data transmitted from ground-based instruments. In the following sections, the meteorological instruments are described in detail.

4.1.1 The Advanced Very High Resolution Radiometer

The Advanced Very High Resolution Radiometer (AVHRR) is a *scanning radiometer*, which means that it makes calibrated measurements of upwelling radiation from small areas (*scan spots* or *pixels*) that are scanned across the subsatellite track (see Fig. 2.16). Images or pictures are constructed by displaying successive scan lines on photographic film or on a computer display. The operation of the AVHRR is representative of many scanning radiometers on low Earth orbiters.

The AVHRR is built by ITT Aerospace/Communication Division in Fort Wayne, Indiana. Figure 4.2 is a photograph of the instrument, and Table 4.3 lists

[2] Search and Rescue is also abbreviated SAR, but this abbreviation is also used for synthetic aperture radar. To avoid confusion, we use S&R for Search and Rescue.

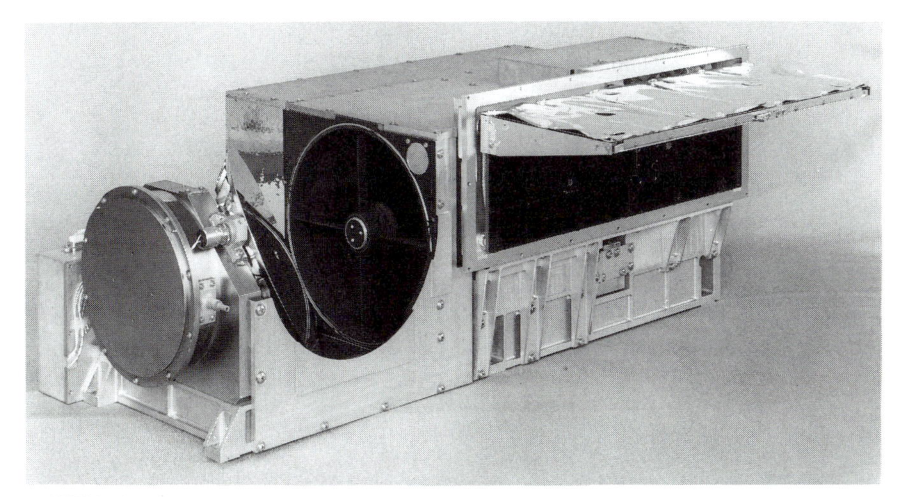

FIGURE 4.2. Photograph of the AVHRR instrument. [Courtesy of NOAA and ITT Aerospace.]

TABLE 4.3. Advanced Very High Resolution Radiometer (AVHRR) System Parameters

Parameter	*AVHRR/1*	*AVHRR/2*
Telescope		
Type	Afocal Mersenne	Afocal Mersenne
Diameter	20.32 cm	20.32 cm
Field of view	1.3 mrad	1.3 mrad
Ground resolution[a]		
Nadir	1.1 km	1.1 km
End of scan	2.3 × 6.4 km	2.3 × 6.4 km
Scan mirror rotation	12π rad s^{-1}	12π rad s^{-1}
Data sampling rate	40 kHz	40 kHz
Scan line		
Angle from nadir	±55.3°	±55.3°
Distance from nadir	±1500 km	±1500 km
Steps	2048	2048
Channels[b] (half-power points)		
1	0.55–0.68 μm	0.58–0.68 μm
2	0.75–1.10 μm	0.725–1.10 μm
3	3.55–3.93 μm	3.55–3.93 μm
4	10.5–11.5 μm	10.3–11.3 μm
5	channel 4 repeated	11.5–12.5 μm
Data rate[c]	750 kbits s^{-1}	750 kbits s^{-1}
Instrument size	27 × 37 × 79 cm	27 × 37 × 79 cm
Instrument mass	30 kg	30 kg
Power consumption	29 W	29 W

[a] From an orbital height of 850 km.
[b] Changes slightly with satellite (see Kidwell, 1986).
[c] Kilobits per second.

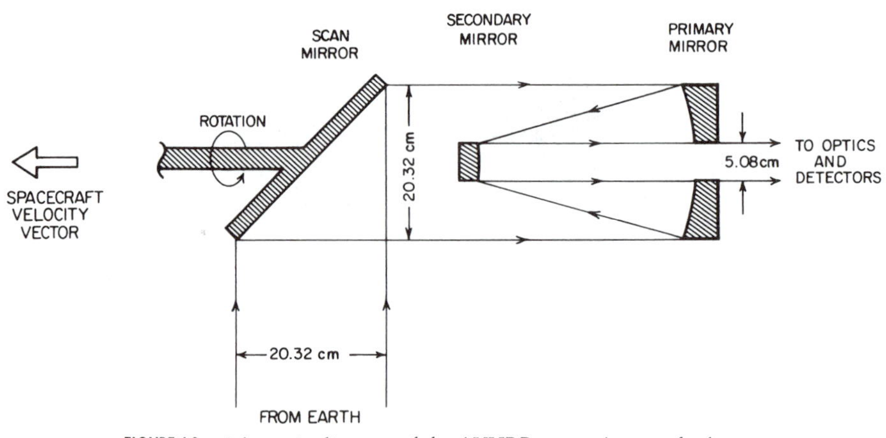

FIGURE 4.3. Schematic diagram of the AVHRR scan mirror and telescope.

the physical properties of the two versions of the instrument that have been flown to date. The AVHRR, like its predecessor, the Very High Resolution Radiometer (VHRR), which flew on the previous generation of NOAA satellites (the ITOS series), consists of a rotating scan mirror, a telescope, internal optics, detectors, and electronics.

The scan mirror is elliptical with a major axis of 29.46 cm and a minor axis of 20.96 cm. It is aligned at a 45° angle to the axis of the telescope, which is nominally parallel to the satellite velocity vector (Fig. 4.3). As the mirror rotates, it scans the field of view of the telescope across the Earth perpendicular to the satellite ground track.[3] The AVHRR scan mirror rotates at a rate of 360 revolutions per minute (rpm). Since the subsatellite point of the TIROS N series satellites moves at the rate of about 392 km min^{-1}, the distance between successive scan lines at the subsatellite point is about 1.1 km. The distance between successive scan lines is one parameter that describes the spatial resolution of a satellite instrument.

The afocal Mersenne telescope consists of parabolic primary and secondary mirrors. The primary mirror is 20.32 cm in diameter, and the beam reflected from the secondary mirror is 5.08 cm in diameter. Unlike other instruments, the reflected beam is collimated (not focused).

After passing through the hole in the primary mirror, the beam encounters an array of beam splitters, mirrors, lenses, and filters (Fig. 4.4) whose job is to deliver a filtered portion of the beam to each of the detectors. The optics are designed to minimize stray radiation and polarization effects that would contaminate the signal. Also, the optics are carefully aligned so that each of the channels views the same spot on Earth within an angle of 0.1 mrad.

Two versions of the AVHRR have been flown. The AVHRR/1 has 4 channels, and the AVHRR/2 has 5 channels. Channels 1 (0.6 μm) and 2 (1.1 μm) on both instruments use silicon detectors that are 2.54 mm on each side. In front of the

[3] Since the spacecraft is moving, however, a plot of the scan spots will make a slight angle with the ground track.

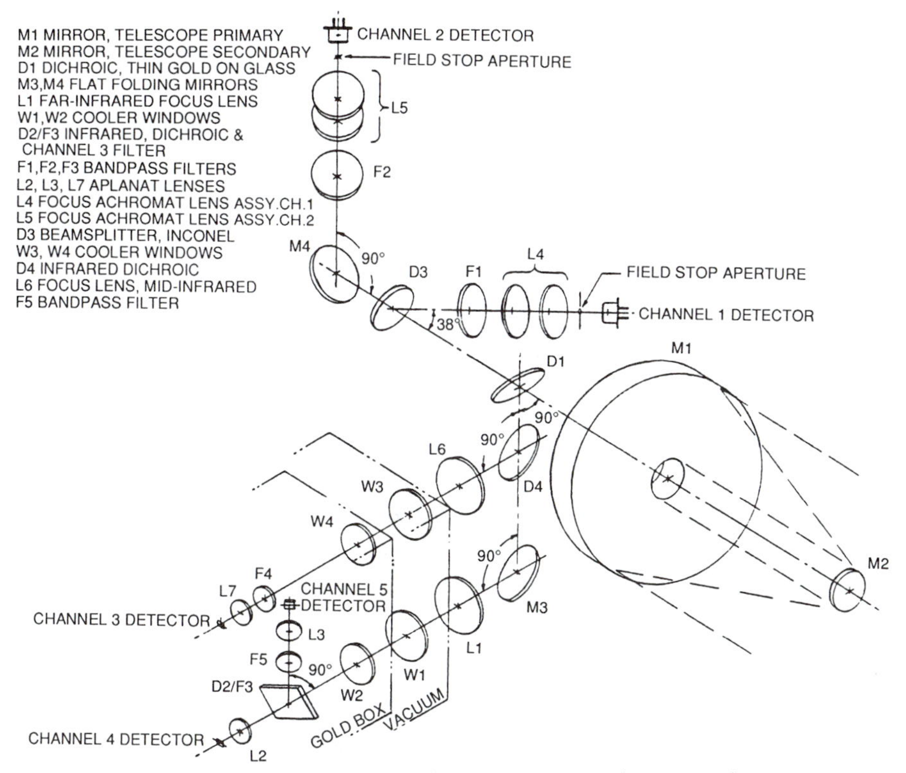

FIGURE 4.4. The AVHRR posttelescope optical system. [Courtesy of NOAA and ITT Aerospace.]

detector is a 0.6-mm-square field stop, which blocks radiation except for that coming from a rectangular scan spot on Earth. Channel 3 (3.7 μm) uses an indium antimonide (InSb) detector that is 0.173 mm square. Channel 4 on the AVHRR/1 (11 μm) and channels 4 (11 μm) and 5 (12 μm) on the AVHRR/2 use mercury cadmium telluride (HgCdTe) detectors that are also 0.173 mm square.[4] To lessen the amount of thermal noise, the detectors for channels 3, 4, and 5 are cooled to 105 K by exposing them to space (2.7-K equivalent blackbody temperature) through the side of the instrument housing. NOAA K will have a third version of the AVHRR (AVHRR/3) on board. AVHRR/3 will add a sixth channel (called *channel 3A*) sensitive to radiation between 1.58 and 1.64 μm. Channel 3A will operate during daylight, and channel 3 will operate at night.

The properties of the detectors and the optics in front of them determine the spectral response of each channel. Figure 4.5 shows the relative response functions for the AVHRR/2 which flew on NOAA 9. The AVHRR/1 response functions are very similar, but the AVHRR/1 lacks channel 5. These response functions are slightly different for the instrument on each satellite. They are measured in the

[4] For more information on the properties of various types of detectors, the reader is referred to Chen (1985).

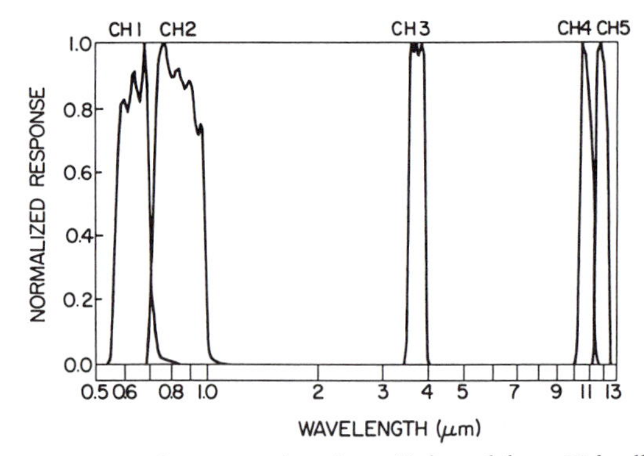

FIGURE 4.5. AVHRR spectral response functions. [Adapted from Kidwell (1986).]

laboratory before launch and are published by NOAA's National Environmental Satellite, Data, and Information Service [NESDIS, Kidwell (1986)].

Radiometer channels are commonly described by either one or two wavelengths. One-wavelength descriptions refer to a center wavelength, frequently the wavelength of peak response. Two-wavelength descriptions indicate the range of wavelengths to which a radiometer is sensitive. By convention, these wavelengths are specified by the half-power points of the response functions. For example, examination of Fig. 4.5 shows that the half-power points of channel 4 on the NOAA 9 AVHRR are approximately 10.3 and 11.3 μm. For some applications it is important to remember that the response of the instrument is not zero outside of this range, it is merely less than 50%.

The output of each sensor is a voltage that is proportional to the energy striking the sensor per unit time. When the area of the sensor, the spectral response function of the optics, and the field of view of the instrument are considered, the output voltage is proportional to the radiance (weighted by the channel spectral response function) of the scene that the instrument is viewing. The analog voltage is digitized for transmission to the ground. In the AVHRR, the digitization has 10 bits, that is, a number between 0 and 1023 is transmitted to the ground to represent the radiance for each pixel.

To make accurate measurements, the output of each channel must be calibrated, that is, compared with a known input. AVHRR channels 1 and 2 are calibrated in the laboratory before launch. Twelve matched, quartz-iodide lamps, which emit known radiance, are used. By turning on combinations of the lamps, a plot of radiance versus output digital count is constructed for each channel.[5] The radiance in this plot is expressed as a percentage (0–100) of the mean annual solar irradiance at the top of the atmosphere weighted by the spectral response function of the channel. The slope and intercept of the straight line that best fits this plot is used to linearly convert digital counts returned by the satellite into

[5] Twelve points are used for channel 1, but for channel 2, only 6–7 points (lamps) are necessary to reach 100% of the solar irradiance.

percent of the mean annual spectrally weighted solar irradiance at the top of the atmosphere.[6] A spectrally averaged radiance[7] can be calculated by multiplying this percentage by the mean annual spectrally averaged solar irradiance and dividing by π. Typically, channels 1 and 2 have a 10 : 1 signal-to-noise ratio for a very dark scene which reflects only 0.5% of the solar irradiance.

Until now, channels 1 and 2 have not been calibrated after launch; the laboratory slope and intercept were used throughout the lifetime of the instrument. As of this writing, however, NESDIS is planning to implement a system whereby AVHRR observations of Earth targets such as White Sands, New Mexico, of known or independently measured reflectance are used to update the slope and intercept every few weeks. This system should alleviate the inevitable postlaunch degradation of the sensor system and improve radiation budget estimates, for example.

AVHRR channels 3, 4, and 5 are calibrated in flight by viewing hot and cold objects. During each rotation of the mirror, the telescope views both cold space and the instrument housing, which is painted black and equipped with platinum resistance thermometers to accurately measure the housing temperature (roughly 290 K). The radiance of the instrument housing (calculated from its measured temperature) plus the digital counts for the housing and for space (essentially zero radiance) allow digital counts for Earth scenes to be linearly converted to radiance (Lauritson *et al.*, 1979). The equivalent blackbody temperature of the scene can then be determined with the Planck function.

NOAA specifications for AVHRR channels 3–5 require the equivalent blackbody temperature of a 300 K scene to be determined within ± 0.12 K. This is equivalent to radiance errors of 2.1, 16.9, and 14.6 mW m^{-2} sr^{-1} μm^{-1} at 3.7, 11, and 12 μm, respectively. Because the sensors basically measure radiance, the radiance error for each channel is approximately constant. The temperature error, however, varies with scene temperature as specified by the Planck function. At 250 K, for example, the temperature errors will be approximately 1.1, 0.20, and 0.18 K at 3.7, 11, and 12 μm, respectively.

The output of each sensor is continuous; a voltage can be measured at any time. This continuous output is discretized by electronically sampling (measuring) the voltage at the rate of 40 kHz (once each 25 μs). Each pixel in the data stream represents one such sample. Since the scan mirror rotates at the rate of 12π rad s^{-1}, it turns about 942 μrad between samples. From a height of 850 km, a scan spot at the subsatellite point moves about 0.80 km between samples. This distance between adjacent scan spots is a second parameter that is used to describe the spatial resolution of a satellite instrument. For each rotation of the scan mirror, the AVHRR records 2048 samples centered on the subsatellite point. This means

[6] In the documentation (e.g., Kidwell, 1986) this value is often called *albedo*, but it takes into account neither solar zenith angle, variation of the Earth–sun distance, nor anisotropic reflectance of the scene; therefore, it is not an albedo as defined in Chapter 3.

[7] Alternately, the albedo of the scan spot can be calculated (see Chapter 10). If the scan spot being observed is assumed to be a Lambertian reflector, its albedo can be calculated by dividing the percentage by the cosine of the solar zenith angle at the scan spot.

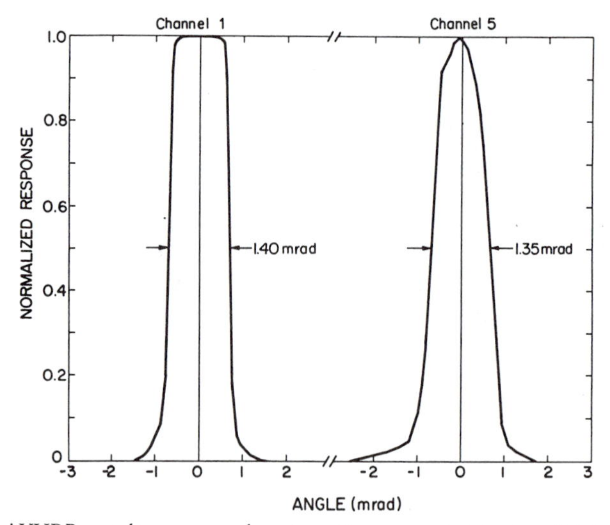

FIGURE 4.6. AVHRR angular response functions. [Courtesy of NOAA and ITT Aerospace.]

that it scans ±55.3° from nadir. From a height of 850-km, it scans about 1500 km from the subsatellite track. All five channels are measured simultaneously.

The final parameter that is used to describe the resolution of a satellite instrument is the *field of view* (FOV, also called the *instantaneous field of view* or IFOV) of the instrument.[8] In the case of the AVHRR, the active area of the detectors themselves (or the field stops in front of the sensors) determines the FOV. All channels of the AVHRR have fields of view of 1.3 ± 0.1 mrad. From 850 km altitude, this translates to a scan spot 1.1 km across at the subsatellite point. At either end of a scan line, each pixel measures about 2.3 km in the direction of satellite motion (*along-track*) and about 6.4 km in the direction perpendicular to the satellite track (*across-track*).

Just as a radiometric instrument measures radiation covering a range of wavelengths, not just from a single wavelength, so an instrument measures radiation coming from a range of directions, not just from a single direction. The angular response curve for an instrument is the relative response as a function of angle from the central angle. Figure 4.6 shows an example of the angular response curves for two of the AVHRR channels. Before launch, the spectral and angular responses of all AVHRR instruments are measured. Similar to the case of spectral response, the field of view of a satellite radiometer is specified by giving the full width at half maximum of the response curve: 1.40 mrad for channel 1 of the instrument depicted in Fig. 4.6, 1.35 mrad for channel 5.

The angular response curve for the AVHRR visible channel is steep near the half-power points; only about 7% of the area under the curve is outside of these points. The curve for the AVHRR infrared channel is broader; about 16% of the

[8] See, however, the discussion of the effects of motion on scan spot size in Section 4.1.6.2.

area under the curve is outside the half-power points.[9] This broadening of the response curves tends to blur temperature and brightness gradients.

These two curves also reveal something about the shape of the FOV. The sensors themselves are square. Because the angular response function for the visible channel is sharp, the FOV is also nearly square. The infrared response function, on the other hand, indicates that the IR FOV has rounded corners. A circle is probably a better approximation of the IR FOV shape than is a square.

All meteorological data on the TIROS N-series satellites are continually broadcast to Earth by a 5.25-W transmitter on 1698 or 1707 MHz. These direct broadcasts are called *high-resolution picture transmission* (HRPT).[10] These data can be received and processed by anyone with the proper equipment. Because the data rate is rather high, however, the necessary equipment is relatively expensive. For those interested only in acquiring visible and infrared images, the AVHRR data are processed on board the satellite into a much lower data rate data stream that can be received with inexpensive equipment. Two channels of the AVHRR, usually channels 1 and 4, are reduced in resolution to about 4 km by averaging along the scan line and by neglecting some scan lines. These data are transmitted to the ground in an analog (television-like) signal by a separate 5-W transmitter operating at a frequency of 137.50 or 137.62 MHz. These transmissions are called *automatic picture transmission* (APT). Barnes and Smallwood (1982) describe the APT transmissions.

Since APT and HRPT can be received only when the satellite is within range of a receiving station, five pairs of digital tape recorders are used on the satellite to record data for playback to ground stations. NOAA operates two Command and Data Acquisition (CDA) stations, one at Gilmore Creek, Alaska, and the other at Wallops Island, Virginia. In addition, the Centre National d'Etudes Spatiales (CNES) operates a receiving station near Lannion, France. All these data are transmitted to the Satellite Operations Control Center (SOCC) in Suitland, Maryland. Each tape recorder can store either one full orbit of 4-km-resolution data, called *global area coverage* (GAC) data,[11] or 10 min of full-resolution HRPT data, called *local area coverage* (LAC) data.

Some of the LAC data, virtually all of the GAC data, and data from the NOAA satellite instruments discussed below are archived by NESDIS (see Section 4.4).

4.1.2 The High Resolution Infrared Radiation Sounder

The High Resolution Infrared Radiation Sounder 2 (HIRS/2) is derived from the HIRS/1, which flew on the Nimbus 6 satellite. The HIRS/2 is built by ITT Aerospace/Communication Division and operates much like the AVHRR. The primary differences between the HIRS/2 and the AVHRR are (1) the HIRS/2 has many more channels (20) than does the AVHRR (4 or 5), and (2) the HIRS/2

[9] Because of the way in which the response curves are measured, the fraction of *energy* sensed by the detectors that comes from outside the half-power points will be *greater* than these areas.

[10] S&R data are processed and transmitted separately.

[11] See Section 10.2.1 for a description of how GAC data are produced from the full-resolution data.

has much coarser resolution (42 km) than the AVHRR (1.1 km). These differences are due to the different requirements of the two instruments. The AVHRR is designed to make images, in which the horizontal structure of the atmosphere is most important, whereas the HIRS/2 is used for soundings, in which the vertical structure of the atmosphere is most important. Table 4.4 contains the HIRS/2 system parameters, and Table 4.5 summarizes the HIRS/2 channels.

The HIRS/2 utilizes two carbon dioxide bands for temperature sounding: seven channels are located in the 15-μm band, and six channels are located in the 4.3-μm band. Older sounding instruments [such as the Vertical Temperature Profile Radiometer (VTPR), which flew on the NOAA satellites prior to the TIROS N series] had only 15-μm temperature sounding channels. The 4.3-μm channels were added to improve sensitivity (change in radiance for a given change in atmospheric temperature) at relatively warm temperatures. Moisture is sensed with three channels in the 6.3-μm band of water vapor. The 9.7-μm channel is designed to sense ozone. Three channels are in atmospheric windows: the 11.11- and 3.76-μm channels help determine the surface (skin) temperature, whereas the 0.69-μm channel is used to detect clouds. A detailed description of the process by which atmospheric soundings are retrieved from these radiances can be found in Chapter 6.

TABLE 4.4. High Resolution Infrared Radiation Sounder (HIRS/2) System Parameters

Parameter	*Value*
Telescope	
Type	Cassegrainian
Diameter	15.24 cm
Field of view	21.8 mrad (1.25°)
Ground field of view	
Nadir	18.5 km diameter
End of scan	31.8 km along-track × 62.8 km cross-track
Scan line	
Time per scan	6.4 s
Angle from nadir	±49.5°
Distance from nadir	±1115 km
Number of steps	56
Scan step	
Time	100 ms
Angle	1.8°
Distance between scan spots	
Cross-track	26.4 km
Along-track	41.8 km
Gap between consecutive passes at equator	540 km
Number of channels	20
Calibration	2 stable blackbodies and space
Data rate	2880 bits s^{-1}
Instrument mass	20.4 kg
Instrument volume	0.040 m^3
Average power consumption	20 W

TABLE 4.5. HIRS/2 Channels[a]

Channel	*Description*	*Central* *Wavenumber* (cm^{-1})	*Central* *Wavelength* (μm)	*Half-power bandwidth* (cm^{-1})	*Specified* NEΔL[b] $(mW\ sr^{-1}\ m^{-2}\ cm)$	*Mean scene temp.* (K)	*Equivalent* NEΔT[c] (K)
1	15 μm CO_2	669	14.95	3	3.00	235	2.77
2	15 μm CO_2	680	14.71	10	0.67	220	0.74
3	15 μm CO_2	690	14.49	12	0.50	220	0.55
4	15 μm CO_2	703	14.22	16	0.31	250	0.31
5	15 μm CO_2	716	13.97	16	0.21	245	0.18
6	15 μm CO_2	733	13.64	16	0.24	260	0.18
7	15 μm CO_2	749	13.35	16	0.20	275	0.14
8	Window	900	11.11	35	0.10	290	0.06
9	O_3	1030	9.71	25	0.15	270	0.13
10	H_2O	1225	8.16	60	0.16	290	0.17
11	H_2O	1365	7.33	40	0.20	260	0.44
12	H_2O	1488	6.72	80	0.19	245	0.96
13	4.3 μm CO_2	2190	4.57	23	0.006	280	0.10
14	4.3 μm CO_2	2210	4.52	23	0.003	265	0.10
15	4.3 μm CO_2	2240	4.46	23	0.004	250	0.24
16	4.3 μm CO_2	2270	4.40	23	0.002	230	0.31
17	4.3 μm CO_2	2360	4.24	23	0.002	250	0.15
18	4.3 μm CO_2	2515	4.00	35	0.002	300	0.04
19	Window	2660	3.76	100	0.001	300	0.02
20	Vis. window	14500	0.69	1000			

[a] Values in this table will change slightly with satellite (see Kidwell, 1986).
[b] Noise equivalent radiance difference (rms error).
[c] Noise equivalent temperature difference (rms error) at the mean scene temperature.

The scan mirror in the AVHRR rotates continuously; the HIRS/2 scan mirror, however, moves in steps. Between each step, the mirror moves 1.8°. The instrument then stares at a location for approximately 65 ms while all 20 channels are sampled. During the next 35 ms, the scan mirror steps to the next position. The total time between scan spots is 100 ms. Fifty-six scan spots between nadir angles of ±49.5° are sampled during each scan line. The total time for a scan line, including the time for the mirror to rotate back to the first position, is 6.4 s. The instantaneous field of view of the radiometer is 1.25°. This scanning geometry results in the scan pattern shown in Fig. 4.7.

Only three detectors are used on the HIRS/2. The visible channel (20) is sampled by a silicon detector at ambient temperature. Channels 1–12 are called the *longwave channels* and are sampled by a single HgCdTe detector cooled to 105 K. The *shortwave channels* (13–19) are sampled by a single InSb detector also cooled to 105 K. Sampling of multiple channels with a single detector is accomplished with a rotating filter wheel. Figure 4.8 shows a schematic diagram of the HIRS/2 optics. The angular length of each filter and the rotation rate of

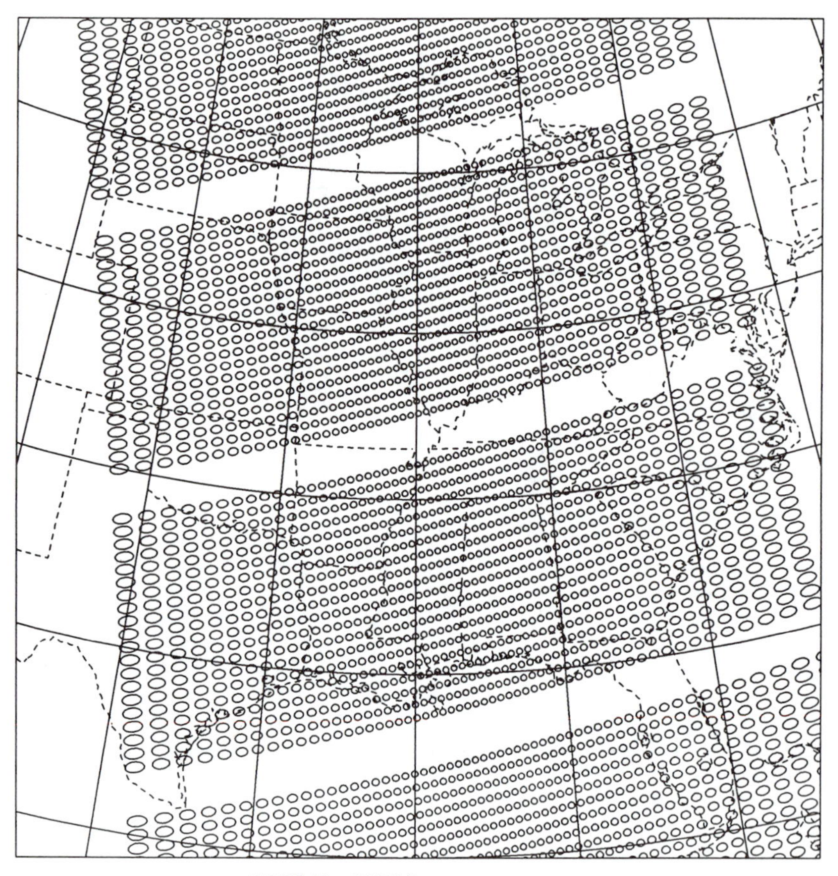

FIGURE 4.7. HIRS/2 scan pattern.

the filter wheel determine the integration time for each channel and are chosen to provide an acceptable signal to noise ratio.

Unlike the AVHRR, the HIRS/2 is not calibrated on each scan line. Instead, every 256 s the instrument goes into calibrate mode. First it looks at space, then at the internal hot and cold calibration sources. Each of these is viewed for a time equivalent to one full scan line. During the calibration period, no Earth data are collected. This results in 3-line gaps in the scan pattern (Fig. 4.7).

Voltages from each of the HIRS/2 channels are digitized using 13 bits and are processed by the TIROS Information Processor (TIP) as are data from the other sounders, the SEM, the DCS, the SBUV/2, and the ERBE. These data are transmitted continuously over the beacon frequencies of 136.77 or 137.77 MHz. The data are also included in the HRPT data. Up to 250 min of these data can be recorded on each of the tape recorders for later playback.

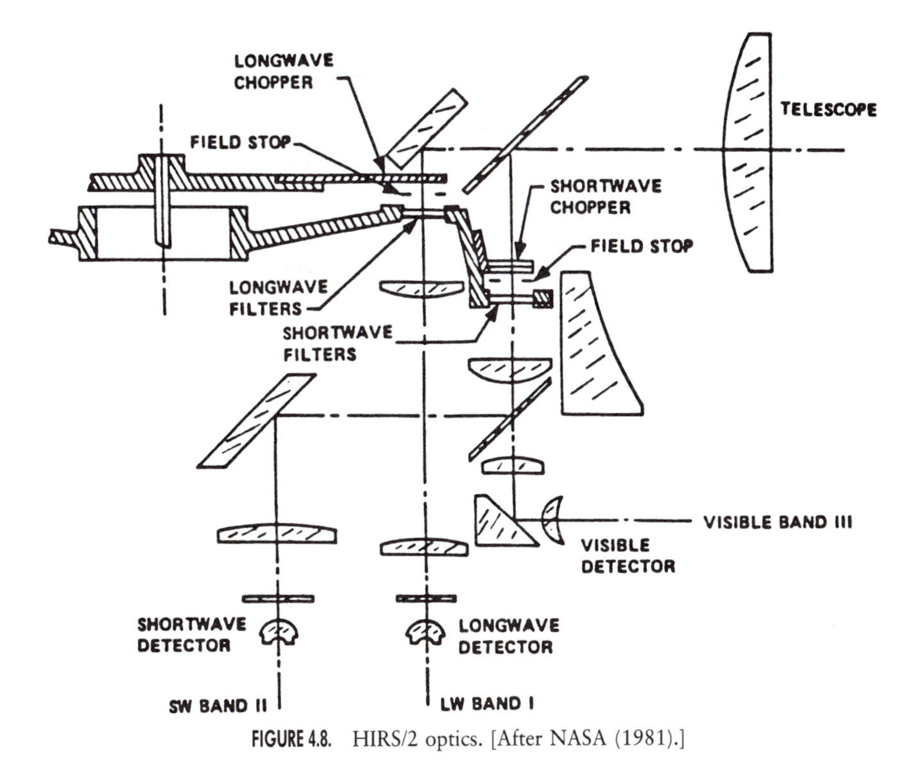

FIGURE 4.8. HIRS/2 optics. [After NASA (1981).]

4.1.3 Microwave Sounding Unit

The second sounding instrument on the NOAA satellites is the Microwave Sounding Unit (MSU). Its primary purpose is to make temperature soundings in the presence of clouds. The instrument is built by the Jet Propulsion Laboratory (JPL) of the California Institute of Technology. Its predecessor was the Scanning Microwave Spectrometer (SCAMS), which flew on the Nimbus 6 satellite. The MSU system parameters are listed in Table 4.6, and a photograph is shown in Fig. 4.9.

Externally the MSU is similar to the HIRS/2 in that it has rotating scan mirrors which step perpendicular to the satellite track. In all there are 11 steps each of 9.47°. The total scan line, therefore covers ±47.35° from nadir. The time for each step is 1.84 s, and the total time for each scan line is 26.6 s. The MSU field of view is 7.5°. These parameters, combined with the orbital parameters, result in the scan pattern shown in Fig. 4.10.

Immediately obvious in Fig. 4.10 is the fact that the MSU resolution is much coarser than that of the HIRS/2. This is a general property of microwave instruments. The Rayleigh criterion for the resolving power of a lens is that two points can just be resolved (distinguished as two points rather than one) when their

TABLE 4.6. Microwave Sounding Unit (MSU) System Parameters

Parameter	*Value*
Instrument type	Dicke Radiometer
Field of view	7.5°
Ground field of view	
Nadir	111 km diameter
End of scan	180 km along-track × 332 km cross-track
Scan Line	
Time per scan	25.6 s
Angle from nadir	±47.35°
Distance from nadir	±1015 km
Number of steps	11
Scan step	
Time	1.84 s
Angle	9.47°
Distance between scan spots (at nadir)	
Cross-track	140 km
Along-track	167 km
Gap between consecutive passes at equator	433 km
Channels	
1	50.30 GHz
2	53.74 GHz
3	54.96 GHz
4	57.95 GHz
RF bandwidth	220 MHz
Noise equivalent ΔT_B	0.3 K
Calibration	Stable blackbody and space
Data rate	320 bits s^{-1}
Instrument mass	20.9 kg
Instrument size	58.4 × 20.3 × 38.1 cm
Average power consumption	30 W

angular separation θ is given by

$$\theta = 1.22\frac{\lambda}{D}, \tag{4.1}$$

where λ is the wavelength of radiation, and D is the diameter of the lens. The same principle applies to a microwave receiving antenna. Consider an infrared instrument ($\lambda \approx 10\ \mu$m) and a microwave instrument ($\lambda \approx 1$ cm) with comparably sized optics. Other factors being equal, the size of a scan spot for the microwave instrument will be approximately 1000 times that of the infrared instrument, due to the ratio of their wavelengths. That the MSU resolution is not 1000 times worse than the HIRS/2 resolution indicates that the optics of the two instruments are not comparable. However, it should be kept in mind that the Rayleigh criterion severely limits the resolution of all microwave instruments. Only by using very large antennas can the resolution of microwave radiometers be made comparable to the resolutions of visible or infrared radiometers.

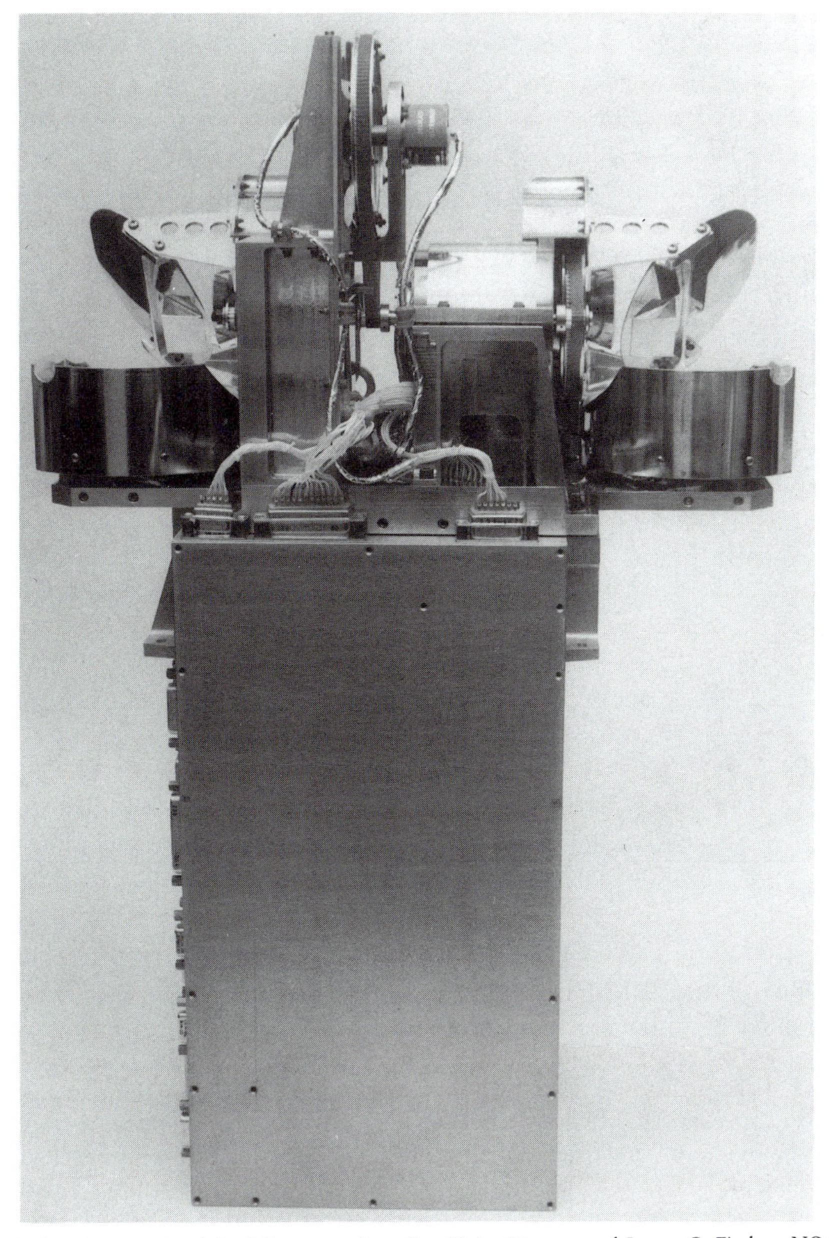

FIGURE 4.9. Photograph of the Microwave Sounding Unit. [Courtesy of James C. Fischer, NOAA/NESDIS.]

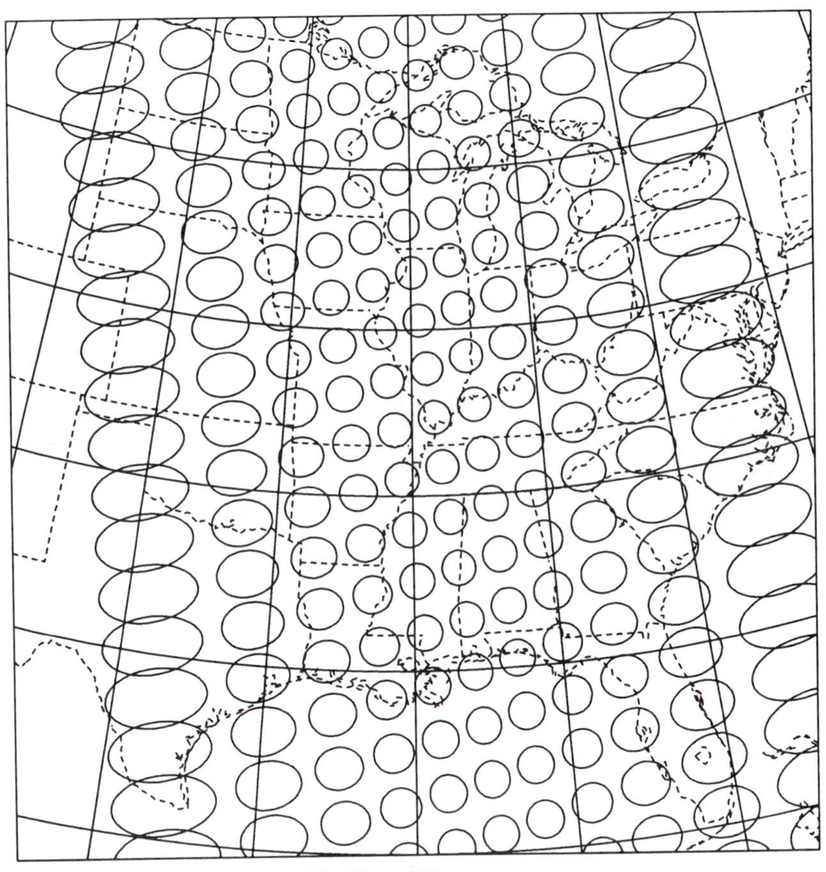

FIGURE 4.10. MSU scan pattern.

Although externally the HIRS/2 and the MSU are similar, internally they are quite different. In the MSU, radiation from each of the two scan mirrors enters a feedhorn where it encounters a transducer that separates the beam into two beams (4 channels total). Each of these beams travels via a waveguide to a device called a *Dicke switch*, which switches between viewing radiation from the feedhorn or from an internal microwave source of known temperature.[12] In the MSU this switching occurs at the rate of 1 kHz. The radiation is then mixed with an internally generated signal at the frequency of the channel and processed by a superheterodyne (radio) receiver. The output of the receiver is proportional to the difference between the brightness temperature of the scene being viewed and the temperature of the internal radiation source.

[12] Note that this detector system essentially amounts to a dipole antenna. The MSU and all other microwave radiometers measure *polarized* radiation. The polarization vector for the MSU rotates with the scan angle.

Calibration of the MSU is accomplished by viewing a blackbody, attached to the instrument housing, and space once each scan. Since no scan lines are devoted exclusively to calibration, the MSU has no breaks in its scan pattern.

The radiance from a typical Earth scene is much lower in the microwave portion of the spectrum than in the infrared portion; however, microwave receivers are much more sensitive than infrared detectors. Brightness temperatures measured by the MSU are accurate to within about 0.3 K rms, which is comparable with the temperature accuracies of the HIRS/2 channels (see Table 4.5). The brightness temperatures are digitized in 12 bits for transmission to Earth.

On the NOAA K,L,M series, both the MSU and the SSU (Stratospheric Sounding Unit) will be replaced with two Advanced Microwave Sounding Units (AMSU-A, and AMSU-B). AMSU will be the primary sounding instrument. The infrared instrument (HIRS/3) will be used to improve the soundings in cloud-free regions (see Chapter 11).

4.1.4 The Stratospheric Sounding Unit

The Stratospheric Sounding Unit (SSU) is designed to retrieve temperatures in the stratosphere. Its immediate predecessor is the Pressure Modulator Radiometer (PMR) flown on Nimbus 6. The SSU is supplied by the United Kingdom Meteorological Office. A diagram of the instrument is shown in Fig. 4.11, and the SSU system parameters are listed in Table 4.7.

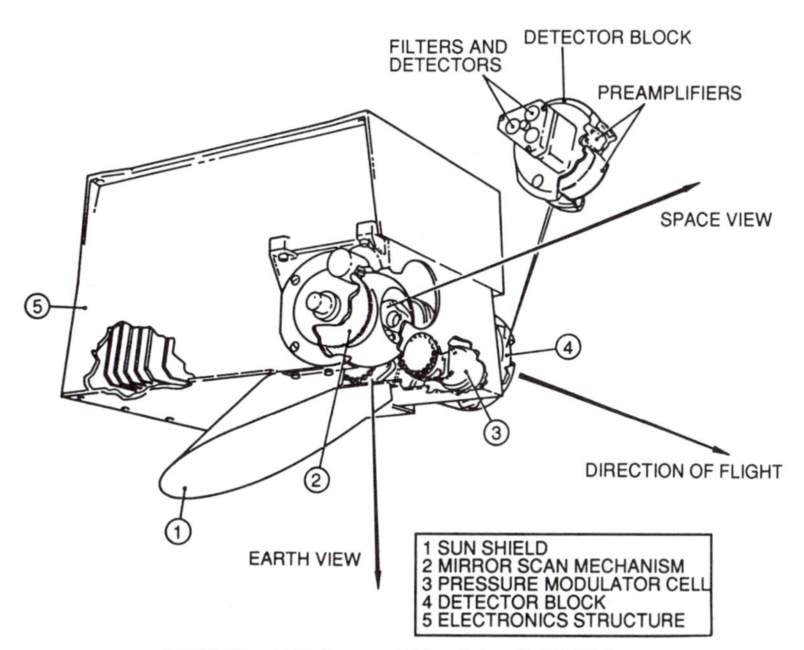

FIGURE 4.11. SSU diagram. [After Schwalb (1978).]

TABLE 4.7. Stratospheric Sounding Unit (SSU) System Parameters

Parameter	*Value*
Telescope	None
Optics	Simple 2-lens, 15 mm diameter, *f*/2.4, imaging system for each channel
Field of view	10°
Ground field of view	
Nadir	148 km diameter
End of scan	187 km along-track × 249 km cross-track
Scan line	
Time per scan	32 s
Angle from nadir	±35°
Distance from nadir	±617 km
Number of steps	8
Scan step	
Time	4 s
Angle	10°
Distance between scan spots (at nadir)	
Cross-track	148 km
Along-track	209 km
Gap between consecutive passes at equator	1330 km
Number of channels	3
Calibration	Stable blackbody and space
Data rate	480 bits s^{-1}
Instrument mass	15 kg
Instrument size	26.5 × 36.5 × 26.3 cm
Average power consumption	18 W

The SSU is a scanning radiometer like all the instruments discussed above. Its scan mirror steps in 10° increments for 8 steps from 35° left to 35° right of the satellite track. Each step takes 4 s, and an entire scan line takes 32 s. Since the field of view of the radiometer is 10°, the scan spots are contiguous in the cross-track direction (Fig. 4.12), but there is an underlap in the along-track direction. Also, the SSU scans only about two thirds as far out as does the HIRS/2 and the MSU. Thus the outermost HIRS/2 and MSU scan spots have no corresponding SSU data. Special handling of the SSU data are necessary in the retrieval process to correct for this (see Chapter 6).

The SSU utilizes an interesting method for filtering the incoming radiation. All three of its channels sense radiation within 50 cm^{-1} of the center of the *Q* branch of the 15-μm CO_2 band. Each channel has a cell containing CO_2 in its optical path. The CO_2 cell acts as a filter with a response function which is very similar to that shown in Fig. 3.17; that is, it is perfectly tuned to sense CO_2. This is called *gas correlation spectroscopy.* The pressure inside the cells is modulated at 40 Hz. The radiation on the detector is a weighted average of the radiance emitted by the atmosphere and by the gas in the cell. As the cell pressure increases, the absorption lines of the CO_2 in the cell broaden and filter radiation from the atmosphere; that is, more of the radiation striking the detector is from the gas

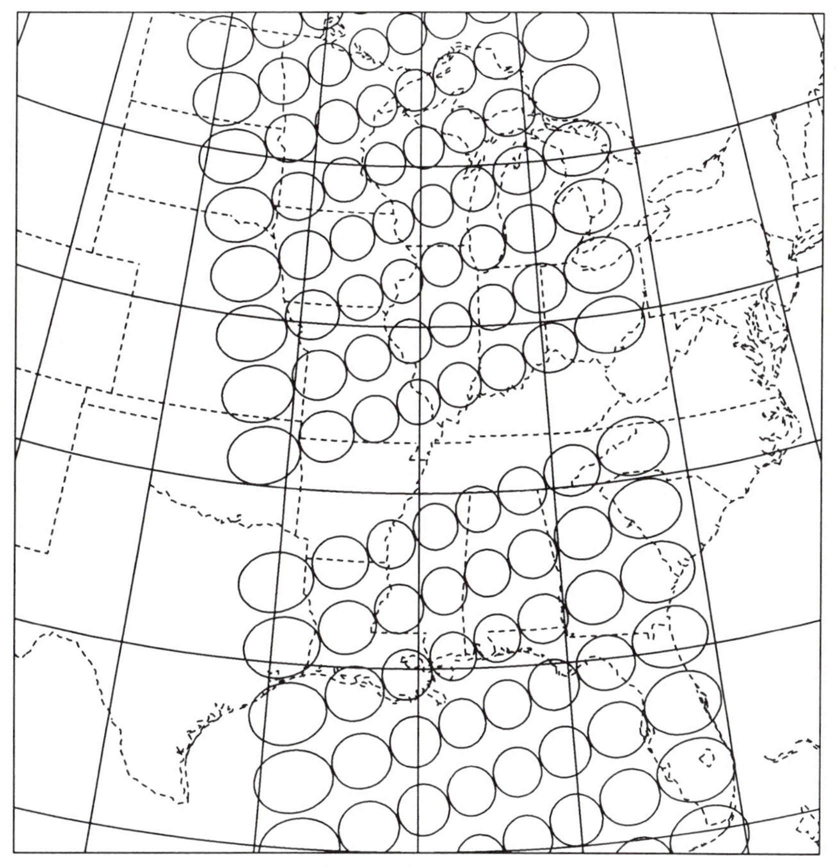

FIGURE 4.12. SSU scan pattern.

in the cell. As the cell pressure decreases, the lines become narrower, and the detector better senses more radiation from the atmosphere. The mean cell pressure determines the level to which the channel is most sensitive (Table 4.8). The amplitude of the 40-Hz signal from each of the three channels yields information on the vertical temperature structure.

TABLE 4.8. SSU Channels

Channel	*Central wavenumber* (cm^{-1})	*Cell pressure* (*hPa*)[a]	*Level of peak sensitivity*		NEΔT (*K*) (*at 273 K*)
			Pressure (*hPa*)	*Height* (*km*)	
1	669.99	100	15	29	0.25
2	669.63	35	5	37	0.50
3	669.36	10	1.5	45	1.25

[a] The hectopascal (hPa) is used throughout the book as the unit of pressure. It is numerically equal to the non-SI unit millibar.

The detector for each of the three SSU channels is a flake of triglycine sulfate.

Once each 256 s, the SSU enters a calibration mode in which it first looks at space, then at an internal blackbody source maintained near 15°C. Calibration takes 32 s, during which time no Earth data are collected. This produces a one-line gap in the scan pattern.

The radiances falling on the detectors are integrated for 3.6 s of each 4-s step and are digitized to 12 bits for transmission to Earth as part of the TIP data stream.

4.1.5 The Solar Backscatter Ultraviolet Radiometer

On board the Advanced TIROS N satellites (NOAA 9 and NOAA 11 to date) is the Solar Backscatter Ultraviolet Radiometer 2 (SBUV/2), which is the successor to the SBUV/1 on the Nimbus 7 satellite. SBUV/2 is built by the Aerospace Systems Division of Ball Corporation in Boulder, Colorado. Its purpose is to estimate the global ozone distribution by measuring backscattered solar radiation in the ultraviolet Hartley–Huggins bands.

SBUV/2 actually consists of two instruments: the Monochrometer is the primary instrument, and the Cloud Cover Radiometer detects clouds that would contaminate the signal (Fig. 4.13). Neither of these instruments scan spatially; they simply view nadir. The field of view of the instruments is 11.3°, which means that the scan spot has a 168 km diameter on the Earth.

Technically, the Monochrometer consists of two Ebert–Fastie monochrometers in series (Fig. 4.14). Before entering the Monochrometer, the radiation is depolarized and chopped at a 20-Hz rate. Radiation enters each half of the Monochrometer through a 0.13 × 3.0 cm entrance slit and reflects from a 25-cm focal length Ebert mirror, which collimates the beam. The beam then encounters a blazed holographic grating[13] that separates the radiation into its constituent wavelengths much like a prism. The radiation again reflects from the Ebert mirror and leaves via an exit slit. The grating is rotated to select which wavelength leaves the Monochrometer. The entrance slit for the second half of the Monochrometer is the same as the exit slit for the first half. The two gratings are mounted on a common shaft to coordinate spectral scanning. The use of two monochrometers in series improves rejection of stray light and sharpens spectral response in comparison to a single monochrometer. The spectral resolution of each channel is 1.1 nm.

The Monochrometer has two operational modes. In *discrete mode*, the gratings are rotated to 12 discrete positions to sample 12 discrete wavelengths (see Table 4.9). Each wavelength is sampled for 1.25 s, and 0.75 s is allowed to step to the next position. All 12 wavelengths are sampled in 24 s. Eight seconds are allowed for the gratings to rotate back to the first position to begin a new wavelength scan. Each wavelength scan thus takes 32 s. In *sweep mode*, the gratings are stepped such that the wavelength is scanned through the range 160–407 nm in 0.148-nm increments. Each step takes 0.1 s, so that the entire wavelength scan

[13] See Halliday and Resnick (1962). Chapter 45 reviews the optical properties of gratings.

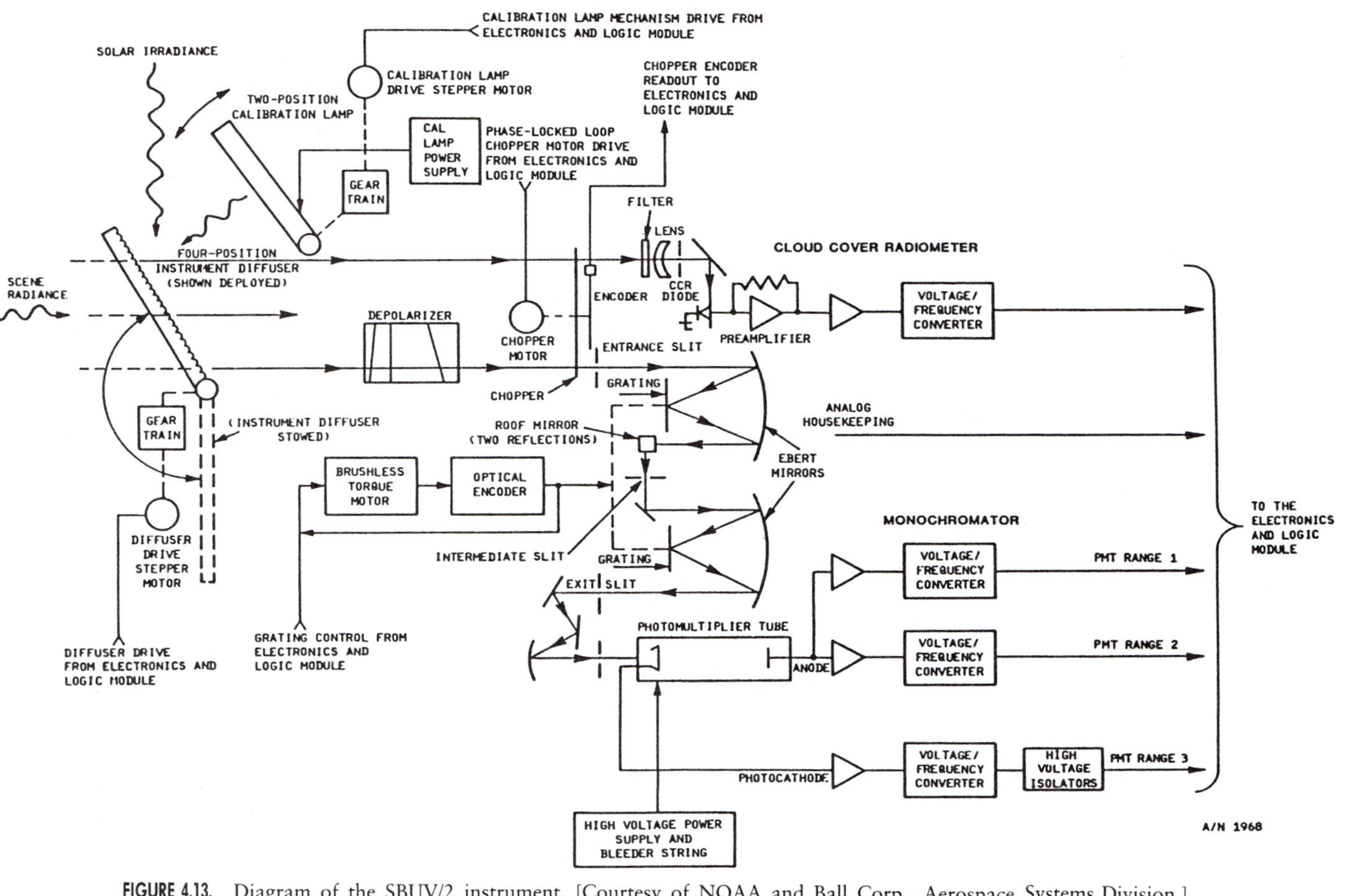

FIGURE 4.13. Diagram of the SBUV/2 instrument. [Courtesy of NOAA and Ball Corp., Aerospace Systems Division.]

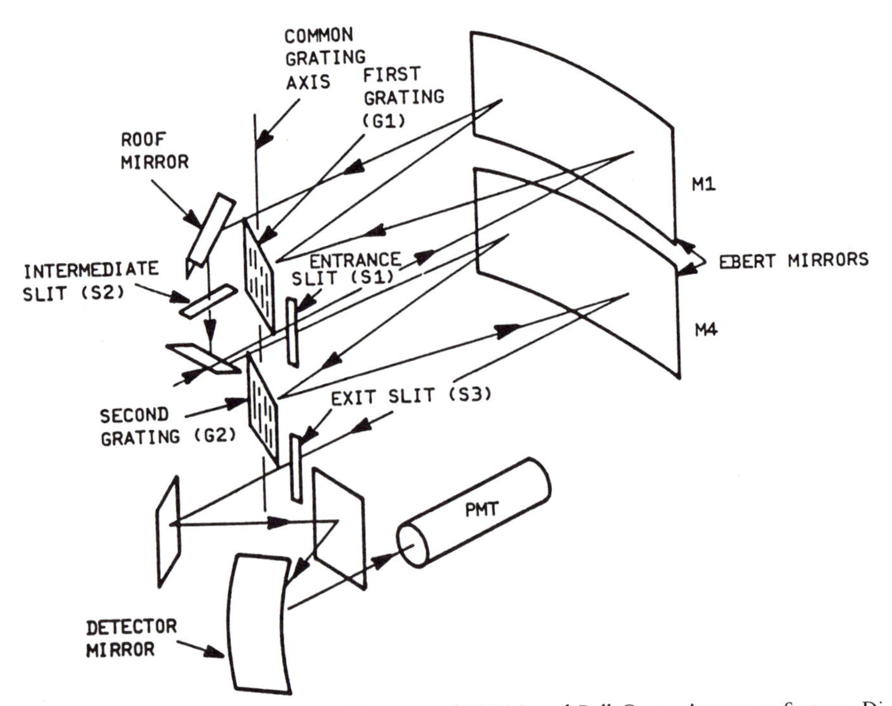

FIGURE 4.14. SBUV/2 Monochrometer. [Courtesy of NOAA and Ball Corp., Aerospace Systems Division.]

takes about 168 s. During the next 24 s, the gratings are rotated back to the initial position.

Retrieval of ozone profiles requires a knowledge of the fraction of solar radiation reflected at each wavelength. Not only must the radiance reflected by the Earth's atmosphere be known, but also the incident solar radiation. Ground-based estimates of the solar radiation at the top of the atmosphere in the ultraviolet portion of the spectrum are not useful because of the opacity of ozone. Therefore, the SBUV/2 measures the ultraviolet solar irradiance itself. A Lambertian diffuser plate painted with $BaSO_4$ (Kodak White) can be swung into place such that once each orbit solar radiation is reflected into the instrument. The Monochrometer can be operated in either discrete mode or sweep mode during this time.

The SBUV/2 is calibrated with an on-board mercury lamp that has eight emission lines in the wavelength range of the instrument: 184.0, 253.7, 302.2, 313.2, 365.0, and 404.7 nm. The lamp can be positioned such that it shines directly into the optics, or such that it first reflects from the diffuser plate, thus checking the properties of the diffuser. During these measurements, the SBUV/2 can be operated in either discrete mode or sweep mode. Normally, discrete mode is used during calibration and 12 wavelengths around the 253.7-nm line are measured.

The Cloud Cover Radiometer (CCR) is a separate radiometer used to detect clouds. It operates at 379 nm using a 3.0-nm-wide filter. The CCR is aligned to

TABLE 4.9. Solar Backscatter Ultraviolet Radiometer (SBUV/2) System Parameters

Parameter	*Value*
Monochrometer	
Instrument type	Spectral scanning Double Ebert–Fastie Monochrometer
Optics	
Depolarizer	4 elements
Chopper wheel	20 Hz
Mirror	Ebert, 25 cm focal length
Entrance slit	0.13 × 3.0 cm
Grating	2400 lines mm^{-1}
Grating step	95.90 μrad
Detector	Photomultiplier tube
Channels	12 (discrete mode): 252.0, 273.5, 283.0, 287.6, 292.2, 297.5, 301.9, 305.8, 312.5, 317.5, 331.2, 339.8 nm
Wavelength Range	160–407 nm (sweep mode)
Field of view	11.3°
Ground resolution (nadir only)	168 km
Calibration	On-board mercury lamp
Cloud Cover Radiometer	
Optics	Chopper wheel (20 Hz) Bandpass filter Singlet lens Spherical folding mirror
Detector	Vacuum photodiode
Channels	1 (379 nm)
Field of view	11.3°
Ground resolution (nadir only)	168 km
Calibration	On-board mercury lamp
Total Instrument Package	
Data rate	320 bits s^{-1}
Instrument mass	
Sensor module	25.6 kg
Electronics and logic module	11.2 kg
Instrument size	
Sensor module	31.1 × 35.6 × 50.5 cm
Electronics and logic module	22.2 × 33.0 × 33.4 cm
Average power consumption	12 W

view the same scene as the Monochrometer, and it has the same 11.3° field of view. The CCR shares the chopper wheel, mercury lamp and diffuser plate with the Monochrometer.

Data from both the Cloud Cover Radiometer and the Monochrometer are digitized to 8 bits for transmission to Earth as part of the TIP data stream.

4.1.6 The Earth Radiation Budget Experiment

Because atmospheric motions are driven by differential absorption of solar radiation and infrared loss to space, the study of the Earth's radiation budget is extremely important. The first successful meteorological instrument in space (on board the Explorer 7 satellite launched 13 October 1959) was an Earth radiation budget instrument. Since then there have been several instruments designed primarily to study the Earth's radiation budget. The latest is the Earth Radiation Budget Experiment (ERBE), which flies on the NOAA 9 and NOAA 10 satellites as well as on the Earth Radiation Budget Satellite (ERBS). The ERBE is designed to make highly accurate (~1%) measurements of incident solar radiation, Earth-reflected solar radiation, and Earth-emitted longwave radiation at spatial scales ranging from global to 250 km and at temporal scales sufficient to generate accurate monthly averages (Luther *et al.*, 1986). It consists of two separate instruments, a scanner and a nonscanner. Both were built by TRW in Redondo Beach, California, in cooperation with NASA's Langley Research Center.

4.1.6.1 The ERBE Nonscanner

The ERBE Nonscanner (ERBE-NS) has four channels which continuously view nadir (except during calibration and solar observations) and one channel which views the sun once during each orbit. The ERBE-NS is discussed by Luther *et al.* (1986). Table 4.10 summarizes the ERBE-NS parameters, and a diagram of the instrument is shown in Fig. 4.15. The radiometers hang from a pedestal that contains the main electronics. The contamination cover remains closed until 29 days (NOAA satellites) after launch to prevent deposition of material from the booster rocket motors on the sensors. The sensor assembly can be rotated both in azimuth and elevation for calibration and direct solar viewing by the four normally Earth-viewing sensors. The Solar Aspect Sensor (SAS) measures solar angles during solar calibration. The Solar Monitor Assembly (SMA) views the sun once during each orbit; it is the primary instrument for solar monitoring.

TABLE 4.10. Earth Radiation Budget Experiment Non-Scanner (ERBE-NS) System Parameters

Parameter	*Value*
Telescope	None
Number of channels	5
Detector	Active cavity radiometer (5)
Calibration	1. Two internal blackbodies 2. On-board tungsten lamp 3. Direct viewing of sun
Digitization	13 bits (each channel)
Data rate	130 bits s^{-1}
Instrument mass	36.7 kg
Instrument size	45.7 × 50.3 × 62.5 cm
Average power consumption	20 W

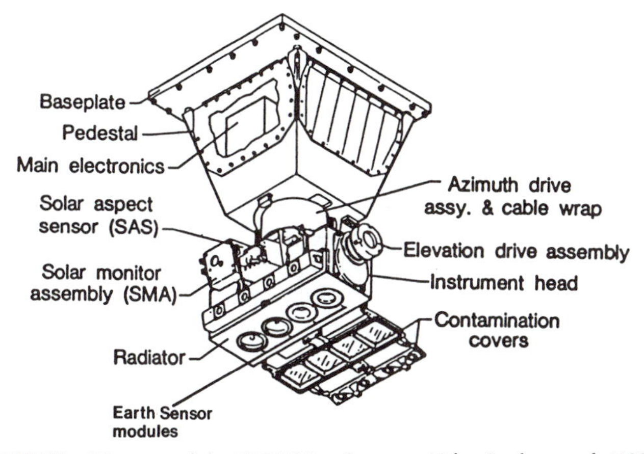

FIGURE 4.15. Diagram of the ERBE Non-Scanner. [After Luther *et al.* (1986).]

The sensors of the ERBE-NS are *active cavity radiometers* similar to those employed on the Solar Maximum Mission (Willson, 1979). Figure 4.16a details the Earth sensor module. It consists of two silver conical cavities wrapped with resistance heater windings on the outside, painted with a black coating on the inside, and mounted inside silver cylinders. The cylinders are wrapped with resistance temperature sensors and soldered back-to-back inside a temperature-controlled heatsink. The heatsink can be commanded to any desired temperature

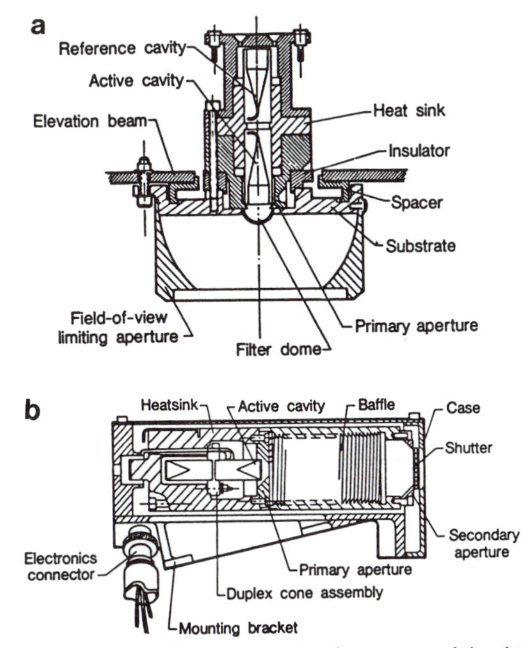

FIGURE 4.16. ERBE-NS active cavity radiometers: (a) Earth sensor module; (b) solar monitor. [After Luther *et al.* (1986).]

between ambient (~15°C) and 60°C within ±0.02°C. The reference cavity views only the heatsink; thus the resistance of its temperature sensor is determined by the heatsink temperature. The active cavity views radiation from the Earth (or sun or internal calibration source), field-of-view-limiting aperture, and filter. The active cavity is maintained at a constant temperature approximately 0.5 K above the temperature of the reference cavity by supplying power to the heater windings around the active cavity. If the irradiance from the Earth increases, less power must be supplied by the heater windings to maintain the desired temperature difference between the active and reference cavities. On the other hand, a decrease in Earth irradiance requires an increase in heating to maintain the temperature difference. The output of the sensor is the voltage across the active cavity heater windings which is required to maintain a constant temperature difference. This voltage (0–10 V, full scale) is digitized using 13 bits for transmission to Earth. The detector time constant is 1–2 s.

The solar monitor (Fig. 4.16b) is very similar to the Earth monitors, except that it has a shutter which covers the sensor for 32 s and then uncovers it for an equal length of time. The shutter eliminates the need for actively controlled heatsink temperature.

The four Earth-viewing sensors are differentiated by filters and field-of-view-limiting apertures. Two sensors have no filters; they absorb essentially all radiation incident on them in the wavelength range of 0.2 μm to beyond 50 μm. These are called *total channels*. The other two sensors are covered by hemispherical, fused-silica (Suprasil-W1) domes that pass radiation in the wavelength range 0.2–3.5 μm (see next section). These are called *shortwave channels*. Longwave radiation emitted by the Earth can be calculated by subtracting the irradiance measured by the shortwave channels from that measured by the total channels. Two of the sensors have wide-enough fields of view (135.8°) that they sense the Earth from horizon to horizon. These are called the *wide-field-of-view* (WFOV) *channels*. They are also called *flat-plate* sensors because they sense essentially the same radiance as would a flat plate, which has a 180° FOV. The other two sensors have fields of view of 66.6°, which allows them to view a spot on Earth with about a 1100-km diameter. These are called *medium-field-of-view* (MFOV) *channels*. The solar monitor has no filter, but has a field of view of 18.0°. The ERBE-NS channels are summarized in Table 4.11.

The ERBE instruments are the most accurately calibrated instruments of their

TABLE 4.11. ERBE-NS Channels

Channel	*Description*	*Filter*	*Wavelength* (*μm*)	*Field of view*		*Precision* (*W m*$^{-2}$)	*Systematic error* (*W m*$^{-2}$)
				Earth	*Solar*		
1	WFOV-T	None	0.2–50+	135.8°	17.6°	±2.6	±4.0
2	WFOV-SW	SW dome	0.2–3.5	135.8°	17.6°	±2.5	±6.0
3	MFOV-T	None	0.2–50+	66.6°	14.0°	±2.0	±3.3
4	MFOV-SW	SW dome	0.2–3.5	66.6°	14.0°	±2.3	±4.5
5	Solar	None	0.2–50+		18.0°	±4.2	±7.0

type ever flown in space. During assembly, all important optical, electrical, and thermal characteristics of the instruments were measured. Before launch, the instruments were calibrated in a specially built vacuum chamber at TRW. Approximately every 14 days in orbit, the normally Earth-viewing sensors are commanded to directly view the sun through a special port. Also approximately every 14 days, the instruments are rotated to the stow position, where they view calibrated blackbodies or tungsten light sources, both of which were carefully calibrated prior to launch and are monitored in flight by calibrated platinum resistance thermometers or photodiodes. In addition, a variety of checks of the electronics can be made. Finally, during data reduction, several consistency checks, which compare the instruments, such as the solar monitor and the Earth-viewing sensors in solar calibration mode, are made. The measurement uncertainty specifications to which each nonscanner sensor was designed and built are listed in Table 4.11.

4.1.6.2 The ERBE Scanner

Like the ERBE Nonscanner, the ERBE Scanner (ERBE-S) is designed to be simple, reliable, and accurate. The ERBE-S instrument is reviewed by Kopia (1986). Table 4.12 summarizes the instrument parameters. The ERBE-S also hangs

TABLE 4.12. Earth Radiation Budget Experiment Scanner (ERBE-S) System Parameters

Parameter	*Value*
Telescope	Cassegrainian, $f/1.84$, 23.03 mm focal length
Field of view	Hexagonal: 4.5° along-track × 3.0° cross-track (see Fig. 4.20)
Ground field of view[a]	
Nadir	66.8 km along-track × 44.5 km cross-track
End of scan	267 km along-track × ∞ cross-track (off Earth)
Rotation rate of scan head	66.7° s^{-1}
Sampling rate	30.303 Hz (1 sample each 33 ms)
Scan line	
Time per scan	4 s
Angle from nadir	Horizon-to-horizon (±61.9°)
Distance from nadir	±3122 km
Distance Between Scan Spots (at nadir)	
Cross-track	32.6 km (2.2°, 48% overlap)
Along-track	26.1 km (60% overlap)
Overlap of consecutive passes at equator	3410 km (55%)
Number of channels	3
Detector	Thermistor bolometer (3)
Calibration	1. Internal blackbody 2. On-board tungsten lamp 3. Indirect viewing of sun
Digitization	12 bits (each channel)
Data rate	960 bits s^{-1}
Instrument mass	32 kg
Instrument size	37.6 × 40.4 × 58.9 cm
Average power consumption	28 W

[a] From a height of 850 km on a NOAA satellite. The resolution of the instruments from the 610-km height of the ERBS is slightly different.

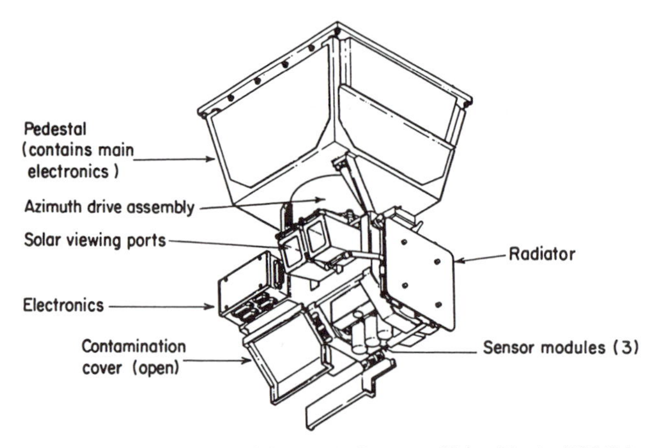

FIGURE 4.17. Diagram of the ERBE Scanner. [After Kopia (1986).]

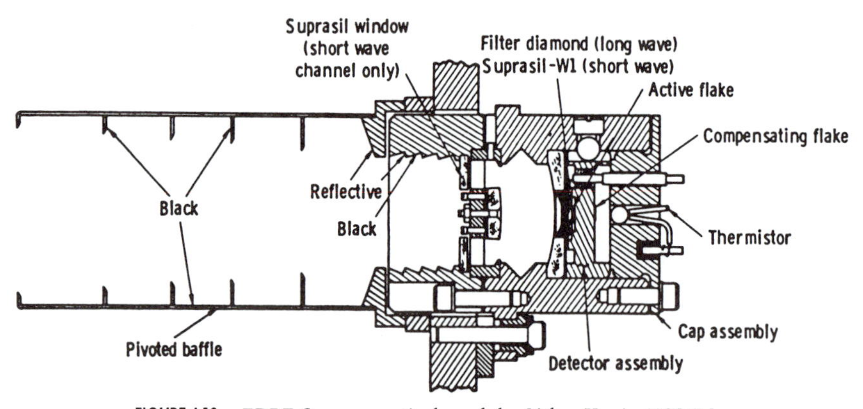

FIGURE 4.18. ERBE Scanner optical module. [After Kopia (1986).]

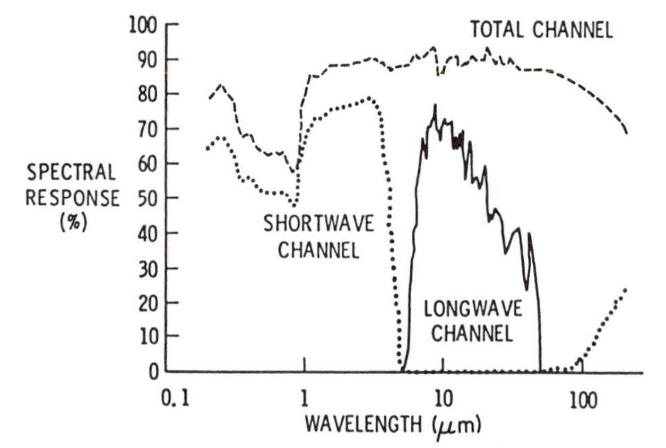

FIGURE 4.19. ERBE Scanner spectral response functions. [After Smith *et al.* (1986).]

from a pedestal that contains the main electronics (Fig. 4.17). The instrument consists of three optical modules that are identical except for filters (Fig. 4.18). Light enters each module through a baffled entrance tube where it encounters a 23.03-mm focal length, *f*/1.84 Cassegrainian telescope. The telescope mirrors are aluminized glass, coated to enhance ultraviolet reflection. Before encountering the detector, the radiation in the shortwave module passes through a suprasil filter, while the radiation in the longwave module passes through a multicoated diamond filter. The total module has no filter. Fig. 4.19 shows the spectral response function of each module, and Table 4.13 describes the channels.

Just above the detector is a hexagonal field stop aperture (Fig. 4.20) that is 3° in the cross-track direction and 4.5° in the along-track direction. This field stop was chosen to minimize problems caused by motion of the scan mechanism during sampling (see below).

The detector is a *thermistor bolometer*. A *bolometer* is a device that measures radiation by measuring the temperature of an object which is warmed by absorption of the radiation. A *thermistor* is a solid-state device whose resistance changes with temperature. Each ERBE-S optical module contains two thermistor flakes about 2 mm square. One is in a temperature-controlled cavity. The other is directly beneath the field stop aperture and is heated by radiation from the telescope. The resistances of the two thermistors are constantly compared in a bridge network. The voltage difference across the bridge, which results when there is an imbalance between the cavity temperature and the active flake temperature, is digitized using 12 bits for transmission to Earth. This voltage is converted to radiance with the help of cold and warm calibration sources of known radiance.

The scan mechanism moves at the rate of 66.7° s^{-1} and is sampled every 33 ms, that is, every 2.2°. Since the scan head moves continuously, the area scanned in 33 ms will be octagonal (Fig. 4.20), and adjacent scan spots will overlap by about 48%. The time between adjacent scan lines is 4 s. Thus adjacent scan lines also overlap (by about 60%) because during the scan period the subsatellite point moves only about 26.1 km, but the along-track field of view is about 66.8 km. The scan head can be rotated azimuthally so that it scans in the along-track direction rather than in the normal cross-track direction. In this mode, each spot along the subsatellite track is viewed once each 4 s for the approximately 10 min that the point is in view of the satellite. These observations can be used to check the anisotropic factors necessary to convert the ERBE-S observations into

TABLE 4.13. ERBE-S Channels

Channel	*Description*	*Filter*	*Wavelength* (μm)	*Precision* (W m^{-2} sr^{-1})	*Systematic error* (W m^{-2} sr^{-1})
1	Shortwave	Suprasil-W	0.2–5.0	±0.3	±0.75
2	Longwave	Diamond	5.0–50	±0.45	±0.75
3	Total	None	0.2–50+	±0.3	±0.5

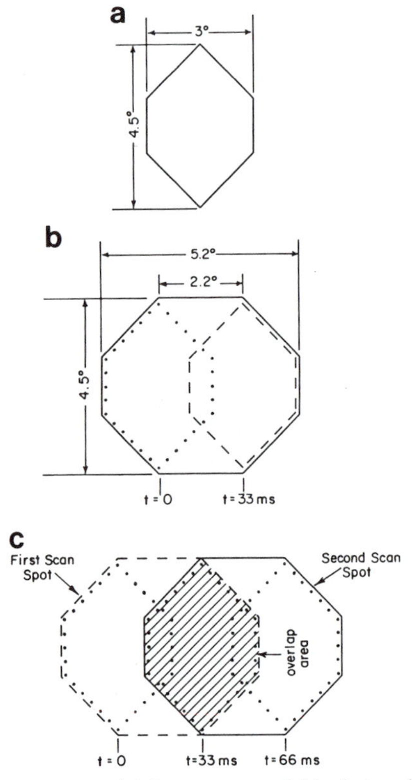

FIGURE 4.20. ERBE Scanner scan spots. (a) Instantaneous field of view from the ERBE-S aperture. (b) A motion-produced ERBE-S scan spot (solid line), that is, the area sensed between two consecutive samples separated by 33 ms as the scanner moves along the scan line. Note that the center is continuously sampled; other areas are sampled for less time. (c) Overlap of two consecutive scan spots (48%).

quantities used in radiation budget studies (see Chapter 10). Unfortunately, this mode has not been used much in practice.

Like the ERBE-NS, the ERBE-S is very carefully calibrated both in space and before launch, and the data are validated after collection. During each scan line, the instrument views cold space and an internal radiance source that includes a temperature-controlled blackbody and a tungsten lamp that is monitored by a photodiode. Approximately once every 14 days, the instrument is rotated to view sunlight reflected from a diffuser plate. Prior to launch, the ERBE-S was carefully calibrated and characterized in the calibration chamber used for the ERBE-NS. Finally, an elaborate system of data validation has been devised (Barkstrom, 1986). As examples, data from the longwave and shortwave channels of the ERBE-S are compared with data from the total channel, and data from the ERBE-S are integrated over the field of view of the ERBE-NS channels and compared. This cross-checking helps ensure that the data are as error-free as possible.

4.1.7 Meteor Satellites

The former Soviet Union has had an active series of operational polar-orbiting meteorological satellites since 1969. There have been three series of Meteor satellites. Through 1991, 31 Meteor-1s, 20 Meteor-2s, and 5 Meteor-3s have been launched (see Appendix A). Both Meteor-2 and Meteor-3 satellites are in current operational use, although the Meteor-2 satellites are being phased out. These satellites are described in the 1989 World Meteorological Organization report, *Information on Meteorological and Other Environmental Satellites.* Tables 4.14 and 4.15 contain descriptions, excerpted from the report, of the Meteor-2 and

TABLE 4.14. Meteor-2 Instrumentation Description[a]

A scanning telephotometer for direct transmission of images in the visible spectrum (0.5–0.7 μm), giving a swath width of 2100 km at a resolution of 2 km
A television-type scanner (spectral range 0.5–0.7 μm), giving a swath width of 2600 km at a resolution of 1 km
A scanning IR radiometer (spectral range 8–12 μm), giving a swath width of 2800 km at a nadir resolution of 8 km
A scanning IR radiometer [spectral ranges (bands) 11.10, 13.33, 13.70, 14.24, 14.43, 14.75, 15.02, and 18.70 μm]; giving a swath width of 1000 kn at an angular resolution of 2°
A device for measuring penetrating (particle) radiation flux densities

[a] From WMO (1989).

TABLE 4.15. Meteor-3 Instrumentation Description[a]

A scanning telephotometer for direct transmission of images in the visible spectrum (0.5–0.7 μm), giving a swath width of at least 2600 km at a resolution of 1.0 × 2.0 km or better at the subsatellite point
A scanning telephotometer for global survey mode in the visible spectrum (0.5–0.7 μm), giving a swath width of at least 3100 km and a resolution at the subsatellite point of 0.7 × 1.4 km or better
A scanning IR radiometer for direct image transmission and global survey mode (spectral range 10.5–12.5 μm), giving a swath width of at least 3100 km; nadir point resolution 3 × 3 km
A scanning IR radiometer for global survey mode (spectral range 8–12 μm), providing a swath width of at least 3100 km; resolution at the nadir point 10 × 10 km
A multispectral scanning IR radiometer (10 channels between 9.6 μm and 15.2 μm and 18 μm) for atmospheric temperature and humidity sensing; swath width 1000 km, surface resolution 50 km
A radiation measuring complex for recording fluxes of protons and electrons with threshold energies from 0.15 MeV to 90 MeV
There is an experimental complex for measuring total ozone content and vertical ozone distribution in individual regions

[a] From WMO (1989).

Meteor-3 instrument capabilities. In summary, they have visible and infrared imaging radiometers, including an APT transmission, an infrared sounder, and an instrument to monitor the space environment. Rao *et al.* (1990) discuss some of the characteristics of the APT transmissions.

The instruments described in Section 4.1 measure radiation in all parts of the electromagnetic spectrum that are useful for satellite meteorology and serve as an introduction to the fundamentals of satellite instrumentation for low Earth orbiters. Next we turn to the instruments on geostationary orbiters.

4.2 OPERATIONAL GEOSTATIONARY SATELLITES

For weather analysis and forecasting, geostationary satellites have an advantage over polar orbiters: images may be made frequently rather than once or twice per day. Thus the motion and rate of change of weather systems may be observed. Meteorological agencies around the world have developed and launched geostationary weather satellites. Normally there are five such satellites positioned nearly equally around the equator (Fig. 4.21).

4.2.1 GMS

The Geostationary Meteorological Satellite (GMS) is positioned at 140° east. GMS 3 and 4 were launched from the Tanegashima Space Center by the Japanese National Space Development Agency (NASDA) and the Japanese Meteorological Agency. GMS has a spin scan radiometer (see below) that returns visible and infrared images very similar to those from GOES. Table 4.16 describes the characteristics of GMS.

4.2.2 Meteosat

Launched by the European Space Agency from Kourou, French Guiana, Meteosat is positioned at the prime meridian. The Meteosat spin scan radiometer returns images in the visible and 11-μm bands as well as the 6.7-μm water vapor band. In fact, Meteosat 1, launched in 1977, provided the first 6.7 μm data from geostationary orbit (Morgan, 1981). Beginning with Meteosat 4, these satellites are also called Meteosat Operational Program (MOP) satellites. Table 4.17 describes the characteristics of Meteosat.

4.2.3 Insat

Insat is positioned at 74° east. Insat 1B was launched from the Space Shuttle as part of the Indian National Satellite System. Insat is the first of a new breed of geostationary meteorological satellites which do not rely on spinning to provide attitude stabilization. They are called *three-axis stabilized*; their instruments always point toward Earth, as do the instruments on the low Earth orbiters. Insat

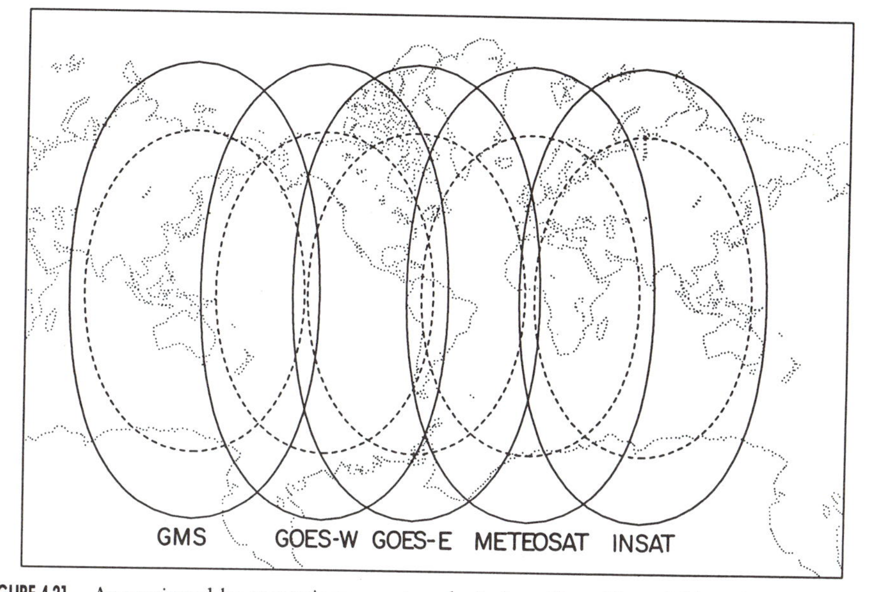

FIGURE 4.21. Areas viewed by geostationary meteorological satellites. The solid line shows the limb; a satellite sees nothing outside this area. The dashed line encloses the area of useful data where the satellite is at least 10° above the horizon.

was built by Ford Aerospace (now Space Systems/Loral), which is building the GOES I–M satellites based on the three-axis-stabilized Insat design. Table 4.18 describes the characteristics of Insat.

TABLE 4.16. Geostationary Meteorological Satellite (GMS) Characteristics[a]

Parameter	*Value*	
	Visible	*Infrared*
Detectors	4 photomultipliers + 4 redundant	1 HgCdTe + 1 redundant
Wavelengths	0.50–0.75 μm	10.5–12.5 μm
Digitization	6 bits (square root)	8 bits
Angular field of view	35 × 31 μrad	140 × 140 μrad
Resolution at nadir	1.25 km	5.0 km
Frame size	10,000 × 10,000 (pixels × lines)	2500 × 2500 (pixels × lines)
Frame time	25 min	25 min
Accuracy	S/N > 6.5 for 2.5% albedo	NEΔT < 0.5 K at 300 K

[a] Information from Rao *et al.* (1990).

TABLE 4.17. Meteosat Characteristics[a]

Parameter	Value		
	Visible	Infrared	Water vapor
Detectors	2 silicon photodiodes + 2 redundant	1 HgCdTe + 1 redundant	1 HgCdTe + 1 redundant
Wavelengths	0.4–1.1 μm	10.5–12.5 μm	5.5–7.1 μm
Digitization	8 bits	8 bits	8 bits
Angular field of view	65 μrad	140 μrad	140 μrad
Resolution at nadir	2.5 km	5.0 km	5.0 km
Frame size	5000 × 5000 (pixels × lines)	2500 × 2500 (pixels × lines)	2500 × 2500 (pixels × lines)
Frame time	25 min	25 min	25 min
Accuracy	S/N > 200 for 25% albedo	NEΔT < 0.4 K at 290 K	NEΔT < 1.0 K at 260 K

[a] Information from Rao *et al.* (1990).

4.2.4 GOES

The series of geostationary satellites operated by NOAA began with the Synchronous Meteorological Satellite 1 (SMS 1) launched 17 May 1974. Like TIROS N, both SMS 1 and SMS 2 were considered experimental satellites and were initially operated by NASA. The Geostationary Operational Environmental Satellites (GOES) were commissioned and are operated by NOAA. SMS 1 and 2 and GOES 1, 2, and 3 were designed and built by Ford Aerospace based on NASA's Applications Technology Satellites 1 and 3 (ATS 1 and 3). GOES 4, 5, 6, and 7 were designed and built by Hughes Aircraft. GOES 8, launched as this book was nearing publication, and its successors (GOES J–M) are built by Space Systems/Loral. Table 4.19 summarizes the physical characteristics of these satellites. Figures 4.22 and 4.23 illustrate the GOES 7 and 8 satellites, respectively.

TABLE 4.18. Insat Characteristics[a]

Parameter	Value	
	Visible	Infrared
Detectors	4	1
Wavelengths	0.55–0.75 μm	10.5–12.5 μm
Digitization	10 bits	10 bits
Angular field of view	78 × 67 μrad	243 × 326 μrad
Resolution at nadir	2.75 × 2.75 km	11 × 11 km
Frame size	20° × 20°	20° × 20°
Frame time	23 min	23 min
Accuracy	S/N > 13 for 2.5% albedo	NEΔT < 0.19 K at 300 K

[a] Information from Rao *et al.* (1990).

TABLE 4.19. Geostationary Operational Environmental Satellite Characteristics

Parameter	*GOES 4–7*	*GOES 8–12*
Satellite mass	835 kg (at launch)	2100 kg (at launch)
Satellite size	Height 3.7 m Diameter 2.1 m	Main body 2 m cube Deployed length 27 m
Solar panel output	At launch: 400 W End of life: 320 W	1050 W at 42 V
Communications frequencies		
UHF	402 and 468 MHz	
S-band	1681 and 2034 MHz	
Instruments	VAS SEM	Imager Sounder SEM
Launch vehicle	Delta 3914	Atlas I/Centaur
Planned lifetime	7 years	5 years, minimum

Normally there are two GOES satellites in operation. GOES–East is stationed at 75° west, and GOES–West is stationed at 135° west. These provide coverage of most of the Western Hemisphere (Fig. 4.21). At times only one GOES is in operation, in which case it is called GOES–Prime, and it is located at 98° west (summer) or 108° west (winter). Table 4.20 lists the positions of the GOES satellites.

The instruments on the GOES are representative of the instruments on most geostationary meteorological satellites; they will be discussed in detail.

4.2.4.1 GOES 4–7 Instruments

The GOES/SMS satellites have three major subsystems: the Space Environment Monitor (SEM), the Telemetry, Tracking, and Command (TTC) subsystem, and a meteorological instrument. The SEM is quite different from the SEM on board the NOAA satellites. It includes a magnetometer, a solar X-ray sensor, and an energetic particle monitor. The instrument package is designed to study solar activity and the Earth's magnetic field. The TTC has several functions including transmission to Earth of data collected by the satellite instruments, satellite command, relay of data from Earth-based data collection platforms (DCPs), and relay of weather facsimile (WEFAX) charts.

Prior to GOES 4 the meteorological instrument on GOES was the Visible and Infrared Spin Scan Radiometer (VISSR). On GOES 4–7, the primary instrument has been the VISSR Atmospheric Sounder (VAS), which has sounding capabilities in addition to the VISSR imaging capabilities. The VAS instrument was designed and built by Santa Barbara Research Center, a subsidiary of Hughes Aircraft. A sketch of the instrument is shown in Fig. 4.24, and the instrument parameters are summarized in Table 4.21.

GOES 4–7 have spun and de-spun portions. The de-spun portion includes the antennas for communication with Earth. The VAS resides in the spun portion of the satellite, and its optical axis coincides with the principal axis of the satellite.

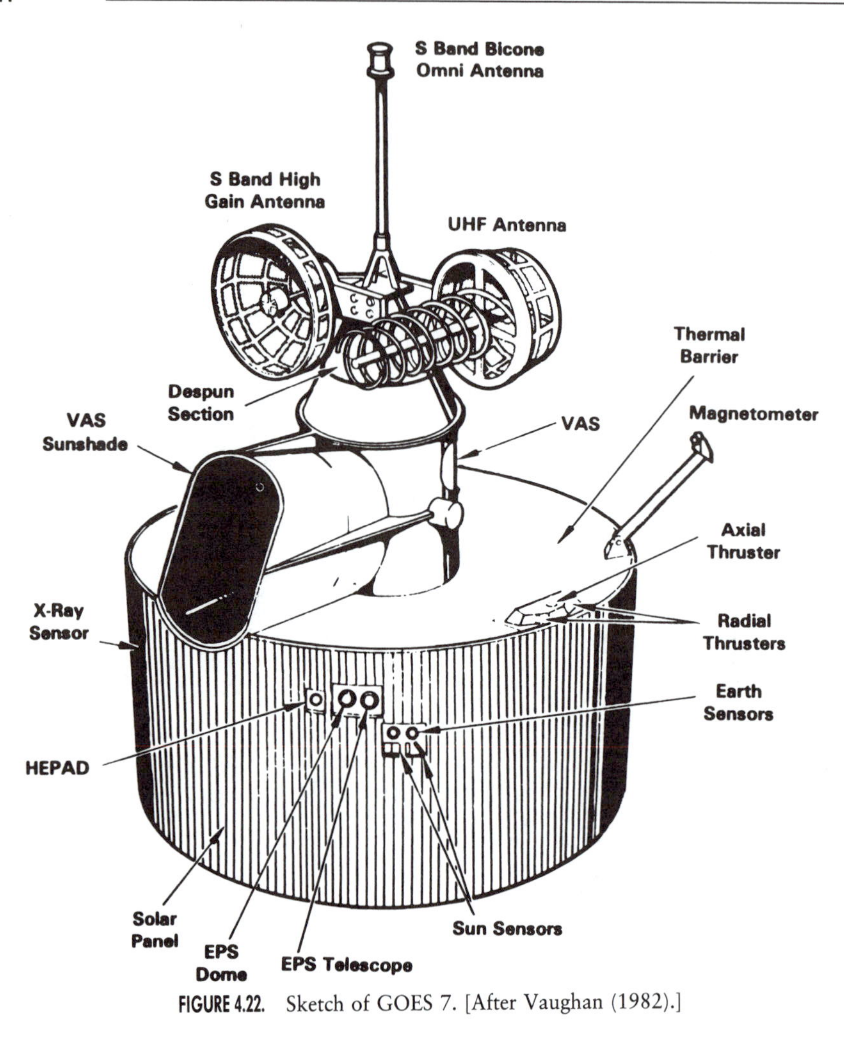

FIGURE 4.22. Sketch of GOES 7. [After Vaughan (1982).]

The spin axis is maintained closely parallel to the Earth's spin axis. The satellite rotates at the rate of 100 rpm (10.4720 rad s^{-1}), which provides east–west optical scanning (thus the name *spin scan radiometer*). North–south optical scanning is provided by a scan mirror that moves in 192-μrad steps. At geostationary altitude, the Earth subtends only 17.4°. Therefore, the mirror need only scan ±10° from the subsatellite point to view the entire hemisphere.[14] A complete hemispherical scan is made in 1821 rotations of the satellite with one mirror step after each rotation. At 100 rpm, a hemispherical image, called a *full-disk* image, is completed in 18.21 min. Usually, full-disk images are terminated short of 1821 scans thus

[14] These figures actually refer to the radiation beam; the mirror itself moves only half this amount, between 40° and 50° from the spin axis.

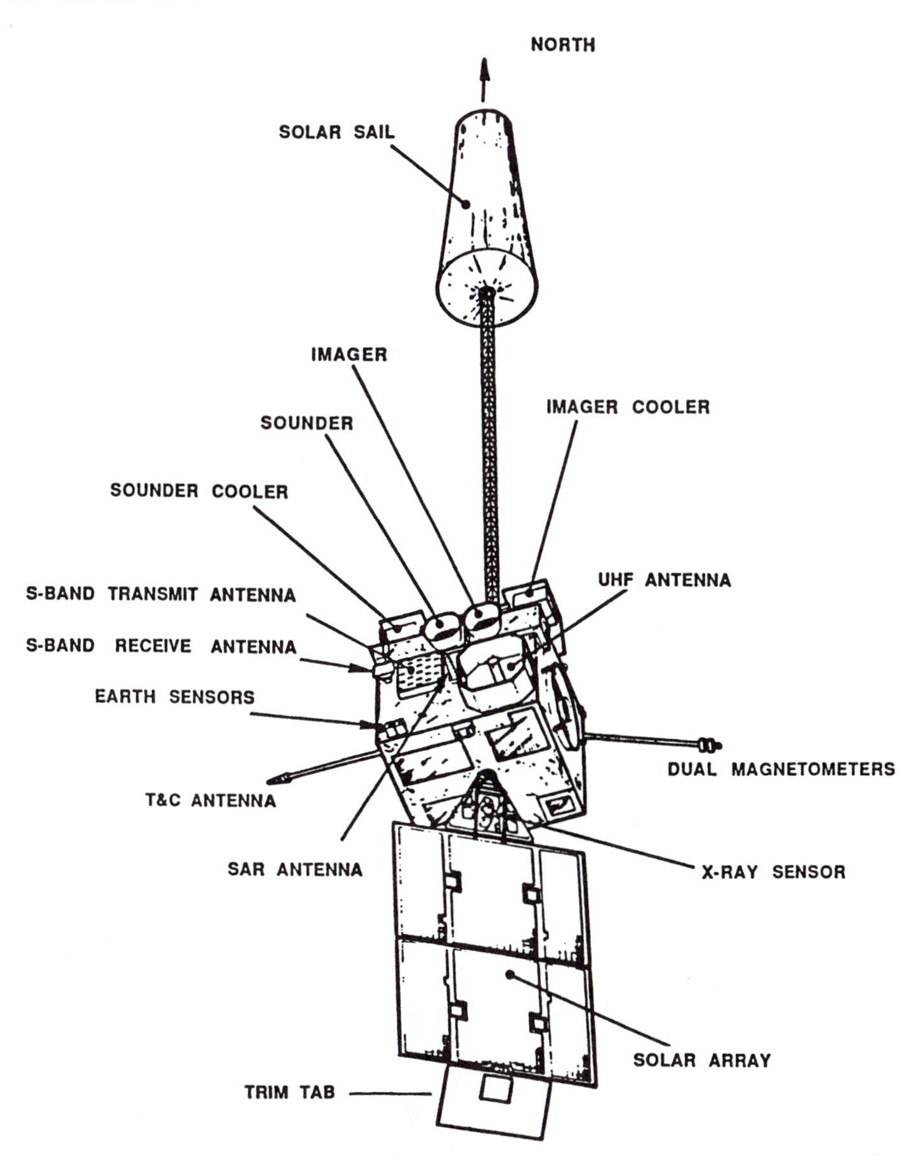

FIGURE 4.23. Sketch of GOES 8. The face of the main body, which contains the imager and sounder telescopes continually points toward Earth. The boom on which the solar sail is mounted is parallel with the Earth's spin axis. [After Komajda and McKenzie (1987).]

eliminating a portion of the Southern Hemisphere. At the end of each image, the mirror rapidly retraces to its northernmost position to start a new image.[15]

The normal GOES 4–7 operating schedule calls for an image to be initiated each 30 min. GOES–East images start on the hour and 30 min past the hour;

[15] Meteosat does not retrace; every other image starts in the south and proceeds northward.

TABLE 4.20. GOES Locations[a]

Satellite	*Instrument*	*Period on station*
	GOES–East (75° **West**)	
SMS 1	VISSR	27 Jun 1974–8 Jan 1976
GOES 1	VISSR	8 Jan 1976–15 Aug 1977
GOES 2	VISSR	15 Aug 1977–26 Jan 1979
SMS 1	VISSR	26 Jan 1979–19 Apr 1979
SMS 2	VISSR	19 Apr 1979–5 Aug 1981
GOES 5	VAS	5 Aug 1981–29 Jul 1984
GOES 7	VAS	25 Mar 1987–21 Jan 1989
	GOES–West (135° **West**)	
SMS 2	VISSR	10 Mar 1975–13 Jul 1978
GOES 3	VISSR	13 Jul 1978–5 Mar 1981
GOES 4	VAS	5 Mar 1981–1 Jun 1983
GOES 6	VAS	1 Jun 1983–1 Aug 1984
GOES 6	VAS	26 Mar 1987–21 Jan 1989
	GOES–Prime (98°–108° **West**)	
GOES 6	VAS	31 Aug 1984–25 Mar 1987
GOES 7	VAS	21 Jan 1989–?
	GOES–Indian Ocean (55° **East**)	
GOES 1	VISSR	23 Oct 1978–24 Nov 1979

[a] Extracted from Gibson (1984).

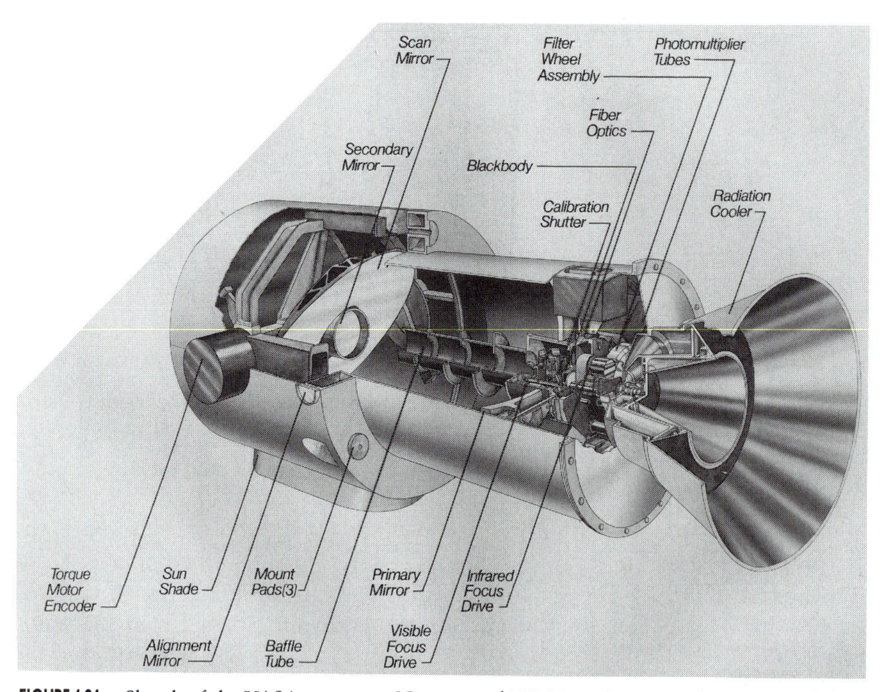

FIGURE 4.24. Sketch of the VAS instrument. [Courtesy of NOAA and Santa Barbara Research Center.]

TABLE 4.21. VISSR Atmospheric Sounder (VAS) Characteristics

Parameter	*Value*
Optics	Ritchey–Chretien telescope 40.64-cm-diameter primary mirror Focal length 292.1 cm Optically flat scan mirror
Detectors	
Visible	8 photomultiplier tubes coupled to focal plane by optical fibers
Infrared	6 detectors cooled to 95 K: 2 small HgCdTe (80 × 80 μm) 2 large HgCdTe (160 × 160 μm) 2 large InSb (160 × 160 μm)
Field of view	
Visible	24 (N–S) × 25 (E–W) μrad
Small IR	192 × 192 μrad
Large IR	384 × 384 μrad
Ground field of view at equator (nadir)[a]	
Visible	0.86 km (N–S) × 0.89 km (E–W)
Small IR	6.9 × 6.9 km
Large IR	13.7 × 13.7 km
Satellite spin rate	100 ± 5 rpm (10.47 rad s^{-1} ± 5%)
Data sampling rate	
Visible	500 kHz (provides E–W
IR	125 kHz sampling)
Scan mirror step	192 μrad (provides N–S sampling)
Distance between scan spots (at nadir)	
Visible	0.75 km (E–W) × 0.86 km (N–S)
IR	3.00 km (E–W) × 6.87 km (N–S)
Channels	
Visible	8 (identical, used simultaneously for high-resolution imaging)
IR	12 (selected by filters inserted in IR optical path)
In-orbit calibration	
Radiometric	IR: temperature-monitored blackbody cyclically reflected into optical axis Visible: optical aperture providing sun signal of 50% Earth albedo
Electronic	IR and visible: linear-ramp voltage into amplifier chain
Digitization	6 bits, visible; 10 bits, IR
Date rate	
Unstretched	28 Mbits s^{-1}[b]
Stretched	2.1 Mbits s^{-1}
Size	
Length	1.5 m
Radial dimension	0.65 m
Mass	
Scanner	64.3 kg
Electronics module	10.5 kg
Average power consumption	40 W

[a] From a height of 35,790 km above the equator.
[b] Megabits per second.

GOES–West images start at 15 and 45 min past the hour. During special *rapid-scan* periods, sectors of limited north–south extent (but full east–west extent) are scanned in less time. Scanning periods of 15, 7.5, 5, and 3 min have been used.

After being reflected by the VAS scan mirror, radiation enters a 40.64-cm diameter, *f*/7.2, Ritchey–Chretien telescope. At the principal focal plane, the radiation is intercepted by an array of eight optical fibers connected to eight photomultiplier tubes sensitive to radiation in the wavelength range 0.54–0.70 μm. This array of optical fibers is arranged in a north–south line, and each fiber has a field of view of 25 μrad (E–W) by 24 μrad (N–S). Thus with each rotation of the satellite, eight lines of visible data are swept out. The 25 $\times$ 24-μrad field of view results in a ground resolution of about 0.9 km at the subsatellite point. The visible channels are sampled at the rate of 500 kHz,[16] which, at a satellite rotation rate of 10.47 rad s^{-1}, means that visible scan spots are separated by 21 μrad or 0.75 km east–west. Visible scan lines are separated by 24 μrad or 0.86 km north–south. Thus visible scan lines are contiguous, but adjacent scan spots overlap slightly.

The VAS visible data are digitized using 6 bits, but the output count is not proportional to radiance; rather, it is approximately proportional to the square root of the radiance. This choice was made because the noise from the photomultiplier tubes is proportional to the square root of the radiance. By making the output proportional to the square root of radiance, the noise level, expressed in counts, is nearly independent of the count value. The visible channels are calibrated before launch, and calibration curves are available from NESDIS.

On each rotation of the satellite, 15,288 samples are taken for each of the eight visible channels. Since a complete full-disk image is made in 1821 rotations, it contains 14,568 lines (8 $\times$ 1821). A single full-disk visible image, therefore, consists of 15,288 $\times$ 14,568 $\times$ 6 = 1,336,293,504 bits (159.3 megabytes), not counting the documentation and navigation information which is normally part of the image file. Archiving all of these data is a challenge.

Since there are eight visible channels, each digitized to 6 bits, and each sampled at the rate of 500 kHz, the visible data are taken at the rate of 24 Mbits s^{-1} (megabits per second). The addition of the infrared data boosts the total data rate to 28 Mbits s^{-1}. This rate only occurs, however, while the instrument is viewing the Earth. Ninety-five percent of the time the instrument views space and collects no data. To decrease the data rate so that users may receive it more easily, the following arrangement is employed. The high-data-rate, raw VAS data are transmitted in real time to the Command and Data Acquisition (CDA) station located at Wallops Island, Virginia. The data are transmitted back to the satellite at a data rate of approximately 2.1 Mbits s^{-1}; which means that the data, which were collected only while the instrument viewed Earth, take one full rotation of the satellite to transmit. These data, called *stretched data*, are retransmitted by the GOES satellite to anyone with the proper receiving equipment.

[16] Actually, the satellite spin rate is allowed to vary by ±5%, but the sampling rate is adjusted accordingly to keep the ground resolution constant.

Real-time ground processing, in addition to stretching the data, allows some useful modifications of the data stream:

- Offsets are added to the visible data to correct for differences in the eight photomultiplier tubes. (This reduces the "striping" that can sometimes be seen.)
- Raw output voltages from the infrared sensors are converted into calibrated, 10-bit counts that are proportional to radiance.
- The data are reordered so that adjacent fields of view are adjacent in the data stream (see below).
- Satellite orbit and attitude information is embedded in the data stream.
- A grid is included in the data stream for those who print images.

The exact format of the retransmitted VAS data stream is detailed in the *Operational VAS Mode AAA Format* (NESDIS, 1987).

In many satellite instruments a beamsplitter is used to separate the visible from the infrared radiation. A beamsplitter allows the visible and infrared detectors to view the same point on Earth at the same time, but it necessarily causes some loss of signal. In the VAS, no beamsplitter is used. The infrared field of view is displaced slightly from the visible field of view. This displacement is corrected in real time by reordering the data during ground processing.

The infrared beam is routed by means of relay optics to a 12-position filter wheel. The filters are described in Table 4.22. The wheel can be rotated to any position during the time that the instrument views space. After being filtered, the infrared beam simultaneously illuminates six detectors. There are four HgCdTe

TABLE 4.22. Typical VAS Infrared Spectral Band Data[a]

Band	*Wave-length* (μm)	*Half-power bandwidth* (cm^{-1})	*Weighting function peak* (*hPa*)	*Description*	*Detector*	*NEΔL* ($mW\ m^{-2}\ sr^{-1}\ cm$) *Imaging*	*Sounding*[b]	*Spin budget*
1	14.71	10	40	CO_2	HgCdTe	4.125	0.583	2
2	14.47	16	70	CO_2	HgCdTe	2.525	0.253	4
3	14.23	16	150	CO_2	HgCdTe	1.763	0.133	7
4	13.99	20	450	CO_2	HgCdTe	1.488	0.112	7
5	13.31	20	950	CO_2	HgCdTe	1.131	0.113	4
6	4.516	45	850	CO_2	InSb	0.028	0.002	7
7	12.67	20	Surface	H_2O	HgCdTe	1.069	0.123	3
8	11.24	140	Surface	Window	HgCdTe	0.119	0.024	1
9	7.248	40	600	H_2O	HgCdTe	1.225	0.082	9
10	6.728	150	400	H_2O	HgCdTe	0.306	0.043	2
11	4.436	40	500	CO_2	InSb	0.026	0.002	7
12	3.940	140	Surface	Window	InSb	0.007	0.001	1

[a] From Hayden (1988).
[b] Obtained from the imaging NEΔ*L* by dividing by the square root of the product of the spin budget and the number of scan spots averaged (typically 25).

detectors, used to monitor wavelengths longer than 6 μm, and two InSb detectors, used for the shorter-infrared wavelengths. Two of the HgCdTe detectors are 80 μm square. They have fields of view of 192 μrad and, therefore, Earth resolutions of 6.9 km. These detectors are called *small detectors*. The two remaining HgCdTe detectors and both of the InSb detectors are 160 μm square, have 384-μrad fields of view and Earth resolutions of 13.7 km. These detectors are called *large detectors*. The infrared detectors are offset from the visible detectors and from each other both in the north–south and the east–west directions to facilitate imaging. The data are properly registered at Wallops Island during the stretching processes. To decrease noise, the infrared detectors are cooled to 94 K by a radiant cooler, which radiatively exhausts heat energy to cold space.

During each rotation of the satellite, two infrared detectors are active and are sampled at the rate of 125 kHz. Since a scan spot near nadir moves only 3 km between samples, the infrared channels are *oversampled*, that is consecutive scan spots overlap. The small detectors are oversampled by about 57%, and the large detectors are oversampled by about 78%. One occasionally hears that the resolution of the small detectors is 3 × 7 km (4 × 8 km on the older VISSR instrument). It should be remembered that two separate aspects of resolution have been combined to make this statement. The instantaneous field of view (half-power points) for the VAS small detectors is 7 × 7 km, but the oversampling causes near-nadir scan spots to be only 3 km apart.

The infrared data are calibrated every scan line by viewing space and an internal blackbody. The data are digitized using 10 bits in a manner such that the counts (0–1023) are linearly proportional to radiance. In the older VISSR instruments, only the 11 μm channel was used, and the data were digitized to 8 bits in a scheme tied to equivalent blackbody temperature rather than to radiance. In this scheme, count 000 corresponded to an equivalent blackbody temperature of 330 K, and the temperature decreased 0.5 K with each count until count 176 (242 K). Thereafter, the temperature decreased 1 K with each count until the maximum count of 255 (163 K) was reached. This digitization scheme has the disadvantages that (1) it is not very sensitive at colder temperatures (1-K steps), and (2) certain count values do not occur due to rounding problems in the conversion of sensor voltage (proportional to radiance) to equivalent blackbody temperature.

The VAS has three operating modes: VISSR, Multispectral Imaging (MSI) and Dwell Sounding (DS). VISSR mode is a backup mode. In VISSR mode, the on-board computer, called the VAS Processor, is turned off, and the instrument operates much like the VISSRs on the first generation of GOES satellites. Normally the filter wheel is set to channel 8 (11 μm, although it can be set to any position), the two small detectors are active, and full-resolution visible and 3 × 7-km IR data are collected.[17]

MSI is the normal VAS imaging mode. Here, full-resolution visible data are collected along with two, three, or four channels of infrared data. Two images

[17] If a filter position other than 8 is chosen, as occasionally happens to make a 6.7-μm image, 3 × 14-km data are collected.

may be collected at 3 × 7-km resolution,[18] one image may be collected at 3 × 7-km resolution and two images at 3 × 14-km resolution, or four images may be collected at 3 × 14-km resolution. In this mode, the filter wheel is under control of the VAS Processor, and the scan mirror steps 192 μrad on each revolution of the satellite.

Dwell sounding mode is used to make soundings; no visible data are collected. In this mode, the filter wheel and the scan mirror are under control of the VAS Processor. All 12 positions of the filter wheel are sampled sequentially. Normally, the small detectors are used for channels 3–5, 7, and 8, and the large detectors are used for the remaining channels.[19] The scan mirror does not step with each rotation of the satellite. In order to reduce noise, the instrument *dwells* on the same scan line for several programmable rotations. The number of rotations of the satellite at each mirror position for each filter wheel position is called the *spin budget*. A typical spin budget is listed in Table 4.22. The data for each channel are averaged over the number of spins. If there are N spins for a particular channel, the noise in the averaged radiance is reduced by a factor of $N^{-1/2}$. The noise is further reduced by averaging the radiance over an area. If M pixels are averaged, the noise in the area-averaged radiance is reduced by a factor of $M^{-1/2}$. Because of the multiple spins on each scan line, only a limited latitude band can be dwell sounded during the 30-min GOES duty cycle. The normal operational schedule calls for a combination of MSI and DS modes to satisfy a large variety of data users (see Gibson, 1984).

Since 1978, digital GOES data have been archived by specially adapted video tape recorders at the Space Science and Engineering Center of the University of Wisconsin under contract to NESDIS. In addition, a number of hard copy prints and negatives are archived by NESDIS. GOES data from these archives may be requested from the Satellite Data Services Division of NESDIS (see Section 4.4).

4.2.4.2 GOES I–M Instruments

The five satellites that constitute the GOES I–M series (GOES I, J, K, L, and M; to be GOES 8, 9, 10, 11, and 12 after launch) are being built by the prime contractor, Space Systems/Loral. Table 4.19 outlines characteristic of the GOES I–M satellites. GOES 8 was launched on 13 April 1994. Menzel and Purdom (1994) describe the GOES I–M satellites and instruments.

The most significant change from previous GOES satellites is that GOES I–M (Fig. 4.23) is a three-axis stabilized satellite designed after Insat. Three-axis stabilization means that the instruments always point toward Earth, whereas the previous GOES spin causing their instruments to sense the Earth only about 5% of the time. The improved duty cycle of the GOES I–M instruments allow better radiometric accuracy and higher spatial resolution than is possible on GOES 4–7. Other improvements include better pointing accuracy (within 4 km at the subsatellite point) and the routine capability to track stars for navigational purposes.

[18] Prior to GOES 7, this two-image option was not available.

[19] The small detectors could also be used for channels 9 and 10.

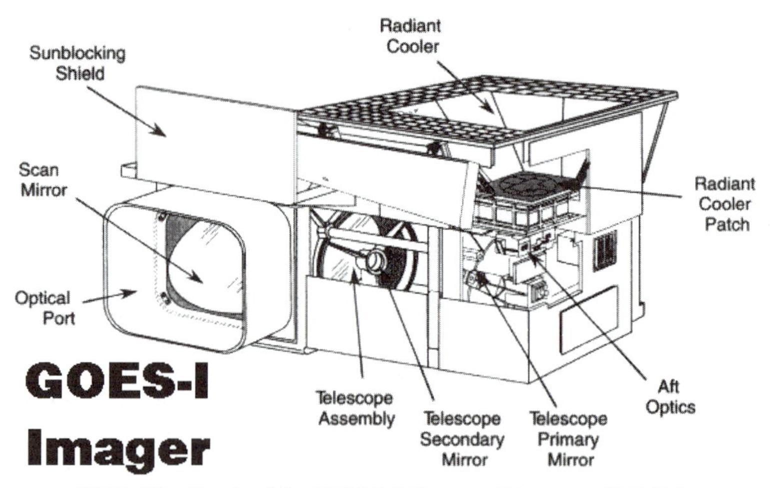

FIGURE 4.25. Sketch of the GOES I–M Imager. [Courtesy of NASA.]

The VAS instrument combines imaging and sounding in one instrument, which means that imaging must be abandoned while soundings are being made. The Imager and Sounder on GOES I–M are separate instruments capable of independent, simultaneous operation. This increases the rate at which soundings are available.

The GOES I–M Imager (Tables 4.23 and 4.24, Fig. 4.25) is built by ITT Aerospace/Communication Division and is similar to the AVHRR. Like AVHRR, the Imager has 5 channels (0.65, 3.9, 6.75, 10.7, and 12.0 μm; see Table 4.24), however, a water vapor channel (6.75 μm) has been substituted for the 0.9-μm AVHRR channel. All five channels are scanned during every Imager frame.

The Imager scans west-to-east on one scan line followed by an east-to-west scan on the next line. Visible data are swept out eight lines at a time with eight silicon photodiode detectors. The 3.9-, 10.7-, and 12.0-μm channels are swept out two scan lines at a time, and the 6.75-μm channel is swept out one line at a time. At the end of each scan line, the mirror steps down (south) 224 μrad (8 km) to begin a new scan.

The ground resolution of some of the channels is improved over that possible with VAS. The 3.9-μm channel has approximately four times the resolution, and the 10.7- and 12.0-μm channels have about twice the resolution, whereas the 0.65- and 6.75-μm channels have about the same resolution.

The radiometric accuracy of some channels also increase (see Tables 4.22 and 4.24). Also, the digitization for the visible channel increases from 6 bits to 10 bits and the output is linear in radiance instead of proportional to the square root of radiance. This should improve the low-light detection capabilities of the visible channel.

In addition, the scan patterns for the GOES I–M Imager are much more flexible than those currently available. For example, it is possible to suspend a full-disk scan to perform a rapid scan of a small area and then to resume the full-disk scan. Currently, rapid scanning must be halted during full disk scanning.

TABLE 4.23. GOES I–M Imager Characteristics[a]

Parameter	*Value*
Optics	Cassegrainian telescope 31.1-cm-diameter primary mirror Focal length 381.2 cm
Earth location accuracy	±30 μrad (rms)
Channel coregistration	±28 μrad
Scanning rate	
Full disk	25 min
3000 × 3000-km area	3.1 min
500 × 500-km area	20 s
Channels	Five (0.65, 3.9, 6.75, 10.7, 12.0 μm)
Digitization	10 bits
Date rate	2.6208 Mbits s^{-1}
In-orbit calibration	
IR (channels 2–5)	Space and internal 290-K blackbody
Visible (channel 1)	None
Size	
Sensor module	115 × 80 × 75 cm
Electronics module	67 × 43 × 19 cm
Power supply	29 × 20 × 16 cm
Mass (3-module total)	120 kg
Power consumption	119 W, daily average

[a] After Koenig (1989).

TABLE 4.24. GOES I–M Imager Channels[a]

Channel	*1*	*2*	*3*	*4*	*5*
Wavelength (μm)	0.55–0.75	3.8–4.0	6.5–7.0	10.2–11.2	11.5–12.5
Detector type	Silicon photodiode	InSb	HgCdTe	HgCdTe	HgCdTe
Noise or NEΔT	±8 counts	0.15 K at 300 K 3.5 K at 230 K	0.3 K at 230 K	0.2 K at 300 K 0.4 K at 230 K	0.2 K at 300 K 0.4 K at 230 K
NEΔL (mW m^{-2} sr^{-1} cm)		0.0058	0.044	0.31	0.35
IFOV (μrad)	28	112	224	112	112
Ground resolution (km)	1.0	4.0	8.0	4.0	4.0
Sampled resolution at subpoint (E–W × N–S, km)	0.57 × 1.0	2.3 × 4.0	2.3 × 8.0	2.3 × 4.0	2.3 × 4.0

[a] From Koenig (1989) and Menzel and Purdom (1994).

Even though three-axis stabilization makes it unnecessary to stretch the data as is done with the VAS instrument, the raw data are still transmitted to Wallops Island, where they are corrected, registered, tagged with ground coordinates, and merged with sounder data before being transmitted back to the satellite. The satellite then retransmits the processed data to users.

The GOES I–M Sounder (Tables 4.25 and 4.26, Fig. 4.26) is also built by ITT Aerospace/Communication Division. It looks much like the Imager because both instruments have nearly identical telescopes. The Sounder has 19 channels including a visible channel for cloud detection.

Like the HIRS/2 instrument, a rotating filter wheel provides channel selection. Four scan spots, on four successive scan lines, are simultaneously sampled during each rotation of the filter wheel (100 ms). The scan mirror steps 280 μrad (10 km at nadir) in the east–west direction with each rotation of the filter wheel. At the end of a scan line, the scan mirror drops 1120 μrad (40 km at nadir) in the

TABLE 4.25. GOES I–M Sounder Characteristics[a]

Parameter	*Value*
Optics	Cassegrainian telescope 31.1-cm-diameter primary mirror Focal length 381.2 cm
IFOV (μrad)	242 (all channels)
Ground resolution (spot diameter at nadir)	8.7 km
Distance between scan spots (at nadir, both E–W and N–S)	10 km
Sampling rate	4 scan spots simultaneously sampled in 100 ms
Earth location accuracy	$\pm$42 μrad (3 σ)
Pixel-to-pixel registration	$\pm$29.72 μrad (3 σ)
Channel coregistration	Within 16 μrad of channel 8
Scanning rate	
Full disk	454 min
3000 × 3000-km area	42 min
1000 × 1000-km area	5.3 min
Channels	19
Digitization	13 bits
Date rate	40 kbits s^{-1}
In-orbit calibration	
IR (channels 1–18)	Space and internal 290-K blackbody
Visible (channel 19)	None
Size	
Sensor module	137 × 80 × 75 cm
Electronics module	67 × 43 × 19 cm
Power supply	29 × 20 × 16 cm
Mass (3-module total)	126 kg
Power consumption	106 W, daily average

[a] After Koenig (1989).

TABLE 4.26. GOES I–M Sounder Channels[a]

Channel	*Detector*	*Central wavelength* (μm)	(cm^{-1})	*Prelaunch* NEΔL ($mW\ m^{-2}\ sr^{-1}\ cm$)	*Bandwidth* (cm^{-1})
1	HgCdTe	14.71	680	1.44–2.42	13
2	HgCdTe	14.37	696	1.23–1.6	13
3	HgCdTe	14.06	711	0.88–1.13	13
4	HgCdTe	13.64	733	0.75–0.92	16
5	HgCdTe	13.37	748	0.74–0.77	16
6	HgCdTe	12.66	790	0.27–0.39	30
7	HgCdTe	12.02	832	0.16–0.23	50
8	HgCdTe	11.03	907	0.10–0.15	50
9	HgCdTe	9.71	1030	0.13–0.24	25
10	HgCdTe	7.43	1345	0.09–0.18	55
11	HgCdTe	7.02	1425	0.06–0.12	80
12	HgCdTe	6.51	1535	0.08–0.13	60
13	InSb	4.57	2188	0.005–0.008	23
14	InSb	4.52	2210	0.004–0.006	23
15	InSb	4.45	2245	0.004–0.007	23
16	InSb	4.13	2420	0.002–0.003	40
17	InSb	3.98	2513	0.002–0.004	40
18	InSb	3.74	2671	0.001–0.004	100
19	Silicon	0.696	14367		

[a] From Koenig (1989) and Menzel and Purdom (1994).

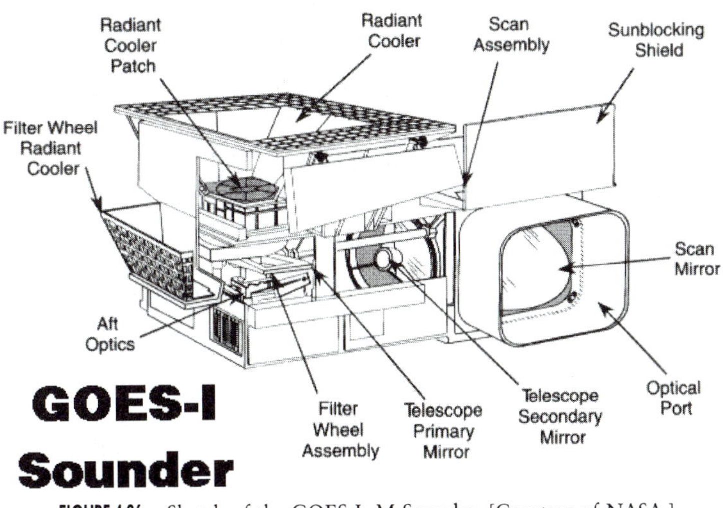

FIGURE 4.26. Sketch of the GOES I–M Sounder. [Courtesy of NASA.]

north–south direction to begin the next scan. The scans are alternately east-to-west and west-to-east.

The Sounder has roughly the same radiometric accuracy as the VAS instrument, but the Sounder does not have to dwell on a point as VAS does to reduce noise. This improves flexibility in scanning. Also, the digitization increases from 10 bits to 13 bits per channel.

A significant improvement in the GOES I–M Sounder is that it has more channels (19 versus 12) to retrieve atmospheric parameters. In particular, it has more surface-sensing channels than VAS (6 versus 3) and more channels in the less-water-vapor-sensitive shortwave IR (<6 μm) channels (6 versus 3). These improvements should allow the GOES I–M Sounder to better separate surface, moisture, and temperature effects (see Hayden, 1989).

The extensive changes in the way the GOES I–M instruments work requires a new data format called GVAR (GOES Variable; see Komajda and McKenzie, 1987). This change requires the modification of the ground equipment for all users who receive the retransmitted data.

An interesting point is that the motion of the mirrors causes perturbations in the satellite attitude. The independent motion of the Sounder mirror would cause the Imager scan lines to be crooked (and vice versa). To prevent this, the GOES I–M satellite has a Mirror Motion Compensation system that automatically anticipates attitude changes of the satellite (caused by the Sounder scan mirror, for example) and makes mirror adjustments (in the Imager, for example) to compensate for them (see Savides and Reseck, 1989).

4.3 OTHER SATELLITE INSTRUMENTS

Although the operational satellites demonstrate many of the principles of meteorological satellite instrumentation, a number of interesting and important instruments are not on board. In this section we briefly describe several of these instruments.

An extremely important series of meteorological satellites is the Nimbus series, which carried experimental instruments (Table 4.27). Many of these instruments were the precursors of the instruments on the NOAA satellites: SIRS, ITPR, and HIRS led to HIRS/2; SCR and PMR led to SSU; NEMS and SCAMS led to MSU; BUV and SBUV/TOMS led to SBUV/2; and ERB led to ERBE. Other Nimbus instruments have not yet seen operational use, but they may lead to the operational instruments of the future. In the remainder of this section, we discuss instruments that we consider to be particularly noteworthy.

4.3.1 Interferometers

The Infrared Interferometer Spectrometer (IRIS) flew on Nimbus 3 and 4. Rather than measure several discrete wavelengths by means of filters, IRIS measured a broad spectrum from 500 to 2000 cm^{-1} (5–20 μm) in 5-cm^{-1} steps.

TABLE 4.27. Nimbus Satellite Instruments and Systems[a]

Nimbus 1	2	3	4	5	6	7
APT	APT	HRIR	BUV	ESMR	ERB	CZCS
AVCS	AVCS	IDCS	FWS	ITPR	ESMR	ERB
HRIR	HRIR	IRIS	IDCS	NEMS	HIRS	LIMS
	MRIR	IRLS	IRIS	SCMR	LRIR	SAM II
		MRIR	IRLS	SCR	PMR	SAMS
		MUSE	MUSE	THIR	SCAMS	SBUV/TOMS
		RTTS	SCR		T&DRE	SMMR
		SIRS	SIRS		THIR	THIR
			THIR		TWERLE	
			RTTS			

[a] Abbreviations:

APT	Automatic Picture Transmission System
AVCS	Advanced Vidicon Camera System
BUV	Backscatter Ultraviolet Spectrometer
CZCS	Coastal Zone Color Scanner
DRIR	Direct Readout Infrared Radiometer
ERB	Earth Radiation Budget Experiment
ESMR	Electrically Scanning Microwave Radiometer
FWS	Filter Wedge Spectrometer
HIRS	High Resolution Infrared Radiation Sounder
HRIR	High Resolution Infrared Radiometer
IDCS	Image Dissector Camera System
IRIS	Infrared Interferometer Spectrometer
IRLS	Interrogation, Recording, and Location System
ITPR	Infrared Temperature Profile Radiometer
LIMS	Limb Infrared Monitor of the Stratosphere
LRIR	Limb Radiance Inversion Radiometer
MRIR	Medium Resolution Radiometer
MUSE	Monitor of Ultraviolet Solar Energy
NEMS	Nimbus E Microwave Spectrometer
PMR	Pressure Modulator Radiometer
RTTS	Real Time Transmission Systems
SAM II	Stratospheric Aerosol Measurement II
SAMS	Stratospheric and Mesospheric Sounder
SBUV	Solar Backscatter Ultraviolet
SCAMS	Scanning Microwave Spectrometer
SCMR	Surface Composition Mapping Radiometer
SCR	Selective Chopper Radiometer
SIRS	Satellite Infrared Spectrometer
SMMR	Scanning Multichannel Microwave Radiometer
T&DRE	Tracking and Data Relay Experiment
THIR	Temperature Humidity Infrared Radiometer
TOMS	Total Ozone Mapping Spectrometer
TWERLE	Tropical Wind Energy Conversion and Reference Level Experiment

The instrument consisted of a Michelson interferometer, the output of which is proportional to the Fourier transform of the spectrum. Figure 4.27 shows a spectrogram retrieved by taking the inverse Fourier transform. Although the 150-km resolution of the instrument is rather coarse, its data are useful for a variety of purposes, among them: (1) temperature and water vapor profiles can be retrieved from them, (2) ozone and other trace-gas concentrations can be retrieved, and (3) the spectral nature of the outgoing longwave radiation can be studied. Because of its great flexibility, IRIS flew on several planetary probes. A similar instrument, but with much higher spectral resolution (0.3 cm^{-1}), called the High-spectral-resolution Interferometer Sounder (HIS), is being tested on aircraft and may be flown in space (Smith *et al.*, 1987).

4.3.2 Limb Scanners

The Limb Radiance Inversion Radiometer (LRIR), the Limb Infrared Monitor of the Stratosphere (LIMS), and the Stratospheric and Mesospheric Sounder (SAMS) are infrared sounders, but they scan the atmosphere at the edge of the Earth (the *limb*) instead of looking down (Fig. 4.28). These instruments have good vertical resolution and can make accurate measurements of the temperature and trace-gas concentration in the mesosphere, stratosphere, and (in the absence of clouds) the upper troposphere (see Chapter 6).

The Stratospheric Aerosol Measurement I Experiment (SAM I, flown on Apollo-Soyuz), SAM II (flown on Nimbus 7), the Stratospheric Aerosol and Gas Experiment (SAGE I, which flew on a dedicated satellite), and SAGE II (which flies on the Earth Radiation Budget Satellite) are called *solar occultation* experiments. They sample the atmosphere by measuring the extinction of sunlight during

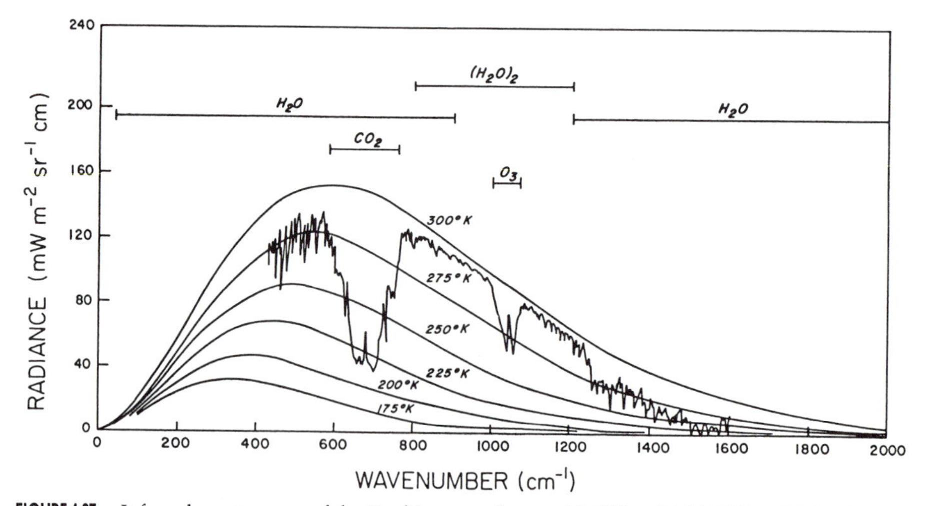

FIGURE 4.27. Infrared spectrogram of the Earth's atmosphere at 15.1°N and 144.7°W on 27 April 1970 taken by the IRIS instrument. [After Liou (1980). Reprinted by permission of Academic Press.]

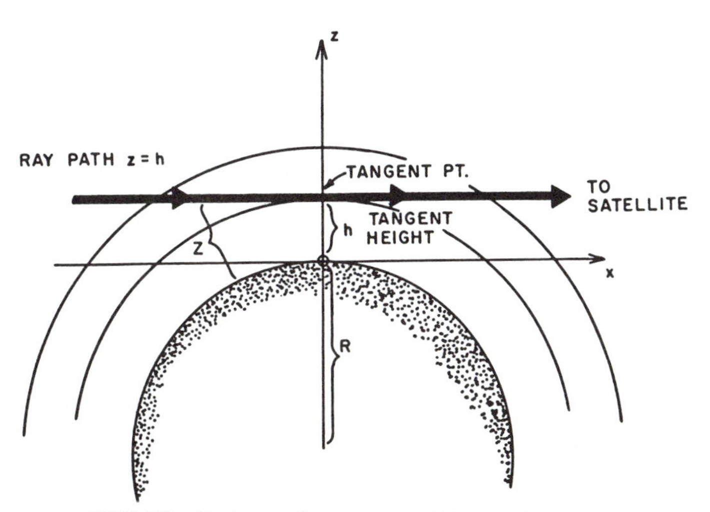

FIGURE 4.28. Limb scanning geometry. [After NASA (1975).]

satellite sunrise and sunset. They can measure the concentrations of aerosols (Chapter 8) as well as gases that have absorption spectra in the solar spectrum, such as ozone, water vapor, and nitrogen dioxide.

4.3.3 Microwave Imagers

A series of microwave imagers flew on the Nimbus satellites, and a new imager (the SSM/I) is currently flying on a Defense Meteorological Satellite Program (DMSP) satellite. The Electrically Scanning Microwave Radiometer (ESMR) flew on both Nimbus 5 and 6. ESMR–5 measured horizontally polarized radiation[20] at 19.35 GHz (1.55 cm). ESMR–6 measured both vertically[21] and horizontally polarized radiation at 37.0 GHz (0.81 cm). These instruments are unique in the fact that they had *phased array antennas*, which means that the beam was scanned electronically; no moving parts were required. ESMR–5 scanned through the satellite subpoint and had a spatial resolution of 25 km at nadir. ESMR–6 scanned in a cone ahead of the satellite in such a way as to keep the zenith angle of the satellite, as measured at the scan spot, nearly constant. ESMR–6 had a spatial resolution of about 20 × 40 km independent of scan position. The ESMRs were meteorologically significant because they can be used to estimate precipitation rate (see Chapter 9). They are also sensitive to sea ice.

The Scanning Multichannel Microwave Radiometer (SMMR) flew on both Nimbus 7 and Seasat. It measured both horizontally and vertically polarized radiation at five wavelengths, 4.54, 2.8, 1.66, 1.36 and 0.81 cm, with resolutions

[20] Electric field vector parallel to the surface at the scan spot and perpendicular to the beam.

[21] Electric field vector perpendicular to both the above horizontal vector and to the beam.

ranging from 20 km (at 0.81 cm) to 100 km (at 4.54 cm). This was accomplished using a rotating scan mirror and six Dicke radiometers much like that of the MSU. SMMR was designed to measure several meteorologically important parameters, among them: precipitation rate, column-integrated atmospheric water vapor and liquid water, sea surface temperature and wind speed, soil moisture, and sea ice concentration.

Finally, the Special Sensor Microwave Imager (SSM/I) is currently flying on some of the DMSP satellites. Similar in construction to the SMMR, the SSM/I measures both horizontally and vertically polarized radiation at 19.35, 37.0, and 85.5 GHz, and vertically polarized radiation only at 22.235 GHz. Measurements at 85 GHz are new, and they should provide improved detection of rain and clouds.

4.3.4 Active Instruments

Most satellite-borne meteorological instruments are *passive*; that is, they measure radiation provided to them by the Earth. *Active* instruments, like radar, transmit their own signal and measure the properties of the returned radiation. Because nature does not necessarily provide optimal amounts of radiation, active instruments can be designed to be more sensitive than passive instruments to certain parameters. Active instruments, however, require much more electrical power to operate than passive instruments. Seasat had three active instruments, one of which was useful for meteorology. The Seasat–A Satellite Scatterometer (SASS) was a 2.05-cm radar that measured the backscatter coefficient from the sea surface. The backscatter coefficient is related to surface wind speed and direction. By observing the same patch of ocean as the satellite approached and as it left an area, these two parameters could be estimated (Jones *et al.*, 1979; see Chapter 7). Unfortunately, Seasat failed due to a short circuit after only 104 days in orbit, but not before demonstrating the feasibility and importance of active instrumentation.

4.3.5 DMSP Instruments

Finally, instruments flown on DMSP satellites (in addition to the SSM/I discussed above) are of interest. The primary DMSP instrument is the Operational Linescan System (OLS), which is a two-channel (visible and 11-μm infrared) imager (Dickinson *et al.*, 1974). It differs from the AVHRR in three interesting ways. First, it has extremely high spatial resolution of about 600 m. Second, it also has a variable scan, which causes the distance between pixels on the same scan line to be the same near nadir and at the ends of the scan line. This makes images easy to display and grid. Third, the visible sensor on the OLS has a much broader bandwidth than the AVHRR visible channel. It covers approximately the same spectral region as AVHRR channels 1 and 2 combined. This greater bandwidth means that more radiation from a given scene is sensed, which means that the instrument has greater sensitivity. Some OLS visible data have been collected at night using reflected moonlight, and city lights are detectable on OLS images.

The data from the OLS are archived mostly in the form of positive transparencies, which can be useful for studies requiring higher resolution than that available from the NOAA satellites or for studies in which digital data on magnetic tape are not appropriate.

DMSP satellites have also carried infrared (SSH/T) and microwave (SSM/T) sounders.

4.4 SATELLITE DATA ARCHIVES

Extensive archives of meteorological satellite data exist. In the United States, two sources are primary. The Satellite Data Services Division of NESDIS [Room 100 Princeton Executive Center, Washington, DC 20233, telephone (301) 763-8400] archives data from the operational satellites. The National Space Science Data Center [NASA/Goddard Space Flight Center, Greenbelt, MD 20771, telephone (301) 286-6695] archives primarily data from NASA research satellites such as the Nimbus satellites.

Meteosat data are archived by EUMETSAT. Information on Meteosat data may be obtained by writing to Meteosat Data Services, Meteosat Exploitation Project, European Space Operations Center, Robert–Bosch Strasse 5, D-6100 Darmstadt, Germany.

GMS data are archived by the National Space Development Agency of Japan, World Trade Center Building 2-4-1, Hamamatsu-cho, Minato-ku Tokyo 105, Japan.

Insat data are archived by the Meteorological Data Utilisation Centre, Satellite Meteorology Division, India Meteorological Department, Lodi Road, New Delhi 110003, India.

Readers should note that access to satellite data and information over the Internet is increasing exponentially. NASA is developing a system of Distributed Active Archive Centers (DAACs) for today's data and for the data to be collected by the Earth Observing System (see Chapter 11). Table 4.28 contains information on how to access the DAACs as of December 1994. Readers who have Gopher software are advised to investigate some of the following Gopher sites:

- eos.nasa.gov
- gopher.gsfc.nasa.gov
- gopher.ngdc.noaa.gov
- gopher.ssec.wisc.edu
- spacelink.msfc.nasa.gov

World Wide Web "Home Pages" are also available for all of the DAACs and for many other interesting sites. We have not listed the Internet addresses for these Home Pages because they are quite likely to change and because up-to-date addresses can be found from the sources listed above. Finally, readers can telnet

TABLE 4.28. Distributed Active Archive Centers (DAACs)

Location	User services		
	Telephone	*E-mail*	*Telnet to*
Alaska SAR Facility	(907) 474-6166	asf@eos.nasa.gov	eosims.asf.alaska.edu 12345
EROS Data Center	(605) 594-6116	edc@eos.nasa.gov	eosims.cr.usgs.gov 12345
Goddard Space Flight Center	(301) 286-3209	gsfc@eos.nasa.gov	eosims.gsfc.nasa.gov 12345
Jet Propulsion Laboratory	(818) 354-9890	jpl@eos.nasa.gov	eosims.jpl.nasa.gov 12345
Langley Research Center	(804) 864-8656	larc@eos.nasa.gov	eosims.larc.nasa.gov 12345
Marshall Space Flight Center	(205) 922-5932	msfc@eos.nasa.gov	eosims.msfc.nasa.gov 12345
NOAA Satellite Active Archive	(301) 763-8402	saainfo@ncdc.noaa.gov	eosims.saa.noaa.gov 12345
National Snow and Ice Data Center	(303) 492-6199	nsidc@eos.nasa.gov	eosims.colorado.edu 12345
Oak Ridge National Laboratory	(615) 241-3952	ornl@eos.nasa.gov	eosims.esd.ornl.gov 12345

directly to the National Space Science Data Center at nssdca.gsfc.nasa.gov and to Spacelink at spacelink.msfc.nasa.gov.

Bibliography

Ball Aerospace Systems Division (1986). *SBUV/2 Final Engineering Report for Flight #1 (S/N002).* Document No. B6802-91, Boulder, CO.

Barkstrom, B. R. (1986). ERBE data and its validation. *Preprints: Second Conference on Satellite Meteorology/Remote Sensing and Applications.* American Meteorological Society, Boston, pp. J1–J3.

Barnes, J. C., and M. D. Smallwood (1982). *TIROS-N Direct Readout Services Users Guide.* NOAA/NESS, Washington, DC.

Chen, H. S. (1985). *Space Remote Sensing Systems: An Introduction.* Academic Press, Orlando.

Clark, J. D. (ed.) (1983). *The GOES User's Guide.* NOAA/NESDIS, Washington, DC.

Dickinson, L. G., S. E. Boselly III, and W. S. Burgmann (1974). *Defense Meteorological Satellite Program (DMSP) User's Guide.* Air Weather Service Tech. Rep. AWS-TR-74-250, Scott Air Force Base, Illinois.

Gibson, J. (ed.) (1984). *GOES Data Users' Guide.* NOAA/NESDIS, Washington, DC.

Halliday, D., and Resnick, R. (1962). *Physics.* John Wiley, New York.

Hayden, C. M. (1988). GOES-VAS simultaneous temperature-moisture retrieval algorithm. *J. Appl. Meteor.*, **27**, 705–733.

Hayden, C. M. (1989). GOES I–M sounding products. *Program: GOES I–M Operational Satellite Conference.* April 3–6, Arlington, VA, pp. 126–130.

Jones, W. L., P. G. Black, D. M. Boggs, E. M. Bracalente, R. A. Brown, G. Dome, J. A. Ernst, I. M. Halberstam, J. E. Overland, S. Peteherych, W. J. Pierson, F. J. Wentz, P. M. Woiceshyn, and M. G. Wurtele (1979). Seasat scatterometer: Results of the Gulf of Alaska workshop. *Science*, **204**, 1413–1415.

Kidwell, K. B. (1986). *NOAA Polar Orbiter Data User's Guide.* NOAA/NESDIS Satellite Data Services Division, Washington, DC.

Koenig, E.W. (1989). Characteristics of the GOES I–M imager and sounder. *Program: GOES I–M Operational Satellite Conference.* NOAA, Washington, DC, pp. 168–175.

Komajda, R.J., and K. McKenzie (1987). *An Introduction to the GOES I–M Imager and Sounder Instruments and the GVAR Retransmission Format.* NOAA Tech. Rep. NESDIS 33, Washington, DC.

Kopia, L. P. (1986). Earth radiation budget experiment scanner instrument. *Rev. Geophys.*, **24**, 400–406.

Lauritson, L. G., Nelson, G. J., and F. W. Porto (1979). *Data Extraction and Calibration of TIROS–N/NOAA Radiometers.* NOAA Tech. Memo. NESS 107, Washington, DC.

Liou, K.-N. (1980). *An Introduction to Atmospheric Radiation.* Academic Press, New York.

Luther, M. R., J. E. Cooper, and G. R. Taylor (1986). The Earth radiation budget experiment nonscanner instrument. *Rev. Geophys.*, **24**, 391–399.

Menzel, W. P., and J. F. W. Purdom (1994). Introducing GOES–I: The first of a new generation of Geostationary Operational Environmental Satellites. *Bull. Am. Meteor. Soc.*, **75**, 757–781.

Morgan, J. (1981). *Introduction to the Meteosat System.* European Space Agency, SP-1041, Paris Cedex, France.

National Aeronautics and Space Administration. *Observations from the Nimbus I Meteorological Satellite.* NASA SP-89, Washington, DC.

National Aeronautics and Space Administration (1966). *Nimbus II Users' Guide.* Goddard Space Flight Center, Greenbelt, MD.

National Aeronautics and Space Administration. *The Nimbus III User's Guide.* Goddard Space Flight Center, Greenbelt, MD.

National Aeronautics and Space Administration (1970). *The Nimbus IV User's Guide.* Goddard Space Flight Center, Greenbelt, MD.

National Aeronautics and Space Administration (1972). *The Nimbus 5 User's Guide.* Goddard Space Flight Center, Greenbelt, MD.

National Aeronautics and Space Administration (1975). *The Nimbus 6 User's Guide.* Goddard Space Flight Center, Greenbelt, MD.

National Aeronautics and Space Administration (1978). *The Nimbus 7 Users' Guide.* Goddard Space Flight Center, Greenbelt, MD.

National Aeronautics and Space Administration (1981). *Handbook of Sensor Technical Characteristics.* Washington, D.C.

National Environmental Satellite, Data, and Information Service (1987). *Operational VAS Mode AAA Format*, Version 2.6, NOAA/NESDIS, Washington, DC.

Rao, P. K., S. J. Holmes, R. K. Anderson, J. S. Winston, and P. E. Lehr (eds.) (1990). *Weather Satellites: Systems, Data, and Environmental Applications.* American Meteorological Society, Boston.

Savides, J., and K. G. Reseck (1989). GOES I-M system characteristics. *Program: GOES I–M Operational Satellite Conference.* NOAA, Washington, DC, pp. 285–314.

Schwalb, A. (1978). *The TIROS-N/NOAA A–G Satellite Series.* NOAA Tech. Memo. NESS 95, Washington, DC.

Schwalb, A. (1982). *Modified Version of the TIROS-N/NOAA A–G Satellite Series (NOAA E–J)—Advanced TIROS-N (ATN).* NOAA Tech. Memo. NESS 116, Washington, DC.

Smith, G. L., R. N. Green, B. A. Wielicki, J. T. Suttles, and L. M. Avis (1986). ERBE data inversion. *Preprints: Second Conference on Satellite Meteorology/Remote Sensing and Applications.* American Meteorological Society, Boston, pp. J9–J13.

Smith, W. L., H. M. Woolf, H. B. Howell, H.-L. Huang, and H. E. Revercomb (1987). The simultaneous retrieval of atmospheric temperature and water vapor profiles—application to measurements with the High spectral resolution Interferometer Sounder (HIS). In A. Deepak, H. Fleming, and J. Theon (eds.), *Advances in Remote Sensing Retrieval Methods*, A. Deepak Publishing, pp. 189–202.

Vaughan, W. W. (ed.) (1982). *Meteorological Satellites—Past, Present, and Future.* NASA Conference Publication 2227, Washington, DC, 60 pp.

Werbowetzki, A. (ed.) (1981). *Atmospheric Sounding Users Guide.* NOAA Tech. Rep. NESS 83. Washington, DC.

Willson, R. C. (1979). Active cavity radiometer type IV. *Appl. Optics*, **18**, 179–188.

World Meteorological Organization (1989). *Information on Meteorological and Other Environmental Satellites.* WMO No. 411, Geneva, Switzerland.

5

Image Interpretation

with Donald L. Reinke

THIS CHAPTER IS an introduction to image interpretation but is by no means a comprehensive treatment of the subject. It would take several volumes, each the size of this book, to show examples of all of the features that can be identified and analyzed, and more are being discovered as we write. Over three decades of weather satellites have produced millions of images—any one of which would provide enough material for a complete seminar. In one chapter we cannot hope to do justice to this broad a subject. Instead, we offer two things: an overview of the concepts that every student of satellite meteorology should know about image interpretation, and a bibliography of works that contain detailed instruction on the subject.

The purpose of meteorological satellite image interpretation is to relate significant features in the image to physical processes that are occurring, or have occurred, in the atmosphere. For example, we want to identify clouds. The shape or texture of the cloud or the proximity to geographic features on the Earth's surface can tell us much about the dynamics occurring in the atmosphere. The simple existence of a cloud on a satellite image can lead us to a number of conclusions, or questions, such as, what was the mechanism for formation of the cloud? The mechanism could be mechanical lifting, it could be convection due to heating, or the cloud could be the residual of a cloud mass that has been advected far from its point of origin. Image interpretation can give us some clues to help identify the mechanism. That knowledge can in turn help us determine the present and future hydrodynamic state of the atmosphere. The objectives of image interpretation are to (1) detect and identify features such as clouds or other obscuring or radiating phenomena on the image and (2) determine the physical mechanisms that produce or sustain those features.

Much of the research in image interpretation being done today involves the development of techniques to automate the analysis of digital imagery. Over the past few decades, the primary analysis tool has been the interpretive skill of the analyst who manually views the satellite image and makes a subjective assessment of the features it contains. The operational meteorological satellites that are currently flying produce more than 10^{13} bytes of image data in a year. With the average imaging system able to display 10^6 bytes at a time, this amounts to more than 10^7 images—far too many to analyze manually. Some progress has been made in the automation of the process of image interpretation, and high-speed digital computers can be used to do a reasonable first-guess analysis of clouds and other features. Total automation of the process, however, still eludes us, and a good understanding of the manual interpretation process is essential to the application of satellite imagery to weather analysis and forecasting.

5.1 SATELLITE IMAGERY

The term *satellite imagery* can be used to describe two similar, yet unique displays: *photo* and *digital*. It is important to differentiate between the two, as some analysis techniques work only on digital images. We refer to a hard-copy image (such as a photograph or laser print copy) as a *photo image*. Satellite imagery that is typically stored on magnetic or optical media is referred to as digital imagery, due to the nature of the collection and archive process. Photo images are easy for an analyst to handle, but digital images can be processed in many more ways.

Most meteorological satellite imaging devices scan the Earth scene in a manner analogous to television raster scans.[1] Radiance measurements are taken along each scan line at time steps which produce a series of elements along that scan (refer to Fig. 2.16). These elements are assigned a unique location within the image by referring to a line (or scan) number and an element (or sample) number. The radiance measured by the satellite sensor at each element location is stored as a single digital value. When displayed on an imaging device such as a color or black-and-white (B/W) monitor, each point on the grid is referred to as a *picture element* or *pixel*. Digital values are used to assign a color or gray shade

[1] Some devices, such as the DMSP SSM/I scanner (Section 5.2), take samples in a helical scan pattern.

to pixels on the final satellite image. The resultant image is displayed either on a monitor or as a photographic (hard-copy) image.

The *resolution* of a satellite instrument can be defined as the size of the smallest element in the Earth scene that can be resolved by the instrument. It is a function of the field of view of the sensor optics and the distance to the Earth surface. One can also think of it as the length of the side of a box on the surface of the Earth from which radiation is gathered during each time step of the scanning process (see Chapter 4.). These boxes are sometimes called "footprints." For a variety of reasons, the size of this footprint varies from one channel to the next. For example, the visible channel on a satellite may have a resolution of 1 km, whereas the infrared channel has a resolution of 8 km. The resolution of a satellite instrument is normally specified separately for each channel.

The resolution of a satellite instrument is usually for the satellite *subpoint* (the point on the Earth directly beneath the satellite). As the satellite scans farther away from the subpoint, the area on the surface of the Earth that is viewed becomes progressively larger because of the curvature of the Earth and the obliquity of view. Geostationary meteorological satellites, because of their altitude, can view the Earth scene out to great distances from the subpoint; thus the resolution varies greatly with distance from the subpoint. Section 5.4 discusses this degradation as a source of error when doing image analysis.

It is important to note that the resolution of the satellite instrument may be quite different from the resolution of an image produced from the data. For example, the resolution of a satellite instrument may be 1 km, but an image constructed from the data might be displayed at a resolution of 10 km by averaging 100 (10×10) of the original pixel values or by selecting (sampling) every tenth pixel and every tenth line.

5.2 SPECTRAL PROPERTIES

By using a filter, placed in front of the satellite sensor, radiation can be measured at specific wavelengths. The filter allows radiation within a narrow segment of the electromagnetic spectrum, called a *spectral interval*, to reach the sensor. These spectral intervals are commonly referred to as *channels* or *bands* (see Chapter 4).

Most meteorological satellites make measurements in the visible and infrared portions of the electromagnetic spectrum. Some also have additional channels which vary from the ultraviolet (100–400-nm band) to the microwave region (0.15–6.0 cm—often listed as a frequency range from 200 to 5 GHz). Chapter 4 contains tables with the spectral bands of a wide range of meteorological satellites. In the remainder of this chapter, we discuss the properties of the more common of these spectral intervals and their application to image interpretation.

5.2.1 Visible Imagery

The first imaging sensors aboard meteorological satellites measured radiation in the *visible band* (0.4–0.7 μm). Figure 5.1 shows a visible image from the GOES

FIGURE 5.1. GOES 7 full-disk visible image.

7 satellite. Visible imagery generally offers the highest spatial resolution and provides us with a view of the Earth that closely matches our senses. Land, clouds, and ocean are easily discernible. One obvious limitation to visible data is that they are available only from the sunlit portion of the Earth. One exception to the daytime-only limitation is a sensitive instrument (the OLS) on board the Defense Meteorological Satellite Program (DMSP) satellites, which can sense features that are illuminated at night by moonlight.

5.2.2 Infrared Imagery

The infrared channels are most often between 1 and 30 μm. The most common infrared band for meteorological satellites is in the 10–12.5-μm window, in which the atmosphere is relatively transparent to radiation upwelling from the Earth's surface. When the word *infrared* is used alone to describe an image, it is nearly always in the 10–12.5-μm window rather than in another portion of the electromagnetic spectrum.

Figure 5.2 shows the infrared image that corresponds to Fig. 5.1. Again, clouds, land, and water are easily discerned. Because infrared radiation can be related to

FIGURE 5.2. GOES 7 infrared image (11 μm).

the temperature of the emitting body, and because the troposphere generally cools with height, this helps us interpret the atmospheric processes occurring within the scene. An important characteristic of the infrared channels is their ability to provide images at night. This provides continuous coverage of cloud evolution over a full 24-hour period.

Normally in image processing, images are displayed such that the greater the radiance, the brighter the pixel. In satellite meteorology, however, infrared images are normally *inverted*; that is, the larger the radiance from an element, the darker the pixel. This way, clouds, which are usually colder than the surface, appear white, and the warmer ground or ocean surface appears darker than clouds, as in visible images.

5.2.3 Water Vapor Imagery

The water-vapor channels are so named because the satellite measures radiation in water-vapor absorption bands. Several wavelengths can be used, but the most common is centered around 6.7 μm. At this wavelength, most of the radiation sensed by the satellite comes from the atmospheric layer between 300 and 600 hPa (Morel *et al.*, 1978); thus it measures middle levels of the troposphere.

Figure 5.3 shows the water-vapor image taken at the same time as Figs. 5.1 and 5.2. Note that the highest clouds can be seen, but surface features cannot be detected because this is not a window channel. Instead, dramatic swirls and eddies of water vapor are seen where high clouds are not present. Water-vapor imagery is commonly looped (see Section 5.3) to display motions in cloud-free regions of the atmosphere.

One way to qualitatively interpret water-vapor imagery is to say that it approximates the relative humidity of the midtroposphere. Consider a bright area and a dark area in a water-vapor image. In the bright area, the satellite measures less radiance than in the dark area. Less radiance means either that the atmosphere is colder at the same level as in the dark area or that there is more water vapor present in the bright area so that the satellite senses a higher, and therefore colder, level. In either case, the relative humidity is likely to be higher in bright areas than in dark areas. Bright and dark areas may also indicate rising and sinking motions, respectively.

FIGURE 5.3. GOES 7 water-vapor image (6.7 μm).

5.2.4 Microwave Imagery

Microwave radiation is sensitive to an array of surface and atmospheric parameters, including precipitation, cloud water, water vapor, water droplet phase, soil moisture, surface temperature, atmospheric temperature, and ocean surface wind speed. In addition, many regions of the microwave spectrum are available for remote sensing. Still further, all microwave instruments measure polarized radiation, and polarization is a factor in the microwave radiation emitted by some of the above atmospheric and surface phenomena. Very roughly, we can say that 19 GHz is a window region and provides a measurement of surface radiance in clear sky areas. The 22-GHz channel is sensitive to water vapor. The 37-GHz channels are attenuated by clouds and rain, and the 85-GHz channels can be used to identify ice clouds and snow. For example, the very cold temperatures seen at 85 GHz in Plate 2 in Indiana, Ohio, and across the South, are probably thunderstorms. These channels are often used in combination to produce products, such as precipitation images (e.g., Spencer, 1986; Spencer *et al.*, 1983, 1988). A great deal of current research is devoted to understanding and fully utilizing the information contained in microwave images.

5.2.5 Multispectral Analysis

Some atmospheric phenomena are more easily identified by using imagery from more than one channel. Figure 5.4 illustrates a simple two-channel scheme for identifying clouds, and Figs. 5.5 and 5.6 are visible and infrared image pairs to which the scheme can be applied. Low-level stratus is often difficult to identify on an infrared image because the cloud may have a radiative temperature that is near that of the surface (see the area north of the hurricane in Fig. 5.5, for example). On a visible image, however, stratus clouds appear bright in contrast to the darker background of land or water. As another example, the detection of thin cirrus clouds can be difficult on a visible image because they are almost transparent to visible light. These same clouds, however, are cold and mostly opaque to infrared radiation from the warmer surface below and thus present a strong cold signal on the infrared image (see the northwest corner of Fig. 5.6).

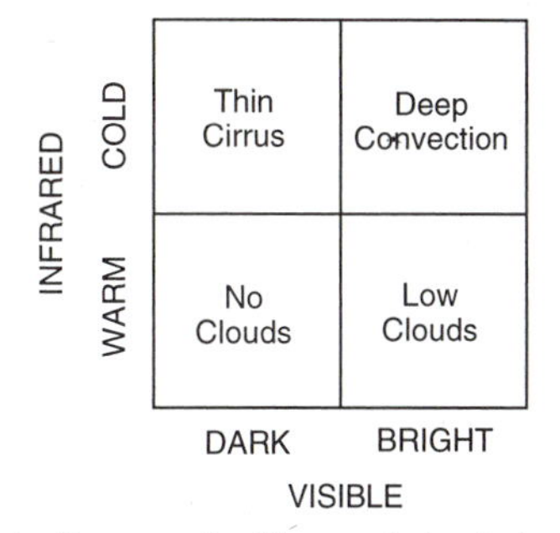

FIGURE 5.4. Schematic diagram of a bispectral cloud classification scheme.

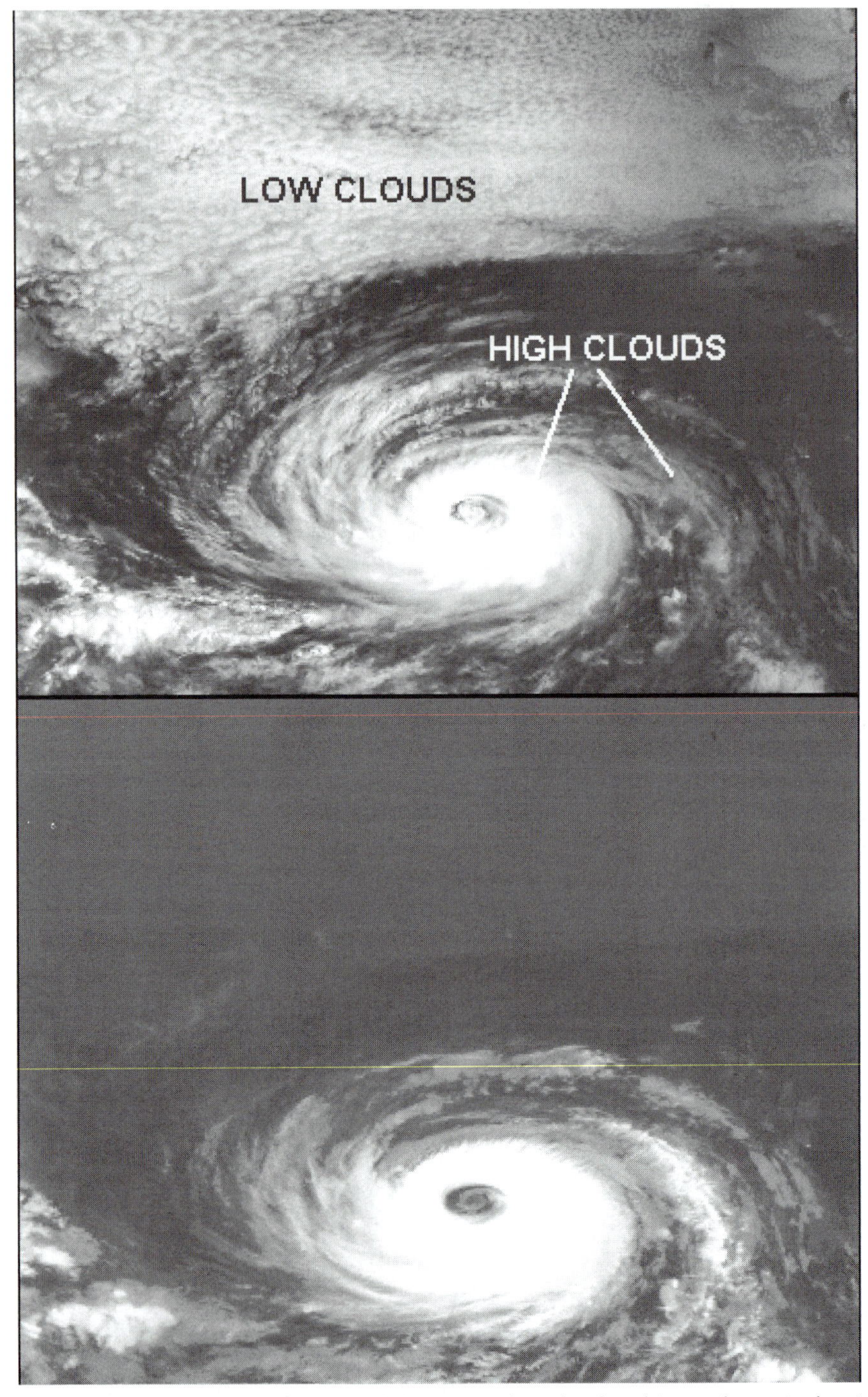

FIGURE 5.5. DMSP visible image (top) sampled concurrently with infrared image (bottom), showing low clouds and cirrus.

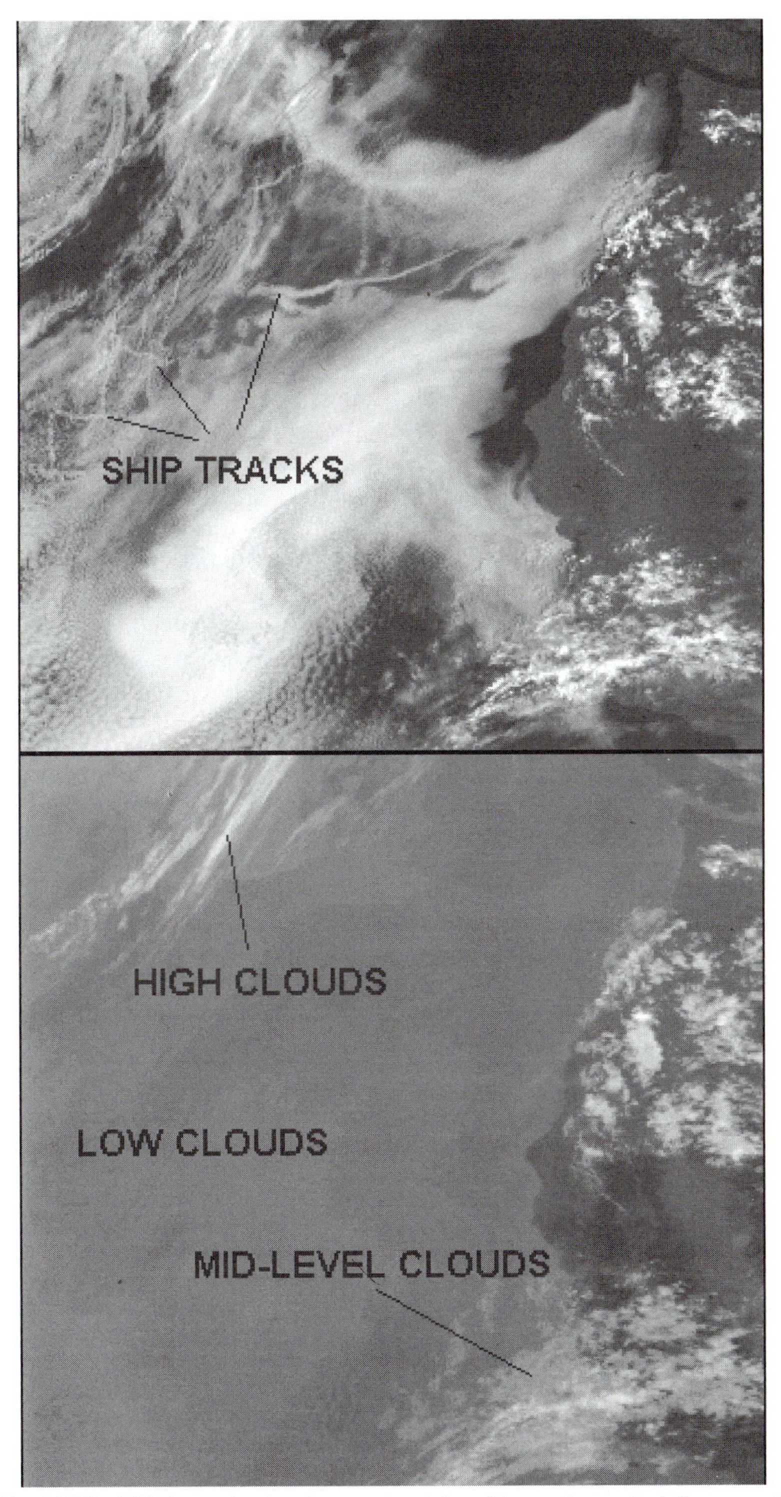

FIGURE 5.6. DMSP visible image (top) and concurrent infrared image (bottom) of the eastern North Pacific Ocean and the western United States, showing low-level stratus and ship tracks.

Deep convection or thick cirrus is both cold and optically thick and therefore is bright in either a visible or an infrared image. A further example of multispectral analysis using 3.7-μm imagery is discussed in Section 5.5.4.4, and the use of multispectral techniques with microwave imagery was mentioned above.

5.3 IMAGE ENHANCEMENT TECHNIQUES

Image enhancement involves highlighting certain values or regions within an image to emphasize and identify meteorological parameters or features and to separate these from land or water. There are three primary methods of enhancement: color enhancement, physical enhancement, and statistical enhancement. Physical enhancement involves the three-dimensional (3D) display of an image, including looping of successive 2D images (the third dimension is time). Statistical enhancement of imagery typically involves the spatial or temporal analysis of adjacent pixels to determine parameters such as the gradient or variance.

5.3.1 Color Enhancement

Color enhancement, as the name implies, involves the assignment of colors (or gray shades) to each pixel of an image based on the value of the pixel. The standard technique for color enhancement of imagery is to apply a *lookup table* to assign a specific color or hue to each possible numerical value that an image pixel can take on. A second technique involves the generation of (red-green-blue) RGB images using multispectral data.

5.3.1.1 Lookup Tables

Since an image is composed of pixels with discrete digital values, we can assign a color or monochromatic gray shade to each value to produce an enhanced image. Today's 8-bit image display systems allow digital values that range from 0 to 255 at each pixel location, so we can employ up to 256 colors or shadings to highlight certain features of the image.

Infrared imagery is well suited to the use of a color enhancement table as shown in Plate 3. Each infrared pixel value corresponds directly to an equivalent blackbody temperature. A lookup table, sometimes referred to as an *enhancement table*, is used to assign a specific color to each possible pixel value or, more commonly, to a range of values. Thus each color represents a temperature or range of temperatures. A temperature scale should be affixed to such images to document the enhancement.

Color enhancement is useful for locating large convective cells that reach from just above ground level to 20 km above the surface, passing through a large range of temperatures. When viewed from the top, the cloud mass around the cell has a range of temperature values from a temperature near that of the surface to one that may be 80 K cooler at the cloud top. Color enhancement is also useful for distinguishing clouds from ground or water surfaces.

Note that enhancement need not be "color." Figure 9.12 shows a gray-scale enhancement called the *MB curve* which has been used for many years to locate convective areas on infrared satellite images.

Color-enhanced images are sometimes referred to as *false color* or *pseudocolor* images because color has been introduce into a monochromatic image.

5.3.1.2 Multispectral Color Enhancement

The production of a multispectral image takes advantage of the red-green-blue (RGB) color display technology available on a color monitor. Each channel to be viewed is assigned to one (or more) of the color guns to highlight multispectral features. To take a two-channel example, if an infrared image is assigned to the red gun, and a visible image is assigned to the blue and green guns, then the features described in Fig. 5.4 become colors on the image. Deep clouds are white, warm cloud-free areas are black, low clouds are cyan (green + blue), and thin cirrus is red. A third channel, such as a water-vapor image, can be added, but the resulting image becomes quite complex.

Multispectral color enhancement technique is not commonly used, but because the human eye is so color-sensitive, it is a powerful way to combine multispectal data to yield information that no single channel can yield. Multispectral images are sometimes referred to as *true color* images, though this is strictly correct only when red, green, and blue images are combined as a color video camera does.

5.3.2 Physical Enhancement

Two common types of physical enhancement are looping and three-dimensional (3D) image displays. Both require some degree of specialized hardware or software to be of value when doing image interpretation. The most frequently used physical enhancement technique is the time-sequencing or looping of imagery. An increasingly popular technique is based on viewing an image which has been rendered three-dimensionally.

5.3.2.1 Looping

Looping of satellite imagery has become an indispensable technique for not only identifying features on a satellite image but also graphically displaying the life cycle of synoptic and mesoscale meteorological features. An image loop is created by rapidly displaying in succession a series of images to either create a sense of movement (a time series of images) or to compare different channels from the same time. Looping of imagery is useful for quantifying the movement of features such as hurricanes, frontal systems, or large thunderstorm complexes. The loop allows an analyst to quickly identify those features that are moving by comparing one image to the next and gives a good overview of both large and small-scale motions. Short-term forecasts, called *nowcasts*, are made by extrapolating ahead in time the feature seen in loops. We cannot show readers a loop in this book, but loops are commonly shown on television weather broadcasts.

Looping is an efficient technique for identifying the misalignment of imagery because even slight movements of land features are easily seen in loops. Looping can also be used to identify features that have a unique spectral signature. For example, a loop of visible and infrared images from the same time can be used to identify cirrus or low stratus clouds.

5.3.2.2 Three-dimensional Display

Although not a common technique, 3D displays of imagery (also called *stereo displays*) can provide information that is not obvious on the standard 2D displays.

There are three types of 3D techniques. The most common is to display an image using a graphical display package that provides shading, light source and perspective views (Vonder Haar *et al.*, 1989). An example is shown in Plate 6, which is discussed in Chapter 7. A similar technique involves viewing offset stereo images (one colored red, the other cyan) with special glasses (one lens red, the other cyan) that separate the views and produce a 3D effect. A third technique employs two separate screens and a stereo viewer by which one eye sees one screen and the other eye sees the second screen.

In addition, there are two types of stereo observations. *True stereo* requires two near-simultaneous images of a scene taken from different platforms or at different locations by a single platform. The two separate images are displayed by any of the above 3D techniques. *False stereo* involves using two spectral images of the same scene. Usually, infrared imagery is used to approximate the height of a pixel, whereas visible imagery is used for the brightness of a pixel. Then each pixel is displayed by one of the above 3D techniques.

Three-dimensional techniques are seldom used because they require special equipment, and true stereo requires near-simultaneous imagery from separate platforms, which is not available with the current generation of meteorological satellites. Future systems, however, will provide frequent revisit times from adjacent satellites and make 3D viewing a more practical analysis tool. The image interpretation techniques discussed in this chapter are confined to 2D images; however, many of the techniques can also be applied to 3D displays.

5.3.3 Statistical Enhancement

Statistical enhancement is an effective technique for automating the identification of features on an image. The most common form of statistical enhancement is the simple comparison of a pixel with its neighbors. This comparison can be in the form of a difference between a pixel and its neighbors, or a gradient or a

Laplacian. Each of these measures is used to identify either the homogeneity of a group of pixels or a boundary or edge passing through a pixel.

One application of this type of measure is to determine whether the pixel is on a land/water or cloud/land boundary (sharp gradient or large difference) or if the pixel is inside a larger homogeneous cloud or land mass (no gradient or small difference). By grouping pixels of like structures, it is possible to locate clouds, land, and water to a reasonable degree of accuracy (Kelly *et al.*, 1989). This technique is sometimes referred to as *edge enhancement*.

5.4 GEOLOCATION AND CALIBRATION

An important aspect of image interpretation is the accurate geographic location of image pixels. Most applications of image interpretation to analysis and forecasting require that features be placed at the proper location within the geographic coordinate system of choice. This is necessary to allow its information content to be transferred to the observation database, which also serves as the primary input to forecast models. In some cases accurate navigation is not necessary, but it is imperative when doing time-sequencing of images.

5.4.1 Navigation

The process of geolocating pixels is also called *navigation* and requires current satellite orbit and attitude parameters plus algorithms that utilize these parameters to calculate the latitude and longitude of each pixel or (the inverse problem) which pixel is closest to a particular latitude-longitude point. This process is described in detail in Chapter 2.

5.4.2 Gridding

Once navigation has been accomplished, and each pixel is assigned a latitude and longitude, the next step is to apply a grid to aid in the visualization of where image features are located geographically. The most common is the geopolitical boundary grid. Another common grid is a latitude—longitude grid, which consists of parallels and meridians placed at the proper location on the image. After navigation has been performed, the image may also be remapped into another coordinate system.

5.4.3 Remapping

Images are often transformed into common map projections (*remapped*) so that image features can be compared to other map or gridded data such as surface or upper air synoptic reports. The original scene viewed by the satellite is called the *satellite projection*. In order to compare the satellite data with standard synoptic weather charts, the images are remapped into a common projection such as polar stereographic or Mercator. Figure 5.7 shows data from four geostationary satellites remapped into a Mollweide projection. The remapping process is done one pixel at a time and can be computer-intensive for large images. (A 1024 × 1024 pixel image contains over 1 million pixels.)

5.4.4 Alignment

The proper navigation or remapping of an image does not guarantee that two images taken at different times will be aligned geographically. This is due to perturbations in the satellite's orbit and attitude parameters. The resulting misalignment is most noticeable when looping images or compositing (overlaying) successive images in time.

Several methods have been used to correct for misalignment with varying degrees of success. One technique relies on the manual navigation of an image by location of known geographic landmarks. Several landmarks are first identified on an image. Because the latitude, longitude, and elevation of the landmarks are known, the orbit and attitude parameters used to navigate the image can be adjusted to force the identified pixels to have the correct latitude and longitude. The new parameters are then used to renavigate the entire image.

Another technique involves *warping* the image to stretch or shrink it until it fits a predefined grid. This technique also requires specific landmarks. By using a predefined grid such as a geopolitical boundary map, the image is warped to fit the graphic display. Warping degrades the integrity of data values due to sampling or averaging. Both of these techniques are less accurate over the oceanic regions where there are few recognizable geographic features. The top portion of

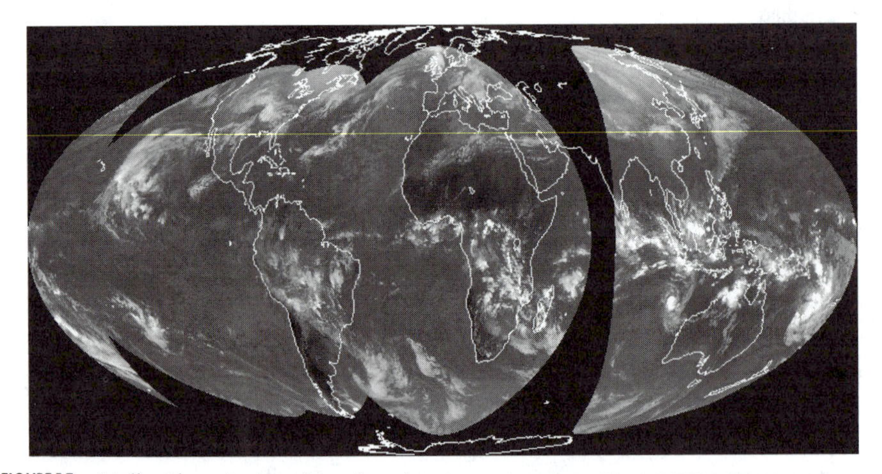

FIGURE 5.7. Mollweide projection of data from four geostationary satellites: GMS 4, GOES 7, Meteosat 3, and Meteosat 5.

Fig. 5.8 shows a poorly aligned GOES 7 image. Note the large error in the geolocation of the image with respect to the graphic. The bottom portion of Fig. 5.8 shows the result after applying a third-order polynomial warping algorithm to the image. The image now fits the graphic that was generated by the original navigation parameters, so the navigation can be applied to the warped image to remap it into any other projection.

Looping of two images during the alignment process, sometimes referred to as *flickering* or *toggling*, can readily identify geographic features that are not aligned. When properly aligned they appear as one stationary image when toggled, but transient cloud, smoke, or airborne sand appear to move. This alignment process is critical for any quantitative movement analysis because the loop should show the movement of cloud features over a fixed geographic background. If the image is not properly aligned, the movement of clouds is biased by the amount of shift in the position of the geographic background.

5.4.5 Merging

One common reason for remapping is the production of a *merged image* product. A cloud system may extend beyond the maximum viewing range of a specific satellite, but is within range of another. Images from the two can be merged to form a single view of the cloud field. Figure 5.7 shows an example of an image that was created by merging four geostationary satellite images. Here the individual images from the GMS 4, GOES 7, Meteosat 3, and Meteosat 5 satellites are remapped onto a Mollweide projection. The image was created by remapping each satellite onto the output Mollweide grid and selecting the pixel from the closest satellite where two of the original images overlap.

5.4.6 Normalization

When merging images from different satellites, it is important to note that differences in the calibration of each individual satellite sensor can cause problems if not taken into account. For example, on a GOES image, a digital count of 200 may represent a temperature of −35°C, whereas the same count on a GMS image may represent a temperature of −25°C. Further, the calibration of the two satellites may differ. To account for these differences, one of the satellites must be *normalized* to the other, so that a digital count value represents the same temperature for both.

The process of normalization requires the extraction of digital values from co-located pixels on overlapping portions of adjacent images. This is normally done on a range of values, such as cool ocean or land points and very cold cloud masses. After a sufficient number of collocated points have been measured, the pairs of brightness temperatures derived from the digital values are fed into a regression scheme to produce a set of normalization coefficients. These coefficients are then applied to one of the images to bring it in line with the other. Because

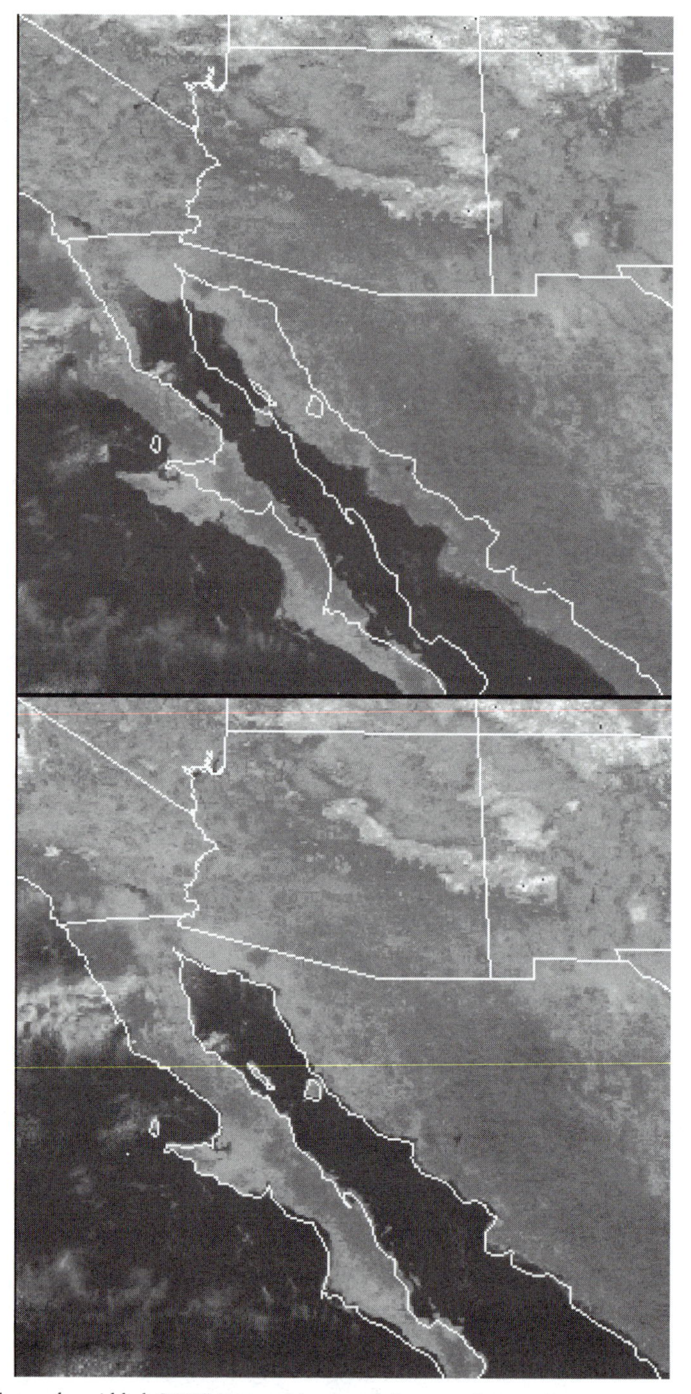

FIGURE 5.8. A poorly gridded GOES 7 image (top) and the same image after remapping to account for navigation error (bottom).

they under-fly all geostationary satellites, NOAA polar orbiting satellites have been used to normalize all of the geostationary satellites that are used in the International Satellite Cloud Climatology Project (ISCCP) global cloud product (see Chapter 8). Image normalization should be done before doing any quantitative analysis of merged imagery.

5.4.7 Errors

There are a number of factors that limit our ability to perform the analysis of satellite imagery. These factors include viewing angle biases, satellite sensor/lens/filter/mirror imperfections, ground station equipment and processing, sensor lag, signal contamination/attenuation, navigation/location errors, and the difficulty in distinguishing land/water/cloud. The severity of a source of error varies with each satellite system, and in some cases, a specific error may not have an adverse affect on the image analysis (e.g., a location error with a magnitude of 2 km has little effect on analysis of data that is sampled at 25-km resolution). Algorithms have been developed to account for several of these errors; however, the analyst should be aware of their presence when making quantitative assessments from satellite imagery.

5.4.7.1 Attenuation

Attenuation is a decrease in the amount of radiation reaching the satellite due to absorption or scattering by the intervening medium. The result is that clouds appear cooler and subsequently higher in the atmosphere. This error is most pronounced in the infrared bands where the primary absorbers are water vapor, carbon dioxide, and ozone. Errors in temperature can be on the order of 10 K too cold, which is approximately a 1.5-km error in height. Attenuation increases with distance from the satellite subpoint and is the cause of the cooler temperatures at the extreme edges of a full-disk geostationary satellite image (referred to as *limb darkening*).

5.4.7.2 Background Contrast

When clouds or other phenomena have a radiative temperature that is close to that of the underlying surface, the analyst may not be able to distinguish them from the background. This source of error is widespread in radiometric measurements from space. The primary technique for accounting for this error is to sample the same scene in multiple spectral intervals (see Section 5.1).

5.4.7.3 Contamination

Contamination is caused by radiation reaching the sensor from beneath a cloud. This problem is most significant with thin clouds, such as cirrus. The additional radiation makes the cloud appear warmer and, therefore, lower in the atmosphere. This error is difficult to account for and, similar to background contrast errors, can most often be detected by sensing in multiple channels.

5.4.7.4 Displacement

Displacement occurs as a result of the viewing-angle geometry and the projection of the image data onto a 2D plane for display as an image or photograph. Figure 5.9 shows how the top of a cloud is displaced to a point away from the satellite subpoint. This error, discussed by Warren (1977) and Weiss (1978), can be corrected if the height of the cloud, the satellite position, and the coordinates of the cloud are known. As one would expect, displacement errors become more significant further from the satellite subpoint. Data from high latitudes collected by the GOES satellites are often difficult to use because of this error.

5.4.7.5 Foreshortening

Foreshortening is due to the effect of the Earth's curvature on the resolution of an image. As an instrument scans from the subsatellite point toward the limb of the Earth, pixels cover larger geographic areas and, with most instruments, are spaced further apart. Figure 5.17 demonstrates this error for a NOAA polar-orbiting satellite. For a geostationary satellite, foreshortening reduces the usefulness of the data beyond about 60° latitude from the subsatellite point. However, some remapping techniques have been developed to ameliorate this problem. The Operational Linescan System (OLS) on board the DMSP satellites has a special (nonlinear) scan mechanism that causes all pixels to be the same distance apart in the cross-track direction (compare Figs. 5.11 and 5.17). It does not change the increase in size of the pixels toward the limb, however.

5.4.7.6 Sensor Lag

Sensor lag produces an error in the radiative temperature assigned to each pixel. As the sensor is moved along a scan line, the sensor "remembers" the radiance it senses for a few microseconds. If L is the scene radiance, and if L_M is the radiance measured by the instrument, then (Hubert, 1974; Adler and Markus, 1982):

$$\frac{dL_M}{dt} = \frac{1}{\tau}(L - L_M) \tag{5.1}$$

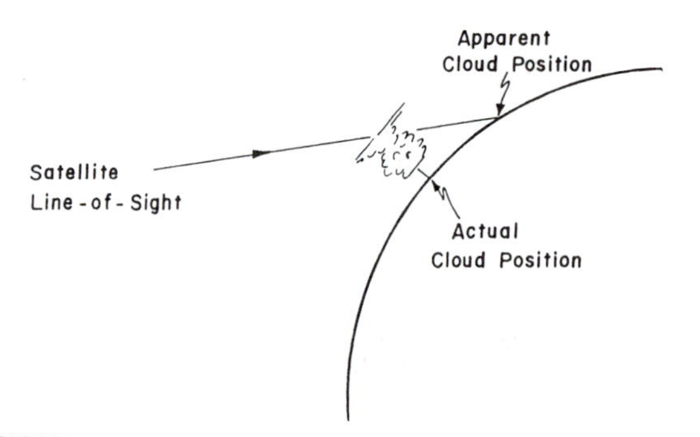

FIGURE 5.9. Displacement due to cloud height. [After Weiss (1978).]

where τ is called the *time constant*, and is about equal to 6 μs for the VISSR or VAS instruments according to Adler and Markus (1982). Since the VAS samples an IR line every 8 μs, a rough approximation is that the output of the sensor is 75% of the radiance from the current pixel, 20% of the previous pixel, and 5% of the pixel before that one.

In addition, the IR sensors on the GOES satellites are over sampled; that is, the east—west distance between pixels is less than the IFOV. Therefore, the instrument senses a cloud, for example, *before* the center of the pixel actually reaches the cloud and *after* the center of the pixel has left the cloud. These two effects together result in a slight smoothing of the IR image, which slightly lessens peaks and increases valleys in the IR data along a scan line. These effects are important only when examining features which are of the same scale as the IFOV of the instrument.

5.4.7.7 Signal Interference

The satellite signal must be transmitted to the ground, and in some cases, such as the GOES satellite, it must be processed, sent back to the satellite, and then retransmitted to Earth stations. During this process there is a potential for interference of the transmitted signal. This interference normally shows up as missing lines of data (dropouts) or garbled data (see Fig. 5.10). Missing lines are easily detected and can be corrected for by a number of averaging schemes that assign values to each pixel in the missing line based on the value of neighboring pixels.

5.4.7.8 Viewing Angle

Viewing-angle errors manifest themselves in several ways. As a satellite samples at increasingly large zenith angles, it looks at the sides of clouds as well as the tops. This is a difficult error to correct. It causes the apparent fraction of the sky covered with clouds to increase, in a similar fashion to the situation for a ground-based observer. Another difficulty caused by view angle is that a larger geographic area is sampled at larger zenith angles, and the corresponding resolution is poorer. A satellite with a ground resolution of 1 km at satellite nadir may have a resolution of 6 km or more at the edge of a scan (see Table 4.3).

Viewing angle must be considered when merging imagery from different satellite platforms. Figure 5.10 shows a sector of an image that was created by merging visible images from Meteosat 3 and DMSP–5D F11. The "seam" between the two images in Fig. 5.10 runs from the lower left to upper center of the image, with the DMSP image to the left of the seam. At the seam, the distance from the subpoint of the Meteosat 3 image is about 7000 km, whereas the distance from the DMSP subpoint to the seam is less than 900 km. At this viewing angle, the resolution of the Meteosat is degraded so that the cloud elements appear "fuzzier" than those from the DMSP image.

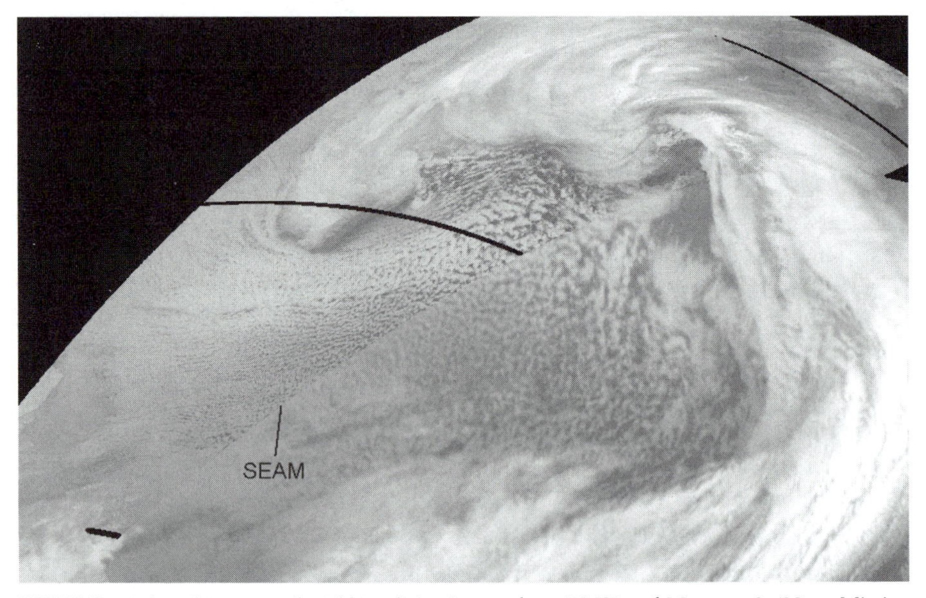

FIGURE 5.10. Infrared image produced by splicing images from DMSP and Meteosat 3. (*Note*: Missing scan lines on DMSP image are included to aid in locating the polar orbiter data.)

5.5 ATMOSPHERIC AND SURFACE PHENOMENA

With an introduction to the characteristics of meteorological satellite imagery and some of the sources for error, we can now look at examples of the phenomena that can be identified from the imagery. Several references are available that give detailed analysis techniques for identifying specific phenomena on a satellite image. These include *Applications of Meteorological Satellite Data in Analysis and Forecasting* (Anderson *et al.*, 1974), *Satellite Meteorology* (Brandli, 1976), *Navy Tactical Applications Guide. Volume I: Techniques and Applications of Image Analysis* (Fett and Mitchell, 1977), *Satellite Interpretation* (Weber and Wilderotter, 1981), *Nowcasting* (Browning, 1982), *Introduction to Weather Satellite Imagery* (Carlson, 1982), *Satellite Image Interpretation for Forecasters* (Parke, 1986), *Weather Satellites* (Rao *et al.*, 1990), and *Images in Weather Forecasting: A Practical Guide for Interpreting Satellite and Radar Imagery* (Bader *et al.*, 1995). In the following sections we list the features that are most commonly identified on meteorological satellite imagery. We refer the reader to the above references for a more detailed description of specific analysis techniques.

5.5.1 Clouds

Clouds are the single most informative feature on the satellite image, and in some cases the most difficult to analyze. The presence of clouds is often the only clue we have to a significant meteorological process occurring in the atmosphere.

Both the character and the extent of the clouds are important, as well as the specific location of the cloud mass. In some cases the absence of clouds is more significant than the occurrence of clouds. We look at some of the characteristics of clouds and examine how we can detect them. We also examine some of the properties of clouds that occasionally make them difficult to detect.

Clouds indicate the presence of moisture and some type of cooling mechanism. The most common cause of cooling is upward vertical motion (convection) that causes adiabatic expansion. Less common, but important for cloud formation, are radiative cooling, and the advection of warm, moist air over a cool surface or cold air over a warm, moist surface. The presence of clouds indicates that one of these mechanisms is, or has recently been, occurring. The type of cloud and its height present further clues about the process that is responsible for it.

We examine the characteristics of clouds in standard height categories. The commonly accepted layers are low, middle, and high clouds, with some clouds, such as cumulonimbus, spanning all three layers. In midlatitudes low clouds are typically found below 3 km, middle clouds between 3 and 7 km, and high clouds between 7 and 12 km, with cumulonimbus occasionally extending beyond 20 km. The placement of clouds in the three layers is normally done using infrared imagery to assign heights based on cloud top temperature, but some cloud types are best observed in visible imagery. We also identify some clouds that require the use of both the visible and infrared imagery for proper identification, and we look at the use of other channels, such as the water vapor absorption bands.

5.5.1.1 Low-level Clouds

The three main low cloud types—cumulus, stratus, and stratocumulus—each exhibit characteristics that allow us to identify them on a satellite image. Cumulus clouds most often appear as individual cells or clusters of cells that are on the order of 1 km in diameter, whereas stratus covers a large area, sometimes hundreds of kilometers in extent, and is normally smooth in texture. Stratocumulus is normally found in bands, called *streets*, or in large clusters. The tops of cumulus and stratocumulus also have a more "lumpy" texture than stratiform clouds.

Some low clouds are difficult to identify without the use of a visible image or knowledge of the underlying surface. Because low clouds are found within several kilometers of the surface of the Earth, they have a radiative temperature that is close to that of the underlying surface, making the detection of low clouds difficult on an infrared image (see Figs. 5.5 and 5.6). Because the infrared image is a display of radiative temperatures, the cloud and ground may appear nearly the same temperature on the infrared image (in some cases where a strong temperature inversion exists, the clouds may even appear warmer than the underlying surface). In the visible, however, the clouds most often appear bright because of the high albedo of the cloud tops compared to that of the land or water surface.

5.5.1.2 Middle-level Clouds

Characteristics of middle-level clouds are similar to those of low clouds, except that they are not as difficult to distinguish from the underlying surface on an infrared image because they are much colder than the surface (Fig. 5.6). Exceptions are extremely cold wintertime outbreaks over high and midlatitudes, and mountainous terrain where the surface reaches middle cloud altitudes. Middle clouds can also be either stratiform (altostratus) or cumuliform (altocumulus) and are used to infer processes that extend into the midtroposphere.

5.5.1.3 High-level Clouds

High clouds are composed exclusively of ice crystals, and some exhibit quite unique characteristics. The majority of the high clouds are stratiform in texture (cirrostatus) and of large extent. Some cirrus can be so thin that they do not appear on the visible image. In this case, however, the infrared signature is often distinct because the ice-crystal clouds are very cold and contrast with the warmer background (see Figs. 5.5 and 5.6).

5.5.2 Storms

Storms are some of the most dramatic things captured in satellite data. In this section we present images of several types of storms, but not by any means all types of storms.

5.5.2.1 Tropical Cyclones

With its unmistakable spiral shape and central eye, the tropical cyclone, including hurricanes and typhoons, is the most memorable feature on any satellite image (see Fig. 5.5). Indeed, if weather satellites detected nothing else besides these monster storms, they would be well worth the money invested in them. A number of techniques have been developed to estimate the movement and intensity of tropical cyclones from satellite images. One of the most widely accepted is the Dvorak technique (Dvorak, 1984), which assigns an intensity based on the size and shape of the dense cloud shield adjacent to the center of the storm (see Chapter 7). Loops of tropical cyclones are quite interesting. In high-resolution visible imagery, one can see low clouds spiraling inward and high clouds spiraling outward. As mentioned in Chapter 1, no tropical cyclone anywhere on Earth has gone undetected since weather satellites became operational in the mid-1960s.

5.5.2.2 Extratropical Cyclones

Perhaps the second most memorable thing in satellite imagery is the extratropical cyclone. The midlatitude cyclone model first described by Bjerknes in 1918 has been dramatically supported by weather satellites, from the cloud bands at warm and cold fronts to the swirl of clouds at the center of a mature cyclone. Figure 5.11 shows the clouds associated with a large midlatitude cyclone over the central United States. A mature cyclone over southeast Canada with a cold front trailing down the U.S. east coast can be clearly seen in Figs. 5.1 through 5.3. Evidence of a warm front can barely be seen at the limb of these images, but the cold high-pressure system over the central United States behind the cold front

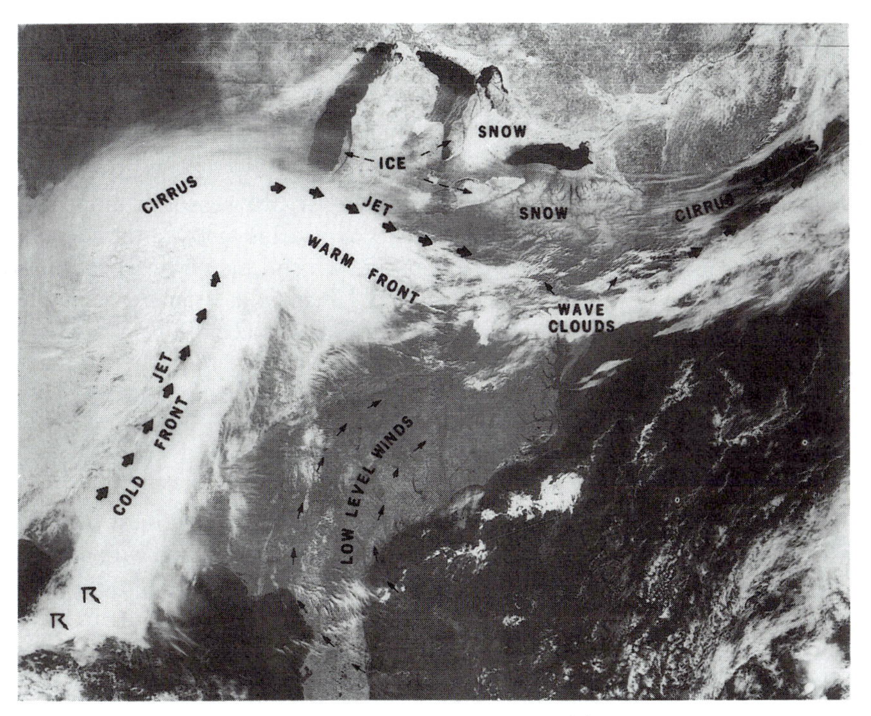

FIGURE 5.11. Midlatitude cyclone and other cloud features (see text).

is quite evident because of the clear weather it has produced. Weldon (1983) reports on the results of a large effort to identify the relationship between cloud patterns and frontogenesis, cyclogenesis, and wind fields in general.

5.5.2.3 Severe Thunderstorms

Severe thunderstorms and potential tornadic thunderstorms can be identified on satellite imagery by several interesting signatures. Severe thunderstorms quickly extend to the top of the troposphere or beyond and are identifiable on infrared images by very cold cloud tops and by rapid areal expansion of the anvil region (Plate 3). By comparing the temperature of the cloud top to the tropopause temperature we can determine whether the top has reached or penetrated the tropopause. These signatures are examined further in Chapter 9.

Lines of thunderstorms, called *squall lines*, often form at or in advance of a rapidly moving cold front, and can easily be detected on satellite imagery. Figure 5.11 shows the bright band of cloudiness associated with thunderstorms along the leading edge of the cold front over the Gulf of Mexico and northward. Satellite loops clearly show the movement of squall lines and most often show the movement of individual cells within the line. Plate 3, which shows the frames of a loop, captures a squall line forming in southern Illinois and central Indiana.

5.5.2.4 Lightning

Although not a storm, lightning is an indicator of thunderstorms. A special low-light nighttime visible sensor on board the DMSP satellite can detect lightning, even though its designers never envisioned that it would do so. The visible sensor on the OLS instrument was designed to be sensitive enough to detect clouds illuminated by moonlight. Because of its sensitivity, the sensor is momentarily saturated when a bright lightning flash occurs as it scans across a thunderstorm. This produces a bright streak along a scan line which lasts until the sensor falls below saturation a few pixels later. Figure 5.12 shows a dramatic example of lightning in an intense squall line over the Gulf of Mexico. In this example, city lights show up over the land, but the bright streaks over the water are clearly caused by lightning from the intense line of thunderstorms that appears as bright (cold) clouds on the corresponding infrared image.

5.5.2.5 Intertropical Convergence Zone (ITCZ)

The ITCZ is an easily recognizable, intermittent band of cloudiness that circles the Earth in the vicinity of the equator. The cloudiness is associated with numerous rain showers and thunderstorms formed by the convergence of the northeast and southeast Trade Winds. In Figs. 5.1 through 5.3 the ITCZ extends in an arch pattern from the central Pacific Ocean to South America. The ITCZ migrates north and south of the equator following the sun with the changing seasons.

5.5.3 Wind Flow

Wind flow can often be inferred by the orientation or character of cloud elements. In this section we discuss several qualitative methods that are in common use. In Chapter 7 we discuss cloud and vapor tracking by which quantitative wind vectors are derived.

5.5.3.1 Low-level Winds

Low-level cumulus clouds often line up in "cloud streets," which are parallel to the low-level wind direction. This pattern is noted in the warm sector of the cyclone in Fig. 5.11, and it can be seen in Fig. 5.15 as well. This pattern is more evident when images are looped.

5.5.3.2 Midlevel Winds

There are relatively few clouds at midlevels to indicate wind flow. Water-vapor imagery, however, shows the midlevels well. The patterns seen in water vapor imagery are nearly like streamlines and can be used to draw midlevel streamline analyses (see Fig. 5.3). Loops of water-vapor imagery are particularly informative.

5.5.3.3 Jet Stream and Jet Streaks

Several distinct cloud patterns are associated with the jet stream. The most common is a sharp edge in a long cirrus cloud shield (see Fig. 5.11). This edge is caused by the counterclockwise rotation in the direction of the jet stream, which

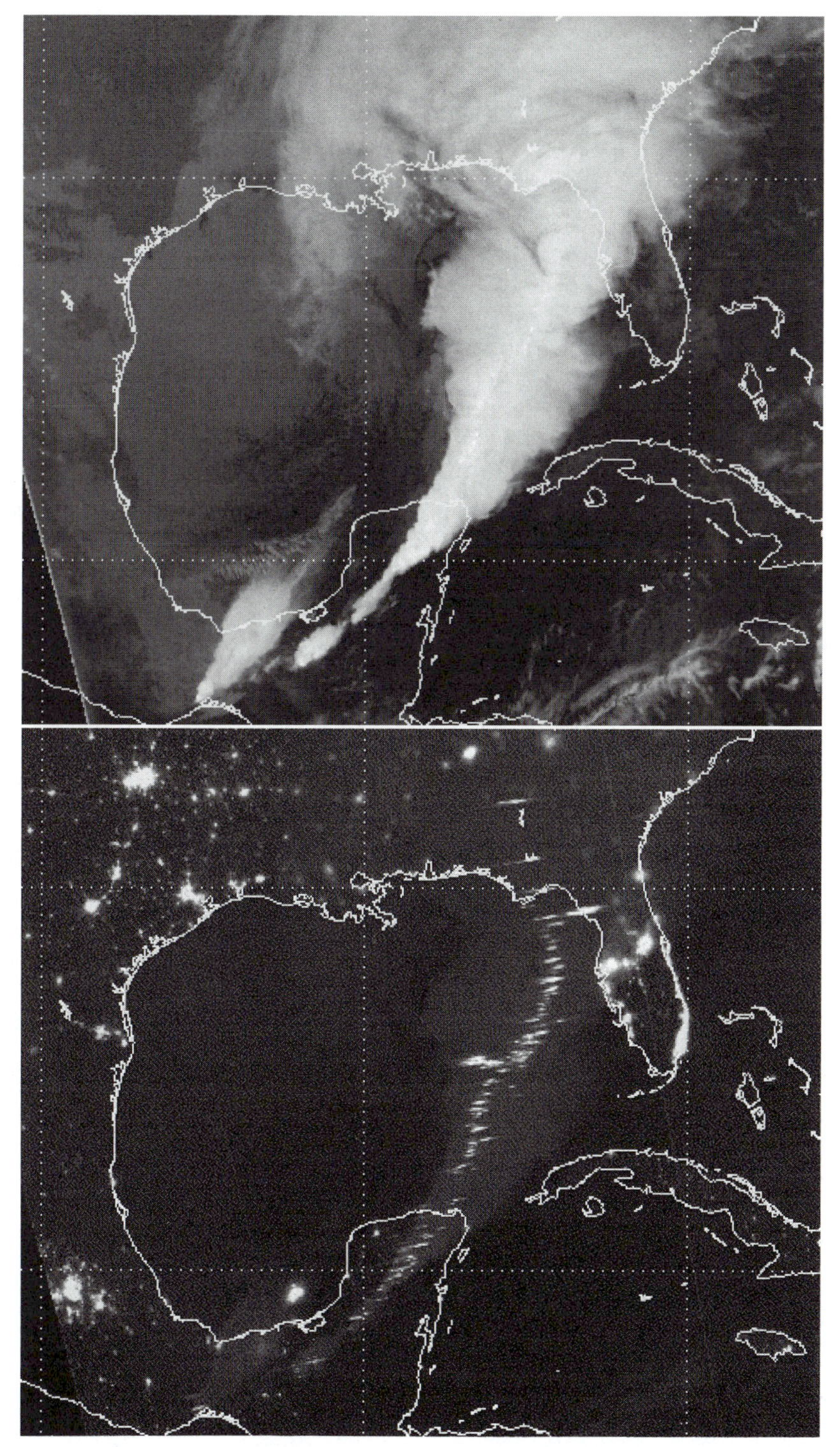

FIGURE 5.12. DMSP infrared image (top) of an intense squall line over the Gulf of Mexico, The companion DMSP low-light visible image (bottom) showing "lightning streaks" in the convective line. [Courtesy of NOAA/National Geophysical Data Center.]

causes upward motion and condensation to the equatorward side of the jet and subsiding air to the poleward side. The axis of the jet stream is at the cloud edge.

Jet streaks are small wind maxima that move through the large-scale circulation patterns. Because they are associated with localized upward vertical motion, they produce clouds that can be used to identify their location. Jet streaks often appear as a thickening of the jet stream cirrus in the vicinity of the maximum winds.

Transverse banding is also associated with the jet stream and is an indication of strong winds. Transverse bands form perpendicular to the axis of the jet stream and have been shown to indicate winds in excess of 40 m s^{-1} (80 kt) (Brandli, 1976).

5.5.3.4 Cyclonic and Anticyclonic Circulations

The detection of cyclonic or anticyclonic flow patterns is an integral part of the global circulation analyses. Most input for the analyses comes from the global rawinsonde sounding network. There are, however, large gaps in the network, especially over oceans. Satellites can be used to identify the location of troughs and ridges and circulation centers. Note the cyclonic circulation system in Figs. 5.1–5.3 off the west coast of South America. In most instances, watching cloud motion in a loop of successive images is the best way to determine the location of circulation centers. The reader is referred to the references listed at the beginning of Section 5.5 for the many techniques to determine cyclonic and anticyclonic circulations from satellite images.

5.5.3.5 Turbulence

Turbulence cannot be detected directly from satellite imagery; however, certain cloud patterns are unmistakably related to turbulence. The most common geographic location for turbulence-induced clouds is in mountainous terrain. The impingement of strong wind flow into a high mountain barrier frequently results in mountain wave turbulence and, if moisture conditions are right, in the formation of mountain wave clouds (Fig. 5.13). Wave clouds are associated with moderate to severe turbulence. Figure 5.14 shows a typical example of mountain wave turbulence over the Appalachians. Wave clouds normally form in bands that are parallel to the mountain barrier that induced them. The spacing between bands

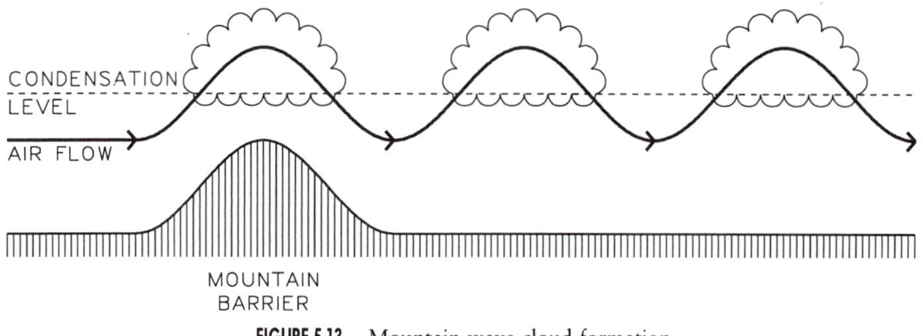

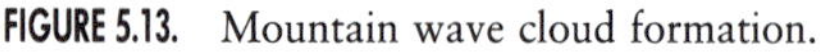

FIGURE 5.13. Mountain wave cloud formation.

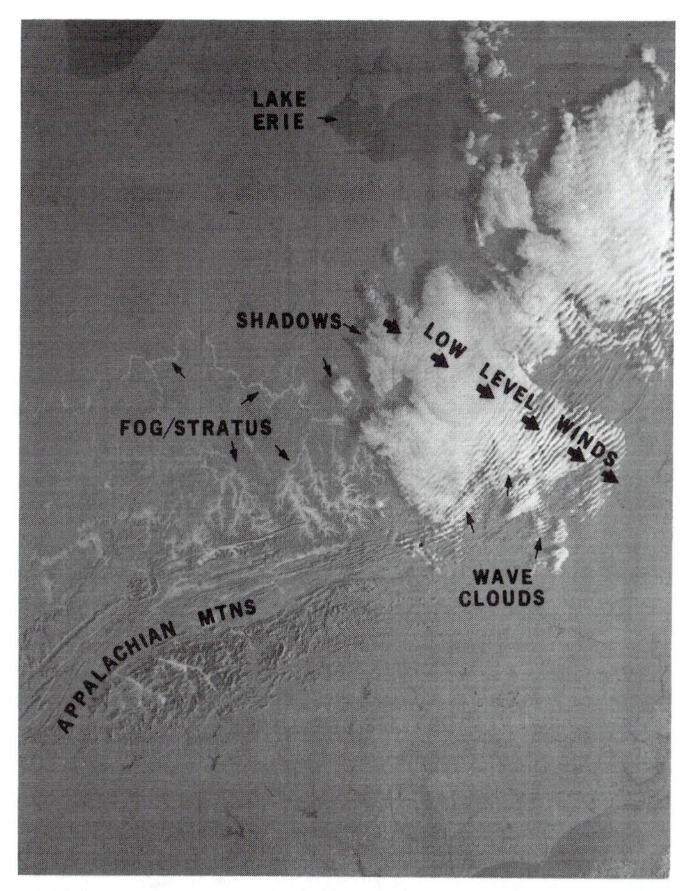

FIGURE 5.14. Visible DMSP image showing "mountain waves" that have formed downwind from the Appalachian Mountains. Also note the characteristic dendritic signature of the fog/stratus that has formed in the valleys, and the cloud shadows to the north and west of the cloud edges.

is linearly related to the normal component of the wind at the barrier, though different authors give different relationships (Brandli, 1976, p. 106; Weber and Wilderotter, 1981, p. 15).

Another form of wave cloud is the *billow cloud* (Ludlam, 1967; Brandli, 1976), which is similar in appearance to the wave cloud but is a high cloud and it is not associated with the topographic barrier. In fact, wave clouds can form when moist air is forced to flow over a cooler (more dense) air mass. The result is a sinusoidal wave pattern propagating downstream and forming clouds in the upper portion of the wave. Billow clouds, because they are high clouds, show up best on an infrared image.

5.5.3.6 Sea Breeze

Another cloud signature that is associated with a wind flow pattern is the sea breeze (or land breeze) cloud boundary. During daylight hours, sinking air suppresses clouds over the cooler water, while rising air forms clouds over the warmer land. Coastlines are therefore regularly outlined in clouds. Figure 5.15 shows an area along the east coast of Florida where sea breeze clouds have been pushed inland by prevailing northeasterly winds. A more localized "lake breeze" is also evident over and downwind of Lake Okeechobee and smaller lakes.

5.5.4 Miscellaneous Phenomena

A wide variety of meteorologically important phenomena in addition to those discussed above can be seen in weather satellite images. We present a few of them in this section.

FIGURE 5.15. Visible DMSP image showing the effect of the "sea breeze" along the east coast of Florida and "lake effect" over Lake Okeechobee.

5.5.4.1 Dust and Sand

Dust, when carried aloft by strong winds, can be detected on satellite imagery. Dust is most easily seen on the visible image when it is advected over a surface of relatively low reflectivity, such as an ocean (Fig 5.16). Dust, or blowing sand, is not as easily distinguishable over brighter regions, such as deserts, because it exhibits a reflectivity that is similar to the surface.

Dust that reaches high altitudes, such as volcanic ash, can be seen in infrared images. It appears warmer near the source of the volcano and, if the dust cloud remains dense enough, farther downstream it blocks the radiation from below the dust cloud and produces a cooler signature similar to a cloud. The presence of dust or sand in the air is an indicator of strong surface winds that are required to carry the dust into the atmosphere.

5.5.4.2 Sun Glint

Sun glint is a bright spot on the visible image caused by the sunlight reflecting from a smooth surface (Fig. 5.16). The signature is most often seen over the ocean, and the intensity of the spot indicates the character of the surface. If surface winds are calm, the ocean is smooth, the reflection is strong, and the spot is well defined. With stronger winds, the sea surface is rough, the reflection is more diffuse, and the spot is less well defined. Note the clouds that are present in the area of the sun glint by comparing the infrared and visible images in Fig 5.16. The bright sun obscures much of the cloud that is evident on the infrared image.

5.5.4.3 Smoke and Fire

Smoke is similar to dust or sand in its appearance on an image, except that it is normally localized and comes from a point source (Fig. 5.17). The most notable feature of smoke is that it appears warm on the infrared image and often has a distinct signature at night when the ground provides a cool background. Most small fires, such as a house or brush fire, are not seen because of the resolution of today's meteorological satellites. However, larger events such as a forest fire are often easily detected. Interestingly, the 3.7 μm band is much more sensitive to fires or hot objects than is the 11-μm band. Images at 3.7 μm have been used to locate steel mills and oil refineries (Kidder and Wu, 1987) and to count forest fires in the Amazon basin.

5.5.4.4 Snow and Ice

Snow and ice can make cloud discrimination difficult. In the visible image, snow and ice can have a reflectivity that is close to that of a cloud, whereas in the infrared the surface is close to that of a low cloud. The primary difference between cloud and snow seems almost too obvious; snow does not move. This simple fact is the basis for distinguishing snow from cloud by using successive images to detect clouds by their movement across a snow field. Even this technique is sometimes difficult because clouds do not always move, and the development or dissipation of clouds can sometimes negate the actual movement of a cloud.

FIGURE 5.16. Visible Meteosat 4 image (top) showing the distinctive signature of airborne dust off the coast of Africa. Also note the brighter sunglint region just southwest of the African coast. Note that the dust and glint are not discernible on the corresponding infrared image (bottom).

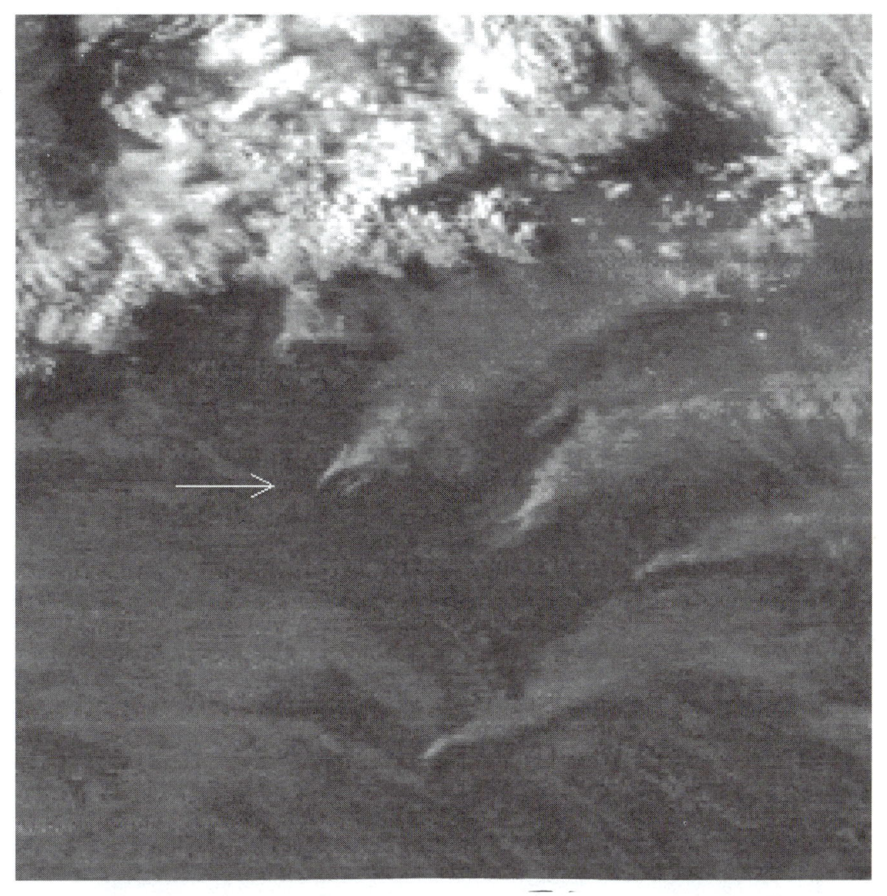

FIGURE 5.17. GOES image showing smoke from the Yellowstone fires of 1988.

Snow and ice can sometimes be identified by the fact that they conform to geographic features. Valleys, rivers, and lakes are often free of snow, whereas surrounding land surfaces are covered, making it easier to distinguish between clouds and snow. This is evident on Fig. 5.11 in the upper Ohio Valley. Also note the ice that covers a portion of Lake Michigan.

A technique using daytime 3.7-μm data has also shown success in discriminating low cloud from snow cover (Kidder and Wu, 1984). Figure 5.18 shows three images of the same scene made by the AVHRR. In the visible and 11-μm infrared images, the clouds in Nebraska and Kansas are nearly indistinguishable from the snow to the east. In the 3.7-μm image, however, there is dramatic contrast between the clouds and the snow due to reflection of sunlight by the clouds but not by the snow.

5.5.4.5 Sea-Surface Temperatures

Sea-surface temperatures, and more importantly temperature gradients, can be analyzed from infrared imagery. The measurement of surface temperatures, of

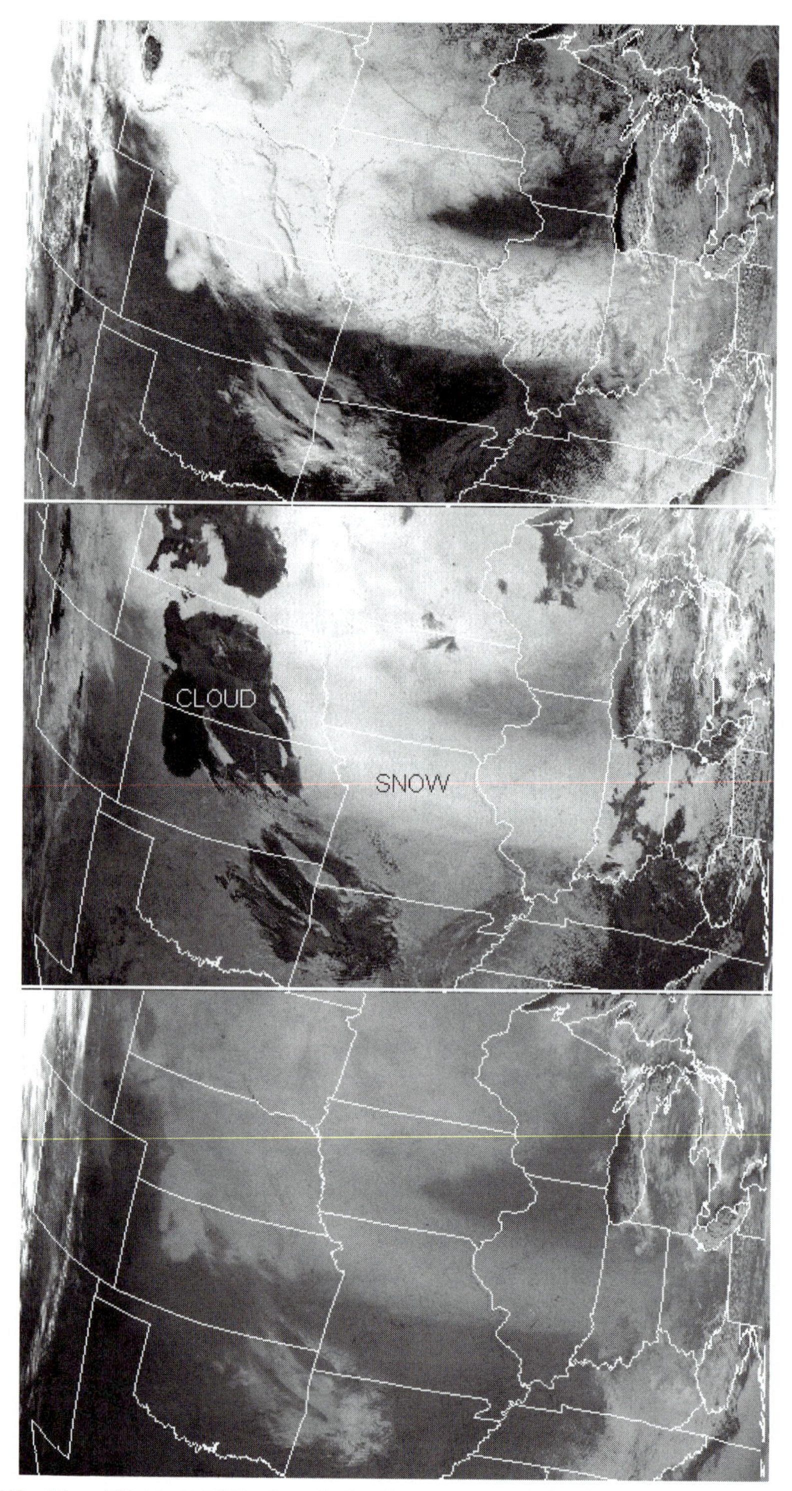

FIGURE 5.18. Three NOAA AVHRR channels for discrimination of cloud from snow: (top) channel 1–visible, (middle) channel 3–3.7 μm, and (bottom) channel 4—11 μm. [Adapted from Kidder and Wu (1984).]

course, requires a cloud-free area. Figure 5.19 is an example of sea-surface temperature gradients in an infrared image. In Chapter 6 we detail schemes for quantitatively retrieving sea-surface temperature, but images such as Fig. 5.19 are useful in themselves, for example, to locate fish.

5.5.4.6 Land and Water Features

The proper interpretation of satellite imagery requires a knowledge of the surface characteristics in the scene being analyzed. There are unique signatures that can be used to identify water, land, snow, ice, and various types of soil and vegetation. Characteristics of these surfaces affect the contribution of that surface to the radiation reaching the satellite. In addition, information about the elevation, latitude, and climate can be used to explain certain features on the image.

The boundary between land and water is important both for identification of clouds and for the proper navigation and gridding of images. Cloud types found over water are often different from those over an adjacent land mass. Land areas can exhibit sharp changes in elevation that are not possible over water. Also, the thermal characteristics of land and water produce varying background radiance properties over the diurnal cycle. The characteristics of land and water are different in each spectral channel and can also vary with the composition of the land and water surfaces.

The manual navigation and gridding of an image requires a recognizable geographic location. The most often used points are land/water boundaries such as coastlines, lakes, or rivers. These features can be striking on both visible and infrared imagery. The quality of gridding is quite evident when the geopolitical boundaries are placed over a coastline.

Water surfaces normally appear darker than land in the visible image. A sea surface with a high sun angle has an albedo of about 0.05–0.10, whereas land (soil) can have an albedo of 0.10 to as high as 0.35 over sand. Forests typically have albedos of 0.03 to 0.10. This contrast is not as great over a rough ocean surface where less of the solar radiation is reflected back toward the satellite.

A useful tool for the analysis of visible imagery is a *radiance background* image. Fig 5.20 is a visible radiance background image that was constructed by processing a month of visible images, each at 1800 UTC, and by selecting the darkest pixel at a given location for the month. This scene represents the image that would be viewed if there were no clouds present. This type of image can be used to determine the presence of clouds, because clouds will increase the brightness.

In the infrared, the difference in the land/water signature is a function of their temperatures. A cold water surface next to a warm land surface gives sharp contrast, whereas a cold water surface next to a cold land surface may be indistinguishable. In addition, water temperatures change more slowly than land, so the contrast may be good at one time of day and poor at a later time. For example, the contrast between Lake Michigan and the shoreline may be poor in the morning in the fall when they are both at nearly the same temperature, but sharp in the afternoon when the land warms.

FIGURE 5.19. GOES visible image (top) of the U.S. east coast (note cloud-free area over the ocean east of the coast) compared with the concurrent infrared image (bottom), which shows a sharp ocean temperature gradient in the vicinity of the gulf stream.

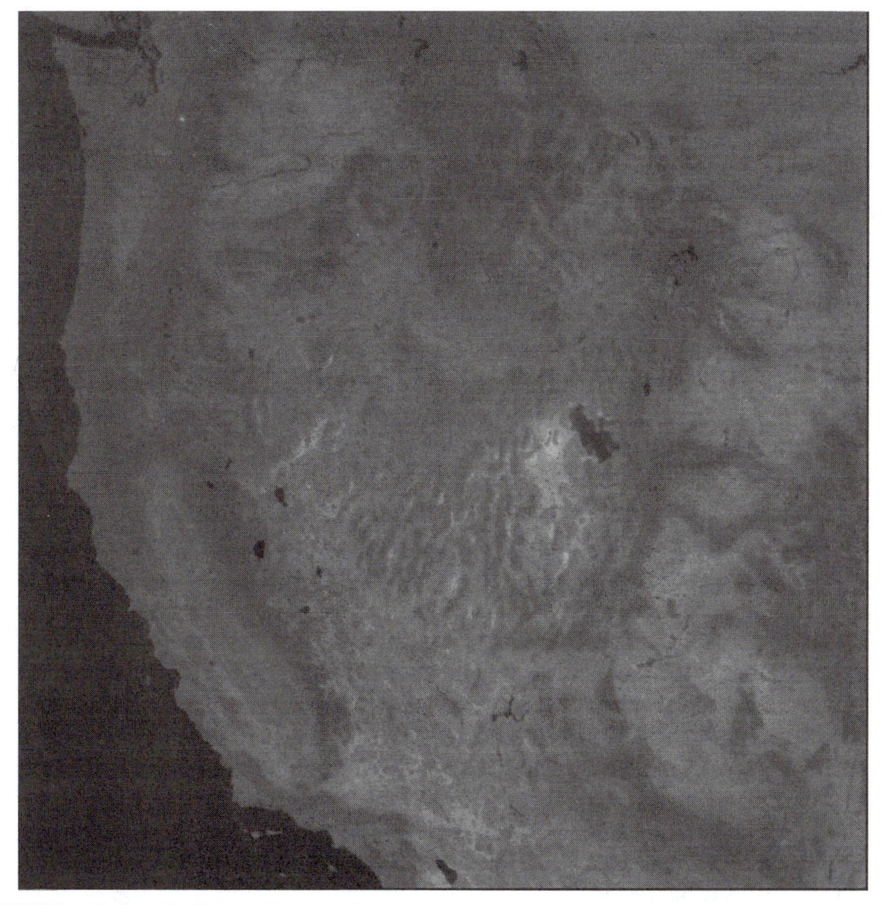

FIGURE 5.20. Visible "background" image that represents the cloud-free radiance for a portion of the northwestern United States at 1800 UTC during the month of July.

5.5.4.7 Sand, Soil, Vegetation

Sandy surfaces have a higher reflectivity than soil or vegetation in the visible channel. Some sandy regions, such as the White Sands in New Mexico, have a reflectivity that is close to snow and cloud. In the infrared, the sandy surface appears cooler at night and warmer during the day than a corresponding soil or vegetation surface.

Agricultural land exhibits variable reflectivity and emittance during different seasons of the year. When the fields are freshly plowed they normally appear darker than a few days later when the ground is dry. The field is normally even darker after crops have grown to cover the bare ground.

A useful tool is the Normalized Difference Vegetation Index (NDVI). It is constructed using visible and near infrared data from a land remote sensing satellite such as Landsat or from the AVHRR on the NOAA satellites. Basically, chlorophyll, the green substance that is responsible for photosynthesis, reflects near infrared radiation more than it reflects visible radiation. This signature is unique to green plants. If L_1 is the radiance in AVHRR channel 1 (visible) and L_2 is the radiance in channel 2, then NDVI is $(L_2 - L_1)/(L_2 + L_1)$. Normalizing by $L_1 + L_2$ removes the effect of varying sun angle and viewing angle. NDVI is high when green plants cover a pixel and low when no green plants are present.

5.6 A FINAL NOTE

In this chapter we have been able to present only a few, mostly black-and-white, images made with data from weather satellites. A new and inexpensive source of satellite imagery has become available in the last few years: images offered on the Internet via file-transfer protcol (ftp) servers such as Gopher or by hypertext-transfer protocol (http) servers such as Mosaic. We strongly encourage readers who have access to the Internet to explore some of the wide variety of still and animated satellite imagery which is available.

Bibliography

Adler, R. F., and M. J. Markus (1982). Determination of thunderstorm heights and intensities from GOES infrared data. *Preprints: Ninth Conf. on Weather Forecasting and Analysis.* American Meteorological Society, Boston, pp. 250–257.

Anderson, R. K., V. J. Oliver, E. W. Ferguson, F. E. Bittner, A. H. Smith, J. F. W. Purdom, G. R. Farr, and J. P. Ashman (1974). *Applications of Meteorological Satellite Data in Analysis and Forecasting,* ESSA Tech. Report NESC 51, U.S. Dept. Commerce, Washington, DC.

Bader, M. J., G. S. Forbes, J. R. Grant, R. B. E. Lilley, and A. J. Waters (eds.) (1995). *Images in Weather Forecasting: A Practical Guide for Interpreting Satellite and Radar Imagery.* Cambridge University Press, New York.

Brandli, H. W. (1976). *Satellite Meteorology,* AWS-TR-76-264, Hq. Air Weather Service, Scott AFB, IL, 188 pp.

Browning, K. A. (ed.) (1982). *Nowcasting,* Academic Press, San Diego.

Carlson, D. L. (1982). *Introduction to Weather Satellite Imagery.* NOAA Tech. Memo. NWS SR-103, NOAA, U.S. Dept. Commerce, 47 pp.

Dvorak, V. F. (1984). *Tropical Cyclone Intensity Analysis Using Satellite Data,* NOAA Tech. Rep. NESDIS 11, NOAA, U.S. Dept. Commerce.

Fett, R. W., and W. F. Mitchell (1977). *Navy Tactical Applications Guide. Volume I Techniques and Applications of Image Analysis.* Naval Environmental Prediction Research Facility, Monterey, CA.

Fett, R. W., P. E. La Violette, M. Nestor, J. W. Nickerson, and K. Rabe (1979). *Navy Tactical Applications Guide. Volume II Environmental Phenomena and Effects.* Naval Environmental Prediction Research Facility, Monterey, CA.

Hubert, L. F. (1974). *A Computation of the Time Constant of SMS IR Sensing System.* NESS Internal Memo., NOAA/NESS, Washington, DC.

Kelly, F. P., T. H. Vonder Haar, and P. W. Mielke, Jr. (1989). Imagery randomized block analysis (IRBA) applied to the verification of cloud edge detectors, *J. Atmos. Oceanic Tech.*, **6**, 671–679.

Kidder, S. Q., and H.-T. Wu (1984). Dramatic contrast between low clouds and snow cover in daytime 3.7 μm imagery. *Mon. Wea. Rev.*, **112**, 2345–2346.

Kidder, S. Q., and H.-T. Wu (1987). A multi-spectral study of the St. Louis area under snow-covered conditions using NOAA—7 AVHRR data. *Remote Sens. Environ.*, **22**, 158–172.

Ludlam, R. H. (1967). Characteristics of billow clouds and their relation to clear air turbulence. *Quart. J. Royal Met. Soc.*, **93**, 419–435.

Morel, P., M. Desbois and G. Szewach (1978). A new insight into the troposphere with the water vapor channel of Meteosat. *Bull. Am. Meteor. Soc.*, **59**, 711–714.

Parke, P. S. (ed.) (1986). *Satellite Imagery Interpretation for Forecasters.* Weather Service Forecasting Handbook No. 6, NOAA, Washington, DC.

Parmenter-Holt, F. C. (1982). Reliability of enhanced infrared (EIR) geostationary satellite data at high latitudes., *Mon. Wea. Rev.*, **110**, 1519–1523.

Rao, K. P., S. J. Holmes, R. K Anderson, J. S. Winston, and P. E. Lehor (eds.) (1990). *Weather Satellites: Systems, Data, And Environmental Applications.* American Meteorological Society, Boston, 503 pp.

Spencer, R. W. (1986). A satellite passive 37-GHz scattering-based method for measuring oceanic rain rates. *J. Climate Appl. Meteor.*, **25**, 754–766.

Spencer, R. W., H. M. Goodman, and R. E. Hood (1988). Precipitation retrieval over land and ocean with the SSM/I, Part 1: identification and characteristics of the scattering signal. *J. Atmos. and Oceanic Tech.*, **6**, 254–273.

Spencer, R. W., W. S. Olson, W. Rongzhang, D. W. Martin, J. A. Weinman, and D. A. Santek (1983) Heavy thunderstorms observed over land by the NIMBUS—7 Scanning Multichannel Microwave Radiometer. *J. Climate Appl. Meteor.*, **22**, 1041–1046.

Vonder Haar, T. H, D. L. Reinke, and S. Naqvi (1989). GEORGE—3D: A minicomputer-based system for 4-dimensional analysis and display of combined satellite, radar, terrain and map data. *Proceedings: Fifth Conference on Interactive Information and Processing Systems*, American Meteorological Society, Boston, pp. 22–25.

Warren, R. (1977) *Displacement Error of Satellite Cloud Tops*, NWS Technical Attach. No. 77-G4, SSD-CRH, 3 pp.

Weber, E. M., and S. Wilderotter (1981). *Satellite Interpretation.* Air Weather Service, 3rd Wea. Wing, Technical Note-81/001, Offutt AFB, NE, 95 pp.

Weiss, C. E. (1978). *Cloud-Location Corrections Near the Horizon of an SMS Image.* Satellite Applications Information Note 78/8, NOAA/NESDIS, U.S. Dept. Commerce, Washington, DC, 8 pp.

Weldon, R. (1983). *Synoptic Scale Cloud Systems.* NESDIS Training Notes, Satellite Applications Laboratory, NOAA, U.S. Dept. Commerce, Washington, DC, 35 pp.

Weldon, R., and S. Holmes (1984). *Characteristics of Water Vapor Imagery.* NESDIS Training Notes, Satellite Applications Laboratory, NOAA, U.S. Dept. Commerce, Washington, DC, 2 pp.

6

Temperature and Trace Gases

ONE OF THE most important applications of satellite measurements is atmospheric sounding, that is, retrieving vertical profiles of temperature and trace-gas concentrations, especially those of water vapor and ozone. In this chapter we discuss the theory, practice, and accuracy of such retrievals, and methods of assimilating these data into numerical models. In these retrievals, radiance from the surface is considered to be noise for which corrections must be made. We also discuss the opposite problem, that is, sensing surface parameters, such as land and sea surface temperature, in which the atmospheric contribution to the radiance is noise that must be eliminated.

6.1 SOUNDING THEORY

There are two basic types of passive atmospheric sounding: vertical sounding, in which the sounding instrument senses radiation coming from the atmosphere and the Earth's surface; and limb sounding, in which only the limb of the atmosphere is sensed. Both sounding methods utilize Schwarzchild's equation. Active atmospheric sounding includes the use of lidar and radar. In this chapter we focus on passive techniques, which are in use today. A discussion of active techniques is postponed to Chapter 11.

6.1.1 Vertical Sounding

Vertical sounding theory starts from the integrated form of Schwarzchild's equation (Eq. 3.38):

$$L_\lambda = L_o \exp\left(-\frac{\delta_o}{\mu}\right) + \int_0^{\delta_o} \exp\left(-\frac{(\delta_o - \delta_\lambda)}{\mu}\right) B_\lambda(T) \frac{d\delta_\lambda}{\mu}, \tag{6.1}$$

which applies to an atmosphere free of scattering. Although optical depth δ is a physically meaningful quantity, it is not necessarily optimal for retrieving soundings. Instead, we substitute h, a generalized height coordinate that can be replaced with height z, pressure p, $\ln(p)$, optical depth, transmittance, or any other function which is monotonic in height; h_o and h_T represent the Earth's surface and the top of the atmosphere, respectively. Recalling that $\tau_\lambda(\delta_1,\delta_2)$, the vertical transmittance between levels δ_1 and δ_2, is given by (Eq. 3.39)

$$\tau_\lambda(\delta_1,\delta_2) \equiv \exp(-|\delta_2 - \delta_1|), \tag{6.2}$$

we have, after some manipulation,

$$L_\lambda = L_o \tau_o^{1/\mu} + \int_{h_o}^{h_T} B_\lambda(T) W_\lambda(h,\mu) dh, \tag{6.3a}$$

$$W_\lambda(h,\mu) = \frac{d}{dh}(\tau_\lambda^{1/\mu}), \tag{6.3b}$$

where L_o is the radiance emerging from the Earth's surface, τ_o is the vertical transmittance from the surface to the satellite, and $\tau_\lambda^{1/\mu}$ is the slant-path transmittance from level h to the satellite. $W_\lambda(h,\mu)$ is called a *weighting function* because it weights $B_\lambda(T)$ in the atmospheric term. Of course, if the scene is overcast, an infrared sounder will see only down to the cloud top, in which case L_o and τ_o should be replaced with the cloud top values L_c and τ_c, respectively, and the atmospheric term should be integrated from h_c to h_T.

If the atmosphere is observed at a number of carefully chosen wavelengths, whose weighting functions sample the atmosphere, it is possible to retrieve $B_\lambda(T)$ (and thus T itself) as a function of h. This idea was first put forward by Kaplan (1959). Figure 6.1 shows τ_λ and W_λ for six channels in the 15-μm band of CO_2 used on the Vertical Temperature Profile Radiometer (VTPR) on the NOAA 2 through NOAA 5 satellites.

In the microwave portion of the spectrum, the weighting function must be modified slightly for soundings made over water. For infrared observations, and for microwave observations over dry ground, the emittance of the surface is approximately one, and there is very little reflection. The surface term is usually approximated as $B_\lambda(T_o)\tau_\lambda^{1/\mu}$, where T_o is the *skin temperature*, that is, the tempera-

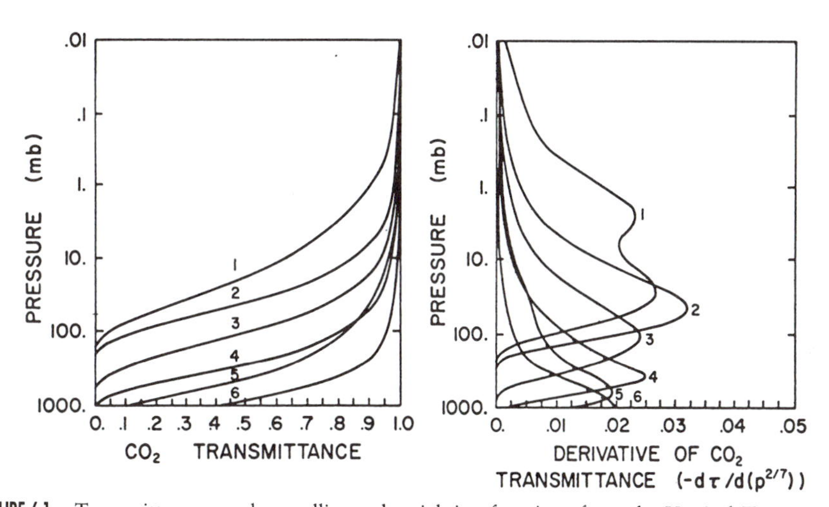

FIGURE 6.1. Transmittances to the satellite and weighting functions from the Vertical Temperature Profile Radiometer on the NOAA 2 through NOAA 5 satellites. In this figure the vertical coordinate is $h = -p^{2/7}$. Note that the weighting functions peak near the level where transmittance is approximately one. [Courtesy of Donald W. Hillger, NOAA/NESDIS.]

ture of the ground itself.[1] Water, however, reflects as much as 60% of the microwave radiation incident on it. In this case, the surface radiance, L_o, is composed of two terms, the radiation emitted by the surface and the reflected sky radiance:

$$L_o = \varepsilon_o B_\lambda(T_o) + (1 - \varepsilon_o) L_{sky}, \tag{6.4}$$

where the water surface has been approximated as a specular reflector with reflectance[2] $1 - \varepsilon_o$. The sky brightness may be determined by applying Schwarzchild's equation:

$$L_{sky} = B_\lambda(T_{space})\tau_o^{1/\mu} + \int_{h_o}^{h_T} B_\lambda(T) \frac{d}{dh}\left[-\left(\frac{\tau_o}{\tau_\lambda}\right)^{1/\mu}\right] dh, \tag{6.5}$$

where $(\tau_o/\tau_\lambda)^{1/\mu}$ is the slant-path transmittance from level h to the Earth's surface. T_{space} is the space temperature (approximately 2.7 K) caused by radiation left over from the Big Bang. Combining Eqs. 6.3 and 6.5 shows that the radiance measured from a satellite is

$$L_\lambda = \varepsilon_o B_\lambda(T_o)\tau_o^{1/\mu} + (1 - \varepsilon_o) B_\lambda(T_{space})\tau_o^{2/\mu} + \int_{h_o}^{h_T} B_\lambda(T) W_\lambda(h,\mu) dh, \tag{6.6}$$

[1] This is not the "surface" temperature as measured by a thermometer in a standard meteorological shelter.

[2] Over water, microwave radiation is polarized, and ε_o is a strong function of polarization (see Fig. 7.9). The following equations apply, but the appropriate value of ε_o, depending on the polarization of the radiometer, must be used.

where

$$W_\lambda(h,\mu) = \left[1 + (1 - \varepsilon_0)\left(\frac{\tau_0}{\tau_\lambda}\right)^{2/\mu}\right]\frac{d\tau_\lambda^{1/\mu}}{dh}. \tag{6.7}$$

This is a general weighting function that approaches Eq. 6.3b as ε_0 approaches one. The differences between these equations and Eqs. 6.3 arises from the fact that reflection affords the satellite the opportunity to see the atmosphere twice, once before reflection, and once after reflection. Figure 6.2 shows W_λ for the three temperature sounding channels of the Nimbus 6 Scanning Microwave Spectrometer (SCAMS).

We note that the radiances measured by satellite radiometers can be useful in themselves; they do not need to be inverted to supply information about the atmosphere. Spencer and Christy (1990), for example, have used brightness temperatures from the Microwave Sounding Unit (MSU) to assemble a global climatology of mean atmospheric temperature beginning in 1979. They used channel 2 of the MSU, which is insensitive to the surface and has a weighting function that peaks in the middle troposphere. The channel 2 brightness temperature, without further processing, is a mean temperature of the troposphere and lower stratosphere as expressed in Eq. 6.6. Since microwave radiation is nearly unaffected by clouds, and since the MSU is calibrated on every scan, the MSU climatology

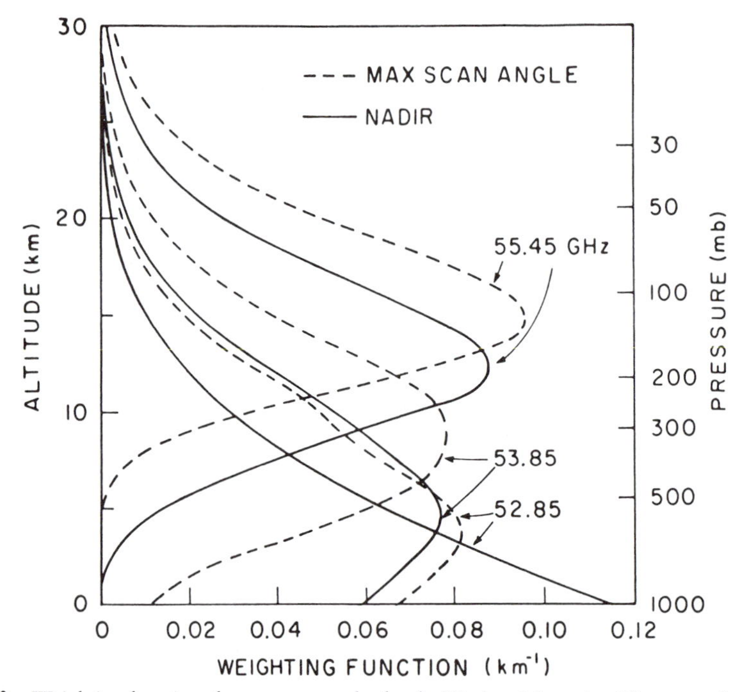

FIGURE 6.2. Weighting functions for two scan angles for the Nimbus 6 Scanning Microwave Spectrometer. [After Staelin *et al.* (1975).]

is potentially a very sensitive indicator of the warming or cooling of the global atmosphere and of changing patterns of atmospheric temperature.

Since all of the radiation sensed by the satellite comes from space, the Earth's surface, or the atmosphere, the weights for these three sources must sum to one. This results in a useful integral constraint on $W_\lambda(h,\mu)$:

$$\int_{h_o}^{h_T} W_\lambda(h,\mu)dh = 1 - \varepsilon_o\tau_o^{1/\mu} - (1 - \varepsilon_o)\tau_o^{2/\mu}. \tag{6.8}$$

As μ increases, the peak of the weighting function moves higher in the atmosphere. For wavelengths in which the weighting function peaks in the troposphere (which cools with height), radiance decreases as the instrument scans away from nadir. This is called *limb darkening*. The complementary effect, *limb brightening*, occurs when the weighting function peaks in a region of the atmosphere in which temperature increases with height. In the typical retrieval process, an empirical correction for limb darkening is applied to raw sounding data before retrieval; all of the scan spots are then treated as if they were at nadir (see Section 6.3.1.1). The original idea for sounding the atmosphere from satellites (King, 1958) employed limb darkening. King suggested that observations at a single wavelength, but at several zenith angles, could be used to retrieve atmospheric temperature profiles.[3]

Transmittance is a function of temperature and the mixing ratio of the absorbing gas. For well-mixed gases such as O_2 and CO_2, we assume that the mixing ratio is known and independent of height. Wavelengths that are sensitive to these gases are used to retrieve atmospheric temperature. For wavelengths sensitive to variable trace gases, such as O_3 and water vapor, we assume that the temperature profile is known and use the radiances to retrieve mixing ratios (or concentrations). Unfortunately, no wavelengths are completely free of absorption by the variable trace gases; this contributes to errors in retrieving temperatures. Also, errors in our knowledge of temperatures result in errors in trace-gas retrievals (see below).

The above equations are strictly applicable only at a single wavelength, but satellite instruments sense a band of wavelengths. The equations must be integrated over wavelength using the spectral response function of each channel as a weight. As discussed in Section 3.4, however, spectral integration is very time-consuming. In the microwave portion of the spectrum, where there are relatively few absorption lines, an *equivalent wavelength* is often employed. The equivalent wavelength is chosen near the center frequency of a satellite channel such that the spectrally integrated radiance sensed by the satellite is approximately the same as the radiance at the equivalent wavelength. When a transmittance must be calculated, a line-by-line calculation is done, but only at one wavelength, the equivalent wavelength.

In the infrared region, line-by-line calculations at even a single equivalent wavelength are too time-consuming. A recursive method is frequently employed

[3] Fleming (1980, 1982, 1985) suggested combining wavelength and zenith angle scanning. Soundings would be retrieved by the method used to invert CAT (computer-assisted tomography; X-ray) scans.

(Smith, 1969; McMillin and Fleming, 1976; Fleming and McMillin, 1977; McMillin *et al.*, 1979). The atmosphere is divided into *predetermined* layers, and the transmittance of each layer, averaged over the response function of the radiometer channel, is calculated line-by-line. In each layer, polynomial functions of temperature, humidity, and (sometimes) zenith angle are then fitted to the results. Given estimates of temperature and humidity (and zenith angle), the polynomial coefficients can be used to quickly calculate transmittance of any layer. The transmittance from a particular level to the satellite is then simply the product of the transmittances of the layers above. This method is fast, easy, and accurate (considering the experimental uncertainties in measurements of line parameters), but it does not let the user change the layers used in the calculations.

6.1.2 Limb Sounding

An important technique in satellite meteorology is *limb sounding* in which the sounder scans the Earth's limb (see Fig. 4.26). The above equations need to be modified slightly for limb sounding because radiation from the Earth's surface is not sensed. Consider a ray traveling in the positive x direction toward the satellite. The height above the surface of the ray at its closest approach to Earth is called the *tangent height*, z_T. The radiance measured by the satellite is given by

$$L_\lambda(z_T) = \int_{-\infty}^{\infty} B_\lambda(T) \frac{d\tau_\lambda(x,z_T)}{dx} dx. \tag{6.9}$$

This can be converted to a vertical integral:

$$L_\lambda(z_T) = \int_{z_T}^{\infty} B_\lambda(T) W_\lambda(z,z_T) dz, \tag{6.10}$$

where

$$W_\lambda(z,z_T) = \left[\left(\frac{d\tau_\lambda(x,z_T)}{dx} \right)_+ + \left(\frac{d\tau_\lambda(x,z_T)}{dx} \right)_- \right] \left| \frac{dx}{dz} \right|, \tag{6.11}$$

and the + and − subscripts indicate values along the positive and negative x axes, respectively.

Limb sounding has several advantages over vertical sounding (Gille and House, 1971):

- The weighting functions peak very sharply at the tangent height because the instrument senses nothing below the tangent height and atmospheric density decreases exponentially above the tangent height (Fig. 6.3).
- The background (space) is cold and uniform in contrast to vertical sounding in which the background (the surface) is hot and variable.
- There is up to 60 times more emitting material along a horizontal path than along a vertical path. This means that temperatures can be measured to higher altitudes than with vertical sounders, and low-concentration gases can be better detected.

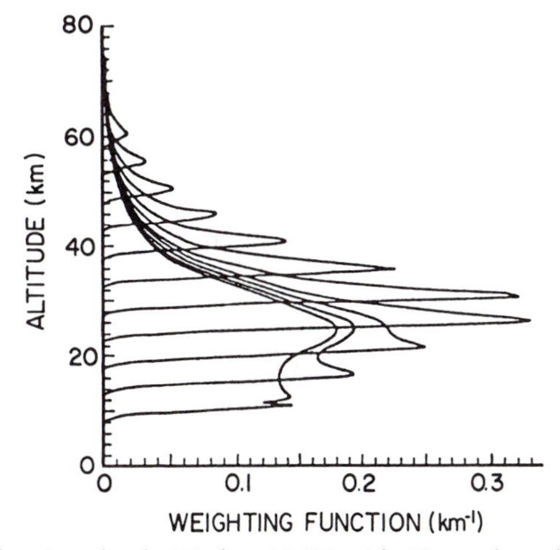

FIGURE 6.3. Weighting functions for the Nimbus 6 LRIR wide 15-μm channel used for temperature sounding. [After Gille *et al.* (1980).]

The disadvantages are

- Clouds and the finite FOV of the instrument (~2 km) limit limb sounding to the upper troposphere and above.
- The geometry limits the horizontal resolution of the soundings, although this is thought not to be a significant limitation in the stratosphere.

6.2 RETRIEVAL METHODS

The physical problem we would like to solve is this: What temperature and trace gas concentration profiles could have produced a set of observed radiances? This is called the *inverse problem* or *retrieval problem*. The opposite problem, called the *forward problem*, is to calculate outgoing radiances given temperature and trace-gas profiles.

Clouds cover approximately 50% of the Earth; thus about 50% of soundings are contaminated with clouds. We defer until Chapter 8 a discussion of methods to detect and correct for clouds in infrared sounding data. In the remainder of this chapter we assume that the radiances are "clear-column radiances" which have been corrected for clouds.

Given clear-column radiances, the forward problem is easy to solve, but the retrieval problem is difficult because the solution is not unique. Even if a noise-free radiometer that measured radiances at all wavelengths could be constructed, a unique solution to the radiative transfer equation would not be guaranteed (Chahine, 1970). When a finite number of wavelengths are observed and the measurements are contaminated with noise, an infinite number of solutions are possible. The retrieval problem becomes one of finding temperature profiles that satisfy the radiative transfer equation *and* approximate the true profiles as closely as possible. Approaches to the retrieval problem can be classified into three general areas:[4] physical retrievals, statistical retrievals, and hybrid retrievals.

6.2.1 Physical Retrievals

In *physical retrieval schemes* the ease of the forward problem is exploited in an iterative process as follows:

1. A first-guess temperature profile is chosen.
2. The weighting functions are calculated.
3. The forward problem is solved to yield estimates of the radiance in each channel of the radiometer.
4. If the computed radiances match the observed radiances within the noise level of the radiometer, the current profile is accepted as the solution.
5. If convergence has not been achieved, the current profile is adjusted (see below).
6. Steps 3 through 5 (or 2 through 5) are repeated until a solution is found.

The two most widely used methods of adjusting the temperature profile are those of Chahine (1970) and Smith (1970). Chahine retrieves temperatures for as many levels as there are channels in the radiometer. Suppose the radiometer has J channels; then the scheme retrieves the temperature at the J levels *located at the peak of the weighting functions.* In this scheme there is a one-to-one correspondence between channel j of the radiometer and level j, where the weighting function of channel j peaks. Let $T_j^{(n)}$ be the nth estimate of the temperature at the jth level and $B_j(T_j^{(n)})$ be the resultant Planck radiance at level j at the wavelength of channel j. Let $L_j^{(n)}$ be the nth estimate of the radiance in channel j, calculated using the $T_j^{(n)}$, and let $\tilde{L}_j$ be the observed radiance in channel j. Chahine iterates the temperature by iterating the Planck radiance:

$$B_j(T_j^{(n+1)}) = B_j(T_j^{(n)}) \left[\frac{\tilde{L}_j}{L_j^{(n)}} \right]. \tag{6.12}$$

The iterated temperature at level j is found using the inverse Planck function. This scheme works because if, for example, the calculated radiance in channel j is greater than the observed radiance, it is reasonable to adjust downward the Planck radiance (and thus the temperature) at the level where the weighting function for channel j peaks. Since the peak of the weighting function is the strongest contributor to the radiance, using the ratio of the observed to the calculated radiance to adjust the Planck radiance is also reasonable.

Smith's (1970) scheme for adjusting temperature profiles is similar to that of Chahine, but he relaxes the requirement that the temperature be retrieved at only

[4] This statement is an oversimplification. The reader is referred to Fritz *et al.* (1972), Deepak (1977, 1980), Houghton *et al.* (1984), Deepak *et al.* (1986), and the current literature for information on the many retrieval techniques.

J levels.[5] Suppose the temperature is to be retrieved at K levels. Let $T_k^{(n)}$ be the nth estimate of the temperature at level k, and $B_j(T_k^{(n)})$ the resulting Planck radiance at the wavelength of channel j. As above, let $L_j^{(n)}$ and $\tilde{L}_j$ be, respectively, the nth estimate of, and the observed value of, the radiance in channel j. At each level, Smith obtains J estimates of an iterated Planck radiance:

$$B_j(T_{jk}^{(n+1)}) = B_j(T_k^{(n)}) + [\tilde{L}_j - L_j^{(n)}], \tag{6.13}$$

where the J estimates of $T_k^{(n+1)}$, obtained by the inverse Planck function, are denoted by $T_{jk}^{(n+1)}$. Not all of these J estimates are equally good because each channel sees some levels better than others. Smith solves for $T_k^{(n+1)}$ as a weighted average of the $T_{jk}^{(n+1)}$ using the weighting functions as weights:

$$T_k^{(n+1)} = \frac{\sum_{j=1}^{J} W_{jk} T_{jk}^{(n+1)}}{\sum_{j=1}^{J} W_{jk}}. \tag{6.14}$$

Since Smith's levels are not restricted to be at the peaks of the weighting functions, he cannot iterate the temperature at a level by using a single channel. At each level, therefore, he obtains a suggested temperature change from each channel and lets the weighting function discriminate among them.

Smith's scheme is more flexible than that of Chahine in that it allows the user to choose the levels at which he will retrieve temperatures (consistent, of course, with the predetermined levels at which the transmittances are calculated). However, from J channels of information, one can calculate independent temperatures at J levels at most. If K is greater than J (as is usually the case) the extra levels are not independent.

To retrieve moisture profiles,[6] it is assumed that the temperature profile, and thus $B_\lambda(T)$, is known. Smith (1970) starts with a first-guess mixing ratio profile $q(p)$ and proceeds in a manner which is formally the same as that for temperature retrievals. The iteration formula is

$$q_{jk}^{(n+1)} = q_k^{(n)} \left[1 + \frac{\tilde{L}_j - L_j^{(n)}}{\Gamma_j^{(n)}} \right], \tag{6.15}$$

where $\Gamma_j^{(n)}$ is a sensitivity factor which estimates the mixing ratio change necessary to correct for a given radiance imbalance. In equation form,

$$\Gamma_j^{(n)} = \int_0^{p_s} U^{(n)}(p) \frac{\partial \tau}{\partial U} \frac{\partial B_j}{\partial T} \frac{dT}{dp} \, dp, \tag{6.16a}$$

$$U^{(n)}(p) = \int_0^{p} q(p) \frac{dp}{g}, \tag{6.16b}$$

[5] Twomey *et al.* (1977) introduced a modification of Chahine's (1970) scheme that also relaxes this requirement (see Section 8.2.1).

[6] An important scheme to retrieve integrated water vapor is discussed in Section 6.6.2.

where $U^{(n)}(p)$ is the nth estimate of the integrated water vapor above pressure level p. For each level, Smith obtains J estimates of the iterated mixing ratio. As above, $q_k^{(n+1)}$ is found as a weighted average of the $q_{jk}^{(n+1)}$ using the weighting functions[7] as weights.

The logic of this scheme can be demonstrated as follows. Note first that since $\partial\tau_j/\partial U$ is always less than zero, the sign of $\Gamma_j^{(n)}$ is opposite that of dT/dp. Consider a channel whose weighting function peaks in the troposphere. Suppose that the calculated radiance in this channel is too high compared with the observed radiance. Since $\Gamma_j^{(n)}$ is a negative number (because $dT/dp > 0$), $q_k^{(n)}$ will be multiplied by a number greater than one, and $q_k^{(n+1)}$ will be greater than $q_k^{(n)}$. Increasing the moisture will move up in the atmosphere the point where the optical depth equals one, and thus the peak of the weighting function for this channel will move up. Since tropospheric temperature decreases with height, $L_j^{(n+1)}$ will be less than $L_j^{(n)}$ and closer to the observed radiance.

Obviously the first-guess profile is important. The closer the first-guess profile is to the actual profile, the better the solution is likely to be. In the past, the first guess has been climatological. A better choice is a nearby radiosonde, but radiosonde releases do not often coincide with satellite overpasses, and few radiosondes are released over the ocean. Forecast soundings generated by numerical models are also being used. Recently, a great deal of work has been done on using the radiances themselves to classify the sounding into a set that has an appropriate first-guess profile and (for the methods listed below) a retrieval matrix (Uddstrom and Wark, 1985; Thompson *et al.*, 1985; McMillin, 1986).

Also obvious is that whatever auxiliary information can be brought to bear on the subject will be helpful. The skin temperature, for example, is an important factor, particularly in moisture retrievals. Sounders usually have window channels which are used in part to measure the skin temperature. Some analysis schemes attempt to incorporate surface observations of air temperature and dewpoint to improve retrievals in the lowest atmospheric layer, in which some of the largest retrieval errors occur. More experimental are attempts to utilize radiosonde-estimated tropopause heights in retrievals (McMillin, 1985, 1986).

None of the retrieval schemes that we discuss is clearly superior to the others. Users who wish to implement a retrieval scheme must weigh the advantages and disadvantages of each. Based on our experience, the principal advantages of the physical retrieval method are

- Physical processes are clearly evident at each stage of the retrieval.[8]
- No large database of coincident radiosonde data is necessary.

The disadvantages are

- The method is computationally intensive.
- It requires accurate knowledge of the transmittances.

[7] Actually, Smith uses a modified weighting function for moisture retrievals.

[8] This has the important implication that changes for such things as elevated terrain or for a failed radiometer channel can be easily implemented.

- Except for information contained in the first guess, it does not utilize known statistical properties of the atmosphere.

6.2.2 Statistical Retrievals

In statistical retrievals, the radiative transfer equation is not directly used. These methods assume that the radiometer has been designed so that its channels (weighting functions) will vertically sample the atmosphere. A set of radiosonde soundings that are nearly coincident in time and space with satellite soundings is compiled. This set, called the *training data set*, is used to calculate a statistical relationship between observed radiances and atmospheric temperatures. These relationships are then applied to other observed radiances to retrieve temperatures. The process can be described as follows.

Suppose there are N sounding pairs in the training data set. Let $\mathbf{l}$ be the $J \times 1$ column vector containing the J radiances observed by the radiometer in one of these soundings. Let $\mathbf{t}$ be the $K \times 1$ column vector, paired with $\mathbf{l}$, which contains the temperatures (and perhaps dewpoints) at K levels in the atmosphere. Finally, let the symbol $\langle \mathbf{x} \rangle$ represent the element-by-element average of the vector $\mathbf{x}$. Statistical retrieval methods require finding the $K \times J$ matrix $\mathbf{C}$ and the vectors $\langle \mathbf{l} \rangle$ and $\langle \mathbf{t} \rangle$ such that

$$\mathbf{t} - \langle \mathbf{t} \rangle = \mathbf{C}(\mathbf{l} - \langle \mathbf{l} \rangle). \tag{6.17}$$

Note that once $\mathbf{C}$, $\langle \mathbf{t} \rangle$, and $\langle \mathbf{l} \rangle$ are known, retrieval of temperatures from observed radiances is a very simple task which involves only vector subtraction, matrix multiplication, and vector addition. This simplicity makes a statistical method attractive for operational retrieval schemes in which numerous soundings must be processed.

There are several ways to find $\mathbf{C}$, $\langle \mathbf{t} \rangle$, and $\langle \mathbf{l} \rangle$. The simplest is the regression solution. Let $\langle \mathbf{t} \rangle$ and $\langle \mathbf{l} \rangle$ be the average of all the vectors $\mathbf{t}$ and $\mathbf{l}$, respectively, in the training data set. Let $\mathbf{T}'$ be the $K \times N$ matrix whose columns are the vectors $\mathbf{t} - \langle \mathbf{t} \rangle$, and let $\mathbf{L}'$ be the $J \times N$ matrix whose columns are the vectors $\mathbf{l} - \langle \mathbf{l} \rangle$. The matrix $\mathbf{C}$ which, in a least-squares sense, minimizes errors in Eq. 6.17 is

$$\mathbf{C} = \mathbf{T}'\mathbf{L}'^{\mathrm{T}}(\mathbf{L}'\mathbf{L}'^{\mathrm{T}})^{-1}, \tag{6.18}$$

where the superscript T indicates that the transpose of the matrix is to be used, and the superscript -1 indicates the matrix inverse. This scheme has been used extensively to retrieve satellite soundings.

One problem with the regression solution is that no "filtering" of noise from the input temperatures or radiances is done. As a result, the $\mathbf{C}$ matrix can be unstable; that is, small radiance errors can produce unacceptably large errors in the retrieved temperatures. Smith and Woolf (1976) developed a technique to filter noise using statistical eigenvectors. Their scheme was used operationally for many years. More recently, Thompson (1992) employed singular value decomposition to improve statistical retrievals.

Just as a good first-guess profile and an accurate knowledge of the transmittances are crucial to the success of the physical retrieval method, so acquiring a representative training data set is crucial to the success of the statistical method. This is more difficult than it might appear. Some of the requirements for these data sets are the following:

- The data sets must be large to ensure that the retrieval matrix will be stable.
- Data sets must be collected for each satellite because of small differences in the radiometers on each.
- The data sets must be updated frequently because, to site an extreme example, a data set collected in the winter cannot be expected to yield accurate retrievals in the summer.
- The data set should be divided into latitude zones, and by surface type (land or ocean).

The last three requirements all tend to decrease the size of the data set, and the last one is particularly troublesome. Of course, few radiosondes are launched over the ocean. Also, because standard radiosondes are launched only near 0000 and 1200 Universal Time, a satellite and a radiosonde will simultaneously observe an area only occasionally. Worse, sunsynchronous satellites view the same area at nearly the same time every day; thus satellite–radiosonde coincidences tend to occur only in narrow longitude zones. To increase the size of the training data set, the spatial and temporal restrictions on which soundings can be accepted as coincident are broadened. Typically, radiosonde soundings within a few hours and a few hundred kilometers of a satellite sounding are considered coincident. This broadened definition introduces the possibility that real atmospheric differences will contaminate the training data set.

Another problem with the statistical retrieval method is that physical processes are embedded in the statistics and are not easily changed. An example of this concerns retrievals over elevated terrain. Relatively few radiosondes are launched over elevated terrain; it is not possible to construct a training data set for elevated locations. The lowest level at which temperature is retrieved in most retrieval schemes is 1000 hPa. If the surface pressure is lower than 1000 hPa, the standard C matrix and mean vectors are applied, then a simple correction is attempted. This approach is likely, however, to introduce biases at the lowest levels.

A clear advantage of the statistical method is that a statistical picture of the structure of the atmosphere is an integral part of the method; that is, the retrieved temperatures cannot be too far from those which have actually been observed in the past. This imposes an even greater requirement that the training data set be appropriately chosen, however; otherwise the *wrong* structure will be built into the retrievals.

To summarize, the principal advantages of the statistical method are

- The actual retrievals are computationally easy.
- It requires no knowledge of the transmittances or use of the radiative transfer equation.

- It extensively utilizes statistical properties of the atmosphere.

The disadvantages of the statistical retrieval method are

- A large training data set of coincident radiosonde and satellite data is necessary.
- Physical processes are embedded in the statistics.

6.2.3 Hybrid Retrievals

Hybrid retrieval methods are in between physical and statistical retrieval methods. They appear much like purely statistical methods, but they do not require the large training data set. They use weighting functions like physical retrievals, but they do not directly involve integration of the radiative transfer equation. Hybrid methods, better known as *inverse matrix methods*, are reviewed by Fritz *et al.* (1972).

To start, the radiative transfer equation is linearized about a standard temperature profile $T_s(h)$:

$$L_\lambda = L_{\lambda s} + \varepsilon_o \frac{\partial B_\lambda}{\partial T} T'_{so}\tau_o + \int_{h_o}^{h_T} W_\lambda(h) \frac{\partial B_\lambda}{\partial T} T'_s(h)dh, \tag{6.19}$$

where $L_{\lambda s}$ is the radiance if $T_s(h)$ were the actual profile, and $T'_s(h)$ is the deviation from T_s at level h. A quadrature (numerical integration) scheme is chosen, which converts Eq. 6.19 into a matrix equation:

$$\mathbf{l}' = \mathbf{A}\mathbf{t}' + \mathbf{e}, \tag{6.20}$$

where $\mathbf{l}'$ is the column vector of radiance deviations from the standard profile, $\mathbf{t}'$ is the vector of temperature deviations from $T_s(h)$, and $\mathbf{e}$ is a column vector that contains the errors of measurement of the radiances. The matrix $\mathbf{A}$ contains the weighting functions, the quadrature weights, and the Planck sensitivity factors $\partial B_\lambda/\partial T$. $\mathbf{A}$ is calculated with a knowledge of the transmittances as in the case of physical retrievals.

Suppose, temporarily, that we have a set of coincident radiosonde and satellite observations. Let the uppercase letters indicate matrices which have N columns for the N sounding pairs in the data set:

$$\mathbf{L}' = \mathbf{A}\mathbf{T}' + \mathbf{E}. \tag{6.21}$$

We seek a matrix C such that

$$\mathbf{T}' = \mathbf{C}\mathbf{L}'. \tag{6.22}$$

Substituting Eq. 6.21 in Eq. 6.18, we have

$$\mathbf{C} = \mathbf{T}'(\mathbf{A}\mathbf{T}' + \mathbf{E})^{\mathrm{T}}[(\mathbf{A}\mathbf{T}' + \mathbf{E})(\mathbf{A}\mathbf{T}' + \mathbf{E})^{\mathrm{T}}]^{-1}. \tag{6.23}$$

Expanding and using the matrix relationship $(\mathbf{AB})^{\mathrm{T}} = \mathbf{B}^{\mathrm{T}}\mathbf{A}^{\mathrm{T}}$,

$$\mathbf{C} = \mathbf{T}'(\mathbf{T}'^{\mathrm{T}}\mathbf{A}^{\mathrm{T}} + \mathbf{E}^{\mathrm{T}})[\mathbf{A}\mathbf{T}'\mathbf{T}'^{\mathrm{T}}\mathbf{A}^{\mathrm{T}} + \mathbf{A}\mathbf{T}'\mathbf{E}^{\mathrm{T}} + \mathbf{E}\mathbf{T}'^{\mathrm{T}}\mathbf{A}^{\mathrm{T}} + \mathbf{E}\mathbf{E}^{\mathrm{T}}]^{-1}. \tag{6.24}$$

Now, we assume that the measurement errors are uncorrelated with temperature deviations. Thus $\mathbf{T'E^T}$ and $\mathbf{ET'^T}$ are negligible in comparison with the other terms. We have, therefore,

$$\mathbf{C} = \mathbf{S_T A^T[A S_T A^T} + \mathbf{S_E}]^{-1}, \tag{6.25}$$

where $\mathbf{S_T}$ is the temperature covariance matrix $(\mathbf{T'T'^T}/N)$ and $\mathbf{S_E}$ is the radiance error covariance matrix $(\mathbf{EE^T}/N)$. This equation, called the *minimum variance method*, has been discussed by many people, including Strand and Westwater (1968) and Rodgers (1970, 1976). The interesting property of Eq. 6.25 is that all of the components can be determined separately: $\mathbf{A}$ can be calculated from the transmittances, $\mathbf{S_T}$ can be determined from a sample of radiosonde soundings, and $\mathbf{S_E}$ can be measured in the laboratory as part of instrument calibration. Thus we can drop the assumption that a data set of coincident radiosonde—satellite soundings has been collected.

A variant of the minimum variance method (due to, e.g., Foster, 1961; Twomey, 1963; Smith *et al.*, 1972) is the *minimum information method,* in which simplifying assumptions about the covariance matrices are made. If it is assumed that temperature at one level is uncorrelated with temperature at any other level, the temperature covariance matrix will be diagonal. If it is further assumed that the temperature variance is independent of level, then

$$\mathbf{S_T} = \sigma_T^2 \mathbf{I}, \tag{6.26}$$

where $\mathbf{I}$ is the identity matrix, and σ_T^2 is the variance of temperature. Similarly, if radiance error in one channel is uncorrelated with radiance error in any other channel, and if the variance of radiance error is independent of channel, then

$$\mathbf{S_E} = \sigma_E^2 \mathbf{I}. \tag{6.27}$$

Equation 6.25 reduces, then, to

$$\mathbf{C} = \mathbf{A^T[AA^T} + \gamma \mathbf{I}]^{-1}, \tag{6.28}$$

where $\gamma \equiv \sigma_E^2/\sigma_T^2$. In practice, γ is often used as a *tuning parameter*, that is, a number that is changed until the best retrievals are obtained. Since the covariance matrices are gone, one is free to use any first-guess $\langle \mathbf{t} \rangle$, rather than the mean of a sample of radiosonde soundings. Also, the method can be used in an iterative fashion to improve the solution. The minimum information method, however, does not utilize the statistical information on atmospheric structure contained in $\mathbf{S_T}$.

The main advantage of the hybrid methods is that they are easier to put into operation than the statistical or physical methods. They share with the physical retrieval method the disadvantage of depending on a knowledge of the transmittances. Most of them share with the statistical retrieval method the advantage of including statistical knowledge of atmospheric structure.

6.3 OPERATIONAL RETRIEVALS

The retrieval method is only a small part of the process by which radiances from operational satellites are converted into temperature and moisture soundings. In this section, we discuss the processes that are used operationally to produce soundings from NOAA and GOES satellites.

6.3.1 The TIROS N Operational Vertical Sounder

On the NOAA satellites there are three sounders: the HIRS/2, the MSU, and the SSU (see Chapter 4). Together they are called the TIROS N Operational Vertical Sounder (TOVS). Figure 6.4 shows the TOVS weighting functions.

The process by which TOVS soundings are retrieved is described by Smith *et al.* (1979a). The reader should be aware, however, that operational sounding retrieval is not a static process. As new techniques are developed, and as problems arise, the system is modified. These modifications can be determined by referring to a periodically updated NESDIS document, the *NOAA Polar Orbiter Data User's Guide.* (e.g., Kidwell, 1986).

The TOVS system consists of four modules: the Preprocessor, the Atmospheric Radiance Module, the Stratospheric Mapper, and the Retrieval Module.

6.3.1.1 The TOVS Preprocessor

The Preprocessor processes data received directly from the Satellite Operations Control Center. The digital counts are converted to radiances, and each scan spot is Earth-located. The SSU data are further processed by the Stratospheric Mapper. A variety of corrections are applied to the HIRS/2 and the MSU data:

- The MSU data are corrected for antenna side-lobes.
- A limb correction is applied to both the MSU and HIRS/2 data so that the other modules can treat the data as if they were all obtained at nadir. The limb correction consists of regression equations, formed from synthesized radiance data, which result in a correction to be *added* to the radiance (Smith *et al.*, 1974). If L_{jm} is the radiance in channel j at scan angle m, then the correction is given by

$$\Delta L_{jm} = \sum_{n=1}^{J} a_{jmn} L_{nm} + b_{jm}. \tag{6.29}$$

- An attempt is made to correct the window channels for water vapor absorption.
- For daytime data, an albedo, estimated using the 3.7-μm channel, is used to correct the 4.3-μm channels for reflected sunlight.
- Since the MSU has coarser spatial resolution than does the HIRS/2, the MSU data are interpolated to the locations of the HIRS/2 scan spots.

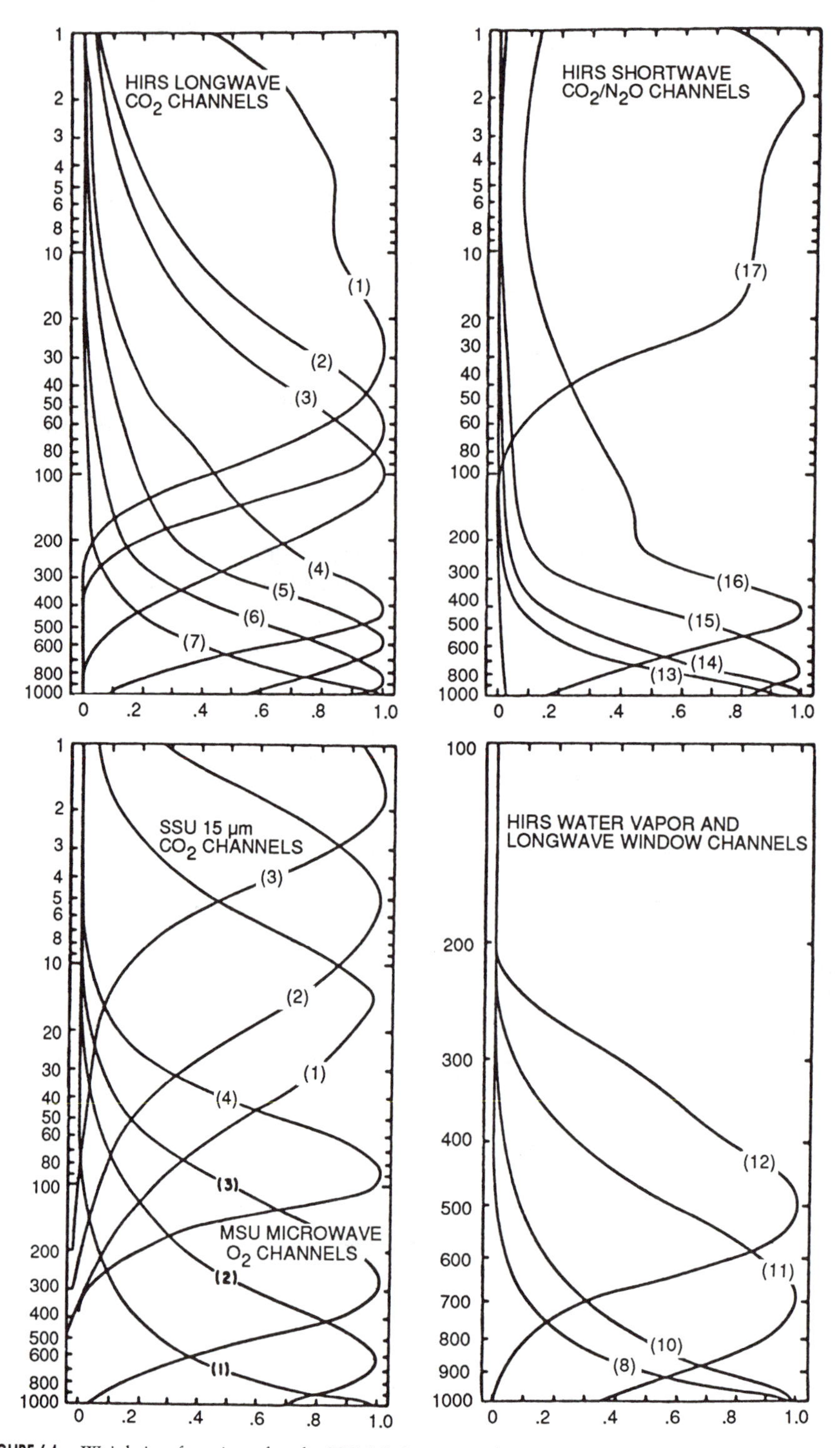

FIGURE 6.4. Weighting functions for the TIROS Operational Vertical Sounder (TOVS). [After Smith *et al.* (1979a).]

Finally, the Preprocessor obtains solar zenith angles, terrain elevations, and initial guess values of skin temperature and surface albedo. All these data are staged to disk to await further processing.

6.3.1.2 The TOVS Atmospheric Radiance Module

The Atmospheric Radiance Module's job is to deliver spatially averaged, clear-column radiances to the Retrieval Module. Perhaps surprisingly, the Atmospheric Radiance Module consumes most of the computer time in the TOVS system.

First, the data are divided into boxes of nine HIRS/2 scan spots cross-track and seven scan spots along-track. One sounding is retrieved to represent this group of 63 scan spots. The nominal resolution of the operationally retrieved TOVS soundings is approximately 250 km.

Next, the 63 scan spots are tested to determine whether they are contaminated by clouds. Several tests are employed:

- In daylight, the albedo measured by HIRS/2 channel 20 is compared with the albedo expected for a cloud-free scene. Clouds, of course, would increase the visible albedo.
- At night, channel 19 (3.7 μm) is less sensitive to clouds than channel 8 (11 μm). During the day, clouds reflect sunlight at 3.7 μm; therefore, channel 19 is much more sensitive to clouds than channel 8. A large difference in the brightness temperatures of these two channels is an indication of clouds.
- The 11-μm brightness temperature is compared with the expected surface temperature. Too low a brightness temperature indicates cloud contamination.
- Finally, the HIRS/2 brightness temperatures are used in regression equations to estimate the MSU brightness temperatures. If clouds contaminate the scan spot, the estimated MSU brightness temperatures will be less than the observed ones.

Finally, the Atmospheric Radiance Module produces clear-column radiances. If four or more of the 63 scan spots are determined to be clear, the clear-column radiances are calculated as a weighted average of the observed radiances from the clear spots. The weighting factor is related to how well the MSU channel 2 brightness temperature is predicted by the HIRS/2 data. Soundings produced from these clear-column radiances are called "clear" or "first path" soundings.

If fewer than four clear scan spots are found, the adjacent-pair or N^* technique of Smith and Woolf (1976) is attempted. This technique, which will be discussed in Chapter 8, assumes that scan spots are partly cloudy and attempts determine clear-column radiances by exploiting differences in cloud amount in adjacent scan spots. If four or more good adjacent pairs can be found, clear-column radiances are estimated. Soundings produced from these radiances are called "partly cloudy" or "second path" soundings.

If the scene is essentially completely overcast (fewer than four good adjacent pairs), the HIRS/2 channels which sense the troposphere are discarded, and the sounding is retrieved using only MSU, SSU, and stratospheric HIRS/2 channels. These soundings are called "cloudy" or "third path" soundings.

In 1986, a higher resolution scheme was implemented (Reale *et al.*, 1986). The 9 × 7 box of HIRS/2 scan spots was divided into six miniboxes, each 3 × 3 scan spots. A clear-column radiance vector is produced for each minibox. In order not to produce six times as many final soundings, each minibox is checked to determine quality and usefulness of the data. The quality flag may be (1) good, (2) redundant, (3) questionable, or (4) bad. "Redundant" means that the minibox has essentially the same radiance vector as its neighbors. In later processing, redundant, questionable, and bad miniboxes can be discarded to produce the appropriate number of soundings.

6.3.1.3 The TOVS Stratospheric Mapper

Because the SSU does not scan as far out as the HIRS/2 and the MSU, HIRS/2 scan spots at the edges of the scan have no corresponding SSU data. This problem is solved by the Stratospheric Mapper, which maintains a global map of SSU radiances on a latitude–longitude grid. The map is updated as new observations arrive. The SSU data are corrected for limb effects before mapping.

6.3.1.4 The TOVS Retrieval Module

The Retrieval Module performs the retrievals using clear-column radiances produced by the Atmospheric Radiance Module and the Stratospheric Mapper. Until September 1988, temperatures and water vapor below 100 hPa were retrieved using statistical eigenvectors (Smith and Woolf, 1976). Above 100 hPa, temperatures and total ozone were retrieved by regression. As of this writing, temperatures and water vapor are retrieved using the Minimum Variance Simultaneous (MVS) method [Eqs. 6.22 and 6.25; see also Fleming *et al.* (1986) and Dey *et al.* (1989)]. The method is "simultaneous" in the sense that temperature and water vapor are retrieved using one matrix.

6.3.1.5 Archived TOVS Data

TOVS sounding data are archived by NESDIS. Users who are interested in ordering historical data, however, should be aware that the archive tapes come in the form of layer-mean virtual temperatures for 15 layers between the surface and 0.4 hPa. Water vapor is archived in the form of layer precipitable water in the three lowest layers (surface–850, 850–700, and 700–500 hPa). The tropopause temperature and pressure, total ozone, cloud-top pressure, cloud amount, and an average value of N^* are saved (Kidwell, 1986).

6.3.2 The VISSR Atmospheric Sounder

VAS soundings have been retrieved operationally at NESDIS's VAS Data Utilization Center (VDUC) since the summer of 1987. The retrieval process, described by Hayden (1988), is similar to that used for TOVS soundings.

First, the raw VAS data are converted to brightness (equivalent blackbody) temperatures and Earth located. As with the TOVS retrieval system, scan spots are blocked before processing. The VAS system is designed to be flexible, so the number of spots in a block is variable. Typically, however, a block of 11 × 11 scans spots is used, which results in a field of retrieved soundings with approximately 75-km resolution.

Next, the data are filtered for clouds. Since visible data are not collected during dwell sounding, the cloud filter relies on three window infrared channels: 3.9, 11.2, and 12.7 μm (channels 12, 7, and 8):

- If the 3.9-μm brightness temperature is significantly different (3–5 K) from the 11.2-μm brightness temperature, the scan spot is assumed to be cloudy.
- If the 11.2-μm brightness temperature (corrected for moisture effects by the "split-window" technique, see below) is significantly colder than the surface air temperature (obtained from analyzed surface observations), the scan spot is classified as cloudy.

If a majority of the scan spots appear to be clear, the clear radiances are averaged, and a sounding is retrieved down to the surface. (During the retrieval process itself, the sounding can be classified as cloudy if the algorithm cools the first-guess surface temperature by more than 6 K.) If most of the scan spots are cloudy, the cloudy spots are averaged, and cloud-top pressure is estimated by comparing the 11.2-μm brightness temperature (corrected for moisture effects) with the first-guess temperature profile. A sounding is retrieved down to the cloud top, and values below cloud top are estimated by interpolating between the cloud-top value and the surface analysis. These cloudy soundings, however, are of questionable accuracy.

The retrieval algorithm is a hybrid scheme described by Smith *et al.* (1986). First-guess temperature and moisture profiles are selected, usually from a model forecast such as the Nested Grid Model (NGM) or the Global Spectral Model. Surface temperature and humidity are obtained from analyzed fields of the corresponding hourly observations. The first-guess estimate of the skin temperature is obtained by regression from the 11.2, 12.7, and (for noncloudy scan spots) 3.9-μm brightness temperatures. Temperatures above the 100-hPa level are obtained from the latest analyses or from climatology. Using the first-guess profiles, the weighting functions and the brightness temperatures in the VAS channels are calculated. The temperature and moisture profiles are calculated by making a single correction to the first-guess profiles, which is why the scheme is known as the "simultaneous" or "one-step" method.

6.3.3 Quality of the Retrieved Soundings

How accurate are soundings retrieved from satellite-measured radiances? Although a great deal of effort has been expended on this question, the answer is still not simple. The basic difficulty is that to assess accuracy, one needs a standard with which to compare. Radiosonde soundings are the most obvious candidates, but two problems occur:

- Radiosonde soundings are not error-free; an rms (root mean square) error of 1 K is often quoted.
- Satellite soundings are fundamentally different from radiosonde soundings.

This latter point is illustrated by example. Figure 6.5 shows two VAS soundings compared with radiosonde soundings. Clearly, the radiosonde soundings exhibit more vertical structure than the satellite soundings. The satellite sounding has trouble where the lapse rate changes abruptly with height, such as at surface inversions or near the tropopause. This is particularly true of the moisture sounding. An examination of the weighting functions (Fig. 6.6) reveals the cause of this difficulty. Current satellite sounders integrate deep layers of the atmosphere. Vertical smoothing of the sounding is unavoidable. Interferometric sounders currently under development (see Section 4.3.1) offer improved vertical resolution

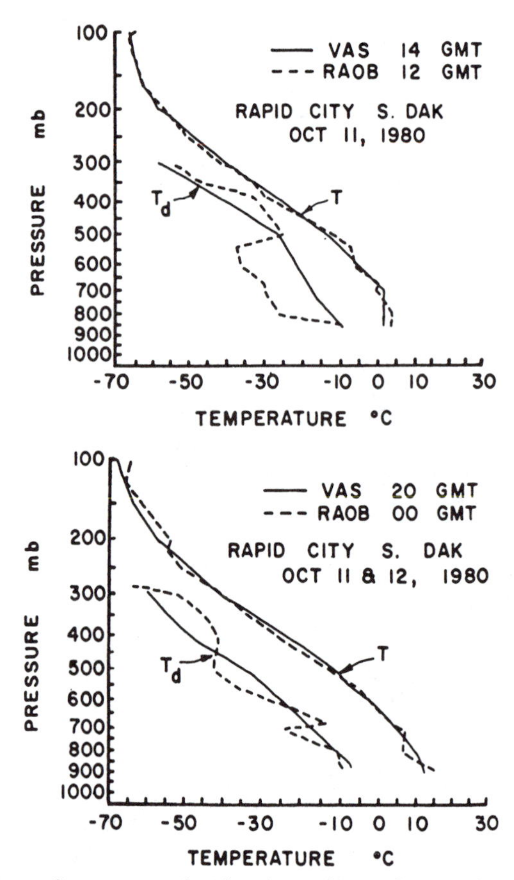

FIGURE 6.5. Two VAS soundings compared with radiosonde soundings. [After Smith *et al.* (1981).]

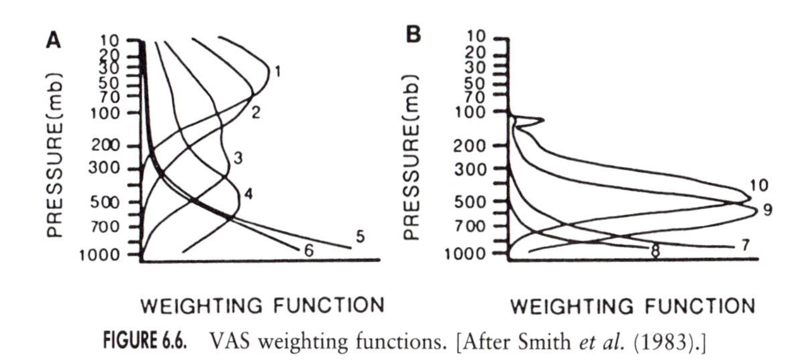

FIGURE 6.6. VAS weighting functions. [After Smith *et al.* (1983).]

(Smith *et al.*, 1979b). Additional improvement in vertical resolution likely will have to wait until active sensors, such as lidar, are feasible in space.

In this section, we discuss the quality of operational retrievals (eigenvector method) from TOVS, which have been extensively studied. The conclusions about TOVS retrievals presented below are representative of VAS retrievals as well.

Gruber and Watkins (1982) have paired a large number of operationally retrieved TIROS N and NOAA 6 soundings with time-interpolated radiosonde soundings. The rms difference between satellite and radiosonde layer-mean virtual temperatures for TIROS N soundings are presented in Fig. 6.7 along with seasonal climatologies of temperature standard deviation, which shows the natural variability of temperature. Since the rms difference is smaller than the natural variability, the satellite soundings explain a substantial fraction of the variance of temperature. It should be remembered that rms differences between satellite and radiosonde soundings overestimate the rms error in the satellite soundings because they contain the radiosonde error as well. Also, the rms difference is the square root of the sum of the squares of the bias (i.e., the mean difference) and the standard deviation. If biases (see below) are removed, rms errors decrease still further.

In general, rms differences between satellite and radiosonde soundings are greatest near the surface and near the tropopause, where lapse rates can vary sharply with height. Cloudy soundings have substantially greater rms differences than do clear soundings. Although rms differences do not vary much with latitude, the natural variability of atmospheric temperature is lower in the tropics than in midlatitudes. Thus satellite soundings explain a smaller fraction of temperature variance in the tropics than they do elsewhere.

Biases (satellite minus radiosonde) in layer-mean virtual temperatures are shown in Fig. 6.8. The biases tend to be small for clear soundings. For cloudy soundings, which rely on the microwave radiances for tropospheric temperatures, the biases are larger in magnitude and tend to be negative (satellite temperatures too cool) in the middle troposphere. The reader should note that apparent seasonal differences in the biases shown in Fig. 6.8 are due partly to real seasonal and latitudinal differences, partly to physical differences between satellite soundings and radiosonde soundings, and partly to changes in the retrieval system. Phillips *et al.* (1979) were the first to report the cool bias in cloudy soundings, and a

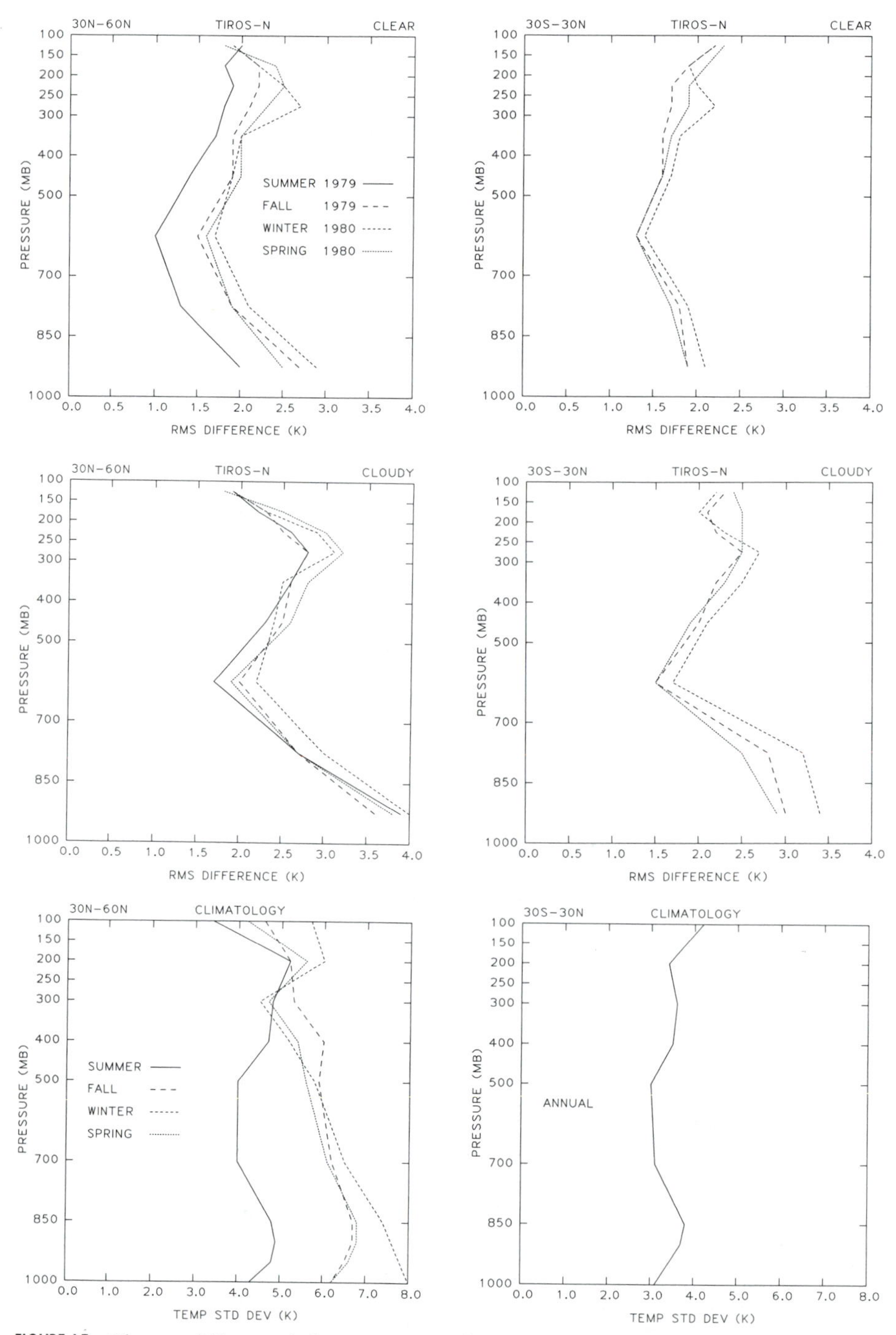

FIGURE 6.7. The rms differences in layer-mean virtual temperature between satellite and radiosonde soundings. [From data in Gruber and Watkins (1982).] Layer-mean values are plotted at the midpoint of the layer. The upper left corner of each plot indicates the latitude band, and the upper right corner indicates the retrieval type (clear or cloudy). Shown for comparison are climatological values of temperature standard deviation; seasonal values are shown in the midlatitudes, and annual values are shown in the tropics.

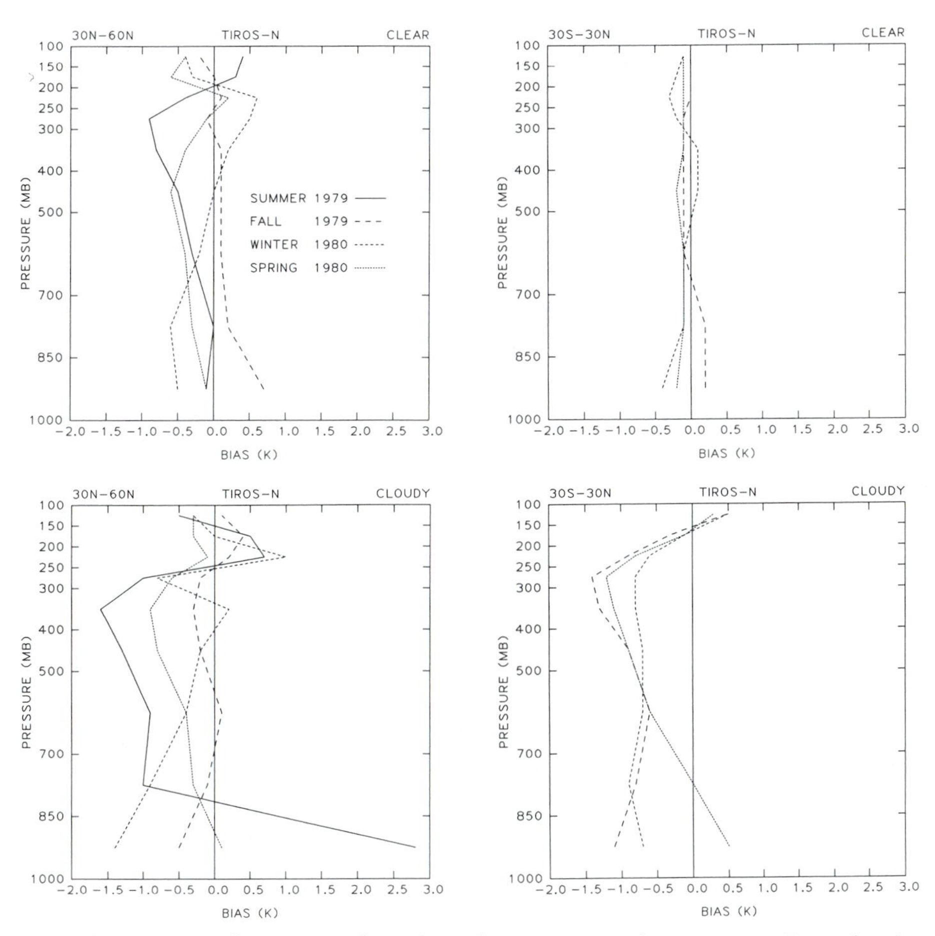

FIGURE 6.8. Biases (satellite minus radiosonde) in layer-mean virtual temperatures. [From data in Gruber and Watkins (1982).]

correction in the operational TOVS retrieval system was implemented in the summer of 1980.

In addition to varying with season and latitude, biases may also vary in other ways. Kidder and Achtemeier (1986) found that TIROS N biases were different day and night. Koehler *et al.* (1983) found that TIROS N and NOAA 6 errors were correlated with synoptic weather patterns. In particular, they found that the temperatures retrieved in troughs were too warm and in ridges too cool. This supports the finding of Schlatter (1981) that gradients of TIROS N temperatures tend to be too weak.

These problems with satellite soundings are pointed out not to dissuade the reader from using satellite soundings, but to illustrate the point that *satellite soundings are not direct substitutes for radiosonde soundings.* Satellite soundings contain a wealth of information on spatial and temporal scales unattainable with radiosondes. However, those who attempt to use satellite soundings without taking into account their differences from radiosonde soundings likely will be disappointed.

6.3.4 Use of Soundings in Models

Satellite soundings are seldom used by themselves; they are used in addition to radiosonde soundings, surface observations, and perhaps other data. How does one blend data from these various sources to yield an accurate picture of the atmosphere? This is known as the *data assimilation problem*. The previous section and the work discussed below suggest the following guidelines. First, it is clear that biases in satellite soundings (and in conventional data) should be removed as part of the assimilation process. Second, the quality or error characteristics of the data should be taken into account during analysis. This can be accomplished by weighting each datum inversely in proportion to its error variance. A third guideline arises from a consideration of atmospheric dynamics. The mass and flow fields of the atmosphere are usually closely in balance. If the mass field is modified by the insertion of a satellite sounding, the flow field probably should be modified to maintain balance. Hayden (1973) suggested that this can be approximated by modifying the wind field by the difference of the geostrophic wind before and after insertion of a temperature from a satellite sounding. Finally, most satellite soundings are not made at synoptic times (0000 and 1200 UTC) when radiosondes are launched and when most analyses are done. Corrections for the asynoptic nature of satellite soundings probably should be attempted.

Much of the work in data assimilation has been done in connection with model impact studies. One hopes, of course, that satellite soundings will improve numerical weather forecasts. Another reason for these studies, however, is that comparing numerical forecasts made with (SAT) and without (NOSAT) satellite data to observations is one way to determine whether inclusion of satellite data has improved the initial analysis. Presumably a better initial analysis will result in a better forecast. Ohring (1979) reviewed early satellite data assimilation experiments. He concluded that satellite soundings had a small positive impact in the data-rich Northern Hemisphere and large positive impact in the data-sparse Southern Hemisphere.

Beginning in the late 1970s there were several model impact studies. Ghil *et al.* (1979) used a global, primitive-equation, general circulation model (Somerville *et al.*, 1974) at NASA's Goddard Laboratory for Atmospheric Sciences (GLAS). The GLAS model had a 4° latitude × 5° longitude horizontal grid and nine sigma levels in the vertical. The purpose of their experiments was to test data assimilation schemes. They used *time-continuous assimilation*. In most numerical models, all data within a window (several hours either side of the time when the model is to be initialized) are analyzed together as if all observations were made at the same time. In time-continuous assimilation, the model is initialized at a time, say 12 h, before the desired initial analysis. The model is run for the intervening period, and new observations are assimilated at the times when they are made.

This process, also called *dynamic initialization*, alleviates the problem of satellite observations not being made at synoptic times. Ghil *et al.* assimilated observations at each model time step (10 min).

At each time step, one is faced with the problem of how to use observations to change the current model fields. Ghil *et al.* tested three methods. In the *direct insertion method*, they essentially replaced the model temperature with the average of the temperature observations within one model grid length. In the *successive correction method*, they did a Cressman-type (1959) analysis of the difference between the observations and temperatures interpolated from model grid points. The weights used depended only on the distance between the observation and a grid point. Three passes with decreasing radii of influence were used. The *statistical assimilation method* was like the successive correction method, except that the weights were based on the error characteristics of the observations. This is similar to the optimum interpolation method of data analysis (Gandin, 1963; Schlatter, 1975). Regardless of the data insertion method used, the winds were geostrophically modified (Hayden, 1973).

Ghil *et al.* used NOAA 4 and Nimbus 6 soundings during the period January through March 1976 from the Data System Test 6 [DST—6; Desmarais *et al.* (1978)]. Forecasts of 48 and 72 h were compared with analyses using S_1 skill scores (Teweles and Wobus, 1954) and rms differences. Surface pressure and 500-hPa height fields over North America and Europe were used for verification. Comparing SAT and NOSAT forecasts, Ghil *et al.* found small but statistically significant positive impact of satellite soundings. This result was highly dependent on the data insertion method, however. The direct insertion method resulted in negligible impact, the successive corrections method resulted in appreciable results, and the statistical assimilation method gave very good results. They concluded that when the better data insertion methods were employed, satellite soundings extended the usefulness of forecasts 8–16 h; thus a 60-h SAT forecast was about as accurate *on the average* as a 48-h NOSAT forecast. They also found that impact increased with quantity of satellite data; when both NOAA 4 and Nimbus 6 soundings were used, the forecasts were better than when data from only one satellite were used.

Tracton *et al.* (1980) did an experiment that was similar to that of Ghil *et al.* (1979); however, their purpose was quite different. Rather than testing different ways to assimilate satellite data, Tracton *et al.* wanted to test the impact that satellite soundings would have on an existing, operational, numerical weather prediction model. They used the six-layer primitive equation (PE) model (Shuman and Hovermale, 1968), which was then operational at the U.S. National Meteorological Center (NMC). The model was hemispheric and had a resolution of 381 km at 60°N. Their data assimilation scheme was that of the operational system: data within 3 h of analysis time were analyzed together; satellite data, in the form of temperatures at mandatory pressure levels, were treated exactly like radiosonde data. Tracton *et al.* used data from DST–6 (as did Ghil *et al.*) and from DST–5. They verified forecasts over North America and Europe using skill scores and rms differences as did Ghil *et al.* They found that the differences between SAT and NOSAT forecasts were "generally small and of inconsistent sign, i.e., beneficial in some cases and harmful in others." On the average, in the DST–6 experiments inclusion of satellite soundings resulted in slightly improved forecasts, but in the DST–5 experiments slightly poorer forecasts resulted.

A possible conclusion from comparing the experiments of Ghil *et al.* (1979) and Tracton *et al.* (1980) is that if Tracton *et al.* had assimilated satellite soundings in a different manner, perhaps using some of the techniques of Ghil *et al.*, they would have had more positive results. Unfortunately the two experiments are not sufficient to draw this conclusion. The difficulty is that the impact which satellite soundings have on numerical forecasts is dependent on the model. Tracton *et al.* (1981) found a negative correlation between model skill and satellite data impact. Poorer models seemed to make better use of satellite soundings. On the other hand, Atlas *et al.* (1982) increased the resolution of the GLAS model, which improved its NOSAT performance, and found increased positive impact of satellite soundings. It seems that the only way to assess the impact of satellite soundings on numerical forecasts made with a specific model is to do a test using that model. This is all the more important considering the fact that both today's models and today's satellite soundings are more sophisticated than those used in the Data System Tests. It is interesting that the current retrieval scheme for the TOVS soundings (the MVS method) was implemented only after two data impact studies at NMC (Dey *et al.*, 1989).

Thomasell *et al.* (1986) tested the impact of TIROS N and NOAA 6 soundings on the forecast model used by the Israel Meteorological Service. Two of their results are interesting. First, satellite soundings were most beneficial when they were used in a two-step analysis procedure as follows. Satellite soundings and surface observations, but no radiosonde data, were analyzed to construct the first-guess fields. Radiosonde soundings were then used to update the first-guess fields to produce the initial analysis for the model. This procedure neatly separates satellite and radiosonde data, and lets each have maximal impact where they will do the most good: satellite data over the oceans, where there are too few radiosondes to produce a good analysis, and radiosonde data over the land, where additional observations from satellites are not needed.[9] Second, Thomasell *et al.* attempted to verify the forecasts of the Israeli model over the ocean as well as over land. They found that satellite soundings improved forecasts of sea level pressure over the ocean to the point that they were almost as good as forecasts over land. (Forecasts over land showed little impact of satellite soundings.) Readers are warned, however, that both the model and the analysis system investigated by Thomasell *et al.* are less sophisticated than those used elsewhere; their results cannot be directly extrapolated.

The consensus opinion (and in the absence of conclusive tests it can only be opinion) of the impact of satellite soundings on numerical forecasts is the following.

[9] A sophisticated analysis system, such as optimum interpolation, should have the same result.

- Satellite soundings in the data-sparse Southern Hemisphere have large positive impact.
- Because the weather is a global phenomenon, Southern Hemisphere soundings will positively impact long-range (5–10-day) Northern Hemisphere forecasts.
- In the Northern Hemisphere, satellite soundings have, on the average, a small positive impact on short- to medium-range forecasts.
- The average positive impact of satellite soundings in the Northern Hemisphere is made up of both positive and negative impacts in individual cases. In most cases, the impact, whether positive or negative, is small. There is evidence (Atlas *et al.*, 1982; Atlas, 1982) that satellite soundings, when properly assimilated, can have large positive impact on numerical forecasts. Whether satellite soundings can have large negative impact on forecasts is controversial.

Satellite soundings are used in today's numerical weather prediction models. As of this writing, operationally retrieved TOVS soundings (up to 20,000 per day) are transmitted over the Global Telecommunications System (GTS) and, thus, are available to forecasters world wide. At NMC, TOVS soundings, in the form of thicknesses, are converted to heights by building on the height of a reference level. Over the data-sparse oceans, the 1000-hPa height is used as a reference level, and all levels of satellite data are assimilated. Over land, the 100-hPa level is used as a reference level, and satellite data above 100 hPa are assimilated. However, different models use different amounts of satellite data. As this book is in preparation, short-range forecasts are made by the Limited Fine Mesh (LFM) model and the Regional Analysis and Forecast System (RAFS), which have data cutoffs of 1 h 15 min and 2 h, respectively, after the time when the models are to be initialized (0000 and 1200 UTC). Because relatively few satellite soundings have been processed by these times, the LFM and the RAFS directly use few satellite soundings. The Medium Range Forecast (MRF) model, which makes forecasts out to 10 days, is initialized at 0000 UTC using observations between 2100 and 0300 UTC. However, the process is not begun until 0600 UTC; thus late-arriving observations, including satellite soundings, are used. Finally, the Global Data Assimilation System (GDAS) is run in delayed mode to produce the NMC "Final" analyses and to provide first-guess fields to initialize the forecast models. GDAS produces analyses four times per day (at 0000, 0600, 1200, and 1800 UTC) using data within ± 3 h of the initial time. To accommodate late-arriving data (including satellite soundings), the data cutoffs are 3.5 to 8.5 h after the initial analyses. A 6-h forecast is made with each of these analyses using the MRF model. The forecasts ending at 0000 and 1200 are used as first guesses for the other models. The forecasts ending at 0600 and 1800 are used as first guesses for the next GDAS run. Thus satellite soundings do affect short-range NMC forecasts, but primarily through their first-guess fields.

O'Lenic (1986) tested the impact of VAS soundings on forecasts made with NMC's Limited Fine Mesh model. In six comparisons of VAS and NOVAS forecasts, VAS soundings over the northeast Pacific Ocean improved the forecast of 500-hPa heights over the United States in four cases and very slightly degraded the forecast in two cases. O'Lenic (1986) reported that the VAS soundings compared favorably with NOAA 7 TOVS soundings. However, VAS soundings are not routinely used by NMC in its forecast models.

6.4 LIMB SOUNDING RETRIEVALS

Limb soundings utilize broadband observations in one or two channels but at many tangent heights rather than narrowband observations in many spectral intervals. Because of the sharpness of the weighting functions, temperature retrievals would seem to be straightforward. Any of the retrieval methods discussed in Section 6.2 should work nicely. The problem, as pointed out by Gille and House (1971), is that a reference level, where both the height and pressure are known, is needed.[10] Suppose that the pressure at reference height z_0 is p_0. Then, using a single wideband channel, a unique temperature profile can be retrieved. A different assumption about the pressure at z_0 yields a different temperature profile. Gille and House solved this problem by utilizing a second, narrower (more opaque) channel. For each channel, a temperature profile is retrieved by assuming a pressure at reference level z_0. If the assumed pressure is incorrect, the two temperature profiles will not agree. Iterating p_0 between inversions yields accurate temperature profiles. However, many inversions are necessary to obtain a single temperature profile, which means that the calculations are lengthy. Fortunately, rapid, accurate retrieval schemes have been developed (Gordley and Russell, 1981; Rodgers, 1976; Rodgers *et al.*, 1984).

Three infrared limb sounders have flown on the Nimbus satellites. The Limb Radiance Inversion Radiometer (LRIR) flew on Nimbus 6, and both the Limb Infrared Monitor of the Stratosphere (LIMS) and the Stratospheric and Mesospheric Sounder (SAMS) flew on Nimbus 7. LRIR had 4 channels: a wide CO_2 channel (WCO_2), 14.4–16.9 μm; a narrow CO_2 channel (NCO_2), 14.9–15.5 μm; an O_3 channel, 8.6–10.2 μm; and an H_2O channel, 23.0–27.0 μm. All channels had 2.0-km vertical resolution and 20-km horizontal resolution at the tangent point except the H_2O channel, which had 2.5 × 25-km resolution. LIMS had six channels: WCO_2, 13.2–17.3 μm; NCO_2, 14.9–15.7 μm; O_3, 8.8–10.8 μm; H_2O, 6.4–7.3 μm; NO_2, 6.1–6.4 μm; and HNO_3, 10.9–11.8 μm. All channels had 1.8 × 18-km resolution except the H_2O and NO_2 channels, which had 3.6 × 28-km resolution. The SAMS was like the SSU in that it utilized filters consisting of gas–filled cells; unlike SSU, however, SAMS was a limb scanner. SAMS had six cells containing CO_2, CO, H_2O, NO, N_2O, and CH_4. Concentra-

[10] Note that this problem is not unique to limb sounding. Vertical soundings are often retrieved down to 1000 hPa because the actual surface pressure is not known. However, vertical sounding retrieval schemes may use estimates of surface pressure from numerical models, whereas this information is not available to limb sounding retrieval schemes.

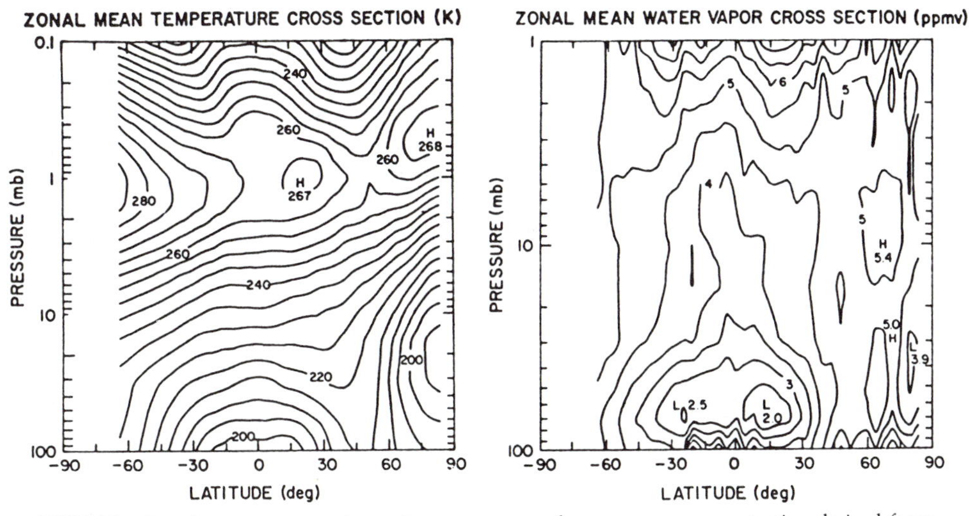

FIGURE 6.9. Zonal mean cross sections of temperature and water vapor concentration derived from LIMS data for 6 January 1979. [After Gille and Russell (1984).]

tions of these gases and temperatures were retrieved from SAMS data. Significantly, SAMS could make measurements above 0.1 hPa in areas which are not in local thermodynamic equilibrium.

Although few measurements in the stratosphere are available for comparison, a substantial effort was made to verify Nimbus 6 measurements. Gille *et al.* (1984a) reported on comparisons between LIMS temperature retrievals and radiosonde and rocketsonde measurements. Temperatures were retrieved from 100 to 0.1 hPa. Typical rms differences of 3 K between LIMS retrievals and in situ observations were found, except for the equatorial tropopause region, where they were higher. Below 1 hPa LIMS biases were generally less than 2 K. Above 1 hPa, LIMS temperatures were systematically cooler than rocketsonde measurements. Russell *et al.* (1984a) reported that LIMS water vapor mixing ratio retrievals were accurate to within 20–30% from 50 hPa to the stratopause (1 hPa) and to within 40% in the lower stratosphere. Figure 6.9 shows zonal mean temperature and water vapor mixing ratios derived from LIMS data. Barnett and Corney (1984) compared retrieved temperatures from SAMS with radiosonde and rocketsondes. They found temperature biases of up to 2 K in the lower stratosphere and 1 K in the upper stratosphere. The rms errors were approximately 1–2 K.

6.5 OZONE AND OTHER GASES

Ozone has absorption lines in all major portions of the electromagnetic spectrum and thus can be measured with a variety of techniques. Krueger *et al.* (1980) reviewed 21 satellite experiments (prior to SBUV/2) to measure ozone. We will discuss a representative sample of the more recent experiments.

Mentioned above is the 9.7-μm channel on the operational HIRS/2 sounder. Planet *et al.* (1984) regressed clear-column radiances, produced during operational TOVS processing, against surface-based observations of total ozone (column-integrated ozone) made with a Dobson spectrophotometer. They found that three HIRS/2 channels yielded an acceptable regression retrieval scheme. Channel 3 (14.49 μm) is sensitive to stratospheric temperature, channel 8 (11.11 μm) is sensitive to surface temperature, and channel 9 (9.71 μm) is sensitive to ozone, stratospheric temperature, and (slightly) surface temperature. The regression retrieval scheme essentially uses channels 3 and 8 to correct channel 9 for temperature dependence. Planet *et al.* showed the retrieval scheme to be accurate within 3–8%. Susskind *et al.* (1984) and Reuter and Susskind (1986) have developed a physical retrieval scheme for total ozone using HIRS/2 and MSU data which also gives good results.

Limb sounders also use the 9.6-μm band to retrieve ozone. Using the emissivity growth approximation of Gordley and Russell (1981), Remsberg *et al.* (1984) retrieved ozone profiles from LIMS data and compared them with balloon-based measurements (ozonesondes) and rocket-based measurements. Ozone was retrieved from ~300 hPa to ~0.1 hPa. Calculated accuracies, based on experiment uncertainties, ranged from 15%, in the region 1–3 hPa, to about 40% at 100 hPa and 0.1 hPa. Comparison with ozonesondes yielded mean differences within 10% in the region 7–50 hPa. The rms differences were ~15%. Comparison with rocket ozone measurements in the region 0.3–50 hPa yielded mean differences less than 16% and rms differences in the range 12–25%. Figure 6.10 shows a zonal mean cross section of ozone concentration derived from LIMS measurements.

Probably the best-known method to retrieve ozone from satellite data utilizes backscattered ultraviolet radiation. Solar ultraviolet radiation, reflected from the

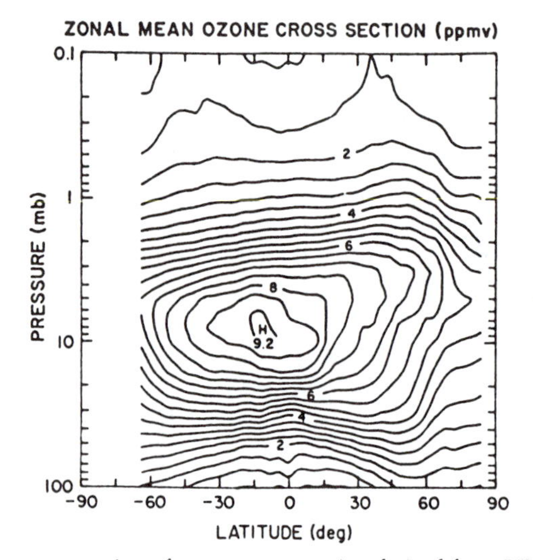

FIGURE 6.10. Zonal mean cross section of ozone concentration derived from LIMS data for 6 January 1979. [After Gille and Russell (1984).]

surface and backscattered by the atmosphere or clouds, provides a radiation source that interacts with ozone in the Hartley–Huggins bands. Observations at several wavelengths can be inverted to produce ozone profiles or total ozone. This was first proposed by Singer and Wentworth (1957). The process for retrieving total ozone (Klenk *et al.*, 1982) illustrates a classical remote sensing technique involving observations at two closely spaced wavelengths. Most of the ozone lies above the tropopause, whereas most of the backscattered ultraviolet solar radiation comes from below the tropopause (Fig. 6.11). Little absorption occurs in the troposphere, and little scattering occurs in the ozone layer. Radiation reaching the satellite must traverse the ozone layer twice, suffering attenuation each time. Schematically, this radiance can be written

$$L_\lambda = E_{\text{sun}}(\lambda)\tau_{\text{ozone}}^{x}\rho(\theta_{\text{sun}},\theta_{\text{sat}},\rho_{\text{sfc}},\rho_{\text{air}}), \tag{6.30}$$

where $E_{\text{sun}}(\lambda)$ is the solar irradiance at the top of the atmosphere, $x = \sec\theta_{\text{sun}} + \sec\theta_{\text{sat}}$, and ρ is a combined surface–troposphere reflectance that depends on scattering from the air (ρ_{air}), reflection from the surface (ρ_{sfc}), and the zenith angles. The problem is to solve for τ_{ozone}, which–ignoring temperature and pressure variation of the absorption coefficients–is related to total ozone:

$$\tau_{\text{ozone}} = \exp[-\beta_{\text{a}}(\lambda)U_{\text{ozone}}], \tag{6.31}$$

where $\beta_{\text{a}}(\lambda)$ is the ozone mass absorption coefficient, and U_{ozone} is the mass of ozone per unit area, that is, the *total ozone*. BUV, SBUV, SBUV/2, and TOMS all make measurements of $E_{\text{sun}}(\lambda)$. Using two closely spaced observations in the Hartley–Huggins bands, one with a higher ozone absorption coefficient than the other, the quantities N_i are formed:

$$N_i \equiv -100\log_{10}\left(\frac{L(\lambda_i)}{E_{\text{sun}}(\lambda_i)}\right). \tag{6.32}$$

Because reflectance changes much less rapidly with wavelength than gaseous absorption, in many cases ρ can be considered nearly constant between λ_1 and

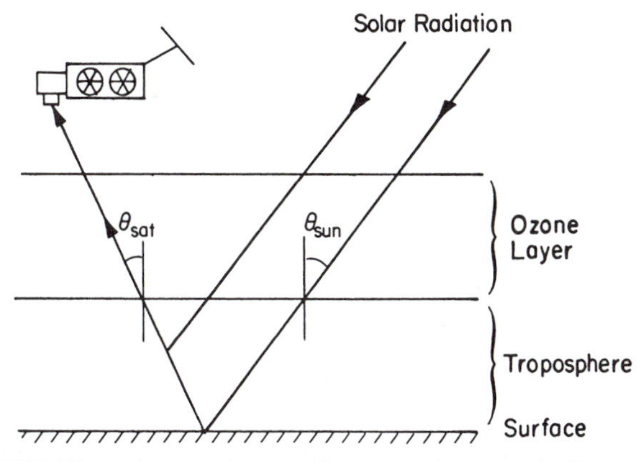

FIGURE 6.11. Schematic diagram illustrating the retrieval of ozone.

λ_2. If ρ did not change, then $N_1 - N_2$ would be directly proportional to U_{ozone} Measurements of N_1 and N_2 plus knowledge of the absorption coefficients and zenith angles would yield total ozone. However, because of the λ^{-4} dependence of Rayleigh scattering, ρ does change between wavelengths λ_1 and λ_2. Rayleigh scattering calculations can account for the influence of the atmosphere on ρ, but not for the surface reflectance. Therefore a third observation at λ_3 outside the ozone band is necessary. This observation establishes the surface reflectance ρ_{sfc}, which is assumed to be the same at λ_1 and λ_2. Scattering calculations then yield ρ at λ_1 and λ_2, which are used to calculate U_{ozone}.

In practice, of course, the retrieval procedure is somewhat more complicated. The BUV that flew on Nimbus 4 operated for 7 years from 1970 to 1977. Klenk *et al.* (1982) used 5 of the 12 BUV wavelengths to retrieve total ozone. Two pairs of wavelengths in the ozone band were used: the *A* pair consisted of 312.60 and 317.63 nm, and the *B* pair consisted of 313.29 and 339.93 nm. These measurements were made with the monochromator. The photometer (similar to the Cloud Cover Radiometer on SBUV/2) made measurements outside the ozone band at 380 nm. Actual retrievals were made by comparing measured radiances with large lookup tables of calculated radiances. The lookup tables were calculated using the method of Dave (1964) and took into account multiple scattering, polarization, the spherical shape of the atmosphere, and the ozone profile shape. Figure 6.12

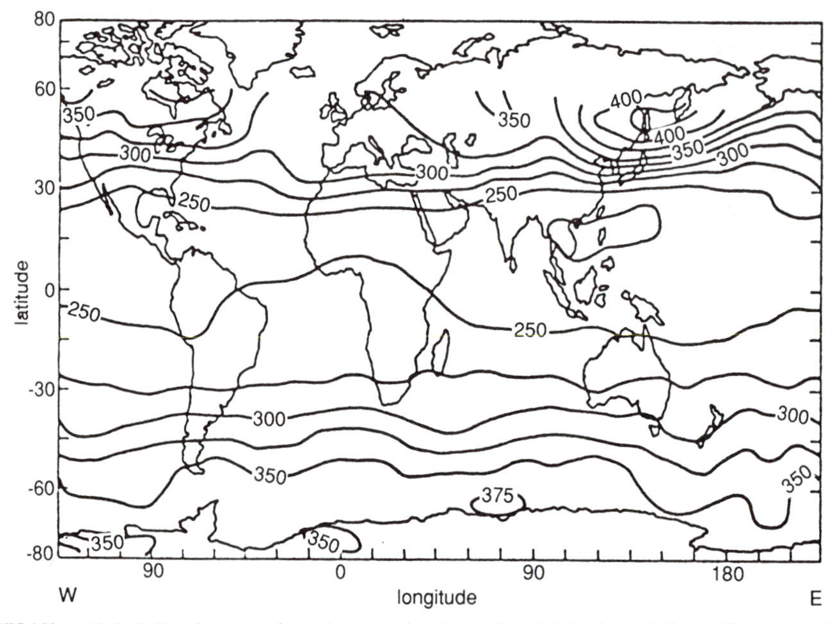

FIGURE 6.12. Global distribution of total ozone for December 1970 derived from Nimbus 4 BUV measurements. The units are Dobson units, that is the depth (in 10^{-3} cm) of pure ozone at standard temperature and pressure (0°C, 1013.25 hPa). To obtain the mass of ozone per unit area (in grams per square meter), multiply by 0.0214. [After Krueger *et al.* (1980).]

PLATE 1 ATS 3 full-disk image. [Courtesy of NASA.]

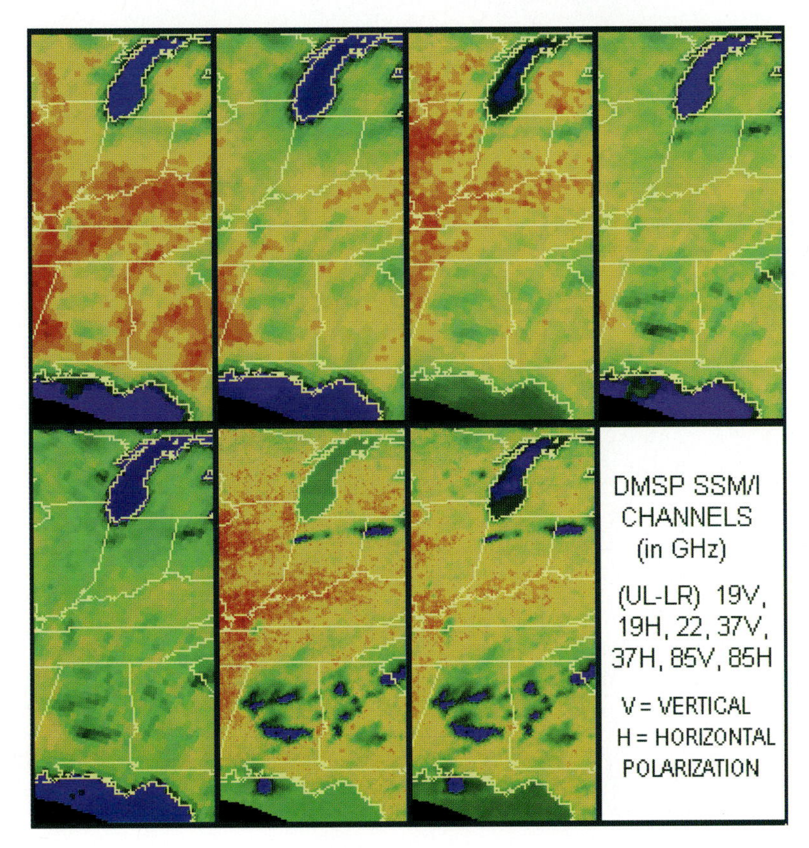

PLATE 2 SSM/I images.

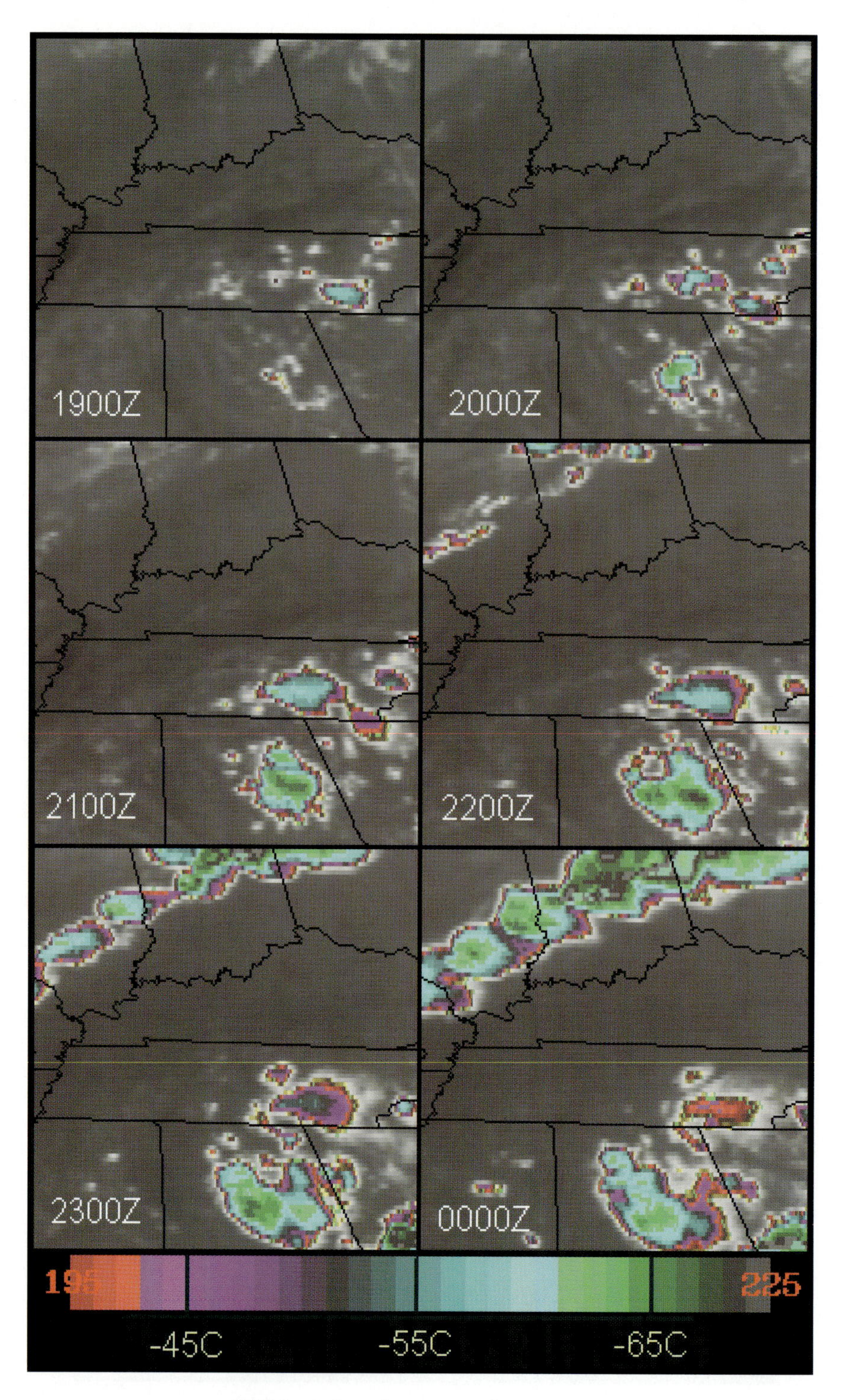

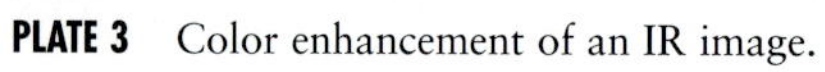

PLATE 3 Color enhancement of an IR image.

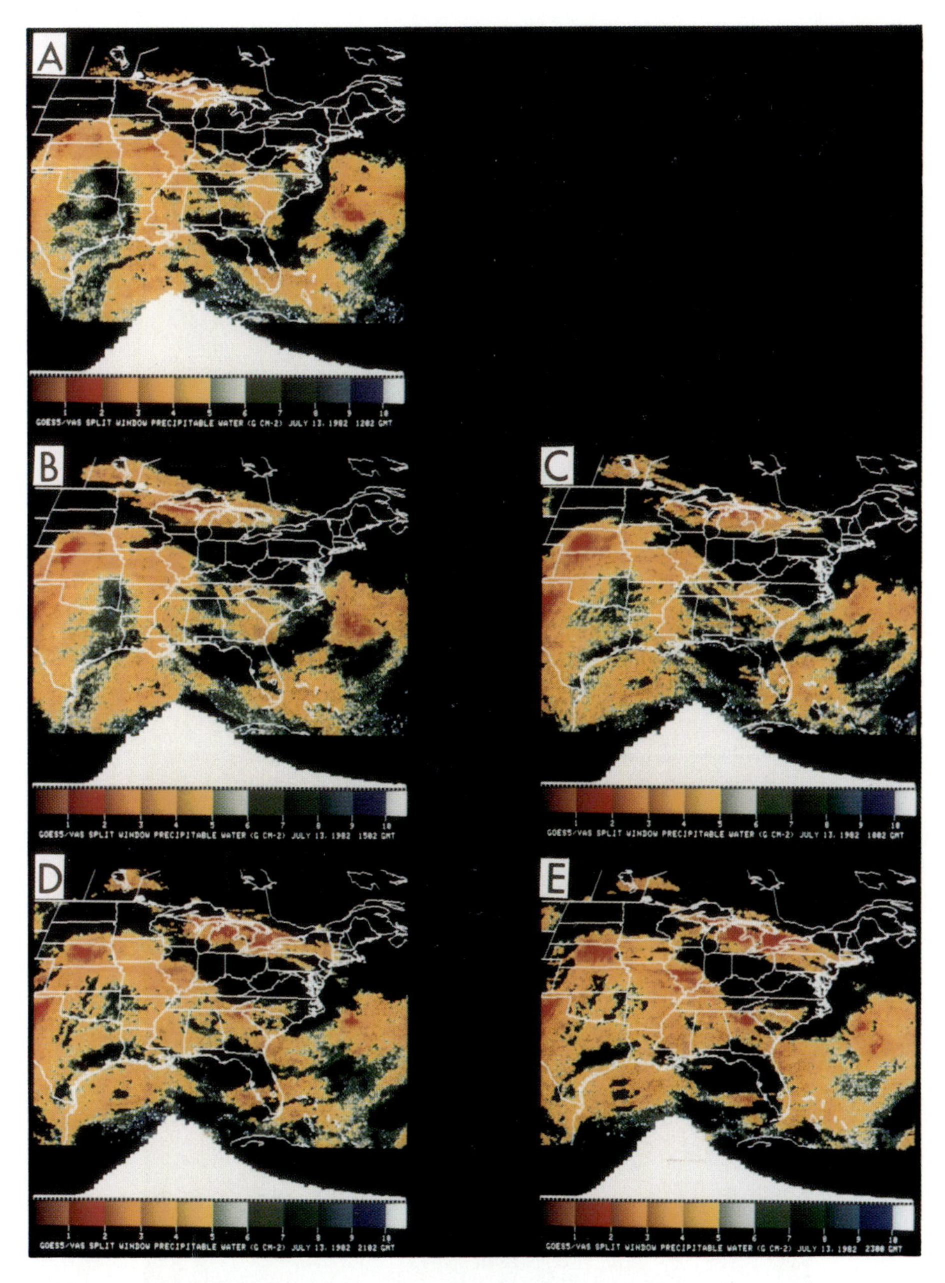

PLATE 4 Precipitable water derived from VAS data using the split window technique. The images are at 3 h intervals (A–E) starting at 1200 UTC 13 July 1981. A histogram of the precipitable water distribution is shown at the bottom of each image. Note that precipitable water exhibits extremely high spatial variability, which can only be captured by a satellite-borne instrument. [After Chesters *et al.* (1983).]

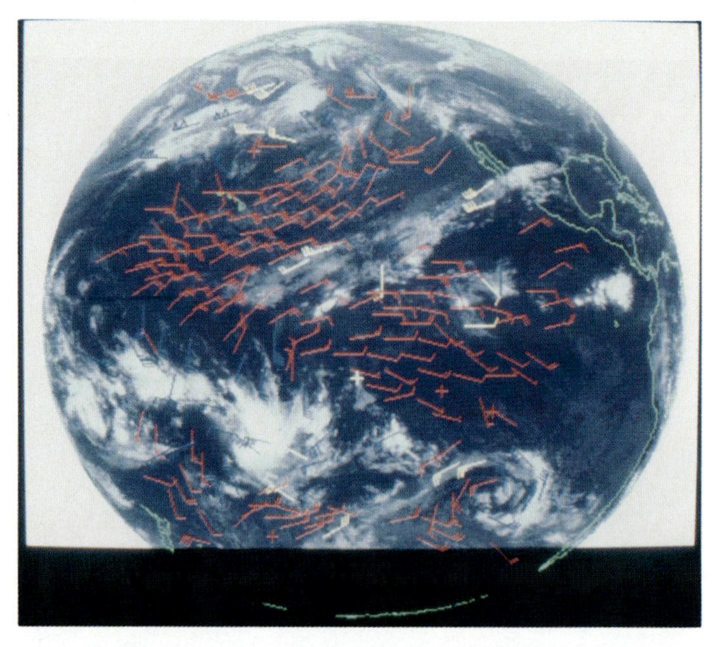

PLATE 5 NESDIS operationally derived winds from the GOES-West satellite on 29 February 1988 at 1545 UTC. The red barbs represent low level winds (surface to 700 hPa), the yellow barbs are mid-level winds (699-400 hPa), and the blue barbs are upper-level winds (above 400 hPa).

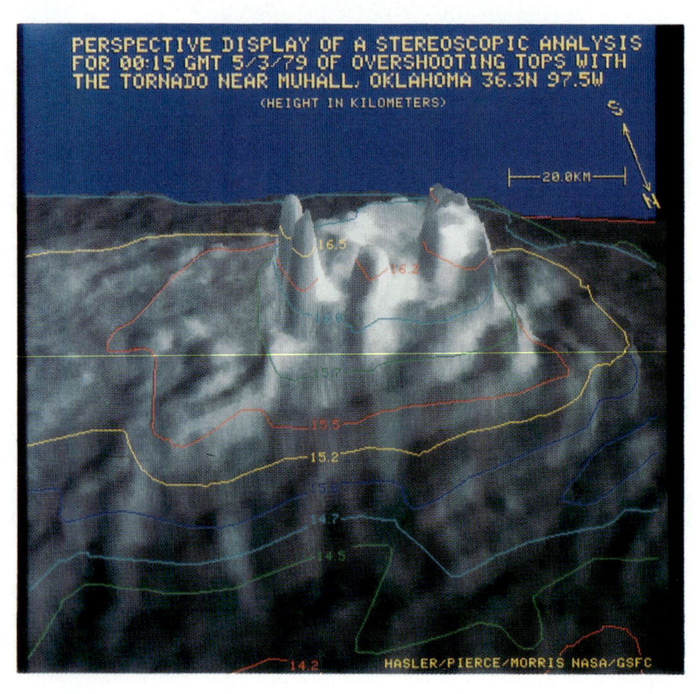

PLATE 6 A perspective view of a thunderstorm cloud top with height contours derived from stereoscopic analysis. [Courtesy of A. F. Hasler, NASA/Goddard Space Flight Center.]

shows the global distribution of total ozone for December 1970 derived from the Nimbus 4 BUV.

Total ozone can be retrieved in the presence of clouds because most of the ozone lies above the cloud tops. However, a change in the retrieval scheme must be made. When the reflectance measured by the photometer was greater than 0.6, it was assumed that clouds were present and that the cloud top was approximately 400 hPa. Separate lookup tables were calculated for cloudy retrievals.

Klenk *et al.* compared their retrievals with surface-based observations and found a bias of 3.5%, which they attribute primarily to differences in ozone absorption coefficients used in the two techniques. The standard deviation of the difference from ground-based measurements was 1.5–3.0%, which results in an rms difference of 3.8–4.6%.

The Total Ozone Mapping Spectrometer (TOMS) on Nimbus 7 is much like BUV or SBUV, except that (1) it scans instead of making measurements only at nadir, and (2) it makes measurements only of total ozone, not ozone profiles. Daily global maps of total ozone at 50–150-km resolution are produced.[11] Bhartia *et al.* (1984a) report that TOMS measures less total ozone than do ground-based instruments. The bias is $-6\% \pm 2\%$. This is thought to be due in large part to a difference in the absorption coefficients used. The standard deviation between TOMS measurements and Dobson (ground-based) measurements averaged 2.6%. This results in an rms difference of about 6.7%. Accuracies of total ozone retrieved from Nimbus 7 SBUV measurements are similar, and these values are probably representative of total ozone retrieved from the SBUV/2 on the NOAA satellites.

Total ozone retrievals utilize measurements in the longer-wavelength, more transparent portion of the Hartley–Huggins band. Estimates of ozone profiles utilize measurements at shorter, more opaque wavelengths where the surface contribution is negligible. The basic equation can be derived as follows (London *et al.*, 1977). Consider a plane-parallel atmosphere as shown in Fig. 6.11. All UV instruments which retrieve ozone profiles only view nadir; therefore, $\theta_{sat} \equiv 0$ ($\mu_{sat} = 1$). Since the atmosphere emits no UV radiation, and since the sun is the only source of UV radiation, the no-emission equations of Section 3.3.2 apply, specifically Eq. 3.45. Assume now that all radiation reaching the satellite is sunlight which has been scattered once and only once from a Rayleigh scatterer. The second term on the right-hand side of Eq. 3.45 may be neglected, and the radiative transfer equation becomes

$$\frac{dL_\lambda}{d\delta_\lambda} = -L_\lambda + \frac{\tilde{\omega}_\lambda}{4\pi} E_\lambda^{sun} \left[\frac{3}{4}(1 + \mu_{sun}^2)\right] \exp\left(-\frac{\delta_o - \delta_\lambda}{\mu_{sun}}\right), \tag{6.33}$$

where Eq. 3.60 has been used for the Rayleigh scattering phase function, and the scattering angle ψ_s has been set equal to θ_{sun} because $\theta_{sat} \equiv 0$. Assuming that the transmittance from the surface is zero, this equation can be integrated to yield

$$\frac{L_\lambda}{E_\lambda^{sun}} = \frac{3(1 + \mu_{sun}^2)}{16\pi} \int_0^{\delta_o} \tilde{\omega}_\lambda \exp\left[-(\delta_o - \delta_\lambda)\left(1 + \frac{1}{\mu_{sun}}\right)\right] d\delta_\lambda. \tag{6.34}$$

[11] TOMS also flew on the Meteor–3 5 satellite.

Recalling that $\tau_\lambda = \exp(\delta_o - \delta_\lambda)$, and choosing p as the vertical coordinate, this equation becomes

$$\frac{L_\lambda}{E_\lambda^{\text{sun}}} = \int_0^{p_o} \tilde{\omega}_\lambda(p) W_\lambda(p,\mu)\, dp, \tag{6.35a}$$

$$W_\lambda(p,\mu) = \frac{3(1 + \mu_{\text{sun}}^2)}{16\pi} \tau_\lambda^{1/\mu_{\text{sun}}} \frac{d\tau_\lambda}{dp}. \tag{6.35b}$$

This is very like Eqs. 6.3, with a slightly different form for W_λ, and with $\tilde{\omega}_\lambda(p)$ in place of $B_\lambda(T)$. With measurements of L_λ and E_λ^{sun} at several wavelengths plus knowledge of the solar zenith angle and ozone absorption coefficients, it should be possible to retrieve $\tilde{\omega}_\lambda$ as a function of p. Now $\tilde{\omega}_\lambda$ is the ratio of the scattering coefficient to the extinction coefficient. Scattering is assumed to be entirely due to Rayleigh scattering by air molecules. The scattering coefficient is given by Eq. 3.58 and has the form $f(\lambda)\rho$, where ρ is atmospheric density. Absorption is assumed to be entirely due to ozone molecules and has the form $\beta_\lambda(T,p)q\rho$, where q is the ozone mixing ratio, and again, ρ is atmospheric density. Therefore, $\tilde{\omega}_\lambda$ is independent of ρ and is a function of q, which is the desired quantity. Mateer (1977) discusses several ways to retrieve ozone concentration from UV measurements.

Using the technique of Schneider *et al.* (1981), Bhartia *et al.* (1984b) retrieved ozone profiles from Nimbus 7 SBUV measurements and compared them with rocket and balloon measurements. They found biases generally less than 10% and rms differences in the 10–20% range.

A third way to measure ozone takes advantage of ozone's absorption of visible light in the Chappuis bands. A solar or stellar occultation experiment is extremely simple. The extinction of radiation is measured as a function of tangent height as the sun or another star rises or sets. The optical depth δ of the atmosphere's limb is simply the product of the number of molecules in the path and the absorption cross section; the extinction is, of course, $\exp(-\delta)$. The Stratospheric Aerosol and Gas Experiment (SAGE) flew on the Applications Explorer Mission 2 (AEM 2) from February 1979 to November 1981. It measured atmospheric extinction in the Chappuis band at 0.6 μm (and three other wavelengths). Approximately 30 observations were made per day, 13,000 over the lifetime of the instrument. Vertical profiles of ozone concentration can be retrieved from the extinction as a function of tangent height [see Section 8.4 and McCormick *et al.* (1979)].

McCormick *et al.* (1984) compared SAGE ozone profiles with balloon and rocket observations. In the altitude range 18–28 km, the bias between SAGE and balloon data (17 comparisons) averaged 9.3% with a standard deviation of 2.8% (rms difference 9.7%). Differences between SAGE and rocket measurements were in the range 11–13.5%.

SAGE II, launched in October 1984 on the Earth Radiation Budget Satellite (ERBS), also had a 0.6-μm channel to measure ozone [and six other channels; Chu (1986)]. Ozone profiles retrieved by Chu *et al.* (1989) were compared against ozonesondes and found to be within 1% on average and within 7% at all altitudes between 20 and 53 km (Cunnold *et al.*, 1989). The SAGE II profiles were found to be more precise than the SAGE profiles, but the SAGE profiles have been revised and Cunnold *et al.* state that the SAGE and SAGE II data can be combined to form a long-term (since 1979) stratospheric ozone data set.

All of the methods for measuring ozone have their advantages and disadvantages. The solar occultation method is simple, both in equipment and in retrieval technique. However, the fact that measurements can be made only at satellite sunrise and sunset limits the number and the spatial distribution of the measurements. Backscatter ultraviolet methods are accurate and measurements have been made over long time periods. However, they cannot be made in darkness, such as over the winter pole. Infrared methods, made with either a downward-looking or a limb sounder are perhaps a little less accurate than UV methods, but they do not depend on reflected sunlight. Retrieval of total ozone from operational sounders is easy, twice-daily global maps of total ozone can be obtained, and measurements have been collected since 1978 (on the NOAA satellites). Limb infrared measurements of ozone profiles have resolution similar to BUV-estimated profiles, but can be retrieved in daylight as well as darkness. They have been made intermittently since 1975.

Several satellite experiments have measured stratospheric gases other than ozone. These measurements are extremely important for the understanding of the chemistry and dynamics of the stratosphere. All are made with infrared limb sounders or solar occultation measurements. These experiments are summarized in Table 6.1. As an example, Fig. 6.13 shows the zonal mean cross section of HNO_3 concentration derived from LIMS data.

TABLE 6.1. Satellite Measurements of Stratospheric Gases Prior to UARS

Gas	*Satellite*	*Instrument*	*Technique*	*Accuracy*	*Reference*
NO_2	Nimbus 7	LIMS	IR Limb Snd	±20%	Russell *et al.* (1984b)
	AEM–2	SAGE I	Solar Occ.	±25%	Chu and McCormick (1979)
	ERBS	SAGE II	Solar Occ.		Chu (1986)
HNO_3	Nimbus 7	LIMS	IR Limb Snd	±20%	Gille *et al.* (1984b)
N_2O	Nimbus 7	SAMS	Limb PMR	6–17% (at 7 hPa) 12–20% (at 20 hPa) 50–100% (at 0.6 hPa)	Jones and Pyle (1984)
CH_4	Nimbus 7	SAMS	Limb PMR	3–10% (at 7 hPa) 5–15% (at 20 hPa) 20–40% (at 0.2 hPa)	Jones and Pyle (1984)
NO	Nimbus 7	SAMS	Limb PMR[a]		
CO	Nimbus 7	SAMS	Limb PMR		
CO_2	Nimbus 7	SAMS	Limb PMR[b]		

[a] Pressure-modulated radiometry, like SSU (see Section 4.1.4).
[b] CO_2 concentration measured above 100 km.

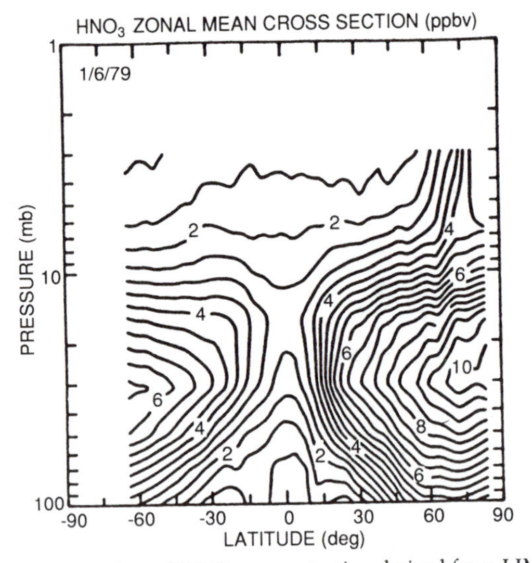

FIGURE 6.13. Zonal mean cross section of HNO_3 concentration derived from LIMS data for 6 January 1979. [After Gille and Russell (1984).]

NASA's Upper Atmosphere Research Satellite (UARS), launched 15 September 1991, is the most important satellite to study the middle atmosphere since the Nimbus program. Its orbit is similar to that of the Earth Radiation Budget Satellite, circular at an altitude of about 600 km with a 57° inclination angle. UARS measures numerous gases (Table 6.2) that are important for stratospheric and mesospheric chemistry, and especially for ozone chemistry. Of particular importance are microwave measurements of chlorine monoxide (ClO), which is thought to be the link between chlorofluorocarbons and ozone destruction.

TABLE 6.2. Stratospheric and Mesospheric Gases Measured by UARS[a]

Instrument	*Gases measured*	*Technique*
Cryogenic Limb Array Etalon Spectrometer (CLAES)	O_3, NO, NO_2, N_2O, HNO_3, N_2O_5, H_2O, CH_4, CF_2Cl_2, $CFCl_3$, HCl, $ClONO_2$	IR spectrometer 3.5–12.7 μm
Improved Stratospheric and Mesospheric Sounder (ISAMS)	O_3, NO, NO_2, N_2O, HNO_3, N_2O_5, H_2O, CH_4, CO	PMR 4.6–16.6 μm
Microwave Limb Sounder (MLS)	O_3, H_2O, ClO	Microwave limb radiometer 63, 183, and 205 mm
Halogen Occultation Experiment (HALOE)	O_3, NO, NO_2, H_2O, CH_4, HCl, HF	Solar occultation 2.43–10.25 μm

[a] From a NASA pamphlet, *UARS: A Program to Study Global Ozone Change*, NASA Headquarters, Washington, DC, 28 pp.

6.6 THE SPLIT-WINDOW TECHNIQUE

Most of the preceding discussion concerns the estimation of atmospheric quantities. The retrieval schemes attempt to correct for any signal from the surface. However, some meteorologically important quantities, such as sea and land surface temperatures, are properties of the surface. To retrieve them, atmospheric modification of the upwelling radiation must be corrected. An extremely important technique used in many of these retrieval schemes is the so-called split-window technique.

The *split-window technique* has its roots in papers by Saunders (1967) and Anding and Kauth (1970). Its best statement, however, is in McMillin and Crosby (1984). Surface observations are made in atmospheric windows. However, these windows are dirty; that is, the radiance leaving the surface is modified by the atmosphere. For infrared windows, the radiative transfer equation can be written

$$L_\lambda = B_\lambda(T_s)\tau_o(\lambda) + B_\lambda(T_A)[1 - \tau_o(\lambda)], \tag{6.36}$$

where $\tau_o(\lambda)$ is the transmittance from the surface to the satellite at wavelength λ, T_s is the surface brightness (equivalent blackbody) temperature, and T_A is a mean brightness temperature of the atmosphere given by

$$B_\lambda(T_A) = \frac{1}{1 - \tau_o(\lambda)} \int_{\tau_o}^{1} B_\lambda(T) d\tau. \tag{6.37}$$

Some windows, particularly the 10.5–12.5-μm window, are wide enough to permit observations in two channels. The split-window technique uses observations at two wavelengths to eliminate the influence of T_A and solve for T_s.

Suppose that radiances at two wavelengths λ_1 and λ_2 are measured. Then

$$L_1 = B_1(T_{B1}) = B_1(T_s)\tau_1 + B_1(T_A)(1 - \tau_1), \tag{6.38a}$$

$$L_2 = B_2(T_{B2}) = B_2(T_s)\tau_2 + B_2(T_A)(1 - \tau_2), \tag{6.38b}$$

where the subscripts 1 and 2 indicate wavelength, T_B is a brightness temperature (equivalent blackbody temperature), and τ is understood to mean τ_o. If the two wavelengths are close together, we can expect three things.

- The weighting functions in the two channels will be similar, and therefore T_A will be the same in both equations. Prabhakara *et al.* (1974) showed that in the 10.5–12.5-μm window, T_A varies by less than 1 K.
- The surface emittance and thus T_s will also be the same in each equation.
- Differences in transmittance will be the result of differences in the absorption coefficient of the same absorbing gas. In the 10.5–12.2-μm window, water vapor is the chief absorber.

If U is the column-integrated water vapor (the precipitable water), then

$$\tau_\lambda \approx \exp\left(-\beta_\lambda \frac{U}{\mu}\right). \tag{6.39}$$

Given measurements of L_1 and L_2, Eqs. 6.38 appear to constitute two equations in three unknowns: T_A, T_s, and U. It will turn out, however, that because the optical depth is small, eliminating T_A also eliminates U. The split-window technique can be used directly to obtain T_s, but not U, from satellite measurements at two wavelengths. U can be retrieved, however, if an independent estimate of T_A is available.

To solve Eqs. 6.38 for T_s, the wavelength dependence must be eliminated. Since the weighting functions peak at the surface, T_s, T_{B1}, and T_{B2} will be close to T_A. Expanding the Planck functions about T_A:

$$B_\lambda(T) \approx B_\lambda(T_A) + \frac{\partial B_\lambda}{\partial T}(T - T_A), \tag{6.40}$$

where the partial derivative is evaluated at $T = T_A$. Writing this equation for each wavelength and eliminating $T - T_A$ between the two equations yields an equation that relates radiance changes at one wavelength to radiance changes at the other:

$$B_2(T) \approx B_2(T_A) + \left(\frac{\partial B_2/\partial T}{\partial B_1/\partial T}\right)(B_1(T) - B_1(T_A)). \tag{6.41}$$

Using Eq. 6.41 to approximate $B_2(T_{B2})$ and $B_2(T_s)$ yields Eq. 6.38b as

$$B_1(T_{B2}) = B_1(T_s)\tau_2 + B_1(T_A)(1 - \tau_2). \tag{6.42}$$

Finally, eliminating $B_1(T_A)$ between Eqs. 6.38a and 6.42 yields the split-window equation

$$B_1(T_s) = B_1(T_{B1}) + \eta[B_1(T_{B1}) - B_1(T_{B2})], \tag{6.43}$$

where

$$\eta \equiv \frac{1 - \tau_1}{\tau_1 - \tau_2}. \tag{6.44}$$

By approximating the Planck function as locally linear, a slightly less accurate but more popular form of the split-window equation can be derived:

$$T_s = T_{B1} + \eta(T_{B1} - T_{B2}). \tag{6.45}$$

Two important points must be made about the split-window equation. First, the split-window technique is a correction technique. The difference between observations at two wavelengths is used to correct one of the observations for atmospheric effects to yield an improved estimate of surface radiance or temperature. Second, the factor η does not depend on the amount of absorber U. Since the transmittance is relatively large in atmospheric windows, Eq. 6.39 can be approximated as

$$\tau_\lambda \approx 1 - \beta_\lambda \frac{U}{\mu}, \tag{6.46}$$

and, therefore

$$\eta \approx \frac{\beta_1}{\beta_2 - \beta_1}. \tag{6.47}$$

The most accurate way of calculating η is not via Eq. 6.47 but by simulating satellite-measured radiances using a variety of model atmospheres with differing absorber amounts. If AVHRR channels 4 (11 μm) and 5 (12 μm) are used for the first and second wavelengths, respectively, then $\eta \approx 3$. (Note that η is positive because channel 5 is more sensitive to water vapor than is channel 4.)

Channels in separate windows, (e.g., the 3.7- and the 10.5–12.5-μm windows), which are both clouded by the same gas, can be used to estimate surface temperature, but one should not expect the equation to take the exact form of the split-window equation. In general, the surface temperature will be a linear combination of the measured brightness temperatures plus a constant term. When channels in different windows are used, the technique is referred to as a *multiple-window technique.*

In practice the split-window technique is rarely used directly. Rather, it provides theoretical justification for regressing the surface temperature of interest against measured or simulated brightness temperatures.

6.6.1 Sea and Land Surface Temperatures

Since 1970, NESDIS has operationally produced and archived global sea surface temperatures (SSTs), which are important for a variety of meteorological and oceanographic purposes (McClain *et al.*, 1985). Until the launch of NOAA 7, SST was based on the 11-μm brightness temperature. A variety of correction schemes were used depending on what instruments were in orbit. Prior to the launch of NOAA 2, only crude corrections based on climatological values of precipitable water were possible. NOAA 2, launched in 1972, had the Vertical Temperature Profile Radiometer (VTPR), an operational sounder. Beginning in 1973, 11-μm Scanning Radiometer (SR) measurements were corrected using retrieved VTPR temperature and moisture profiles. Because of problems with the retrieved soundings, after 1976 NESDIS retrieved SSTs with regression equations using 11-μm SR measurements and VTPR radiances. With the launch of TIROS N, AVHRR and HIRS/2 measurements were substituted for SR and VTPR measurements. All of these SST retrieval algorithms are known as global operational SST computation (GOSTCOMP). McClain (1979) has detailed these algorithms.

With the launch of NOAA 7 in 1981, 11-, 12-, and 3.7-μm channels became simultaneously available. New, more accurate SST retrieval algorithms using the split- and multiple-window techniques were derived. The equations were based on regression between calculated radiances and SSTs, but they were corrected based on comparisons between NOAA 7 data and SSTs measured by drifting buoys (Strong and McClain, 1984).

The current operational retrieval scheme[12] is detailed by McClain *et al.* (1985). The following discussion summarizes the major aspects of the scheme. AVHRR data are first blocked into arrays as small as 8 km on a side, near coastlines, to 25 km on a side, in the open ocean. Daytime and nighttime retrievals have separate algorithms. An important part of the algorithm is the detection and elimination of clouds. In the daytime this is accomplished by taking advantage of the fact that at 0.9 μm the sea surface is nearly black and very uniform. Blocks in which 0.9-μm reflectance varies by more than 0.32% or in which a pixel exceeds an empirically determined bidirectional reflectance are rejected as cloudy. At night reflected solar radiation is not available; less accurate techniques based on thermal infrared observations must be used. One of these tests relies on the uniformity of the sea surface. The 11-μm brightness temperature of the pixels in a block must agree to within 0.2 K. A second test relies on the fact that over the clear ocean, the 11-, 12,- and 3.7-μm brightness temperatures are correlated with each other. Regression equations, derived from cloud-free data sets, are used to predict the 3.7-μm brightness temperature from the 11-μm brightness temperature and to predict the 11-μm brightness temperature from the 12-μm brightness temperature. The predicted and observed temperatures must closely agree. Unfortunately, low uniform stratus clouds can pass both of these tests. A final test, which eliminates most stratus clouds requires that the 3.7- and 11-μm brightness temperatures differ by less than 0.7 K.

After cloud scenes have been eliminated, the sea surface temperatures are retrieved. In the daytime, the 3.7-μm brightness temperatures are contaminated by reflected sunlight. Therefore only the 11- and 12-μm brightness temperatures are used in a regression equation:

$$SST = 1.0346T_{11} + 2.5779(T_{11} - T_{12}) - 283.21. \tag{6.48}$$

In this equation the SST is in degrees Celsius, and the brightness temperatures are in kelvins. At night, all three channels may be used; three independent SSTs are calculated:

$$SST_1 = 1.0088T_{3.7} + 0.4930(T_{3.7} - T_{11}) - 273.34, \tag{6.49a}$$

$$SST_2 = 1.0350T_{11} + 2.5789(T_{11} - T_{12}) - 283.18, \tag{6.49b}$$

$$SST_3 = 1.0170T_{11} + 0.9694(T_{3.7} - T_{12}) - 276.58. \tag{6.49c}$$

All three SSTs must agree to within 1 K. SST_3, called the *triple-window solution*, is usually chosen as the SST. However, the 3.7-μm AVHRR channel has had problems with electrical noise. When the noise is too great at 3.7 μm, SST_2 is used instead, and the consistency test is not applied. Two final tests are applied to the retrieved SSTs, either day or night. First, the SST must be between −2.0 and +35.0°C, and second, the SST must be within 7.0 K of the monthly climatological value for its location. SSTs retrieved with this scheme are called *multichannel sea surface temperatures* (MCSSTs). They are archived by NESDIS on a variety of space and time scales (McClain *et al.*, 1985). An example is shown in Fig. 6.14.

[12] As with any operational retrieval system, the algorithm is occasionally updated. Users of the data should ascertain the exact algorithm used.

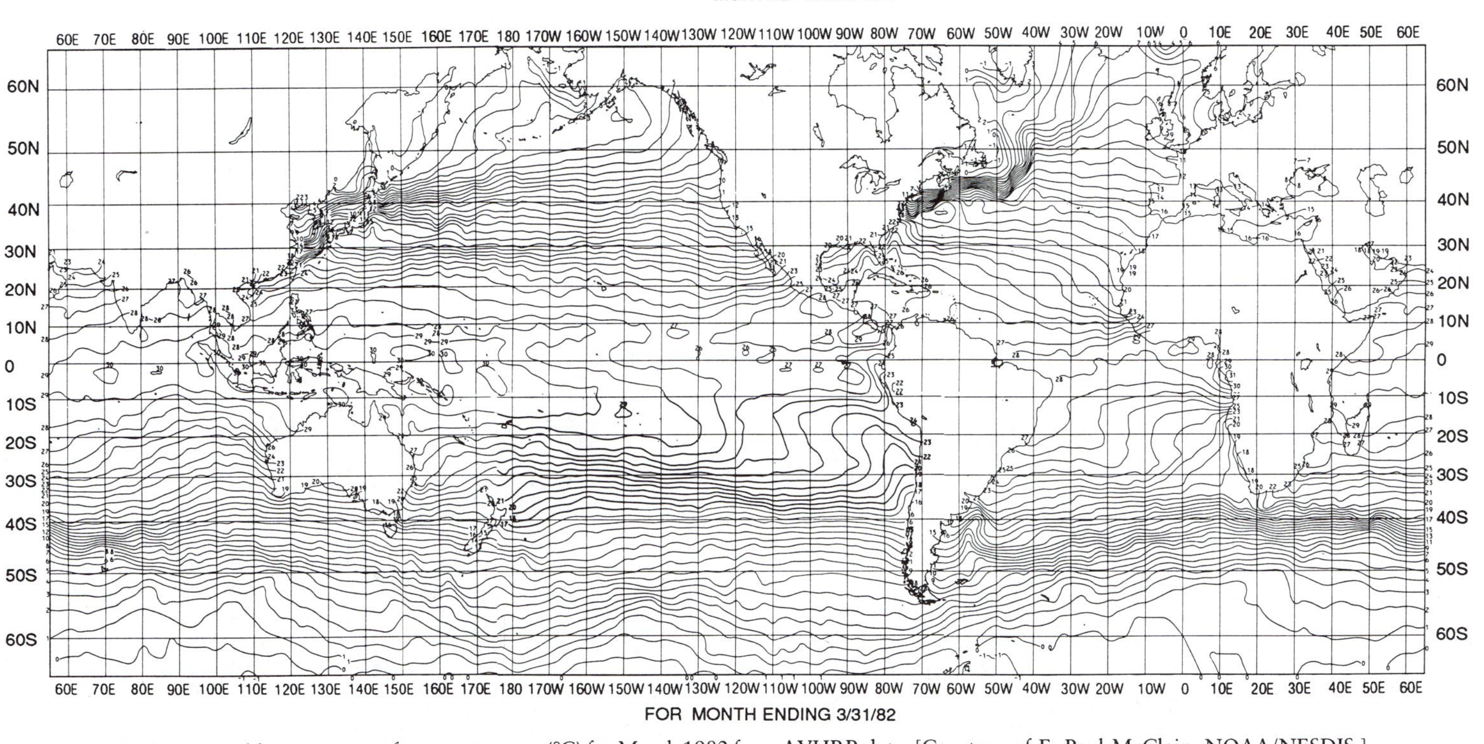

FIGURE 6.14. Monthly mean sea surface temperature (°C) for March 1982 from AVHRR data. [Courtesy of E. Paul McClain, NOAA/NESDIS.]

One problem with the MCSST algorithm is that it is affected by stratospheric aerosols produced by volcanic eruptions. Using atmospheric transmittance models, Walton (1985) showed that MCSSTs could be in error by several kelvins in the presence of realistic aerosol distributions. He developed a three-channel algorithm that could be used to derive SSTs even in the presence of aerosols. Since it depends on the 3.7-μm channel, however, it can be used only at night and only when the 3.7-μm channel is not too noisy.

Atmospheric sounding retrievals usually also retrieve skin temperature. Thus soundings over ocean produce sea surface temperatures essentially as a byproduct. SSTs have been retrieved by Susskind and Reuter (1985) using HIRS/2 and MSU data and by Bates and Smith (1985) using VAS data. The above three techniques only retrieve SSTs under cloud-free conditions. A fourth technique utilizes data from the Scanning Multichannel Microwave Radiometer (SMMR) on board the Nimbus 7 satellite (Milman and Wilheit, 1985). This technique, which is based on regression between calculated microwave brightness temperatures and SSTs, is largely unaffected by clouds. However, the Nimbus 7 SMMR was an experimental instrument that was operated only every other day. Also, the instrument was affected by heating and cooling in orbit. Corrections for these and other problems were found to be necessary.

Assessing the accuracy of satellite-estimated sea surface temperatures is not an easy task, in part because satellites sense only perhaps the top millimeter of water. Comparative measurements from buoys sample at least the top few centimeters, and ship-based measurements sample the top few meters. Significant gradients can exist between the levels sensed by satellites and those sensed by in situ sensors. There is also the problem of precisely matching the location of the satellite observation with the in situ measurement, particularly in high-gradient coastal areas. To discuss these problems and to compare SST retrieval systems with each other and with carefully quality-checked in situ measurements, a series of three workshops were held between January 1983 and February 1984 (Njoku, 1985; Njoku *et al.*, 1985). Table 6.3 contains comparisons between the four SST techniques mentioned above and ship observations. The operational AVHRR technique performed better than the three experimental techniques.

Finally, land surface temperatures can also be estimated with satellite data. Price (1984) has successfully estimated land surface temperatures using a split-

TABLE 6.3. Comparison of Satellite-Estimated SSTs with Ship Observations[a]

	Instrument			
Comparison	*AVHRR*	*HIRS/2–MSU*	*VAS*	*SMMR*
Bias (satellite–ship) (K)	−0.36	0.30	0.90	−0.21
Standard deviation (K)	0.51	0.92	0.56	1.11
Root-mean-square difference (K)	0.62	0.98	1.06	1.13

[a] After Njoku (1985).

window-type technique. Land surface temperatures are useful for a variety of applications including frost detection and warning.

6.6.2 Precipitable Water

The split-window equations also allow precipitable water to be estimated. Assuming that the Planck function is locally linear, Eqs. 6.38 become

$$T_{B1} - T_A = (T_s - T_A)\tau_1, \tag{6.50a}$$

$$T_{B2} - T_A = (T_s - T_A)\tau_2. \tag{6.50b}$$

This time eliminating T_s, we get

$$\frac{\tau_1}{\tau_2} = \frac{T_{B1} - T_A}{T_{B2} - T_A}. \tag{6.51}$$

Assuming again that τ is a function only of the integrated amount of a single absorbing gas,

$$\frac{\tau_1}{\tau_2} = \exp\left[-(\beta_2 - \beta_1)\frac{U}{\mu}\right]. \tag{6.52}$$

Eliminating τ_2/τ_1 from Eqs. 6.51 and 6.52,

$$U = \frac{\mu}{\beta_2 - \beta_1}\ln\left(\frac{T_{B1} - T_A}{T_{B2} - T_A}\right). \tag{6.53}$$

Thus, if one has measurements of split-window brightness temperatures plus an estimate of T_A, one can retrieve U. Chesters *et al.* (1983) have used VAS split-window images plus estimates of T_A derived from radiosonde measurements to estimate precipitable water over the United States (Plate 4). The success of this technique rests on the fact that atmospheric temperature is much less variable spatially than water vapor. Robinson *et al.* (1986) have used all 12 VAS channels plus surface observations in regression equations to retrieve precipitable water. Mostek *et al.* (1986) found these estimates useful in analyzing the clear, preconvective environment of thunderstorms.

The above infrared techniques for estimating precipitable water may be used over either land or ocean, but they may not be used in the presence of clouds. A similar technique employing two microwave frequencies (22.235 and 31 GHz) can be used to retrieve precipitable water over the ocean (but not over land) in the presence of clouds (Grody, 1976). This same technique also yields estimates of column-integrated cloud liquid water content, again over the ocean. This technique is discussed in Chapter 8.

Bibliography

Achtemeier, G. L., H. T. Ochs III, S. Q. Kidder, and R. W. Scott (1986). Evaluation of a multivariate variational assimilation of conventional and satellite data for the diagnosis of cyclone systems. In Y. K. Sasaki (ed.), *Variational Methods in Geosciences*. Elsevier, New York, pp. 49–54.

Anding, D., and R. Kauth (1970). Estimation of sea-surface temperature from space. *Remote Sens. Environ.*, **1**, 217.

Atlas, R., M. Ghil, and M. Halem (1982). The effect of model resolution and satellite sounding data on GLAS model forecasts. *Mon. Wea. Rev.*, **110**, 662–682.

Atlas, R. (1982). The growth of prognostic differences between GLAS model forecasts from SAT and NOSAT initial conditions. *Mon. Wea. Rev.*, **110**, 877–882.

Baker, W. E., R. Atlas, M. Halem, and J. Susskind (1984). A case study of forecast sensitivity to data and data analysis techniques. *Mon. Wea. Rev.*, **112**, 1544–1561.

Barnett, J. J., and M. Corney (1984). Temperature comparisons between the NIMBUS 7 SAMS, rocket/radiosondes and the NOAA 6 SSU. *J. Geophys. Res.*, **89**, 5294–5302.

Bates, J. J., and W. L. Smith (1985). Sea surface temperature: Observations from geostationary satellites. *J. Geophys. Res.*, **90**, 11609–11618.

Bhartia, P. K., K. F. Klenk, C. K. Wong, and D. Gordon (1984a). Intercomparison of the NIMBUS 7 SBUV/TOMS total ozone data sets with Dobson and M83 results. *J. Geophys. Res.*, **89**, 5239–5247.

Bhartia, P. K., K. F. Klenk, A. J. Fleig, C. G. Wellemeyer, and K. Gordon (1984b). Intercomparison of NIMBUS 7 solar backscattered ultraviolet ozone profiles with rocket, balloon, and Umkehr profiles. *J. Geophys. Res.*, **89**, 5227–5238.

Chahine, M. T. (1968). Determination of the temperature profile of an atmosphere from its outgoing radiance. *J. Optical Soc. Am.*, **58**, 1634–1637.

Chahine, M. T. (1970). Inverse problems in radiative transfer: Determination of atmospheric parameters. *J. Atmos. Sci.*, **27**, 960–967.

Chahine, M. T. (1974). Remote Sounding of cloudy atmospheres. I. The single cloud layer. *J. Atmos. Sci.*, **31**, 233–243.

Chahine, M. T. (1975). An analytical transformation for remote sensing of clear-column atmospheric temperature profiles. *J. Atmos. Sci.*, **32**, 1946–1952.

Chang, H. C., P. H. Hwang, T. T. Wilheit, A. T. C. Chang, D. H. Staelin, and P. W. Rosenkranz (1984). Monthly distribution of precipitable water from the Nimbus 7 SMMR data. *J. Geophys. Res.*, **89**, 5328–5334.

Chesters, D., L. W. Uccellini, and W. Robinson (1983). Low-level water vapor fields from the VISSR Atmospheric Sounder (VAS) "split window" channels. *J. Climate Appl. Meteor.*, **22**, 725–743.

Chu, W. P., M. P. McCormick, J. Lenoble, C. Brogniez, and P. Pruvost (1989). SAGE II inversion algorithm. *J. Geophys. Res.*, **94**, 8339–8352.

Chu, W. P. (1986). Inversion of SAGE II measurements. *Preprints: Second Conference on Satellite Meteorology/Remote Sensing and Applications*, American Meteorological Society, Boston, pp. J49–J51.

Chu, W. P., and M. P. McCormick (1979). Inversion of stratospheric aerosol and gaseous constituents from spacecraft solar extinction data in the 0.38–1.0-μm wavelength range. *Appl. Optics*, **18**, 1404–1413.

Cram, J. M., and M. L. Kaplan (1985). Variational assimilation of VAS data into a mesoscale model; assimilation method and sensitivity experiments. *Mon. Wea. Rev.*, **113**, 467–484.

Cressman, G. P. (1959). An operational objective analysis system. *Mon. Wea. Rev.*, **85**, 367–374.

Cunnold, D. M., W. P. Chu, R. A. Barnes, M. P. McCormick, and R. E. Veiga (1989). Validation of SAGE II ozone measurements. *J. Geophys. Res.*, **94**, 8447–8460.

Curran, R. J., R. R. Nelms, F. Allario, R. A. Bindschadler, E. V. Browell, J. L. Bufton, R. L. Byer, S. Cohen, J. J. Degnan, W. B. Grant, R. Greenstone, W. S. Heaps, B. M. Herman, A. Jalink, D. K. Killinger, L. Korb, J. B. Laudenslager, M. P. McCormick, S. H. Melfi, R. T. Menzies, V. Mohnen, J. Spinhirne, and H. J. Zwally (1987). *LASA–Lidar Atmospheric Sounder and Altimeter*. Earth Observing System Instrument Panel Report, Vol. IId, NASA, Washington, DC.

Dave, J. V. (1964). Meaning of successive iteration of the auxiliary equation in the theory of radiative transfer. *Astrophys. J.*, **140**, 1292–1303.

Deepak, A. (ed.) (1977). *Inversion Methods in Atmospheric Remote Sounding.* Academic Press, New York.

Deepak, A. (ed.) (1980). *Remote Sensing of Atmospheres and Oceans.* Academic Press, New York.

Deepak, A., H. Fleming, and M. Chahine (eds.) (1986). *Advances in Remote Sensing Retrieval Methods.* A. Deepak Publishing, Hampton, VA.

Desmarais, A. J., S. Tracton, R. McPherson, and R. van Haaren (1978). The NMC report on the Data Systems Test. NOAA/National Meteorological Center, Washington, DC.

Dey, C. H., R. A. Peterson, B. A. Ballish, P. M. Caplan, L. L. Morone, H. J. Thiebaux, H. H. White, H. E. Fleming, A. L. Reale, D. G. Gray, M. D. Goldberg, and J. M. Daniels (1989). *An Evaluation of NESDIS TOVS Physical Retrievals Using Data Impact Studies.* NOAA Tech. Memo. NWS NMC 69, Washington, DC.

Fishman, J., C. E. Watson, J. C. Larsen, and J. A. Logan (1989). Distribution of tropospheric ozone determined from satellite data. *J. Geophys. Res.*, **95**, 3599–3617.

Fleming, H. E., and L. M. McMillin (1977). Atmospheric transmittance of an absorbing gas. 2: A computationally fast and accurate transmittance model for slant paths at different zenith angles. *Appl. Optics*, **16**, 1366–1370.

Fleming, H. E. (1980). Application of computerized axial tomography techniques to satellite remote sensing of the atmosphere. Extended Abstract Volume, International Radiation Symposium, Fort Collins, CO, 11–16 August, pp. 87–89.

Fleming, H. E. (1982). Satellite remote sensing by the technique of computed tomography. *J. Appl. Meteor.*, **21**, 1538–1549.

Fleming, H. E. (1985). Temperature retrievals via satellite tomography. In Deepak, A., H. E. Fleming, and M. T. Chahine (eds.), *Advances in Remote Sensing Retrieval Methods*, A. Deepak Publishing, Hampton, VA, pp. 55–67.

Fleming, H. E., M. D. Goldberg, and D. S. Crosby (1986). Minimum variance simultaneous retrieval of temperature and water vapor from satellite radiance measurements. *Preprints: Second Conference on Satellite Meteorology/Remote Sensing and Applications.* American Meteorological Society, Boston, pp. 20–23.

Foster, M. R. (1961). An application of Wiener-Kolmogorov smoothing theory to matrix inversion. *J. Soc. Indust. Appl. Math.*, **9**, 387–392.

Fritz, S., D. Q. Wark, H. E. Fleming, W. L. Smith, H. Jacobowitz, D. T. Hilleary, and J. C. Alishouse (1972). *Temperature Sounding from Satellites.* NOAA Tech. Rep. NESS 59, Washington, DC.

Gandin, L. S. (1963). *Objective Analysis of Meteorological Fields.* Gidrometeor. Izd., Leningrad. English translation by Israel Program for Scientific Translations, Jerusalem, 1965. [NTIS N6618047.]

Ghil, M., M. Halem, and R. Atlas (1979). Time-continuous assimilation of remote-sounding data and its effect on weather forecasting. *Mon. Wea. Rev.*, **107**, 140–171.

Gille, J. C., and F. B. House (1971). On the inversion of limb radiance measurements. I: Temperature and thickness. *J. Atmos. Sci.*, **28**, 1427–1442.

Gille, J. C., P. L. Bailey, and J. M. Russell III (1980). Temperature and composition measurements from the l.r.i.r. and l.i.m.s. experiments on Nimbus 6 and 7. *Phil. Trans. Roy. Soc. London*, **A296**, 205–218.

Gille, J. C., J. M. Russell III, P. L. Bailey, L. L. Gordley, J. H. Lienesch, W. G. Planet, F. B. House, L. V. Lyjak, and S. A. Beck (1984a). Validation of temperature retrievals obtained by the Limb Infrared Monitor of the Stratosphere (LIMS) experiment on Nimbus 7. *J. Geophys. Res.*, **89**, 5147–5160

Gille, J. C., J. M. Russell III, P. L. Bailey, E. E. Remsberg, L. L. Gordley, W. F. J. Evans, H. Fischer, B. W. Gandrud, A. Girard, J. E. Harries, and S. A. Beck (1984b). Accuracy and precision of the nitric acid concentrations determined by the Limb Infrared Monitor of the Stratosphere experiment on Nimbus 7. *J. Geophys. Res.*, **89**, 5179–5190.

Gille, J. C., and J. M. Russell III (1984). The Limb Infrared Monitor of the Stratosphere: Experiment description, performance, and results. *J. Geophys. Res.*, **89**, 5125–5140.

Gordley, L. L., and J. M. Russell III (1981). Rapid inversion of limb radiance data using an emissivity growth approximation. *Appl. Optics*, **20**, 807–813.
Grody, N. C. (1976). Remote sensing of atmospheric water content from satellites using microwave radiometry. *IEEE Trans. Antennas Propag.*, **AP-24**, 155–162.
Gruber, A., and C. D. Watkins (1982). Statistical assessment of the quality of TIROS–N and NOAA–6 satellite soundings. *Mon. Wea. Rev.*, **110**, 867–876.
Halem, M., E. Kalnay, W. E. Baker, and R. Atlas (1982). An assessment of the FGGE satellite observing system during SOP-1. *Bull. Am. Meteor. Soc.*, **63**, 407–426.
Hanel, R. A., B. Schlachman, F. D. Clark, C. H. Prokesh, J. B. Taylor, W. M. Wilson, and L. Chaney (1970). The Nimbus III Michelson interferometer. *Appl. Optics*, **9**, 1767.
Hanel, R. A., B. Schlachman, D. Rodgers, and D. Vanous (1971). The Nimbus 4 Michelson interferometer. *Appl. Optics*, **10**, 1376.
Hayden, C. M. (1973). Experiments in the four-dimensional assimilation of Nimbus 4 SIRS data. *J. Appl. Meteor.*, **12**, 425–436.
Hayden, C. M. (1979). Remote soundings of temperature and moisture. In J. S. Winston (ed.), *Quantitative Meteorological Data from Satellites.* World Meteorological Organization Tech. Note No. 166 (WMO No. 531), Geneva, pp. 1–32.
Hayden, C. M., W. L. Smith, and H. M. Woolf (1981). Determination of moisture from NOAA polar orbiting satellite sounding radiances. *J. Appl. Meteor.*, **20**, 450–466.
Hayden, C. M. (1988). GOES–VAS simultaneous temperature-moisture retrieval algorithm. *J. Appl. Meteor.*, **27**, 705–733.
Hillger, D. W., and T. H. Vonder Haar (1981). Retrieval and use of high resolution moisture and stability fields from Nimbus 6 HIRS radiances in pre-convective situations. *Mon. Wea. Rev.*, **109**, 1788–1806.
Houghton, J. T., F. W. Taylor, and C. D. Rodgers (1984). *Remote Sounding of Atmospheres.* Cambridge University Press, Cambridge.
Jones, R. L., and J. A. Pyle (1984). Observations of CH_4 and N_2O by the NIMBUS 7 SAMS: A comparison with in situ data and two-dimensional numerical model calculations. *J. Geophys. Res.*, **89**, 5263–5279.
Kaplan, L. D. (1959). Inference of atmospheric structure from remote radiation measurements. *J. Optical Soc. Am.*, **49**, 1004–1007.
Kelley, G. A. M., G. A. Mills, and W. L. Smith (1978). Impact of Nimbus 6 temperature soundings on Australian region forecasts. *Bull. Am. Meteor. Soc.*, **59**, 393–405.
Kidder, S. Q., and G. L. Achtemeier (1986). Day-night variation in operationally-retrieved TOVS temperature biases. *Mon. Wea. Rev.*, **114**, 1775–1778.
Kidwell, K. B. (1986). *NOAA Polar Orbiter Data User's Guide.* NOAA/NESDIS Satellite Data Services Division, Washington, DC.
King, J. I. F. (1958). The radiative heat transfer of planet Earth. In J. A. Van Allen (ed.), *Scientific Uses of Earth Satellites.* Univ. of Michigan Press, Ann Arbor, pp. 133–136.
Klenk, K. F., P. K. Bhartia, A. J. Fleig, V. G. Kaveshwar, R. D. McPeters, and P. M. Smith (1982). Total ozone determination from the backscattered ultraviolet (BUV) experiment. *J. Climate Appl. Meteor.*, **21**, 1672–1684.
Koehler, T. L., J. C. Derber, B. D. Schmidt, and L. H. Horn (1983). An evaluation of soundings, analyses and model forecasts derived from TIROS–N and NOAA–6 satellite data. *Mon. Wea. Rev.*, **111**, 562–771.
Krueger, A. J., B. Guenther, A. J. Fleig, D. F. Heath, E. Hilsenrath, R. McPeters, and C. Prabhakara (1980). Satellite ozone measurements. *Phil. Trans. Roy. Soc. London*, **A296**, 191–203.
London, J., J. E. Frederick, and G. P. Anderson (1977). Satellite observations of the global distribution of stratospheric ozone. *J. Geophys. Res.*, **82**, 2543–2556.
Mateer, C. L. (1977). Experiments with the inversion of Nimbus–4 BUV measurements to retrieve the ozone profile. In A. Deepak (ed.), *Inversion Methods in Atmospheric Remote Sounding.* Academic Press, New York, pp. 577–595.
McClain, E. P. (1979). Satellite derived Earth surface temperatures. In J. S. Winston (ed.), *Quantitative Meteorological Data from Satellites.* World Meteorological Organization Tech. Note No. 166 (WMO No. 531), Geneva, 60–86.

McClain, E. P., W. G. Pichel, and C. C. Walton (1985). Comparative performance of AVHRR-based multichannel sea surface temperatures. *J. Geophys. Res.*, **90**, 11587–11601.

McCormick, M. P., P. Hamill, T. J. Pepin, W. P. Chu, T. J. Swissler, and L. R. McMaster (1979). Satellite studies of the stratospheric aerosol. *Bull. Am. Meteor. Soc.*, **60**, 1038–1046.

McCormick, M. P., T. J. Swissler, E. Hilsenrath, A. J. Krueger, and M. T. Osborn (1984). Satellite and correlative measurements of stratospheric ozone: Comparison of measurements made by SAGE, ECC Balloons, Chemiluminescent, and Optical Rocketsondes. *J. Geophys. Res.*, **89**, 5315–5320.

McMillin, L. M., and H. E. Fleming (1976). Atmospheric transmittance of an absorbing gas: A computationally fast and accurate transmittance model for absorbing gases with constant mixing ratios in inhomogeneous atmospheres. *Appl. Optics*, **15**, 358–363.

McMillin, L. M., H. E. Fleming, and M. L. Hill (1979). Atmospheric transmittance of an absorbing gas. 3: A computationally fast and accurate transmittance model for absorbing gases with variable mixing ratios. *Appl. Optics*, **18**, 1600–1606.

McMillin, L. M., and D. S. Crosby (1984). Theory and validation of the multiple window sea surface temperature. *J. Geophys. Res.*, **89**, 3655–3661.

McMillin, L. M. (1985). The use of Baysian classifier to retrieve tropopause heights from satellite measurements. *Preprints: Ninth Conference on Probability and Statistics in Atmospheric Sciences*. American Meteorological Society, Boston, pp. 358–359.

McMillin, L. M. (1986). The use of classification procedures to improve the accuracy of satellite soundings of temperature and moisture. *Preprints: Second Conference on Satellite Meteorology/Remote Sensing and Applications*. American Meteorological Society, Boston, pp. 1–4.

Milman, A. S., and T. T. Wilheit (1985). Sea surface temperatures from the Scanning Multichannel Microwave Radiometer on Nimbus 7. *J. Geophys. Res.*, **90**, 11631–11641.

Mostek, A., L. W. Uccellini, R. A. Peterson, and D. Chesters (1986). Assessment of VAS soundings in the analysis of a preconvective environment. *Mon. Wea. Rev.*, **114**, 62–87.

Njoku, E. G. (1985). Satellite-derived sea surface temperature: workshop comparisons. *Bull. Am. Meteor. Soc.*, **66**, 274–281.

Njoku, E .G., T. P. Barnett, R. M. Laurs, and A. C. Vastano (1985). Advances in satellite sea surface temperature measurements and oceanographic applications. *J. Geophys. Res.*, **90**, 11573–11586.

O'Lenic, E. (1986). The effect of VISSR Atmospheric Sounder data on some LFM analyses and forecasts. *Mon. Wea. Rev.*, **114**, 1832–1846.

Ohring, G. (1979). Impact of satellite sounding data on weather forecasts. *Bull. Am. Meteor. Soc.*, **60**, 1141–1147.

Phillips, N., L. McMillin, A. Gruber, and D. Wark (1979). An evaluation of early operational temperature soundings from TIROS–N. *Bull. Am. Meteor. Soc.*, **60**, 1188–1197.

Planet, W. G., D. S. Crosby, J. H. Lienesch, and M. L. Hill (1984). Determination of total ozone amount from TIROS radiance measurements. *J. Climate Appl. Meteor.*, **23**, 308–316.

Price, J. C. (1984). Land surface temperature measurements from the split window channels of the NOAA–7 Advanced Very High Resolution Radiometer. *J. Geophys. Res.*, **89**, 7231–7237.

Prabhakara, C., G. Dalu, and V. G. Kunde (1974). Estimation of sea surface temperature from remote sensing in 11 to 13 μm window region. *J. Geophys. Res.*, **79**, 5039–5044.

Reale, A. L., D. G. Gray, M. W. Chalfant, A. Swaroop, and A. Nappi (1986). Higher resolution operational satellite retrievals. *Second Conference on Satellite Meteorology/Remote Sensing and Applications,* American Meteorological Society, Boston, pp. 16–19.

Remsberg, E. E., J. M. Russell III, J. C. Gille, L. L. Gordley, P. L. Bailey, W. G. Planet, and J. E. Harries (1984). Validation of Nimbus 7 LIMS measurements of ozone. *J. Geophys. Res.*, **89**, 5161–5178.

Reuter, D., and J. Susskind (1986). Determination of water vapor profiles and total ozone burden from HIRS2/MSU sounding data. *Preprints: Second Conference on Satellite Meteorology/Remote Sensing and Applications*, American Meteorological Society, Boston, pp. 30–35

Robinson, W. D., D. Chesters, and L. W. Uccellini (1986). Optimized retrievals of precipitable water fields from combinations of VAS satellite and conventional surface observations. *J. Geophys. Res.*, **91**, 5305–5318.

Rodgers, C. D. (1970). Remote sounding of the atmospheric temperature profile in the presence of cloud. *Quart. J. Roy. Meteor. Soc.*, **96**, 654–666.

Rodgers, C. D. (1976). Retrieval of atmospheric temperature and composition from remote measurements of thermal radiation. *Rev. Geophys. Space Sci.*, **14**, 609–624.

Rodgers, C. D., R. L. Jones, and J. J. Barnett (1984). Retrieval of temperature and composition from NIMBUS 7 SAMS measurements. *J. Geophys. Res.*, **89**, 5280–5286.

Russell, J. M. III, J. C. Gille, E. E. Remsberg, L. L. Gordley, P. L. Bailey, H. Fischer, A. Girard, S. R. Drayson, W. F. J. Evans, and J. E. Harries (1984a). Validation of water vapor results measured by the Limb Infrared Monitor of the Stratosphere experiment on Nimbus 7. *J. Geophys. Res.*, **89**, 5115–5124.

Russell, J. M. III, J. C. Gille, E. E. Remsberg, L. L. Gordley, P. L. Bailey, S. R. Drayson, H. Fischer, A. Girard, J. E. Harries, and W. F. J. Evans (1984b). Validation of nitrogen dioxide results measured by the Limb Infrared Monitor of the Stratosphere (LIMS) experiment on Nimbus 7. *J. Geophys. Res.*, **89**, 5099–5107.

Saunders, P. M. (1967). Aerial measurements of sea surface temperature in the infrared. *J. Geophys. Res.*, **72**, 4109–4117.

Schlatter, T. W. (1975). Some experiments with a multivariate statistical objective analysis scheme. *Mon. Wea. Rev.*, **103**, 246–257.

Schlatter, T. W. (1981). An assessment of operational TIROS–N temperature retrievals over the United States. *Mon. Wea. Rev.*, **109**, 110–119.

Schneider, W. H., P. K. Bhartia, K. F. Klenk, and C. L. Mateer (1981). An optimum statistical technique for ozone profile retrieval from backscattered UV radiances. *Preprints: Fourth Conference on Atmospheric Radiation*, AMS, Boston, pp. 33–37.

Shuman, F., and J. Hovermale (1968). An operational six-layer primitive equation forecast model. *J. Appl. Meteor.*, **7**, 525–547.

Singer, S. F., and R. C. Wentworth (1957). A method for the determination of the vertical ozone distribution from a satellite. *J. Geophys. Res.*, **62**, 299–308.

Smith, W. L. (1969). A polynomial representation of carbon dioxide and water vapor transmission. *ESSA Tech. Rep. NESC 47*, Washington, DC.

Smith, W. L. (1970). Iterative solution of the radiative transfer equation for temperature and absorbing gas profile of an atmosphere. *Appl. Optics*, **9**, 1993–1999.

Smith, W. L., H. M. Woolf, and H. E. Fleming (1972). Retrieval of atmospheric temperature profiles from satellite measurements for dynamical forecasting. *J. Appl. Meteor.*, **11**, 113–122.

Smith, W. L., H. M. Woolf, P. G. Abel, C. M. Hayden, M. Chalfant, and N. Grody (1974). Nimbus 5 sounder data processing system. Part I: Measurement characteristics and data reduction procedures. *NOAA Tech. Memo. NESS 57*, Washington, DC.

Smith, W. L., and H. M. Woolf (1976). The use of eigenvectors of statistical covariance matrices for interpreting satellite sounding radiometer observations. *J. Atmos. Sci.*, **33**, 1127–1140.

Smith, W. L., H. M. Woolf, C. M. Hayden, D. Q. Wark, and L. M. McMillin (1979a). The TIROS–N operational vertical sounder. *Bull. Am. Meteor. Soc.*, **60**, 1177–1187.

Smith, W. L., H. B. Howell, and H. M. Woolf (1979b). The use of interferometric radiance measurements for sounding the atmosphere. *J. Atmos. Sci.*, **36**, 566–575.

Smith, W. L., V. E. Suomi, W. P. Menzel, H. M. Woolf, L. A. Sromovsky, H. E. Revercomb, C. M. Hayden, D. N. Erickson, and F. R. Mosher (1981). First sounding results from VAS–D. *Bull. Am. Meteor. Soc.*, **62**, 232–236.

Smith, W. L. (1983). The retrieval of atmospheric profiles from VAS geostationary radiance observations. *J. Atmos. Sci.*, **40**, 2025–2035.

Smith, W. L., H. M. Woolf, C. M. Hayden, and A. J. Schreiner (1985). The simultaneous retrieval export package. *Technical Proceedings of the Second International TOVS Study Conference*, Cooperative Institute for Meteorological Satellite Studies, University of Wisconsin, Madison, pp. 224–273.

Smith, W. L., H. M. Woolf, and A. J. Schreiner (1986). Simultaneous retrieval of surface atmospheric parameters: A physical and analytically correct approach. In Deepak, A., H. E. Fleming, and

M. T. Chahine (eds.), *Advances in Remote Sensing Retrieval Algorithms*. A. Deepak Publishing, Hampton, VA, pp. 221–232.

Smith, W. L., H. M. Woolf, H. B. Howell, H. E. Revercomb, and H.-L. Huang (1988). High resolution interferometer–the retrieval of atmospheric temperature and water vapor profiles. *Preprints: Third Conference on Satellite Meteorology and Oceanography*, American Meteorological Society, Boston, pp. 266–271.

Smith, W. L., H. E. Revercomb, H. B. Howell, H.-L. Huang, R. O. Knuteson, E. W. Koenig, D. D. LaPorte, S. Silverman, L. A. Sromovsky, and H. M. Woolf (1990). GHIS—the GOES high-resolution interferometer sounder. *J. Appl. Meteor.*, **29**, 1189–1204.

Somerville, R. C. J., P. H. Stone, M. Halem, J. E. Hansen, J. S. Hogan, L. M. Druyan, G. Russell, A. S. Lacis, W. J. Quirk, and J. Tenenbaum (1974). The GISS model of the global atmosphere. *J. Atmos. Sci.*, **31**, 170–194.

Spencer, R. W., and J. R. Christy (1990). Precise monitoring of global temperature trends from satellites. *Science*, **247**, 1558–1562.

Staelin, D. G., A. H. Barrett, P. W. Rosenkranz, F. T. Barath, E. J. Johnson, J. W. Waters, A. Wouters, and W. B. Lenoir (1975). The Scanning Microwave Spectrometer (SCAMS) Experiment. In *The Nimbus 6 User's Guide*. NASA, Goddard Space Flight Center, Greenbelt, MD, pp. 59–86.

Strand, O. N., and E. R. Westwater (1968). Statistical estimation of the numerical solution of a Fredholm integral equation of the first kind. *J. Assoc. Comp. Mach.*, **15**, 100–114.

Strong, A. E., and E. P. McClain (1984). Improved ocean surface temperatures from space—comparisons with drifting buoys. *Bull. Am. Meteor. Soc.*, **65**, 138–142.

Suomi, V., T. Vonder Haar, R. Krauss, and A. Stamm (1971). Possibilities for sounding the atmosphere from a geosynchronous spacecraft. *Space Research XI*, Akademie–Verlag, Berlin, pp. 609–617.

Susskind, J., and D. Reuter (1985). Retrieval of sea surface temperatures from HIRS2/MSU. *J. Geophys. Res.*, **90**, 11602–11608.

Susskind, J., J. Rosenfield, D. Reuter, and M. T. Chahine (1984). Remote sensing of weather and climate parameters for HIRS2/MSU on TIROS–N. *J. Geophys. Res.*, **89**, 4677–4697.

Teweles, S., and H. Wobus (1954). Verification of prognostic charts. *Bull. Am. Meteor. Soc.*, **35**, 455–463.

Thomasell, A., Jr., A. Gruber, H. Brodrick, N. Wolfson, and Z. Alperson (1986). The impact of satellite soundings on numerical forecasts of the Israel Meteorological Service. *Mon. Wea. Rev.*, **114**, 1251–1262.

Thompson, O. E. (1992). Regularizing the satellite tamperature-retrieval problem through singular-value decomposition of the radiative transfer physics. *Mon. Wea. Rev.*, **120**, 2314–2328.

Thompson, O. E., M. D. Goldberg, and D. A. Dazlich (1985). Pattern recognition in the satellite temperature problem. *J. Climate Appl. Meteor.*, **24**, 30–48.

Tracton, M. S., A. J. Desmarais, R. J. van Haaren, and R. D. McPherson (1980). The impact of satellite soundings on the National Meteorological Center's analysis and forecast system—the data systems test results. *Mon. Wea. Rev.*, **108**, 543–586.

Tracton, M. S., A. J. Desmarais, R. J. van Haaren, and R. D. McPherson (1981). On the system dependency of satellite sounding impact—comments in recent impact test results. *Mon. Wea. Rev.*, **109**, 197–200.

Twomey, S. (1963). On the numerical solution of Fredholm integral equations of the first kind by the inversion of the linear system produced by quadrature. *J. Assoc. Comp. Mach.*, **10**, 97–101.

Twomey, S., B. Herman, and R. Rabinoff (1977). An extension to the Chahine method of inverting the radiative transfer equation. *J. Atmos. Sci.*, **34**, 1085–1090.

Uddstrom, M. J., and D. Q. Wark (1985). A classification scheme for satellite temperature retrievals. *J. Climate Appl. Meteor.*, **24**, 16–29.

Walton, C. (1985). Satellite measurements of sea surface temperature in the presence of volcanic aerosols. *J. Climate Appl. Meteor.*, **24**, 501–507.

7

Winds

TO SPECIFY THE current state of the atmosphere and to predict its future state, one must know the flow field in addition to the mass field. In this chapter we discuss the ways in which winds (the flow field) can be derived from satellite measurements. The chapter begins with cloud- and vapor-tracking techniques, proceeds to methods for deriving winds from the soundings discussed in the last chapter, presents some special techniques for determining winds over the ocean, and concludes with a new Doppler wind-estimation method introduced on the Upper Atmosphere Research Satellite.

7.1 CLOUD AND VAPOR TRACKING

The great advantage of geostationary satellites is that they can make frequent images of the same area. Since the launch of ATS 1 in 1966, meteorologists have watched the movement of clouds and attempted to use them to estimate winds. The concept is very simple; the vector difference of the location of a cloud in two successive images divided by the time interval between images is an estimate of the horizontal wind at the level of the cloud. Winds estimated by this method are called *cloud-track* or *cloud-drift* winds. Hubert (1979) reviews the techniques used in cloud tracking. The two basic methods for tracking clouds, manual tracking and automatic tracking, are discussed in the next two sections.

7.1.1 Manual Tracking

In *manual tracking*, an analyst individually locates the clouds to be tracked on each of two or more consecutive images (Hubert and Whitney, 1971). This is usually done by positioning a cursor at the cloud center on a video display device. A computer takes the cloud positions and calculates the wind vectors using navigation algorithms. Often three images are used. The wind vector at the center time is the average of the vectors before and after the center time. If the two vectors are not consistent, the analyst is asked to recheck the positions.

Manual tracking has two advantages over automatic tracking. First, people are very good at choosing appropriate clouds to track. Low clouds can even be

FIGURE 7.1. (a) SMS 2 visible image at 2058 UTC 25 April 1975. (b) Low-level winds derived by manual cloud tracking using the image in (a) and a companion image at 2105 UTC. [After Negri and Vonder Haar (1980).]

tracked through thin cirrus overcast. Second, in theory more cloud vectors can be obtained manually than with automatic methods because individual clouds (or cloud systems) are tracked rather than areas of clouds. However, a disadvantage of manual tracking is that it is extremely tedious, which effectively limits the number of clouds that can be manually tracked.

Figure 7.1 shows a visible SMS 2 image and the wind vectors manually produced from it and its companion image. Note one deficiency with cloud-track winds: If there are no clouds, no winds can be estimated. A similar problem occurs with hurricanes and other storms; low-level winds in the most intense part of the storm often cannot be tracked because of obscuration by thick cirrus overcast.

It is interesting that the scanning frequency is an important factor in determining the number of clouds that can be tracked. Rodgers *et al.* (1979) investigated the effects of time and space resolution on cloud tracking in hurricanes. They found that if 3 min or 7.5 min rapid-scan GOES data were used in place of the normal

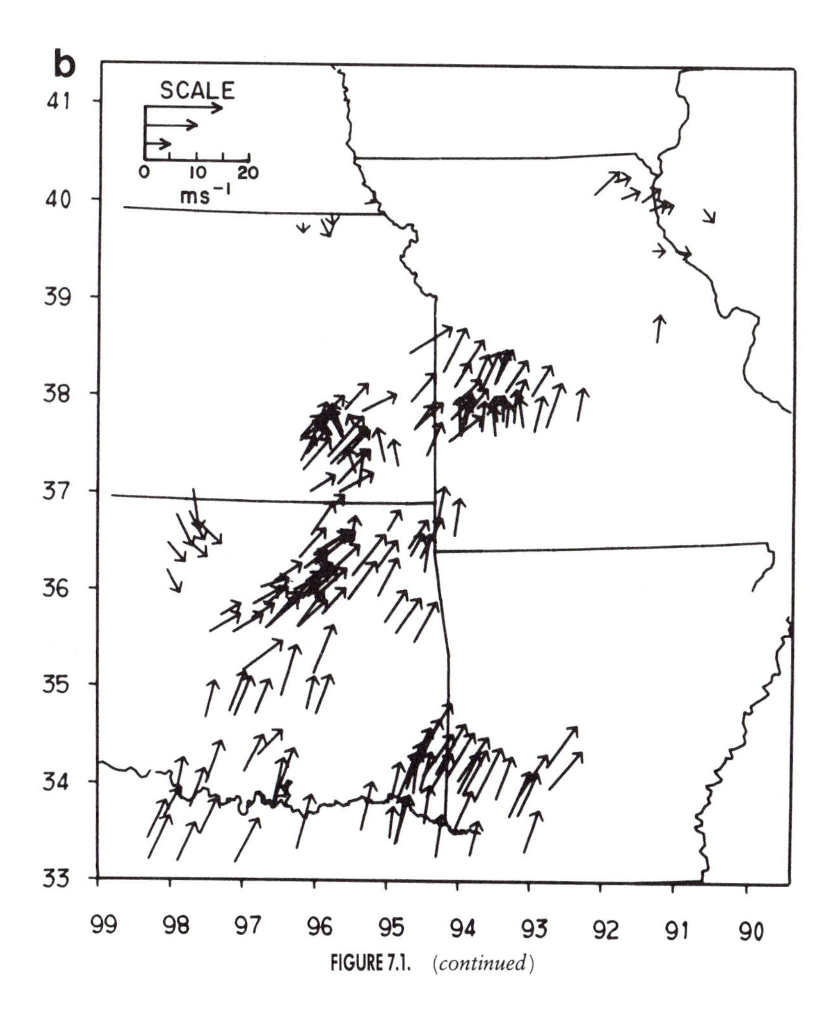

FIGURE 7.1. *(continued)*

30-min data, 10 times as many clouds could be tracked. To achieve this increase, however, it was necessary to use full resolution (1 km) visible images. Coarser resolution images prevented sufficiently accurate cloud location.

7.1.2 Automatic Tracking

Automatic tracking is done primarily using the cross-correlation method[1] of Leese *et al.* (1971) as improved by Smith and Phillips (1972) and Philips *et al.* (1972). In this technique, tracking of individual cloud features, which is difficult for a computer, is abandoned in favor of a computationally intensive method in which the average motion of an area of clouds is calculated. Figure 7.2 illustrates this process. In the first of two images, a target array of typically 16 × 16 pixels is selected. The problem is to locate this area in the second image, assuming that the clouds have moved, but changed little, during the time interval between images. A search array of typically 32 × 32 pixels, *centered* on the location of the target array, is chosen in the second image. The target array can be positioned either vertically or horizontally to any of 17 lag positions ($-8 \le m,n \le +8$) in the search array. A correlation coefficient[2] is calculated for each lag position:

$$r_{m,n} = \frac{\sum_{i=1}^{16}\sum_{j=1}^{16}[S_{i+m+8,j+n+8} - \bar{S}(m,n))](T_{i,j} - \bar{T})}{\left\{\sum_{i=1}^{16}\sum_{j=1}^{16}[S_{i+m+8,j+n+8} - \bar{S}(m,n))]^2 \sum_{i=1}^{16}\sum_{j=1}^{16}(T_{i,j} - \bar{T})^2\right\}^{1/2}}, \tag{7.1}$$

where $T_{i,j}$ is the pixel at the ith row and jth column in the target array ($1 \le i,j \le 16$), $S_{k,l}$ is the pixel at the kth row and lth column in the search array ($1 \le k,l \le 32$), $\bar{T}$ is the average of pixels in the target array, and $\bar{S}(m,n)$ is the average of the pixels in the 16 × 16 subarea of the search array at lag position (m,n):

$$\bar{S}(m,n) = \frac{1}{256}\sum_{i=1}^{16}\sum_{j=1}^{16} S_{i+m+8,j+n+8}. \tag{7.2}$$

The lag position (m,n) where the correlation coefficient is maximum is assumed to be the final position of the target array. Navigation algorithms are used to translate the initial and final positions into a wind vector.

The cross-correlation method works well, but the following should be considered. Computations based on Eq. 7.1 are lengthy even for a fast computer. In practice the correlation coefficient is calculated using fast Fourier transforms (see Leese *et al.*, 1971). Often the volume of computation is further reduced by using a first-guess wind from a recent analysis to estimate where to begin searching in the search array. Calculations of the correlation coefficient are performed in a subarea of the search array centered on the first-guess position. If a suitable peak

[1] Pattern matching techniques have been successfully implemented. See Wolf *et al.* (1977) and Endlich and Wolf (1981).

[2] A least-squares or least-absolute-value approach can also be used. See Smith (1975).

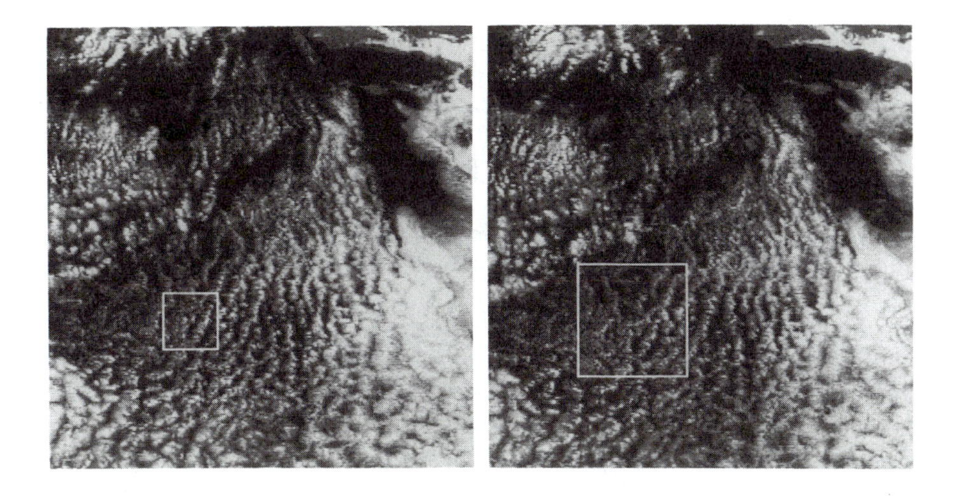

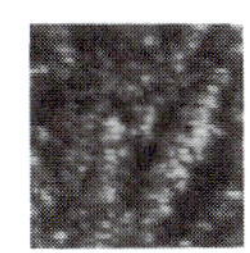

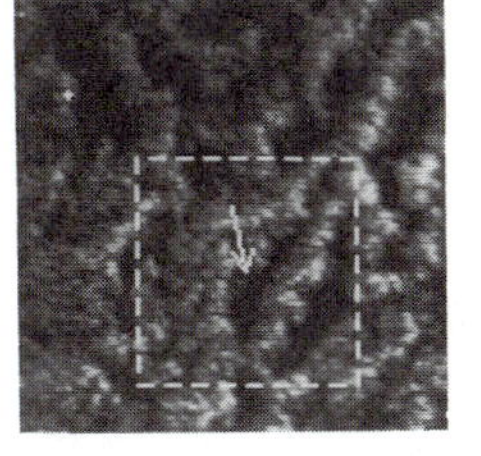

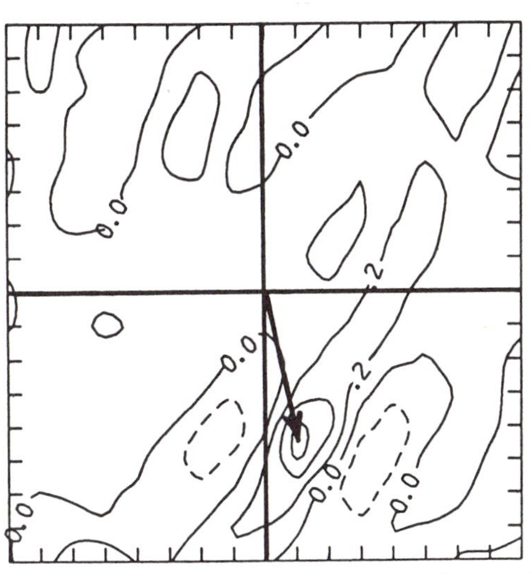

FIGURE 7.2. Top: Visible GOES—West images for 17 July 1987 at (left) 1916 UTC and (right) 1945 UTC. The boxes show the target array in the 1916 UTC image and the search array in the 1945 UTC image. Middle: Blown-up images of the target and search arrays. The dashed box shows the location of best fit of the target area, and the arrow represents the wind. Bottom: The correlation matrix. The wind is determined by the location of the maximum correlation coefficient.

can be found, the process is terminated. If not, calculations for the remainder of the search array are performed.

Occasionally the algorithm fails to find a single, well-defined peak of the correlation coefficient in the search array. Sometimes all the correlation coefficients are small, and there is no important maximum. This usually means that the cloud feature changed substantially between images. Sometimes the correlation coefficients are all large and there is no important maximum. This usually means that the cloud field was nearly uniform so that many locations produce high correlation coefficients. Finally, there may be more than one maximum. This occurs when the cloud field is repetitive, such as stratocumulus, so that more than one location in the search array matches the target array. Occasionally two levels of clouds moving with different velocities are contained in the arrays. In this case the array of correlation coefficients often has a single peak at zero lag.

Several strategies are available for dealing with these cases. Infrared data can be used to reject arrays in which more than one cloud layer is present. Alternately, an analyst can locate places which are suitable for tracking and reject those which will confuse the algorithm.

The great advantage of the automatic method is that wind vectors can be calculated by computer; it eliminates requiring a person to perform repetitive, error-prone tasks. However, it has two disadvantages: (1) fewer wind vectors may be produced, since only one wind vector is calculated for each target array; and (2) those vectors which are produced may be in error. Manual editing of the final set of wind vectors, which is desirable for manually produced wind vectors, is essential for automatically produced vectors.

Editing of wind vectors is a complex task having both internal (computer-done) and external (manually done) components. The internal components may consist of comparing each vector with its neighbors, with an existing analysis, or with an analysis of the entire set of wind vectors. If a vector deviates excessively from the analysis, it is rejected or flagged for manual inspection. Also, if the correlation coefficient matrix fails to find an unambiguous peak, the vector is rejected or flagged. The external component consists of inspection by an analyst, who looks at the plotted vectors, especially any flagged vectors, to determine whether they appear to be meteorologically meaningful. Depending on the applications intended for the vectors, a balance must be achieved between the internal and external editing process. In general, it is better to have the internal editor flag rather than reject vectors because many meteorologically interesting situations produce "anomalous" winds. An analyst can decide better than a computer which vectors are significant. On the other hand, the analyst should not have to examine too many vectors, else the advantage of the automatic method is lost.

7.1.3 Quality

Very accurate navigation and image registration are required to produce accurate cloud-track winds. Navigation errors which do not change between images do not degrade the accuracy of the wind calculation, but they cause the location of the wind vector to be misplaced horizontally. Usually this type of error is not serious. Far more serious are navigation errors which change between images. These result in errors, sometimes large, in the wind vector itself. Fortunately, these errors are infrequent today. When they occur, the pair of pictures with which the winds are produced is usually abandoned. However, one of the images can be *reregistered* by shifting it until landmarks are lined up in both images. The process of retrieving winds then proceeds as above. If reregistration is used, however, cloud vectors which are far from a landmark will be less accurate than those which are closer because the navigation algorithms are nonlinear.

Not all clouds move passively with the wind. Clouds with large vertical extent such as cumulonimbus clouds must be avoided. So too must orographically generated clouds, clouds associated with gravity waves, and clouds such as thunderstorm anvils which are evaporating or growing in a way not associated with the ambient wind. One must also be careful about clouds at frontal boundaries.

Studies have been done to determine which types of clouds are suitable for tracking. Fujita *et al.* (1975) concluded that cumulus clouds less than 0.5 km in diameter are too small to represent the wind. The clouds that best represented the subcloud layer wind are cumulus between 0.5 and 4 km in diameter. Larger cumulus clouds were also found to be suitable tracers if their vertical extent is not too large. Hasler *et al.* (1979) used in situ aircraft observations to determine that low-level cumulus clouds and cirrus clouds move essentially with the wind and thus are good candidates for tracking. Few clouds can be tracked at midlevels.

A related question is, What types of clouds can be tracked with satellite imagery? Visible images from the GOES satellites have 1-km resolution. Thus clouds a few kilometers in diameter can be tracked. If infrared imagery is used, as it must be at night, the 3 $\times$ 7-km resolution limits the size of what can be tracked. As discussed in Section 5.4.7.6, a cloud that fills only a single pixel is not likely to be well positioned. Thus, clouds or cloud systems several tens of kilometers across—larger than the ideal cloud—must be used for tracking in the infrared.

Finally, the height assignment of the calculated wind vectors is difficult. The motion of small cumulus clouds best represents the wind at cloud base, and the motion of cirrus best represents the mean wind in the cloud. How does one estimate these levels?

Several methods for using the satellite data themselves to make a height estimate are discussed in Chapter 8. These include observations of cloud shadows, comparison of the equivalent blackbody temperature of the cloud top with a sounding, and stereoscopic estimates made using two different satellites to view the same cloud. Unfortunately, shadows and stereoscopy can be used only occasionally, shadows because only low sun angles are useful, and stereoscopy because the cloud must be viewed by each satellite *at nearly the same time* (normally GOES–East and GOES–West are scheduled 15 min apart). IR height estimates may be unreliable because trackable clouds have either unknown emittance (cirrus) or unknown cloud fraction (low cumulus). Other techniques include (1) assigning the vector to the level that deviates least from the latest analysis and (2) assigning the vector to the level that in the past has agreed best with rawinsonde observations. Often the later is used. Two multichannel infrared techniques for estimating cloud height are discussed in Sections 7.1.5 and 7.1.6.

Unfortunately, there is more uncertainty in the height of the cloud than we would like. This translates into an error in the wind because placing the correct wind at the wrong level is just as bad as having the wrong wind at the correct level.

7.1.4 Operational Winds

Winds from cloud motion are produced operationally by all of the geostationary satellite operating agencies. NESDIS produces winds from GOES–East and GOES–West data, the European Space Agency (ESA) produces winds from Meteosat data, the Indian Space Research Organization (ISRO) produces winds from Insat data, and the Meteorological Satellite Center (MSC) of Japan produces winds from GMS data. The algorithms used in these different centers differ slightly. We will use the NESDIS algorithms as an example.

7.1.4.1 Wind Algorithms

At NESDIS, winds are produced on the VAS Data Utilization Center (VDUC) system four times daily nominally at 00, 06, 12 and 18 UTC. A sequence of three half-hourly images ending 1–2 h before the desired time is assembled. Because 8-km-resolution infrared data are are available around the clock, they are most frequently used. However, 8-km-resolution visible data may be used at the discretion of the analyst. With the exception of high-level winds over South America, NESDIS usually tracks clouds only over the ocean.

Low-level winds (also called *picture-pair* winds) are computed by automatic tracking. A pair of images is chosen and automatically reregistered using cross correlation on preselected landmarks. These images are usually 30 min apart, but can be 60 min apart. Approximately every 315 km, a square area of 16 × 16 pixels (roughly 125 × 125 km) is selected from the first image. A 32 × 32-pixel search window is extracted from the second image. The pixels in these arrays are examined to determine whether middle or high clouds are present. If its equivalent blackbody temperature is colder than the 700-hPa temperature, a pixel is not used in the calculations. Further, if more than 70% of the pixels are too cold, no wind is calculated. A first guess wind is obtained from the latest 850-hPa analysis. A cross-correlation matrix is computed, and the location of the maximum correlation coefficient is interpolated between 8-km steps in the cross-correlation matrix. Navigation software applied to the beginning and ending locations of the center of the target array yields a wind vector. The resultant wind vectors are compared (1) with their neighbors, (2) with the first guess, and (3) with an analysis of all the winds (Thomasell, 1979). Vectors that deviate too severely are eliminated or flagged for manual inspection. All of these computations are performed on NOAA's supercomputer. The low-level winds are all assigned to the 900-hPa level.

All winds other than the low-level winds are derived manually by NESDIS on the VDUC system,[3] and manual tracking is used at the discretion of the analyst to supplement the picture-pair winds at low levels. If three usable images are available, two wind vectors are calculated, one between the first and second images, and one between the second and third images. The two vectors are compared. If they differ too much, they are discarded. If they are acceptable, a vector between the first and last images is used as the wind. All vectors are subjectively assigned to a pressure level by the analyst based on (1) the equivalent blackbody temperature of the cloud, (2) recent analyses, and (3) experience. The vast majority of these vectors are high-level winds, but middle- and even low-level winds are occasionally produced manually.

All winds, both automatic and manual, are inspected by an analyst to ensure consistency. The analyst is free to delete any vector that seems unjustified. It takes about 3 h to produce a set of wind vectors from the two GOES satellites. Plate 5 shows one set of winds from GOES—West. The average set contains about 500 vectors, including all levels from both GOES satellites. These vectors are transmitted over GTS as are the vectors produced by ESA, ISRO, and MSC. NESDIS has archived GOES cloud-track winds since 1974. ESA, MSC and ISRO also archive the winds which they produce.

7.1.4.2 Accuracy

For several years the World Meteorological Organization (WMO) Committee on Coordination for Geostationary Meteorological Satellites (CGMS) has organized the collection of wind data sets from satellites and rawinsondes for comparison. Satellite winds are compared with (1) winds from other satellites (in the region where satellite coverage overlaps) and (2) rawinsonde winds. NESDIS performs comparisons between winds from GOES, Meteosat, and GMS and between GOES winds and rawinsonde winds. These comparisons are similar to those discussed in Chapter 6 between temperature and moisture soundings made by satellites and rawinsondes. Observations that are nearly collocated in time and space are compared. Recent comparisons between winds from different satellites have yielded rms vector differences of about 6 m s^{-1} for high-level winds and about 4 m s^{-1} for low-level winds. Comparison between satellite winds and rawinsonde winds is worse: rms vector differences of 10–15 m s^{-1} for high-level winds and 5–6 m s^{-1} for low-level winds.

There are several sources of differences between satellite winds and rawinsonde winds, among them: improper height assignment, inexact collocation in either time or space, errors of measurement and tracking, and unrepresentative rawinsonde winds. Height assignment appears to be a primary source of error. One would expect errors in height assignment in a sheared environment to increase rms vector differences. When satellite winds are compared with rawinsonde winds

[3] ESA uses automatic tracking for all of its winds (Bowen *et al.*, 1979). Histograms of the IR and visible pixels are used to determine whether an area is trackable, and if so, to what level to assign the vector.

not at the assigned level (as above) but at the level where the satellite wind differs least from the rawinsonde wind (the *level of best fit*), rms vector differences decrease by 40–50%. Also, satellite—satellite differences are smaller than satellite—rawinsonde differences. Since both satellites track virtually the same clouds, height assignment is not a factor in this comparison. If height assignment can be improved, cloud-drift wind errors will likely decrease. Sections 7.1.5 and 7.1.6 describe techniques that offer improved height assignment.

7.1.4.3 Use in Numerical Models

Cloud-drift winds are used in numerical models throughout the world. NMC uses winds from GMS and Meteosat as well as the two GOES satellites. As of the time of this writing, winds from INSAT are not yet in use at NMC. In contrast to the satellite soundings, satellite winds are used in both short- and medium-range forecasts at NMC. About half of the wind vectors that go into the GDAS run are used in the LFM and RAFS runs. Impact studies, such as those conducted for satellite soundings, have not, to our knowledge, been performed for cloud-drift winds, although some studies have included cloud-drift winds with satellite soundings and other satellite data in their SAT/NOSAT comparisons.

7.1.5 CO_2 Slicing

Menzel *et al.* (1983) have used VAS images in the 15-μm CO_2 band to track clouds. Figure 7.3 shows images in VAS channels 3, 4, and 5 (14.2, 14.0, and 13.3 μm). Note that as one moves closer to the 15-μm center of the CO_2 band, low clouds disappear. Clouds that appear only in channel 5 are low clouds. Clouds that appear in channels 4 and 5 but not in channel 3 are middle clouds. Clouds that appear in all channels have high cloud tops. This *slicing* of the atmosphere using CO_2 channels results in winds that are more likely to be assigned to the correct height. In addition, using ratios of the radiances, a quantitative estimate of the cloud top can be made (see Chapter 8). Cloud heights can be estimated within 50 hPa rms by this method.

7.1.6 Vapor Tracking

Although 6.7-μm water-vapor images have been available for many years on the polar-orbiting Nimbus satellites, the advent in 1977 of the high-resolution (5 km) water vapor channel on Meteosat made possible the tracking of water vapor features (Eigenwillig and Fischer, 1982). Vapor tracking can also be accomplished using lower-resolution (14-km) VAS data (Stewart *et al.*, 1985). The advantage of *vapor-track winds* is that they may be estimated in areas that are free of clouds. In addition, since the weighting function of the water vapor channels peaks near the 450-hPa level (depending on moisture), the derived winds are representative of the middle troposphere, a region in which there are few trackable clouds. Thus vapor-track and cloud-track winds complement each other.

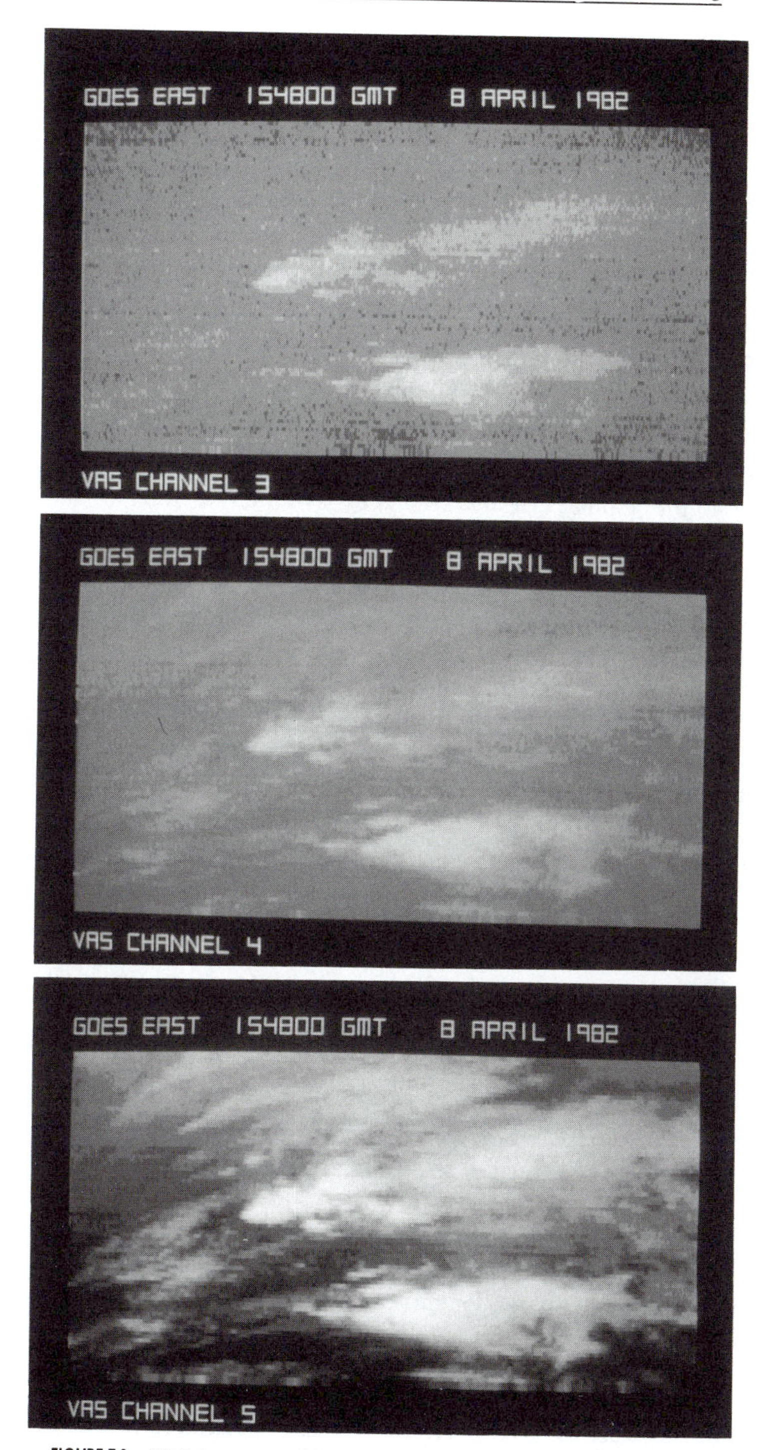

FIGURE 7.3. VAS images used in CO_2 slicing. [After Menzel *et al.* (1983).]

Vapor-track winds suffer from the same difficulties as cloud-track winds, in particular the task of assigning a height to the derived wind vector. Stewart *et al.* (1985) have found some success in assigning the wind to the level where the 6.7-μm brightness temperature equals the environmental temperature, determined from an analysis. A problem unique to vapor tracking is that water vapor images are "fuzzy." It is sometimes difficult to identify trackable features. High-pass filtering and edge enhancement have been used to alleviate this problem.

To date, the accuracy of vapor-track winds has been somewhat less than that of cloud-track winds. Eigenwillig and Fischer (1982) found rms speed differences of about 5 m s^{-1} and rms direction differences of about 20°. The accuracy of vapor-track winds, which are still experimental, will likely improve.

High clouds are visible in water vapor images. Szejwach (1982) has shown that the emittance of cirrus clouds is about the same at 6.7 and 11 μm. Simultaneous observation of thin cirrus at both wavelengths allows the temperature (and thus the height) of the cloud to be estimated. The use of both water vapor and window infrared data to track high clouds may improve the accuracy of upper-level wind estimation by improving the height assignment.

7.2 WINDS FROM SOUNDINGS

In general, the flow field is not independent of the mass field. Given temperature soundings and an estimate of the sea-level pressure, one can calculate the height of any pressure surface. In the extratopics, winds are related to the height field. Three relationships come immediately to mind. The *geostrophic wind* is related to the gradient of the geopotential:

$$\mathbf{V}_g = \frac{1}{f}\,\hat{\mathbf{k}} \times \nabla\Phi, \tag{7.3}$$

where Φ is the geopotential (gz), f is the Coriolis parameter [$= 2(d\Omega_e/dt)\sin\Theta$, $d\Omega_e/dt$ = angular velocity of rotation of Earth, Θ = latitude], and $\hat{\mathbf{k}}$ is a unit vector in the vertical direction. The *gradient wind* blows in the same direction as the geostrophic wind, but with magnitude

$$V_{gr} = \frac{2V_g}{1 + \left(1 + 4\dfrac{V_g}{fR_T}\right)^{1/2}}, \tag{7.4}$$

where R_T is the radius of curvature of the trajectory of an air parcel. The gradient wind is a better approximation for curved flow than the geostrophic wind. Finally, the *quasi-geostrophic approximation* (see Holton, 1992) may be used to estimate the wind. The question is: How accurate are winds calculated from satellite temperature soundings?

Hayden (1984) used surface pressure reports and station elevations to calculate the 1000-hPa height field. He then used VAS soundings and the hypsometric equation to calculate the heights of pressure surfaces aloft. He estimated winds from the heights using the three approximations mentioned above.[4] For 2 days in 1982, he compared these winds with rawinsonde observations (sample size = 174). The gradient wind estimate was the best of the three. At 300 hPa the comparison between the satellite-estimated gradient wind and the rawinsonde-measured wind was good. The bias was less than 1 m s^{-1}, and the rms error was under 10 m s^{-1}. The correlation coefficient was 0.87. At 850 hPa, the results were not as good. Although the rms errors were lower than at 300 hPa, the correlation coefficient was much lower, essentially zero on one of the days. These results make physical sense because satellite soundings are least accurate near the surface. On the other hand, the mean temperature of a deep layer of the troposphere is accurately retrieved from radiance data. Therefore, the winds aloft ought to be accurately estimated by this technique, assuming that the gradient wind equation is capable of approximating the actual wind. Hayden's results are supported by those of Lord *et al.* (1984), who compared winds derived from satellite soundings with aircraft observations.

It is interesting to note that winds derived from VAS data complement cloud-track winds because VAS winds can only be calculated in clear areas.

7.3 OCEAN SURFACE WINDS

Unique properties of the ocean surface or of oceanic storms have allowed the development of several techniques to measure surface winds over the ocean.

7.3.1 Active Microwave Winds

Since the invention of radar, there has been an interest in the radar return from the sea surface (sea clutter). An extensive series of observations using aircraft-mounted radars during the 1960s indicated that there is a relationship between surface wind and radar return. NASA then began development of a specialized radar known as a *scatterometer*, which very accurately measures the backscattering cross section per unit sea surface area. This is a unitless parameter called the *normalized radar cross section*, $\sigma°$. Scatterometers were flown on aircraft and on Skylab. The results were sufficiently good that a scatterometer was planned for Seasat.

Seasat's mission was to demonstrate the feasibility of global ocean monitoring from space. Its instrument complement consisted of a Visible and Infrared Radiometer (VIRR); one passive microwave instrument, the Scanning Multichannel Microwave Radiometer (SMMR, identical to the one flown on Nimbus 7); and three active microwave instruments (radars): an altimeter (ALT), a synthetic aperture radar (SAR), and a scatterometer, the Seasat–A Satellite Scatterometer (SASS).

[4] In the gradient wind relationship, R_T was approximated as the radius of curvature of the geostrophic streamline.

Seasat was launched on 28 June 1978, but it suffered a power supply failure on 10 October 1978.

SASS had bar-shaped, dual-polarization, fan-beam antennas extending horizontally at angles of 45°, 135°, 225°, and 315° from the satellite velocity vector. The instrument was designed to measure $\sigma°$ in three swaths, one left, one right, and one centered on the satellite ground track (Fig. 7.4). The orientation of the antennas caused the instrument to scan at a 45° angle to the ground track. The antenna footprint was electronically divided by Doppler filters into cells approximately 18 × 70 km. To estimate both wind direction and wind speed, each spot of ocean had to be observed at least twice from different azimuth angles. SASS accomplished this by observing a spot with the forward antennas, then observing it at right angles 1–3 min later with the aft antennas.

The relationship between $\sigma°$ and surface wind is complex and not completely understood. We present only a simplified treatment; the reader is referred to Stewart (1985) for a more thorough discussion. Wind induces waves of all wavelengths on the sea surface. Waves that are near the wavelength of the radar backscatter radiation through *Bragg scattering*: reflected radiation from one wave reinforces (or cancels) that from the successive wave depending on wave spacing and viewing geometry. Since the SASS wavelength was about 2 cm, centimeter-size *capillary waves* are responsible for the backscatter. Several parameters influence $\sigma°$: primarily wind speed, zenith angle, relative azimuth between the radar beam and the wind, and radiation polarization; but also water temperature, viscosity, the spectrum of larger waves, and surface tension. The largest backscatter occurs in the upwind and downwind directions. Backscatter minima are in the vicinity of the crosswind directions.

To account for the large wind shear that occurs in the boundary layer, the wind speed U employed in these calculations is taken to be the *19.5 m neutral*

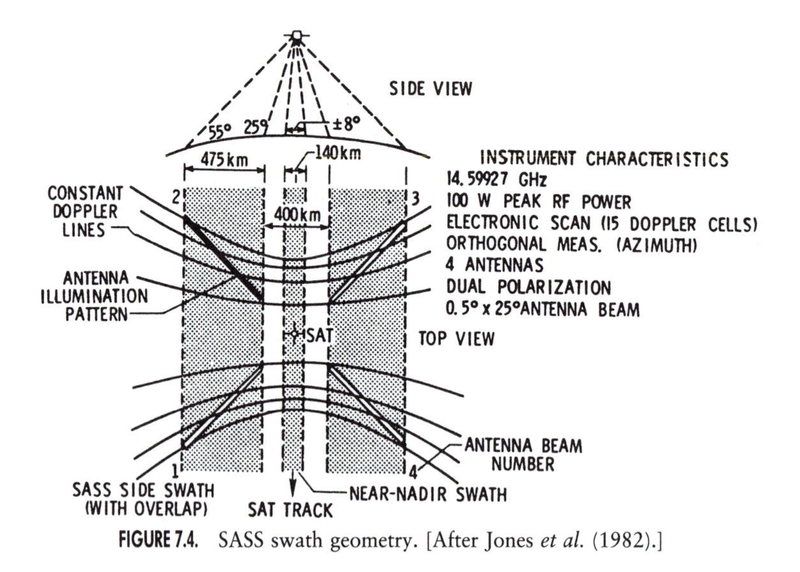

FIGURE 7.4. SASS swath geometry. [After Jones *et al.* (1982).]

stability wind, which is the wind speed that would occur at a height of 19.5 m above the surface if the atmosphere had neutral (dry adiabatic) stability. U is calculated from *in situ* observations by applying a boundary-layer model which uses the height of the anemometer (among other things) as an input parameter.

The algorithm for retrieving wind speed and direction from SASS observations of σ° is called the *SASS I model function* (Schroeder *et al.*, 1982; Jones *et al.* 1982). The model assumes that

$$\sigma^\circ = GU^H, \tag{7.5}$$

where G and H are functions of zenith angle (ζ), relative azimuth angle (χ), and polarization (P, vertical or horizontal). Tabulated values of G and H were determined empirically by fitting prelaunch aircraft and postlaunch SASS observations of σ° to in situ observations of wind speed and direction. Suppose that SASS has made two measurements of σ° at 90° angles. Then we have two equations:

$$\sigma_1^\circ = G(\zeta_1, \chi_1, P_1)U^{H(\zeta_1, \chi_1, P_1)}, \tag{7.6a}$$

$$\sigma_2^\circ = G(\zeta_2, \chi_1 + 90^\circ, P_2)U^{H(\zeta_2, \chi_1 + 90^\circ, P_2)}. \tag{7.6b}$$

These are two nonlinear equations in two unknowns: U and χ_1. They can be solved by plotting U versus χ_1 (Fig. 7.5). The two equations yield two curves. Since σ° is maximum in the upwind and downwind directions and minimum in the crosswind directions, the curves have two maxima and two minima. Since the two observations are made 90° apart, the two curves are shifted 90° so that the minima of one are superimposed on the maxima of the other. The points where the curves intersect are possible solutions, called *aliases*. Unfortunately most of the time there are four aliases, only one of which represents the wind. Sometimes two or three aliases are found. When more than two SASS cells overlap,

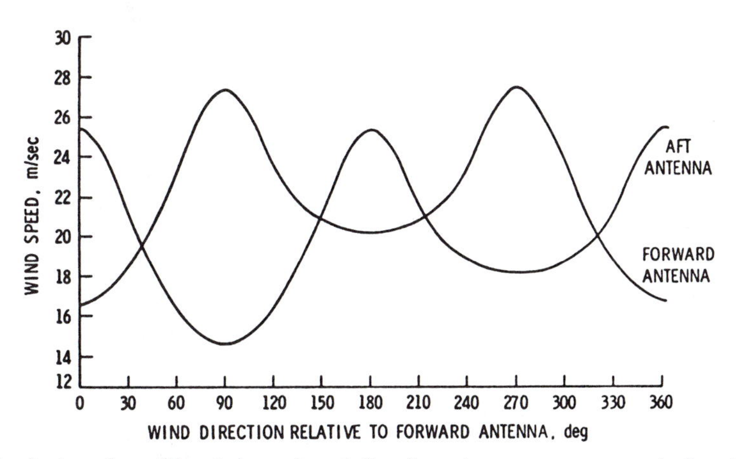

FIGURE 7.5. A plot of possible wind speeds and directions given measurements of σ° made by the forward and aft antennas. The intersections of the two curves are the aliases, one of which represents the true 19.5-m neutral stability wind. [After Jones *et al.* (1982).]

a least-squares estimator is used to derive the aliases. In the narrow swath under the satellite, the wind speed, but not the wind direction, can be retrieved.

The SASS retrievals must be dealiased using auxiliary meteorological information such as ship wind reports or model first-guess fields. Figure 7.6 shows an example of SASS aliases, a model first-guess wind analysis, and a dealiased analysis produced by an objective dealiasing procedure (Baker *et al.*, 1984). This dealiasing procedure is complicated, and it has been suggested that future scatterometers have two additional antennas to reduce the number of aliases.

The goal for SASS was to retrieve ocean surface winds with rms errors of 2 m s^{-1} in speed and 20° in direction. Lame and Born (1982) reported wind speed accuracy of 1.3 m s^{-1} and direction accuracy of 16°. Woiceshyn *et al.* (1986), however, report that a substantial number of the retrieved SASS winds may not have met the accuracy goals. They found several areas where the SASS I model function was deficient, such as in neglecting viscosity and surface tension, and they suggest that an improved model is both necessary and possible.

In addition to establishing accuracy, it is important to determine the contribution which a satellite instrument may make to meteorological analyses and forecasts. In September 1978, an explosively developing cyclone (dubbed the *QEII storm*) off the east coast of the United States damaged the ocean liner *Queen Elizabeth II* (Gyakum, 1983). Duffy and Atlas (1986) found that SASS surface winds had little impact on forecasts of the storm. However, when upper-level winds were made to agree with the SASS surface winds, a large positive impact was found on forecasts. In other words, surface winds alone were insufficient to improve forecasts. Duffy *et al.* (1984) found a small, inconsistent impact of SASS winds in forecasts made with the U.S. Navy's operational numerical forecast model. They cite incorrectly dealiased winds as a possible reason for this low impact. Baker *et al.* (1984) found a small positive impact of SASS winds, but this positive impact disappeared when satellite temperature soundings were included. This indicates that temperature soundings and surface winds may be partially redundant in their ability to improve weather forecasts.

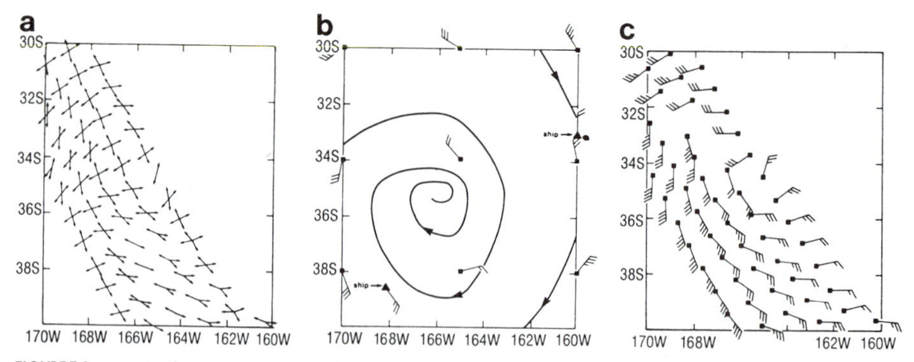

FIGURE 7.6. (a) Ambiguous SASS wind vectors (aliases) at approximately 2200 UTC 7 September 1978. (b) Model first guess 1000-hPa wind vectors and streamlines for 0000 UTC 8 September 1978. (c) Objectively dealiased SASS winds. [After Baker *et al.* (1984).]

7.3.2 Passive Microwave Winds

Microwave radiation emitted from the sea surface is a function of wind speed for two reasons. First, the same wind-generated waves that alter the radar backscatter coefficient also change the surface emittance. Figure 7.7 shows the calculated emittance at 19.4 GHz of a wind-roughened ocean surface as a function of incidence (viewing) angle. The emittance change is strongly dependent on polariza-

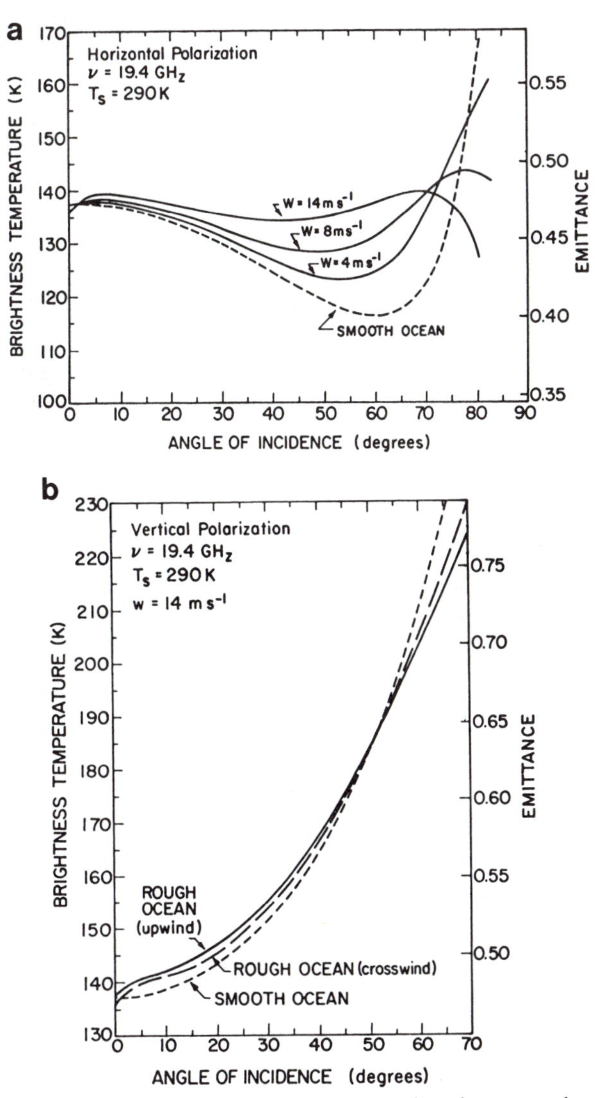

FIGURE 7.7. Calculated emittance at 19.4 GHz of a wind-roughened ocean surface as a function of incidence angle (measured from vertical): (a) horizontal polarization, (b) vertical polarization. [After Stogryn (1967). © 1967 IEEE.]

tion and viewing angle. For nadir viewing, wind-induced roughness has little effect on emittance. For nonnadir viewing, roughness increases emittance for horizontally polarized radiation; roughness decreases emittance for vertically polarized radiation out to 55° and increases it thereafter. Note that the angle between the viewing azimuth and the wind direction affects the emittance slightly. The second reason that sea surface emittance is a function of wind speed is that at wind speeds above about 7 m s^{-1} foam begins to form on the sea surface. The fraction of the surface covered by foam increases with wind speed. Since foam is essentially black, emittance increases rapidly, and nearly linearly, with wind speed above 7 m s^{-1}. Wilheit (1979) has proposed a sea surface emittance model which includes the effects of foam and roughness.

The change of sea surface emittance with wind speed is the basis for wind speed retrieval algorithms using Scanning Multichannel Microwave Radiometer (SMMR) data. Wilheit and Chang (1980) developed such algorithms by regressing simulated brightness temperatures (or simple functions of brightness temperature) against wind speed. Wilheit *et al.* (1984) tested and improved these algorithms using SMMR observations. Several problems were encountered with the real data (including manmade radio frequency interference in the 6.6-GHz channel). However, the best SMMR-estimated winds were within about 2.7 m s^{-1} rms of winds measured by NOAA Environmental Data Buoys or ships.[5] The SSM/I instrument on the DMSP satellites can also estimate surface wind speed over the ocean. Since surface emittance is only weakly dependent on wind direction, passive microwave techniques yield only wind speed.

7.3.3 Tropical Cyclone Winds

The strongest and most dangerous winds over the ocean are the result of tropical cyclones. One of the chief contributions of meteorological satellites is simply the detection of these storms. Since the 1960s, no tropical cyclone anywhere on Earth has gone undetected. However, while shipping interests are pleased to know about the location of tropical cyclones, they also need to know the wind speed so that they can keep their ships at a safe distance. Not all tropical cyclones are equally dangerous in terms of wind speed and the ocean waves which result. There is a great deal of interest, therefore, in using satellite data to estimate hurricane winds.

7.3.3.1 The Dvorak Technique

The most important technique for estimating winds in tropical cyclones was developed by Dvorak (1973, 1975, 1984). Each tropical cyclone goes through a life cycle which may be classified into one of several types by its appearance in visible or infrared satellite images. Figure 7.8 shows examples of six storm types identified by Dvorak. (Two types are not shown.) In addition to typing the storm, its current strength can be determined from the satellite images. *T–numbers*

[5] The 19.5-m neutral stability wind was used in these comparisons.

(T1–T8) are used for this purpose. The T–numbers are determined by detailed examination of the images following a decision tree (Fig. 7.9) and by guidelines on the expected day-to-day change in the T–number of a storm. Once-daily images are generally used in this technique, and it is desirable for the images to be at the same time each day to avoid confusion caused by diurnal changes in tropical storm cloudiness. The T–number is related to the minimum sea level pressure (central pressure), although the relationship is different in the different oceans. A *current intensity* (CI) number can be derived from the T–number. The CI number is related to the storm's intensity (maximum sustained surface wind speed). Comparison with aircraft-observed winds shows that the Dvorak technique has an rms error of approximately 6 m s^{-1} in tropical cyclone intensity.

In 1984 Dvorak introduced a variant of the above technique, called the *enhanced IR technique*, which uses specially enhanced infrared images instead of visible images. This, of course, allows intensity estimates to be made at night. A decision tree similar to that in Fig. 7.9 is used in the analysis.

The visible and enhanced IR Dvorak techniques are subjective and require training to be used effectively, but they yield good wind estimates. They are in use throughout the tropical regions of the world.

Partly to lessen the subjective nature of the Dvorak technique, Dvorak (1984) introduced a digital IR technique for estimating tropical cyclone strength. The digital IR technique uses two parameters: the equivalent blackbody temperature of the eye of a storm (T_{eye}), and the average equivalent blackbody temperature of the surrounding cloud shield (T_{cloud}). The warmer T_{eye} and the colder T_{cloud}, the more intense the storm. T–numbers can be estimated from these two parameters using a table (see Dvorak, 1984).

7.3.3.2 Microwave Sounding

Microwave sounders are nearly unaffected by clouds; thus they can probe the interior of tropical cyclones. The surface pressure drop in the center of a tropical cyclone is the result of warming in the storm's core. This warming is maximum in the upper troposphere but below the tops of the clouds. It is therefore not detectable with infrared sounders. Rosenkranz *et al.* (1978) first noticed that the 55.45-GHz channel of the Nimbus 6 Scanning Microwave Spectrometer (SCAMS) showed a warm anomaly (difference between temperatures inside and outside the storm) above a strong western Pacific typhoon. This warm anomaly is explained by the fact that the 55.45 GHz weighting function peaks in the upper troposphere in the region of the maximum tropical cyclone temperature anomaly. Kidder *et al.* (1978) found that the 55.45-GHz brightness temperature anomaly is correlated with tropical cyclone central pressure, and Kidder *et al.* (1980) succeeded in relating the radial gradient of 55.45-GHz brightness temperature to surface pressure gradient and then to surface wind speed at outer radii, specifically to the radii of 15.4-m s^{-1} (30-knot) and 25.7-m s^{-1} (50-knot) winds, which are important for ship routing.

Velden and Smith (1983) extended this technique in two ways. First, they used data from the Microwave Sounding Unit (MSU) on board the NOAA satellites.

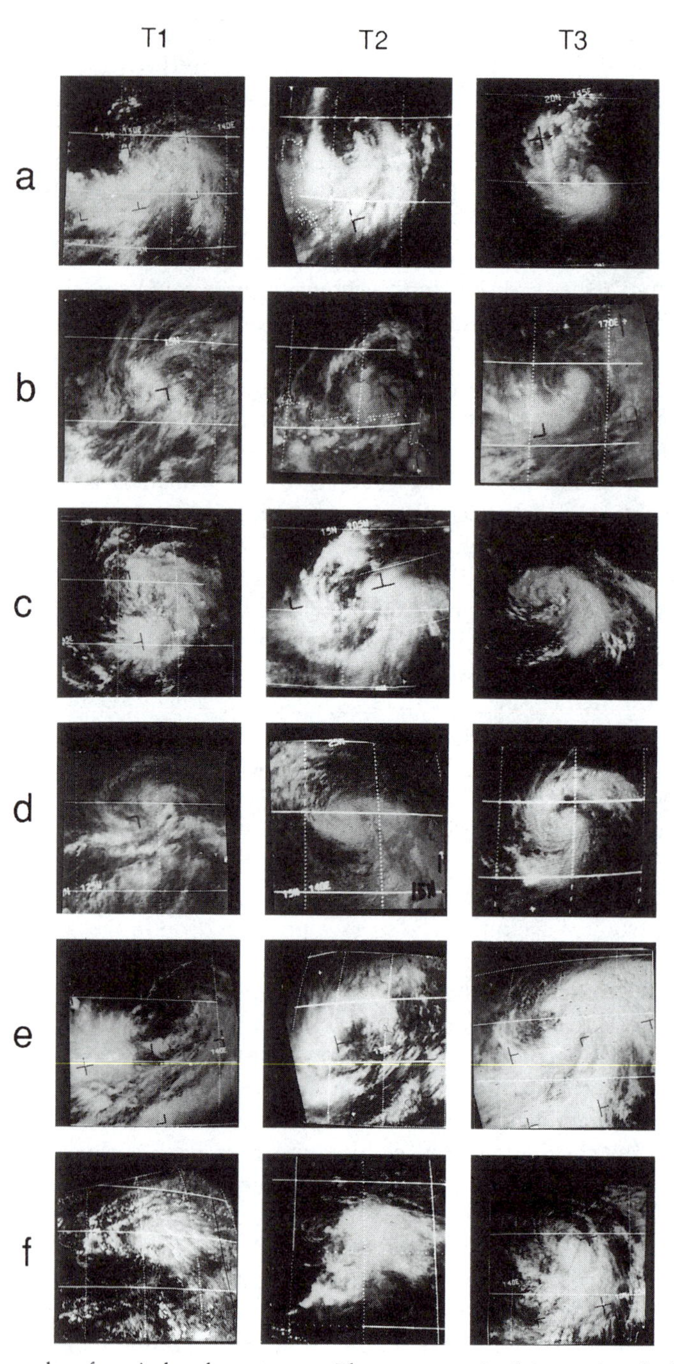

FIGURE 7.8. Examples of tropical cyclone patterns. The storm type (a–f) is noted at the left (types g and h are not shown). The storm intensity as indicated by the T–number is noted at the top. [After Dvorak (1975).]

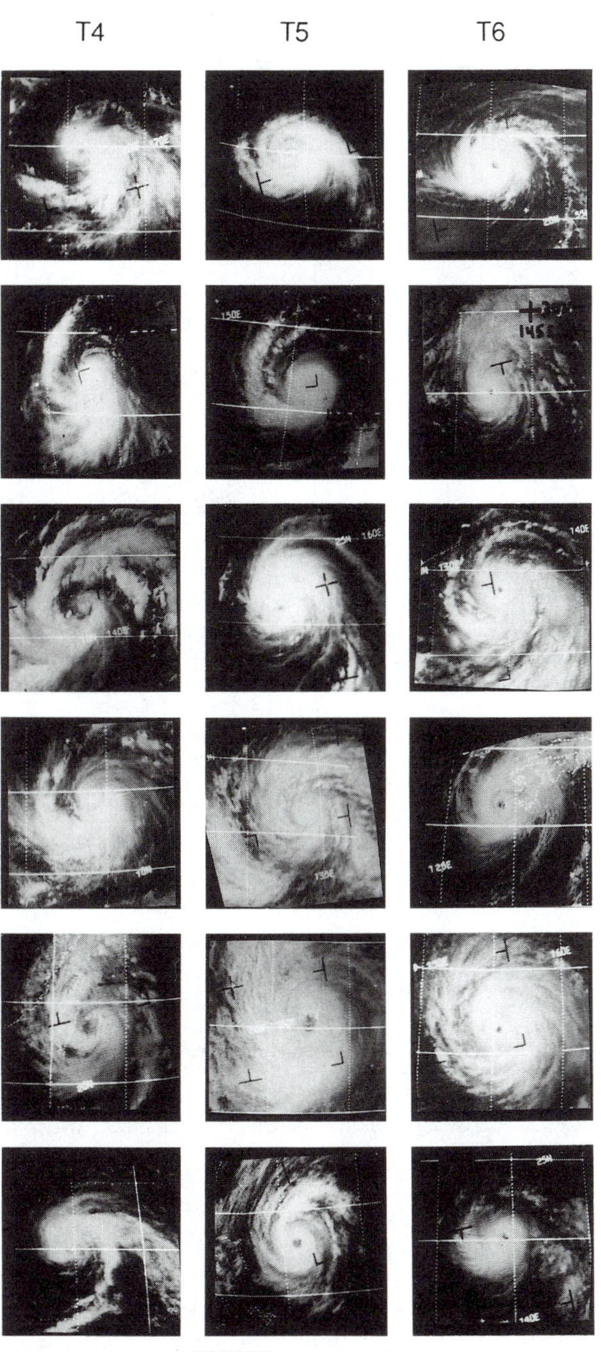

FIGURE 7.8. *(continued)*

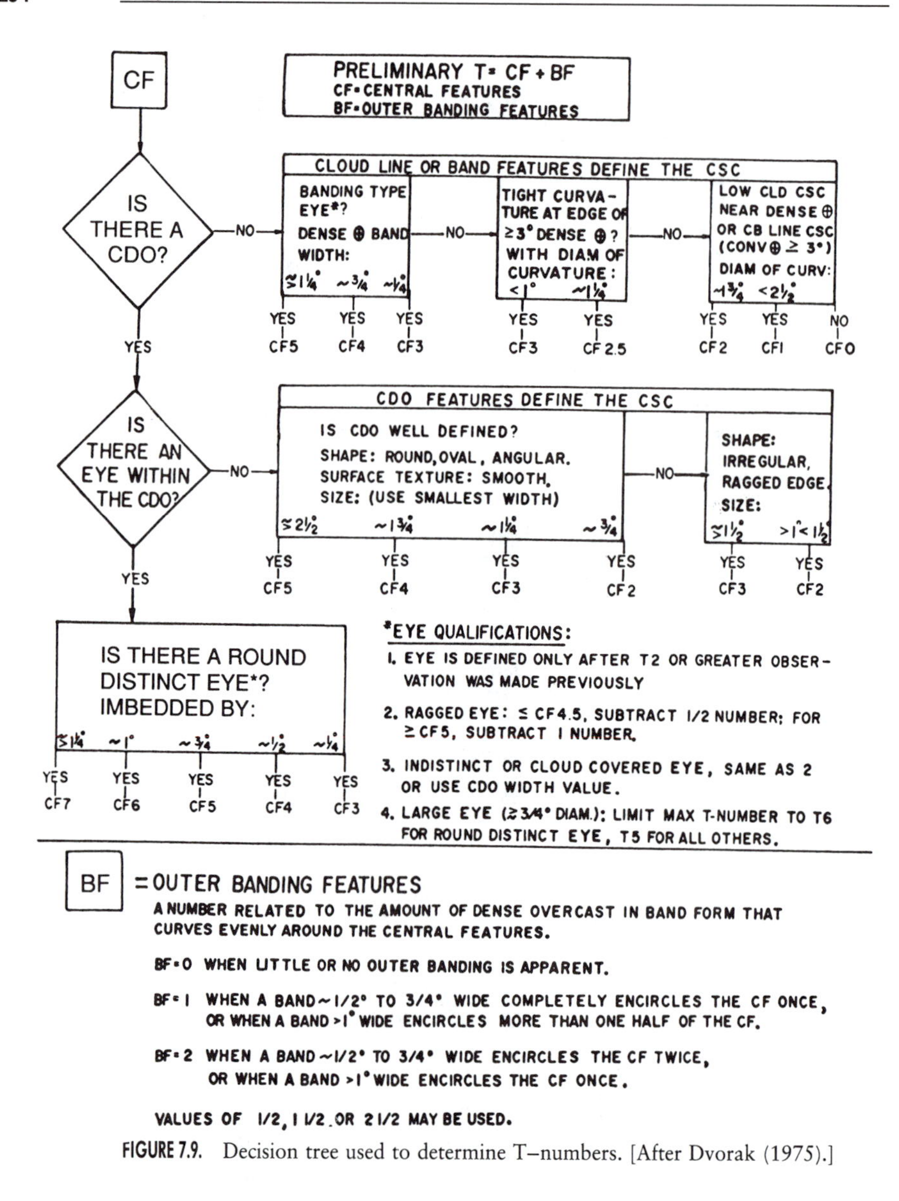

FIGURE 7.9. Decision tree used to determine T–numbers. [After Dvorak (1975).]

The MSU has better spatial resolution than did the SCAMS, and it measures the brightness temperatures more accurately. Second, they retrieved atmospheric temperature profiles inside the tropical cyclones. The retrieved temperature anomaly at 250 hPa is correlated with both the central pressure (minimum sea-level pressure) and with intensity (maximum sustained sea-level wind speed). They found rms errors of about 6 hPa in central pressure estimates and about 6 m s^{-1} in intensity estimates. This technique has been used semioperationally at the U.S. National Hurricane Center.

A difficulty with the technique is that none of the microwave sounders launched to date have provided contiguous coverage of the Earth; gaps exist between consecutive orbits. Furthermore, the outermost scan spots must be discarded because the resolution becomes too poor. This widens the gaps between consecutive orbits to the point that there is only about a 50% probability that a particular storm will be viewed sufficiently well to apply these techniques. This difficulty is one of the reasons that placing a microwave sounder on a geostationary satellite would be extremely valuable.

7.4 DOPPLER WIND MEASUREMENTS

The Upper Atmosphere Research Satellite (UARS) has two instruments devoted to measuring horizontal winds, which are important because the dynamics and chemistry of the upper atmosphere are inextricably linked. The instruments are the High Resolution Doppler Imager (HRDI) and the Wind Imaging Interferometer (WINDII). Both view the limb of the atmosphere and measure the Doppler shift in emission and absorption lines in the visible and near-infrared portion of the spectrum. The chemical species detected are molecular oxygen, atomic oxygen, and the OH molecule. The Doppler shift determines the relative velocity between the satellite and the volume of air being sampled. Correcting for the motion of the satellite and the rotation of the Earth yields one component of the horizontal wind. By observing the same location both as the satellite approaches and as it departs from an area, one can estimate the horizontal wind vector. The instruments scan vertically to give a vertical profile of the wind with about 4-km vertical resolution. It is expected that winds will be measured to accuracies of between 5 and 15 m s^{-1}, depending on altitude. Since UARS was launched as this book was being written, the reader is referred to the literature for results of these wind measurements.

Bibliography

Atlas, R., E. Kalnay, W. E. Baker, J. Susskind, D. Reuter, and M. Halem (1985). Simulation studies of the impact of future observing systems on weather prediction. *Preprints: Seventh Conference on Numerical Weather Prediction*, American Meteorological Society, Boston, pp. 145–151.

Baker, W. E., R. Atlas, E. Kalnay, M. Halem, P. M. Woiceshyn, S. Peteherych, and D. Edelmann (1984). Large-scale analysis and forecast experiments with wind data from the Seasat A scatterometer. *J. Geophys. Res.*, 89, 4927–4936.

Bengtsson, L., M. Kanamitsu, P. Kallberg, and S. Uppala (1982). FGGE 4-dimensional data assimilation at ECMWF. *Bull. Am. Meteor. Soc.*, 63, 29–43.

Bowen, R. A., L. Fusco, J. Morgan, and K. O. Roska (1979). *Operational Production of Cloud Motion Vectors (Satellite Winds) from METEOSAT Image Data: Use of Data from Meteorological Satellites.* European Space Agency Report No. 143, 65–75.

Curran, R. J., D. E. Fitzjarrald, J. W. Bilbro, V. J. Abreu, R. A. Anthes, W. E. Baker, D. A. Bowdle, D. Burridge, G. D. Emmitt, S. E. D. Ferry, P. H. Flamant, R. Greenstone, R. M. Hardesty, K. R. Hardy, R. M. Huffaker, M. P. McCormick, R. T. Menzies, R. M. Schotland, J. K. Sparkman, Jr., J. M. Vaughan, and C. Werner (1987). Laser Atmospheric Wind Sounder Instrument Panel Report.

Earth Observing System, Vol. IIg, National Aeronautics and Space Administration, Washington, DC.

Desbois, M., G. Seze, and G. Szejwach (1982). Automatic classification of clouds on METEOSAT imagery: Application to high-level clouds. *J. Appl. Meteor.*, **21**, 401–412.

Duffy, D. G., R. Atlas, T. Rosmond, E. Barker, and R. Rosenberg (1984). The impact of Seasat scatterometer winds on the Navy's operational model. *J. Geophys. Res.*, **89**, 7238–7244.

Duffy, D. G., and R. Atlas (1986). The impact of Seasat–A Scatterometer data on the numerical prediction of the *Queen Elizabeth II* storm. *J. Geophys. Res.*, **91**, 2241–2248.

Dvorak, V. F. (1973). *A technique for the analysis and forecasting of tropical cyclone intensities from satellite pictures.* NOAA Tech. Memo. NESS 45, U.S. Department of Commerce, Washington, DC.

Dvorak, V. F. (1975). Tropical cyclone intensity analysis and forecasting from satellite imagery. *Mon. Wea. Rev.*, **103**, 420–430

Dvorak, V. F. (1984). *Tropical Cyclone Intensity Analysis Using Satellite Data.* NOAA Tech. Rep. NESDIS 11, Washington, DC.

Eigenwillig, N., and H. Fischer (1982). Determination of midtropospheric wind vectors by tracking pure water vapor structures in Meteosat water vapor image sequences. *Bull. Am. Meteor. Soc.*, **63**, 44–58.

Endlich, R. M., and D. E. Wolf (1981). Automatic cloud tracking applied to GOES and Meteosat observations. *J. Appl. Meteor.*, **20**, 309–319.

Fujita, T. T., E. W. Pearl, and W. E. Shenk (1975). Satellite-tracked cumulus velocities. *J. Appl. Meteor.*, **14**, 407–413.

Green, R., G. Hughes, C. Novak, and R. Schreitz (1975). The automatic extraction of wind estimates from VISSR data. In C. L. Bristor (ed.), *Central Processing and Analysis of Geostationary Satellite Data.* NOAA Tech. Memo. NESS 64, Washington, DC.

Grody, N. C., C. M. Hayden, W. C. C. Shen, P. W. Rosenkranz and D. H. Staelin (1979). Typhoon June winds estimated from scanning microwave spectrometer measurements at 55.45 GHz. *J. Geophys. Res.*, **84**, 3689–3695.

Gyakum, J. R. (1983). On the evolution of the QEII storm: Synoptic aspects. *Mon. Wea. Rev.*, **111**, 1137–1155.

Halem, M., E. Kalnay, W. E. Baker, and R. Atlas (1982). An assessment of the FGGE satellite observing system during SOP—1. *Bull. Am. Meteor. Soc.*, **63**, 407–426.

Hasler, A. F., W. C. Skillman, W. E. Shenk, and J. Steranka (1979). *In situ* aircraft verification of the quality of satellite cloud winds over oceanic regions. *J. Appl. Meteor.*, **18**, 1481–1489.

Hasler, A. F. (1981). Stereographic observations from geosynchronous satellites: An important new tool for the atmospheric sciences. *Bull. Am. Meteor. Soc.*, **62**, 194–212.

Hayden, C. M. (1984). *Estimating the Wind Field from VAS Temperature Soundings.* Unpublished manuscript. Available from the author at NOAA/NESDIS, 1225 W. Dayton St., Madison, WI 53706.

Holton, J. R. (1992). *An Introduction to Dynamic Meteorology*, 3rd ed., Academic Press, San Diego.

Hubert, L. F., and L. F. Whitney, Jr. (1971). Wind estimation from geostationary-satellite pictures. *Mon. Wea. Rev.*, **99**, 665–672.

Hubert, L. F., and L. F. Whitney, Jr. (1974). *Compatibility of Low-Cloud Vectors and Rawinsondes for Synoptic Scale Analysis.* NOAA Tech. Rep. NESS 70, Washington, DC.

Hubert, L. F. (1979). Wind derivation from geostationary satellites. In J. S. Winston (ed.). *Quantitative Meteorological Data from Satellites.* World Meteorological Organization Tech. Note No. 166 (WMO No. 531), Geneva, pp. 33–59.

Hubert, L. F., and A. Thomasell, Jr. (1979). *Error Characteristics of Satellite-Derived Winds.* NOAA Tech. Rep. NESS 79. Washington, DC.

Jones, W. L., L. C. Schroeder, D. H. Boggs, E. M. Bracalente, R. A. Brown, G. J. Dome, W. J. Pierson, and F. J. Wentz (1982). The SEASAT–A satellite scatterometer: The geophysical evaluation of remotely sensed wind vectors over the ocean. *J. Geophys. Res.*, **87**, 3297–3317.

Kastner, M., H. Fischer, and H.-J. Bolle (1980). Wind determination from Nimbus 5 observations in the 6.3 μm water vapor band. *J. Appl. Meteor.*, **19**, 409–418.

Kidder, S. Q., W. M. Gray, and T. H. Vonder Haar (1978). Estimating tropical cyclone central pressure and outer winds from satellite microwave data. *Mon. Wea. Rev.*, **106**, 1458–1464.

Kidder, S. Q., W. M. Gray, and T. H. Vonder Haar (1980). Tropical cyclone outer surface winds derived from satellite microwave sounder data. *Mon. Wea. Rev.*, **108**, 144–152.

Lame, D. B., and G. H. Born (1982). Seasat measurement system evaluation: Achievements and limitations. *J. Geophys. Res.*, **87**, 3175–3178.

Leese, J. A., C. S. Novak, and B. B. Clark (1971). An automatic technique for obtaining cloud motion from geostationary satellite data using cross correlation. *J. Appl. Meteor.*, **10**, 118–132.

Lord, R. J., W. P. Menzel, and L. E. Pecht (1984). ACARS wind measurements: An intercomparison with radiosonde, cloud motion, and VAS thermally derived winds. *J. Atmos. Ocean. Tech.*, **1**, 131–137.

Mack, R. A., A. F. Hasler, and R. F. Adler (1983). Thunderstorm cloud top observations using satellite stereoscopy. *Mon. Wea. Rev.*, **111**, 1949–1964.

Menzel, W. P., W. L. Smith, and T. R. Stewart (1983). Improved cloud motion-vector and altitude assignment using VAS. *J. Climate Appl. Meteor.*, **22**, 377–384.

Negri, A. J., and T. H. Vonder Haar (1980). Moisture convergence using satellite-derived wind fields: A severe local storm case study. *Mon. Wea. Rev.*, **108**, 1170–1182.

Njoku, E. G., and L. Swanson (1983). Global measurements of sea surface temperature, wind speed and atmospheric water content from satellite microwave radiometry. *Mon. Wea. Rev.*, **111**, 1977–1987.

Phillips, D. R., E. A. Smith, and V. E. Suomi (1972). Comment on "An automated technique for obtaining cloud motion from geosynchronous satellite data using cross correlation." *J. Appl. Meteor.*, **11**, 752–754.

Rodgers, E. B., R. C. Gentry, W. Shenk, and V. Oliver (1979). The benefits of using short-interval satellite images to derive winds for tropical cyclones. *Mon. Wea. Rev.*, **107**, 575–584.

Rodgers, E. B., R. Mack, and A. F. Hasler (1983). A satellite stereoscopic technique to estimate tropical cyclone intensity. *Mon. Wea. Rev.*, **111**, 1599–1610.

Rosenkranz, P. W., D. H. Staelin, and N. C. Grody (1978). Typhoon June (1975) viewed by a scanning microwave spectrometer. *J. Geophys. Res.*, **83**, 1857–1868.

Schroeder, L. C., D. G. Boggs, G. Dome, I. M. Halberstam, W. L. Jones, W. J. Pierson, and F. J. Wentz (1982). The relationship between wind vector and normalized radar cross section used to derive SEASAT–A satellite scatterometer winds. *J. Geophys. Res.*, **87**, 3318–3336.

Smith, E. A., and D. R. Phillips (1972). Automated cloud tracking using precisely aligned digital ATS pictures. *IEEE Trans. Computers*, **C-21**,715–729.

Smith, E. A. (1975). The McIDAS system. *IEEE Trans. Geosci.*, **GE-13**, 123–128.

Stewart, R. H. (1985). *Methods of Satellite Oceanography*. Univ. California Press, Berkeley.

Stewart, T. R., C. M. Hayden, and W. L. Smith (1985). A note on water-vapor wind tracking using VAS data on McIDAS. *Bull. Am. Meteor. Soc.*, **66**, 1111–1115.

Stogryn, A. (1967). The apparent temperature of the sea at microwave frequencies. *IEEE Trans. Antennas Propag.*, **AP-15**, 278–286.Suchman, D., and D. W. Martin (1976). Wind sets from SMS images: An assessment of quality for GATE. *J. Appl. Meteor.*, **15**, 1265–1278.Szejwach, G. (1982). Determination of semi-transparent cirrus cloud temperature from infrared radiances: Application to METEOSAT. *J. Appl. Meteor.*, **21**, 384–393.

Thomasell, A., Jr. (1979). *Wind Analysis by Conditional Relaxation*. NOAA Tech. Rep. NESS 77, Washington, DC.

Velden, C. S., and W. L. Smith (1983). Monitoring tropical cyclone evolution with NOAA satellite microwave observations. *J. Climate Appl. Meteor.*, **22**, 714–724.

Velden, C. S., W. L. Smith, and M. Mayfield (1984). Applications of VAS and TOVS to tropical cyclones. *Bull. Am. Meteor. Soc.*, **65**, 1059–1067.

Whitney, L. F. (1983). International comparison of satellite winds—an update. *Adv. Space Res.*, **2**, 73–77.

Wilheit, T. T. (1979). *A Model for the Microwave Emissivity of the Ocean's Surface as a Function of Wind Speed*. NASA Tech. Memo. 80278, Goddard Space Flight Center, Greenbelt, MD.

Wilheit, T. T. and A. T. C. Chang (1980). An algorithm for retrieval of ocean surface and atmosphere parameters from the observations of the scanning multichannel microwave radiometer. *Radio Sci.*, **15**, 525–544.

Wilheit, T. T., J. A. Greaves, J. A. Gatlin, D. Han, M. Krupp, A. S. Milman, and E. S. Chang (1984). Retrieval of ocean surface parameters from the scanning multifrequency microwave radiometer (SMMR) on the Nimbus 7 satellite. *IEEE Trans. Geosci. Remote Sensing*, **GE-22**, 133–143.

Woiceshyn, P. M., M. G. Wurtele, D. H. Boggs, L. F. McGoldrick, and S. Peteherych (1986). The necessity for a new parameterization of an empirical model for wind/ocean scatterometry. *J. Geophys. Res.*, **91**, 2273–2288.

Wolf, D. E., D. J. Hall, and R. M. Endlich (1977). Experiments in automatic cloud tracking using SMS-GOES data. *J. Appl. Meteor.*, **16**, 1219–1230.

8

Clouds and Aerosols

CLOUDS COVER ROUGHLY half the Earth, and aerosols are always present. They constitute major factors in determining the radiation budget of Earth, and they play crucial, if as yet not completely understood, roles in modulating the climate. They also contaminate radiometric observations necessary for sounding the atmosphere and remotely sensing surface parameters. Space-based instruments are the only means by which the global distribution of clouds and aerosols can be adequately sampled.

In the first part of the chapter, we discuss methods for determining cloud parameters from three sources: sounders, imagers, and microwave radiometers. In the second part of the chapter, we detail methods for retrieving stratospheric and tropospheric aerosols.

8.1 CLOUDS FROM SOUNDERS

Many cloud properties are of interest to the meteorologist. The ones most often retrieved are cloud-top temperature, cloud-top height (or pressure), and *cloud amount* (the fractional area covered by cloud). Useful also are reflectance, emittance, optical depth, phase (ice or water), liquid water content, and drop size distribution. Attempts to retrieve cloud properties can be divided into schemes that use sounding instruments, schemes that use visible and infrared imaging instruments, and schemes that use microwave instruments.

Most efforts to detect clouds using sounder data have been directed toward removing cloud contamination so that accurate, clear-column radiances can be calculated. It is also desirable, however, to retrieve the properties of the clouds.

If we assume that clouds reflect little radiation at the wavelengths used by sounders,[1] then the radiation sensed by a radiometer can be written as the sum of three terms:

$$L(\lambda) = (1 - N)L_{\mathrm{clr}}(\lambda) + N\varepsilon L_{\mathrm{cld}}(\lambda) + N(1 - \varepsilon)L_{\mathrm{clr}}(\lambda), \tag{8.1}$$

where $L(\lambda)$ is the measured radiance of a scene, $L_{\mathrm{clr}}(\lambda)$ is the radiance from the scene if clouds were not present (the clear-column radiance), $L_{\mathrm{cld}}(\lambda)$ is the radiance that would come from the cloud if it were black, N is the cloud amount, ε is the cloud emittance, and $1 - \varepsilon$ is the cloud transmittance. The three terms represent, respectively, the radiation from the clear area, the radiation emitted by the cloud, and the radiation transmitted through the cloud. Combining the first and third terms results in

$$L(\lambda) = (1 - N')L_{\mathrm{clr}}(\lambda) + N'L_{\mathrm{cld}}(\lambda), \tag{8.2}$$

where $N' \equiv N\varepsilon$ is called the *effective cloud amount*.[2] Equation 8.2 is the basis of efforts to determine clear-column radiances for sounding retrievals and to determine the cloud properties N' and p_c, the pressure of the cloud top. We turn our attention first to the problem of determining clear-column radiances, which was deferred from Chapter 6.

8.1.1 Clear-Column Radiances

The basis for determining clear-column radiances was first described by Smith (1968). It applies Eq. 8.2 to two "adjacent" scan spots:

$$\tilde{L}_1(\lambda) = (1 - N_1')L_{\mathrm{clr}}(\lambda) + N_1'L_{\mathrm{cld}}(\lambda), \tag{8.3a}$$

$$\tilde{L}_2(\lambda) = (1 - N_2')L_{\mathrm{clr}}(\lambda) + N_2'L_{\mathrm{cld}}(\lambda), \tag{8.3b}$$

(A tilde indicates an observed quantity.) The scan spots do not have to be strictly adjacent, but they have to be close enough that the same values of $L_{\mathrm{clr}}(\lambda)$, $L_{\mathrm{cld}}(\lambda)$, and $\varepsilon(\lambda)$ describe each spot. This means that (1) the clouds have basically the same height and microphysical composition in both spots, and (2) the land surface

[1] Chahine (1982) treats the case of channels in the 4.3-μm band in which clouds reflect some solar radiation.

[2] The reader should be aware that the work of Harshvardhan and Weinman (1982), Weinman and Harshvardhan (1982), and others calls into question whether this approximation, linear in N', is adequate in a field of broken clouds.

and overlying clear atmosphere are indistinguishable in both spots. The only difference between the scan spots is assumed to be the cloud amount N.

Strictly speaking, Eqs. 8.3 describe only a single layer of clouds. If there is only one layer of clouds, N is independent of wavelength. If there are two cloud layers, N will seem to decrease as the height of the peak of the weighting function increases because channels that peak higher in the atmosphere are less sensitive to lower clouds (McMillin and Dean, 1982). Terms can be added to Eq. 8.2 to account for additional cloud levels (Chahine, 1977), but multiple cloud layers will not be treated here.

Assuming only one cloud layer, $L_{cld}(\lambda)$ can be eliminated between Eqs. 8.3 to yield the clear-column radiance:

$$L_{clr}(\lambda) = \frac{\tilde{L}_1(\lambda) - N^* \tilde{L}_2(\lambda)}{1 - N^*}, \tag{8.4}$$

where N^* (pronounced *N star*) is

$$N^* \equiv \frac{N_1'}{N_2'} = \frac{L_{clr}(\lambda) - \tilde{L}_1(\lambda)}{L_{clr}(\lambda) - \tilde{L}_2(\lambda)}. \tag{8.5}$$

If $N_1' \neq N_2'$, an estimate of N^* plus observations $\tilde{L}_1(\lambda)$ and $\tilde{L}_2(\lambda)$ will yield $L_{clr}(\lambda)$.

There are several ways to estimate N^*. Chahine (1974) treats N^* as an additional unknown to be retrieved iteratively in the same scheme used to retrieve temperatures. This technique is useful when the sounder makes measurements in a single band, such as the 15-μm band.

If an infrared sounder makes measurements in both the 15- and 4.3-μm bands, the difference in response to clouds in the two bands can be used to estimate N^* (Chahine, 1974; McMillin and Dean, 1982). First, a regression equation is developed, based on synthesized (not observed) radiances, to estimate the clear radiance in a relatively transparent 4.3-μm channel (wavelength $\lambda_{4.3}$) in terms of the clear radiances in the 15-μm channels:[3]

$$\hat{L}_{clr}(\lambda_{4.3}) = a_0 + \sum_i a_i \hat{L}_{clr}(\lambda_i). \tag{8.6}$$

(A caret over a variable indicates that it has been estimated using other variables.) Second, measured radiances in two adjacent scan spots are used to calculate the sensitivity parameters

$$S(\lambda_i) \equiv \frac{\tilde{L}_1(\lambda_i) - \tilde{L}_2(\lambda_i)}{\tilde{L}_1(\lambda_{4.3}) - \tilde{L}_2(\lambda_{4.3})}. \tag{8.7}$$

[3] Note that McMillin and Dean (1982) use brightness temperatures rather than radiances in this equation. Their scheme is more accurate than the one outlined here, but this scheme contains the essential physics and is easier to explain. Also note that 4.3-μm radiances must be corrected for reflected solar radiation.

If a single layer of clouds[4] affects the two scan spots, Eqs. 8.3 imply that the parameters $S(\lambda_i)$ are independent of cloud amount:

$$S(\lambda_i) = \left[\frac{\varepsilon(\lambda_i)}{\varepsilon(\lambda_{4.3})}\right]\left[\frac{L_{\text{clr}}(\lambda_i) - L_{\text{cld}}(\lambda_i)}{L_{\text{clr}}(\lambda_{4.3}) - L_{\text{cld}}(\lambda_{4.3})}\right]. \tag{8.8}$$

Therefore, the sensitivity parameters can be used in combination with one of the scan spots to relate the radiance in the 4.3-μm channel to the radiance in each 15-μm channel:

$$\hat{L}(\lambda_i) = \tilde{L}_2(\lambda_i) + S(\lambda_i)[\hat{L}(\lambda_{4.3}) - \tilde{L}_2(\lambda_{4.3})]. \tag{8.9}$$

In particular, the clear radiances are related:

$$\hat{L}_{\text{clr}}(\lambda_i) = \tilde{L}_2(\lambda_i) + S(\lambda_i)[\hat{L}_{\text{clr}}(\lambda_{4.3}) - \tilde{L}_2(\lambda_{4.3})]. \tag{8.10}$$

Eliminating $L_{\text{clr}}(\lambda_i)$ between Eqs. 8.6 and 8.10 yields an estimate of the clear radiance in the 4.3-μm channel:

$$\hat{L}_{\text{clr}}(\lambda_{4.3}) = \frac{a_0 + \sum a_i\tilde{L}_2(\lambda_i) - \tilde{L}_2(\lambda_{4.3})\sum a_iS(\lambda_i)}{1 - \sum a_iS(\lambda_i)}. \tag{8.11}$$

Finally, N^* is given by

$$N^* = \frac{\hat{L}_{\text{clr}}(\lambda_{4.3}) - \tilde{L}_1(\lambda_{4.3})}{\hat{L}_{\text{clr}}(\lambda_{4.3}) - \tilde{L}_2(\lambda_{4.3})}. \tag{8.12}$$

Since N^* is independent of wavelength, it can be used in Eq. 8.4 to calculate clear radiances in the other channels.

If a microwave sounder is flown with an infrared sounder, N^* is easily obtained. Since microwave radiation is nearly unaffected by non-raining clouds, microwave observations are essentially clear radiances. Smith and Woolf (1976), McMillin and Dean (1982), and Susskind *et al.* (1984) use microwave observations in their cloud filters. McMillin and Dean develop a regression equation that relates MSU channel 2 (weighting function peak ~700 hPa) to radiances in the 15-μm band:

$$\hat{L}(\lambda_{\text{MSU2}}) = b_0 + \sum_i b_i\hat{L}_{\text{clr}}(\lambda_i). \tag{8.13}$$

Only clear radiances are used to develop Eq. 8.13. When the 15-μm channels are contaminated by clouds, $\hat{L}(\lambda_{\text{MSU2}})$ will be less than the observed value, $\tilde{L}(\lambda_{\text{MSU2}})$. Using observed 15-$\mu$m radiances in scan spots 1 and 2 to calculate $\hat{L}_1(\lambda_{\text{MSU2}})$ and $\hat{L}_1(\lambda_{\text{MSU2}})$, N^* is given by

$$N^* = \frac{\tilde{L}(\lambda_{\text{MSU2}}) - \hat{L}_1(\lambda_{\text{MSU2}})}{\tilde{L}(\lambda_{\text{MSU2}}) - \hat{L}_2(\lambda_{\text{MSU2}})}. \tag{8.14}$$

[4] The cloud microphysics must be the same in the two spots so that $\varepsilon_1(\lambda_i) = \varepsilon_2(\lambda_i) = \varepsilon(\lambda_i)$ and $\varepsilon_1(\lambda_{4.3}) = \varepsilon_2(\lambda_{4.3}) = \varepsilon(\lambda_{4.3})$.

The cloud filter used operationally to process TOVS soundings from NOAA satellites[5] is described by McMillin and Dean (1982). Since no one test or technique is sufficient to detect and correct for clouds in all situations, several empirically determined tests are employed. These are briefly mentioned here. Recall that TOVS soundings are processed in blocks of 7 × 9 or 3 × 3 HIRS/2 scan spots. The first task is to determine whether any of the scan spots are clear. Seven tests are applied:

1. The frozen sea test uses MSU channel 1 to determine whether the ocean is ice covered. A cold brightness temperature (<236 K) indicates water, not ice. If the surface is water, and the HIRS/2 channel 8 (11 μm) brightness temperature is less than 270 K (the temperature at which sea water freezes), the scan spot is assumed to be cloud covered.
2. In the adjacent spot test, the HIRS/2 channel 8 brightness temperature of a scan spot is compared with its eight neighbors. If the center spot is more than 1.5 K colder than any of the surrounding spots, the spot is assumed to be cloudy.
3. The albedo test is designed to detect low clouds, which are difficult to detect with infrared observations. The HIRS/2 channel 20 albedo is compared with an expected value for that location. If the albedo exceeds the expected albedo by more than a threshold value, the spot is cloudy. This test can be applied only during daylight hours, obviously, and it is not applied if the surface temperature is thought to be below freezing so that ice and snow will not be mistaken for clouds.
4. The window channel tests compare brightness temperatures in HIRS/2 channels 8, 18 (4.0 μm) and 19 (3.7 μm). If they differ by too great an amount,[6] the spot is cloudy. Because of reflected solar radiation, the difference threshold must be greater (and the test is therefore less sensitive) during the day than at night.
5. In the interchannel regression test, HIRS/2 channel 7 (13.4 μm) and MSU channels 2 and 3 are estimated by regression equations from the other channels. If the estimated and observed brightness temperatures differ by too great an amount, the spot is assumed to be cloudy.
6. The surface temperature test is designed to detect clouds by estimating a surface temperature from the measured radiances and comparing with yesterday's retrieved surface temperature, a shelter temperature, or the sea surface temperature. If the estimated surface temperature is significantly colder than the expected surface temperature, the spot is cloudy.
7. The maximum value test classifies as cloudy any scan spot whose HIRS/2 channel 8 brightness temperature is more than 4 K colder than the maximum channel 8 brightness temperature for the block.

[5] The much simpler cloud filter used for VAS soundings is discussed in Section 6.3.2.

[6] Several empirically determined thresholds are used. The reader is referred to McMillin and Dean (1982) for the details.

These tests do a good job of detecting clouds, but they are not perfect. Clouds over snow can go undetected, as can low clouds at night. Also, areas with very low cloud amount or thin cirrus are always a problem. The tests can err in the opposite direction by classifying snow-covered areas and areas behind cold fronts as cloudy. If four or more spots in a block pass all tests, their radiances are averaged and returned as clear-column radiances.

If enough clear scan spots are not found, the N^* technique is attempted. Estimates using both the 4.3/15-μm method (N_{H}^*) and the microwave method (N_{M}^*) are made. A large number of tests are performed during the calculations to ensure that an accurate N^* is produced. Among the tests are the following:[7]

- N^* is not calculated if the HIRS/2 channel 8 radiances for two adjacent spots differ by less than 7.5 W m^{-2} sr^{-1} cm. This, plus other tests, ensures that N^* is not calculated if $N_1 \approx N_2$.
- An indication that multiple cloud layers are present (and therefore that the N^* technique is inappropriate) is that the clear radiances are too cold. A test to eliminate these cases is made.
- Estimates of N^* must yield reasonable values within empirically determined limits. Different tests are applied to N_{H}^* and to N_{M}^*. If only one of N_{H}^* or N_{M}^* survives the above tests, it is taken as N^*. If both survive, they are averaged using the formula

$$\frac{N^*}{N^* - 1} = \frac{1}{2}\left[\frac{N_{\mathrm{H}}^*}{N_{\mathrm{H}}^* - 1} + \frac{N_{\mathrm{M}}^*}{N_{\mathrm{M}}^* - 1}\right], \tag{8.15}$$

  and clear-column radiances are calculated.
- Finally, the clear-column radiances are subject to window channel tests and interchannel regression tests.

If the N^* calculation succeeds for a sufficient number of scan spot pairs, the calculated radiances are averaged (along with radiances from any spots identified as clear) and are returned as clear-column radiances. If the N^* technique fails (too few pairs), retrievals are done using only microwave channels in the troposphere.

The N^* technique consumes substantial computer time. It is important because microwave-only (third path) retrievals are not as accurate as retrievals that use both infrared and microwave channels. The testing done as part of the N^* technique is crucial because in the cases when the single-cloud-layer model is inappropriate or when the technique produces unreasonable radiances, the accuracy of the retrieved soundings is lower.

8.1.2 Cloud Amount and Cloud-Top Pressure

Methods for retrieving cloud amount and cloud-top pressure from sounder data may be roughly divided into two related classes: radiance residual methods

[7] Details may be found in McMillin and Dean (1982), page 1010.

and radiance ratio methods. *Radiance residual methods* are applied after a sounding has been retrieved, and therefore, the clear-column radiances and profiles of temperature and moisture are known. Several papers discuss the development of these methods (e.g. Smith *et al.*, 1970; Smith and Woolf, 1976; Chahine, 1975, 1982). We present a simplified version of the procedure developed by Chahine (1975).

1. Choose a set of possible cloud-top pressures, which range from a level above the surface to the tropopause.
2. For each p_c in the set, calculate L_{cld} for a channel that peaks near the surface:[8]

$$L_{\text{cld}}(\lambda) = B_\lambda(T_c)\tau_\lambda(p_c,0) + \int_0^{p_c} B_\lambda(T)\frac{\partial\tau_\lambda(p,0)}{\partial p}\,dp. \tag{8.16}$$

   Clouds are treated as black because the emittance is included in the effective cloud amount.

3. For each p_c, calculate N' using the just-calculated L_{cld}, the value of L_{clr} calculated during sounding retrieval, and the observed $\tilde{L}$

$$N' = \frac{\tilde{L} - L_{\text{clr}}}{L_{\text{cld}} - L_{\text{clr}}}. \tag{8.17}$$

   Assuming one cloud level and channels close enough together so that $\varepsilon(\lambda_i)$ is constant, N' is independent of channel.

4. For the remainder of the channels (λ_i) that sense clouds, use Eq. 8.16 to calculate $L_{\text{cld}}(\lambda_i)$, then use Eq. 8.2 to calculate $\hat{L}(\lambda_i)$. This step produces a set of radiances $\{\hat{L}(\lambda_i)\}$ for each p_c.
5. Compare the calculated radiances with the observed radiances. The value of p_c that minimizes a measure of the radiance residuals, perhaps

$$R(p_c) \equiv \left\{\sum_i \left[\frac{\hat{L}(\lambda_i) - \tilde{L}(\lambda_i)}{\tilde{L}(\lambda_i)}\right]^2\right\}^{1/2}, \tag{8.18}$$

   is accepted as the cloud-top pressure. The associated N' is accepted as the effective cloud amount.

The radiance residual method has been used by Susskind *et al.* (1984), among others. Smith *et al.* (1970) and Chahine (1982) discuss multiple-cloud-layer retrievals.

Radiance ratio methods have been studied by several groups including Smith *et al.* (1974), McCleese and Wilson (1976), Smith and Platt (1978), Chahine (1982), and Menzel *et al.* (1983). They all apply Eq. 8.2 to two sounding channels:

$$L(\lambda_1) = (1 - N')L_{\text{clr}}(\lambda_1) + N'L_{\text{cld}}(\lambda_1), \tag{8.19a}$$

$$L(\lambda_2) = (1 - N')L_{\text{clr}}(\lambda_2) + N'L_{\text{cld}}(\lambda_2). \tag{8.19b}$$

[8] For simplicity, p has been used as the vertical coordinate, though $\ln p$ is more often used.

If the two channels are close enough together so that $\varepsilon(\lambda_1) = \varepsilon(\lambda_2)$, then N' will be the same in both equations. Eliminating N gives

$$\frac{L(\lambda_1) - L_{\mathrm{clr}}(\lambda_1)}{L(\lambda_2) - L_{\mathrm{clr}}(\lambda_2)} = \frac{L_{\mathrm{cld}}(\lambda_1) - L_{\mathrm{clr}}(\lambda_1)}{L_{\mathrm{cld}}(\lambda_2) - L_{\mathrm{clr}}(\lambda_2)} \equiv f(\lambda_1,\lambda_2,p_c). \quad (8.20)$$

As long as there are no temperature inversions present, $f(\lambda_1,\lambda_2,p_c)$ is monotonic in p_c and only weakly dependent on the atmospheric temperature profile (Fig. 8.1).

To retrieve p^c for a particular scan spot, one calculates the left-hand side of Eq. 8.20 and finds the level at which the radiance ratio agrees with $f(\lambda_1,\lambda_2,p_c)$. N' is then calculated from Eq. 8.19. The procedure for doing these calculations depends on what information is available. Three approaches have been utilized:

- If a sounding has just been retrieved from the radiances, L_{clr} will be known for both channels, and $f(\lambda_1,\lambda_2,p_c)$ can be calculated from Eqs. 8.16 and 8.20 (Smith *et al.*, 1974).
- $f(\lambda_1,\lambda_2,p_c)$ can be calculated from a nearby rawinsonde sounding, a forecast sounding, or even from climatology, and the values of L_{clr} can be ei-

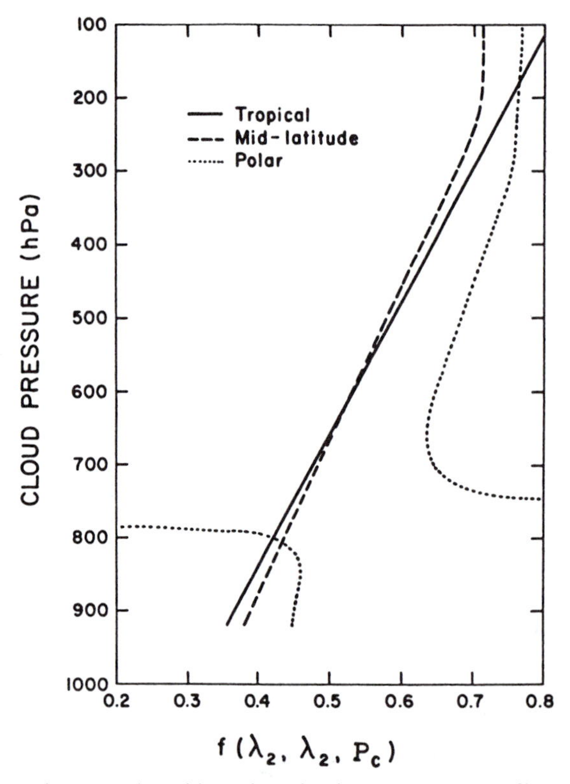

FIGURE 8.1. $f(\lambda_1,\lambda_2,p_c)$ for tropical, mid-latitude and polar temperature profiles. HIRS/1 channels 6 (λ_1 = 13.35 μm) and 7 (λ_2 = 13.65 μm) were used. The discontinuity in the polar plot is caused by a temperature inversion. [After Wielicki and Coakley (1981).]

ther calculated from the sounding or obtained from a nearby (clear) scan spot (Smith and Platt, 1978; Menzel *et al.*, 1983).

- If it can be assumed that p_c, $L_{clr}(\lambda_1)$, and $L_{clr}(\lambda_2)$ are constant in a local area, but that N' changes, then a plot of $L(\lambda_1)$ versus $L(\lambda_2)$ will be a straight line with slope $[L_{cld}(\lambda_1) - L_{clr}(\lambda_1)]/[L_{cld}(\lambda_2) - L_{clr}(\lambda_2)]$. Comparing this slope with climatologically derived $f(\lambda_1,\lambda_2,p_c)$, gives p_c (McCleese and Wilson, 1976).

The radiance ratio method is the basis of CO_2 slicing, which Menzel *et al.* (1983) used to assign heights to wind vectors (see Section 7.1.5).

Wielicki and Coakley (1981) performed an error analysis to determine how accurately p_c and N' can be retrieved from sounder data. Simulated radiances for the HIRS/1 instrument (which is similar to the HIRS/2 instrument that flies on the NOAA satellites) were calculated for 10 cloud amounts (0.1–1.0), 12 cloud-top pressures (920–100 hPa), and 2 climatological mean profiles (tropical and midlatitude). Cloud-top pressure and effective cloud amount were retrieved using both the radiance ratio (Smith and Platt, 1978) and the radiance residual (Chahine, 1975) techniques. The retrieved parameters were compared with the parameters used to calculate the radiances.

Wielicki and Coakley first investigated errors resulting from instrument noise. In these calculations, temperature and humidity profiles (and thus L_{clr}, L_{cld}, and $f(\lambda_1,\lambda_2,p_c)$) were assumed to be exactly correct. Realistic instrument noise was added to the synthesized instrument observations. Figure 8.2 shows the bias and standard deviation in cloud amount and cloud-top pressure retrieved by the radiance ratio method using synthesized HIRS/1 channels 6 and 7 (13.35 and 13.65-μm) radiances. (Note these results depend strongly on the channels used. Other channel combinations gave better results.) As expected, the method has trouble with low cloud and with small cloud amounts because the clear-column radiances are not very different from the observed radiances. The radiance residual technique produces similar biases, but standard deviations are reduced aloft (by 50% at 200 hPa), due to the addition of channels that better sense high clouds. Unfortunately, the additional channels slightly degrade retrieval accuracy at low levels because the channels do not sense low cloud, yet the technique attempts to minimize their radiance residuals. In other words, the additional channels add noise, but no information, to low cloud retrievals.

Bias errors are uncomfortably large. Wielicki and Coakley found that nearly all of the bias error can be explained by nonlinear constraints placed on the retrievals. These constraints consist of requiring cloud amounts to be in the range 0.0–1.0 and cloud-top pressures to be between 920 hPa and the tropopause pressure. When a value is out of range, it is reassigned to the closest point in the range, except that when p_c is greater than 920 hPa, clear sky is assumed. In the presence of instrument noise, the retrieved cloud-top pressure varies in approximately a Gaussian fashion about the true pressure. As the difference $\hat{L} - L_{clr}$ approaches the noise level of the instrument, either because the clouds are lower or the cloud amount is smaller, the distribution of retrieved cloud-top pressures

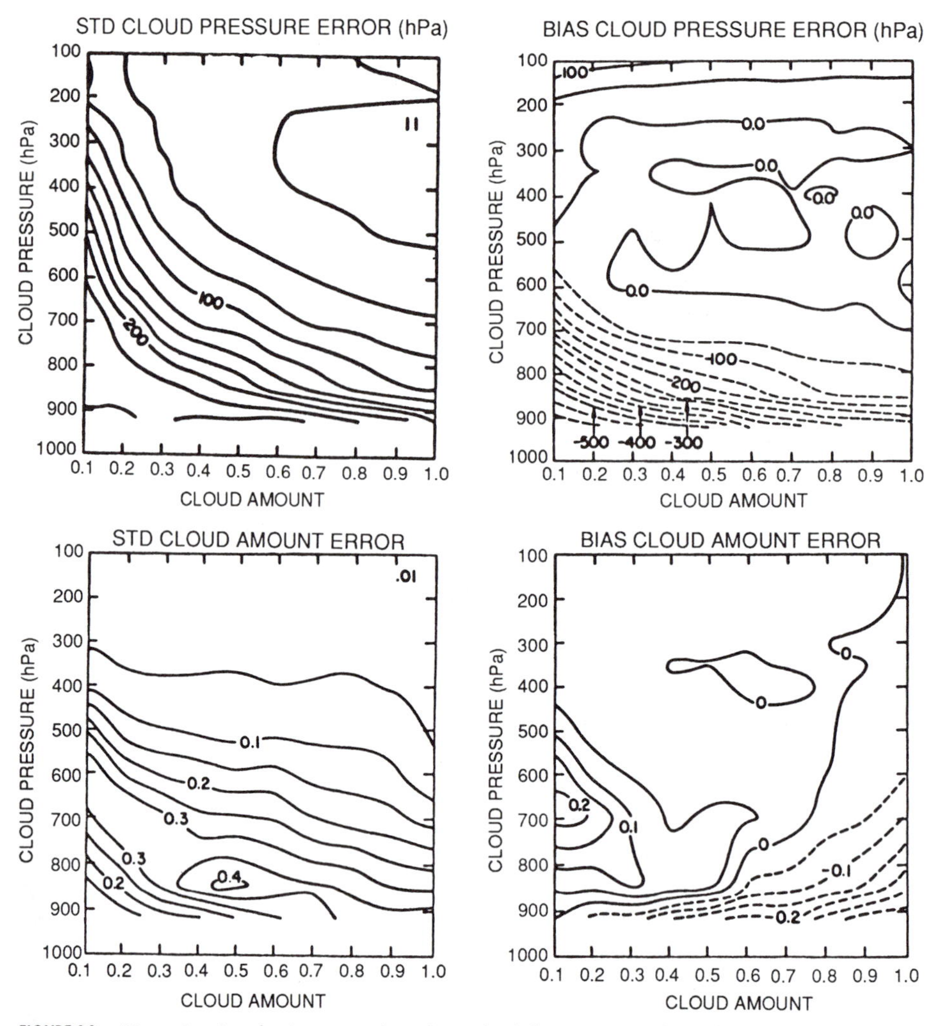

FIGURE 8.2. Biases (retrieved minus actual) and standard deviations resulting from instrument noise in retrieved cloud amount and cloud-top pressure (hPa) using the radiance ratio method on synthesized HIRS/1 channels 6 and 7 (13.35 and 13.64 μm) radiances. The rms error is the square root of the sum of the squares of the bias and standard deviation. [After Wielicki and Coakley (1981).]

broadens (i.e., the standard deviation of cloud-top pressure error increases). For scenes with true cloud-top pressures less than 920 hPa, instrument noise causes a fraction of the retrieved pressures to be greater than 920 hPa and thus causes the scene to be misclassified as clear. Removal of this lower tail of the distribution results in a bias toward cloud-top pressures that are too low. High clouds are similarly biased toward cloud-top pressures that are too high. Instrument noise also produces a distribution in retrieved cloud amounts. If the true cloud amount is near zero, some of the retrievals will be less than zero and will be assigned to zero cloud amount. When the cloud amount bias is calculated, this results in a high bias for low cloud amounts. Similarly, too few scenes are classified as completely cloudy. Misclassification of scenes covered with low cloud as clear causes cloud amounts in the lower troposphere to be biased too low. Wielicki and Coakley tested modifications to the constraints, such as using a visible channel to detect low cloud and estimating clear skies when $\hat{L} - L_{clr}$ is less than twice the instrument noise. These modifications resulted in substantially reduced biases in their simulations.

Wielicki and Coakley also tested the effects of errors in temperature and moisture retrievals on cloud amount and cloud-top pressure accuracy. The errors for 2-K temperature error and 50% moisture error were roughly comparable to those for instrument error.

In summary, cloud amounts and cloud-top pressures retrieved from sounder data using either the radiance ratio or radiance residual techniques have reasonable accuracy in the mid- and upper-troposphere for moderate to high cloud amounts. Clouds in the lower troposphere or those which have small cloud amounts are poorly retrieved.

8.2 CLOUDS FROM IMAGERS

Most of the work on cloud estimation has been done using window channels on imaging instruments. These instruments have the advantage of much higher spatial resolution than sounders. Perhaps the most often used technique for cloud characterization is simple manual inspection of satellite photographs or video images. These analyses are used daily for weather analysis and forecasting and in the construction of *nephanalyses*, which are maps showing chiefly cloud type and amount at various atmospheric levels (Fye, 1978). Some of these techniques were discussed in Chapter 5. In this chapter we concentrate on objective techniques.

8.2.1 Threshold Techniques

The oldest (Arking, 1964; Koffler *et al.*, 1973), simplest, and still most frequently used technique for objectively extracting cloud information from digital satellite images is the threshold technique. Here a visible brightness or infrared temperature threshold is set such that if a pixel is brighter or colder than the threshold, the pixel is assumed to be cloud covered. The fractional area covered with cloud is simply the ratio of the number of cloudy pixels to the total number of pixels. The cloud height can be determined by comparing the infrared temperature of the pixel with a sounding. The simplicity of this technique makes it attractive. Thresholding is one of the means of filtering clouds from data to be used in sea surface temperature (SST) retrievals (McClain *et al.* 1985). In practice, however, two complications arise: (1) the problem of clouds smaller than the satellite scan spot and (2) the problem of how to set the threshold.

Shenk and Salomonson (1972) investigated the effects of sensor spatial resolution on determination of cloud amount using the threshold technique. Consider an ideal scene in which only a uniform background (radiance L_{clr}) and uniform cloud (radiance L_{cld}) appear. Pixels that are partly filled with cloud will have radiances between L_{clr} and L_{cld}. If the threshold is set at nearly L_{clr}, then any cloud in a pixel will cause it to appear 100% cloudy, which will result in an overestimate of cloud amount. Figure 8.3a shows cloud amount retrieved from simulated satellite data as a function of the ratio (R) of the mean cloud area to the area of a satellite scan spot. The line labels are the true cloud amount. For $R = 100$ (cloud diameter approximately 10 times the diameter of a scan spot) and beyond, there is good agreement between the retrieved and true cloud amounts. For clouds approximately the same size as the scan spot, however, a scene with 40% true cloud amount appears to be completely overcast. Alternately, if the threshold is at nearly L_{cld}, only completely cloudy pixels will be classified as cloudy, partly cloudy pixels will be classified as clear, and cloud amount will be underestimated.

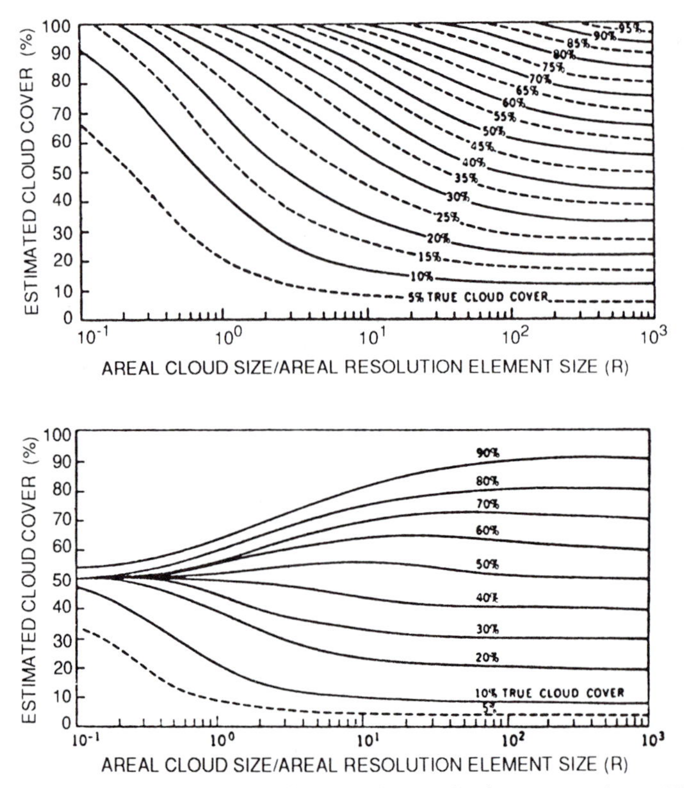

FIGURE 8.3. Retrieved cloud amount versus the ratio of mean cloud area to pixel area. The line labels indicate true cloud amount. (a) The threshold is nearly L_{clr}. (b) Two thresholds, with pixels falling between the thresholds being counted as 50% cloud covered. [After Shenk and Salomonson (1972).]

A less biased estimate of cloud amount can be obtained by setting two thresholds, one at nearly L_{clr} and the other at nearly L_{cld}. Pixels that are between the two thresholds are counted as 50% cloud covered. Figure 8.3b shows the effects of this scheme. The large positive bias in cloud amount has disappeared, and for R greater than about 10 (cloud diameter more than $\sqrt{10}$ times the scan spot diameter), reasonable cloud amounts are retrieved. Obviously, high-spatial-resolution radiometers are desirable for cloud detection.

A related problem, which has not as yet received adequate treatment in satellite meteorology, concerns the estimation of cloud amount in a field of broken clouds. Because broken clouds radiatively interact with each other, and because for nonnadir viewing the satellite sees the sides as well as the tops of the clouds, effective cloud amount in fields of broken clouds is not necessarily proportional to the fractional area covered with cloud (Harshvardhan and Weinman, 1982; Weinman and Harshvardhan, 1982).

Setting the threshold is the chief difficulty in threshold techniques. The problem is that the threshold is a function of many variables: surface type (land, ocean, ice), surface conditions (snowcover, vegetation, soil moisture), recent weather (which changes surface temperature and reflectance), atmospheric conditions (haze, temperature inversions), season, time of day, and even satellite–Earth–sun geometry (bidirectional reflectance, sun glint). In the past, thresholds have been set manually; that is, an analyst views each image to be processed and picks an area that appears clear. Pixels that are a few percent brighter or a few degrees colder than the background are classified as cloud-covered. Manual setting of the threshold is, however, incompatible with automated processing.

Today, thresholds are more likely to be set by examining the temporal or spatial variance of observed radiances. Spatial variance methods will be discussed below. Here we discuss the temporal variance approach.

Determination of clear-sky radiances is critical to the success of the threshold technique because clouds are assumed to perturb the clear-sky radiances. Suppose that images for the last several ($\sim$15) days are analyzed to select the minimum visible albedo and maximum infrared radiance during the period.[9] Except in relatively unusual circumstances, such as regions which are cloud-covered for the entire period, regions in which snow has fallen, or regions that experience large day-to-day variations in surface temperature, these extremes come close to clear-sky values. Spatial smoothing of the results will reduce noise caused by choosing extreme values, but care must be taken in high gradient areas such as near coastlines. Minnis and Harrison (1984) have used a sophisticated version of this approach.

It must be noted that in some situations, it is difficult to detect clouds with visible and 11-μm data using thresholding (or other) techniques. Low clouds over ice or snow, for example, present no contrast in either visible imagery, because

[9] Bidirectional reflectance and limb darkening corrections should be made to the data before this analysis.

both ice and snow have similar albedos, or in infrared imagery, because both have similar temperatures (see Chapter 5). Cirrus is generally easier to detect with infrared than with visible data, yet cirrus can be thin enough to be warmer than an infrared threshold. Small broken clouds can be widely scattered enough to escape detection. Finally, multiple levels of cloud present problems for all cloud retrieval techniques because of obscuration of the lower clouds by the upper clouds and because some pixels will contain both levels and perhaps the surface as well.

Except in combination with radiative transfer techniques (see below), threshold techniques generally are limited to determining cloud amount (by calculating the fraction of pixels that are cloudy) and to estimating the cloud-top height by comparing the infrared equivalent blackbody temperature with a sounding.

8.2.2 Histogram Techniques

Histogram techniques serve as alternates to threshold techniques. The basic idea is that a histogram of the pixels in an area will show clusters of pixels that represent cloud or surface types. The histograms have as many dimensions as the number of channels of data.[10] Figure 8.4, for example, shows a two dimensional histogram of visible and 11-μm infrared radiances. Tall convective clouds are cold and bright; they form a cluster in the upper right of the histogram. The sea surface is warm and dark; it forms a cluster in the lower left corner. Other cloud and surface types are noted in the figure. Since some pixels sample two or more cloud or surface types (partly cloudy pixels, for example), the histogram does not consist of isolated islands, but rather in connected islands. Unfortunately, some histograms are much more difficult to interpret than this one in part because some cloud or surface types are too variable to form a local maximum in the histogram.

Setting a threshold for the detection of clouds is equivalent to drawing a vertical or horizontal line in Fig. 8.4. As expected due to small temperature differences, it would be difficult to set an infrared threshold (draw a horizontal line) that distinguishes low cumulus clouds from land or sea. It is interesting to note that it would also be difficult to set a visible threshold (draw a vertical line) that successfully distinguishes cloud from land. This is in part because the Meteosat image from which the histogram was formed included the bright Saharan sands, but it illustrates the care that must be taken in setting thresholds.

Histogram techniques attempt to locate clusters of pixels as opposed to drawing vertical or horizontal lines. There are several methods for identifying clusters. The dynamic cluster method (Desbois *et al.* 1982; cf. Simmer *et al.*, 1982; Phulpin *et al.*, 1983) randomly chooses a maximum number of points in the histogram (roughly 15) to serve as cluster centers. (The coordinates of each point are (C_1, C_2, ...), where C_i represents the count (related to radiance) measured in channel *i*.) Each point in the histogram is assigned to the closest cluster. Since randomly selected clusters are likely to be misplaced, the next step results in a correction.

[10] Or more, if texture parameters are included. (See Section 8.2.3.)

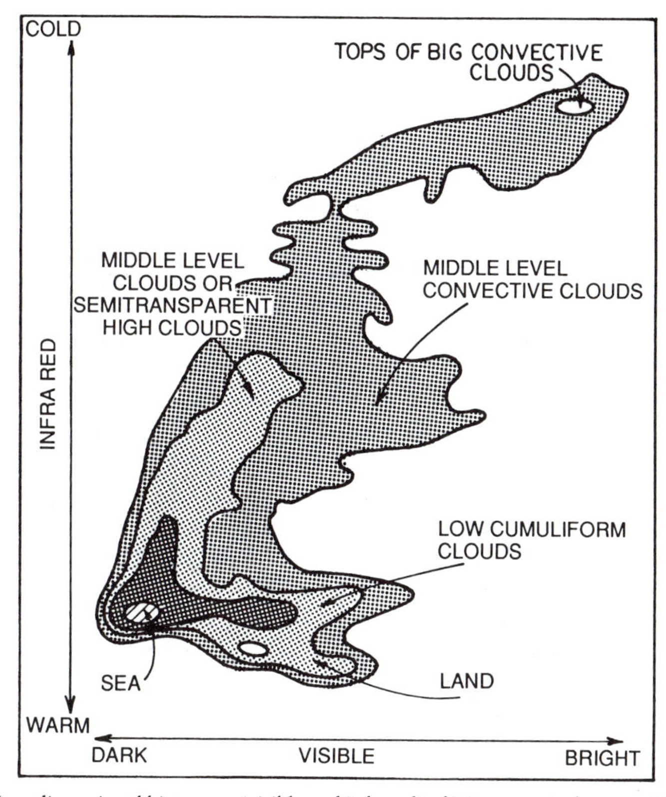

FIGURE 8.4. Two-dimensional histogram (visible and infrared) of Meteosat pixel counts. [After Desbois *et al.* (1982).]

The center of gravity (using the number of observed pixels at each point as a weight) and standard deviations (about each axis) are calculated for each cluster. The center of gravity serves as the new cluster center. Points are again assigned to the clusters, this time using the standard deviations as distance measures. Any cluster that receives too few points is eliminated and its points are distributed among the remaining clusters. The correction step is repeated until the cluster centers and their standard deviations do not change.

After cluster centers are identified in the histogram, they must be classified. The clusters can be identified manually, or by setting thresholds, or by comparing with previous classifications. Once surface clusters have been identified, cloud amount can be determined by dividing the number of pixels that belong to cloudy clusters (nonsurface clusters) by the total number of pixels. Cloud temperature and brightness can be estimated from the average temperature and brightness of each cloudy cluster. Note that cloud amount defined in this manner applies to an area, not to an individual pixel, which is considered to be completely clear (if it belongs to a clear cluster) or completely cloudy (if it belongs to a cloudy cluster).

8.2.3 Pattern Recognition Techniques

For some applications, it is important to know the type of cloud (or surface). *Pattern recognition techniques* attempt to classify arrays of pixels in a manner analogous to the way a person might perform such a classification.

Consider the following simple hypothetical example. Suppose that an analyst, armed with a visible satellite image of an oceanic area, were charged with the task of determining whether an observer on a ship would report clear, partly cloudy, or cloudy skies. The analyst would examine a small area of the image centered on the location of the ship. To be consistent, the size of the area should be approximately the same as that which would be seen by the surface observer. If the area in the image were uniform and bright, the analyst would classify it as cloudy. If it were uniform and dark, the analyst would classify it as clear. If a portion of it were bright and a portion dark, the analyst would classify it as partly cloudy.

A simple pattern recognition algorithm might simulate this process as follows. Suppose that the mean and standard deviation of the pixels in the area were calculated. Uniform scenes, either clear or cloudy, would have a low standard deviation. Partly cloudy scenes must have a high standard deviation. Cloudy scenes would have high mean brightness, whereas clear (ocean) scenes would have low mean brightness. Partly cloudy scenes would have intermediate values of mean brightness. The algorithm could first check the standard deviation. If it were sufficiently high, the scene would be classified as partly cloudy. For low-standard-deviation scenes, the mean brightness would discriminate between clear and cloudy cases. Note that in pattern recognition, measures of the spatial variability of pixels in a scene (such as standard deviation) are referred to as *texture parameters*. *Spectral parameter* is the term sometimes used to describe nontexture parameters such as mean or extreme radiances or differences of these quantities between channels.

A training data set is necessary to determine the discrimination values used in pattern recognition techniques. In the above simple example, an analyst would classify a set of scenes as clear, cloudy, or partly cloud; then the mean brightnesses and standard deviations for each classification would be examined in order to select threshold values.

Harris and Barrett (1978) used mean brightness and two texture parameters (including standard deviation) to classify 5×5-pixel scenes in DMSP visible images as (1) clear, (2) stratiform cloud, (3) cumuliform cloud, or (4) mixed cloud.

Ebert (1987) attempted the difficult task of classifying cloud and surface types over the Arctic using four-channel AVHRR data (0.6, 0.9, 3.7, and 11 μm). She used eight parameters (five spectral and three texture) to classify 32×32-pixel scenes into 18 cloud, surface, and mixed cloud-surface categories. The maximum likelihood algorithm (Duda and Hart, 1973) was similar to that used in histogram techniques in that scenes were assumed to form clusters in the eight-dimensional parameter space. Means and variances for each cluster were calculated. Standard deviation was used to measure the "distance" of a scene to each cluster. The scene was assigned to the "closest" cluster. A panel of three analysts provided the training and verification data sets. The classification scheme agreed with the analysts more than 80% of the time.

8.2.4 Multispectral Techniques

In this category we group cloud retrieval techniques that rely on radiance measurements at two or more wavelengths and use simple models to make retrievals. Radiative transfer techniques are similar, but they use the results of more complex radiative transfer calculations to make retrievals.

The best-known multispectral technique is the bispectral technique of Reynolds and Vonder Haar (1977), in which visible and 11-μm infrared radiance observations are used to retrieve cloud amount and cloud-top temperature. Each scan spot is assumed to be partly covered with a single cloud layer. The radiance observed in the visible channel is modeled as solar radiation reflected from the cloud tops and from the surface:

$$L_{\text{VIS}} = N\gamma_{\text{cld}}E_{\text{sun}} + (1 - N)\gamma_{\text{clr}}E_{\text{sun}}, \tag{8.21}$$

where N is the cloud amount; γ_{cld} and γ_{clr} are the bidirectional reflectances of the clouds and the surface, respectively, and are assumed known; and E_{sun} is the solar irradiance (integrated over the radiometer spectral response), which is assumed to be known and to be the same at the surface and at cloud top. The radiance observed in the infrared channel is similarly modeled as being emitted by the surface (through the holes in the cloud), emitted by the cloud, and transmitted through the cloud after being emitted by the surface:

$$L_{\text{IR}} = (1 - N)L_{\text{clr}} + N\varepsilon L_{\text{cld}} + N(1 - \varepsilon)L_{\text{clr}}, \tag{8.22}$$

where ε is the cloud emittance (assumed known), L_{cld} is the radiance the cloud would have if it were black, and L_{clr} is the radiance of the clear area (also assumed known). These equations are solved for the two unknowns:

$$N = \frac{L_{\text{VIS}} - \gamma_{\text{clr}}E_{\text{sun}}}{(\gamma_{\text{cld}} - \gamma_{\text{clr}})E_{\text{sun}}}, \tag{8.23a}$$

$$L_{\text{cld}} = \frac{L_{\text{IR}} - (1 - N\varepsilon)L_{\text{clr}}}{N\varepsilon}. \tag{8.23b}$$

After L_{cld} has been retrieved, the cloud-top temperature is calculated using the Planck function. (Of course, if $\varepsilon < 1$, the temperature will represent the temperature of the interior of the cloud rather than the cloud-top temperature.) The cloud-top height may be retrieved by finding the height in an appropriate sounding where the temperature equals the cloud-top temperature.

In practice, retrievals based on single pixels were shown to be rather noisy. An error analysis showed that averaging cloud height and cloud amount over approximately 50 pixels would produce useful results.

Reynolds and Vonder Haar tested the bispectral technique using Scanning Radiometer data from the NOAA 2 satellite. Three rawinsonde stations were chosen for analysis: Denver, Oklahoma City, and White Sands Missile Range. Cloud amount and cloud height were calculated and averaged for each pixel within 75 km of the station. This area approximated the area viewed by a ground-based observer. Between 13 and 15 satellite passes were analyzed for each station. The retrieved parameters were compared with surface observations. The cloud amount rms error was 0.2, and the bias was −0.05. Cirrus caused problems in cloud height retrieval, in part because cloud emittance was assumed to be 0.9 for all clouds. For all cases, rms cloud height error was 4.7 km with a bias of −4.2 km. When cirrus cases were eliminated, the rms cloud height error was 0.5 km with a bias of +0.27 km.

A problem with the bispectral technique is that bidirectional reflectance is a difficult quantity to specify a priori. It is a function of the size, shape, and spacing of the clouds in addition to the optical depth (McKee and Cox, 1974; Reynolds *et al.*, 1978; Weinman and Harshvardhan, 1982).

Platt (1983) explored extensions to the bispectral technique including reflection between the cloud layer and surface, parameterization of the infrared cloud emittance in terms of the cloud visible optical depth, and multiple cloud layers.

Another multispectral technique allows the retrieval of the temperature of thin cirrus. Szejwach (1982) showed both theoretically and experimentally (with an aircraft-mounted radiometer) that cirrus clouds have nearly equal emittances near 6.4 and 11.5 μm in spite of the large wavelength separation. Since collocated scan spots in the two channels will contain the same cloud amount N, they will have the same effective cloud amount N'. The radiances in the two channels must, therefore, be linearly related (see Eqs. 8.19 and 8.20); when observed pairs of radiances are plotted (Fig. 8.5), they fall on a straight line. The point on the line where $N' = 1$ is the point where the cloud acts as a blackbody. This point can be found with the aid of the Planck function by plotting the curve $L_1 = B_1[B_2^{-1}(L_2)]$ The temperature (T_N) where the line (L) intersects the curve (Γ) is the temperature of the cloud. This technique is being used with Meteosat data.

8.2.5 Spatial Coherence

A very precise method for determining cloud amount and cloud-top temperature is the *spatial coherence method* of Coakley and Bretherton (1982). The technique starts with several assumptions that preclude it from being applicable in all situations but that result in precise estimates in those cases to which it does apply. In part the precision results from the fact that the technique is capable of detecting cases to which it is not applicable. Clouds are assumed to exist in layers with uniform cloud top and emittance (the original technique only dealt with a single layer of cloud) and to exist over a uniform background. Portions of the image must be completely clear and portions completely cloudy. The technique is well suited to analyzing marine stratocumulus, but is applicable in other situa-

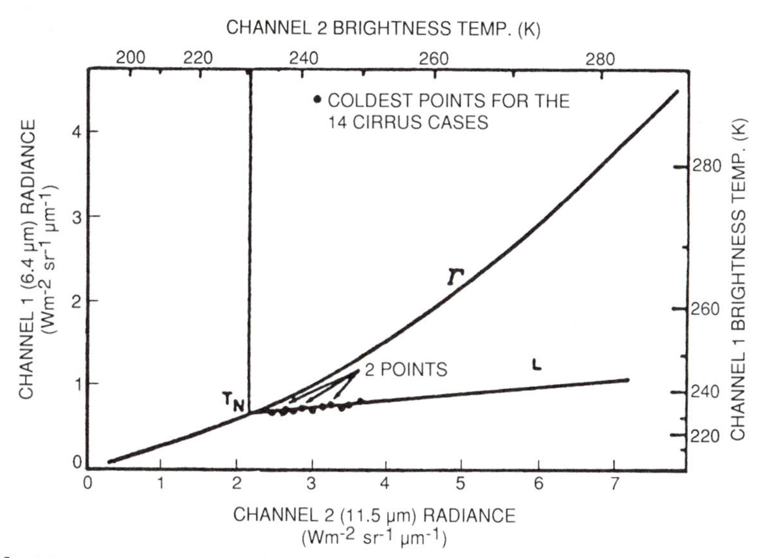

FIGURE 8.5. Meteosat water vapor radiance versus window channel radiance over a thin cirrus cloud. The observed radiances fall on the straight line L. The intersection of this line with the Planck curve ε gives the cloud-top temperature. [After Szejwach (1982).]

tions as well. It is not applicable, for example, near fronts or when cirrus overlies the scene.

The spatial coherence technique rests on the idea that portions of the image that are either completely clear or completely cloudy will exhibit little spatial variability, whereas partly cloudy areas will exhibit large spatial variability. Infrared data are analyzed over a moderately large sector (on the order of $(250\text{ km})^2$). The image is subdivided into blocks of at least 2×2 pixels. For each block, the mean and standard deviation of the radiance are calculated. When plotted (Fig. 8.6), the data form an arch pattern. The feet of the arch (where the spatial variation of radiance in each block is low) are the blocks that are completely clear or completely cloudy. The average of the radiances in a foot represents the radiance of the surface or the cloud top. The width of the foot, as measured by the standard deviation of the radiances in the foot, is a measure of the uncertainty in the background or cloud radiance. No other technique yields these important uncertainty estimates.

Once the clear and cloudy radiances are known, the cloud amount can be calculated for each pixel using Eq. 8.2. Usually, however, the mean cloud amount for the entire sector is calculated:

$$\overline{N} = \frac{\bar{L} - L_{\text{clr}}}{L_{\text{cld}} - L_{\text{clr}}}, \tag{8.24}$$

where $\bar{L}$ is the mean radiance in the sector. The uncertainty in cloud amount can also be calculated. It is composed of instrument noise and uncertainty in L_{clr} and

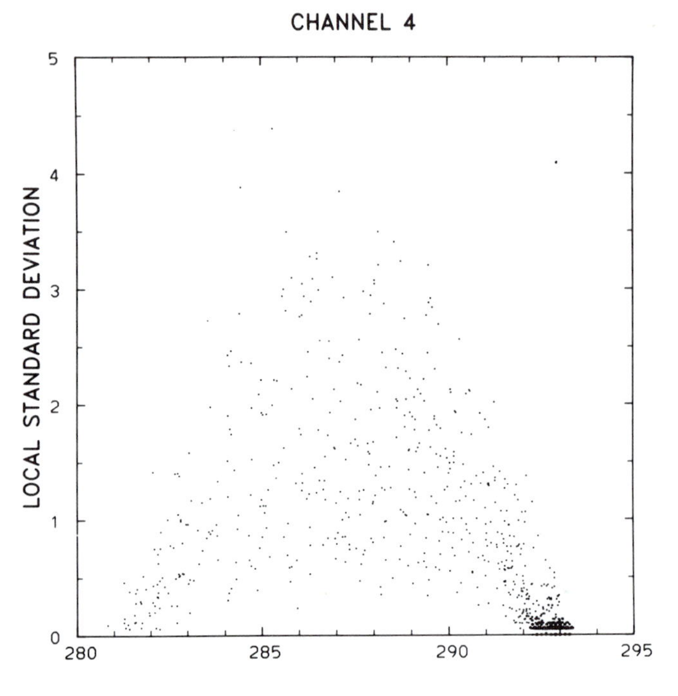

FIGURE 8.6. Spatial coherence diagram of a one-layer cloud system. The ordinate is the standard deviation, and the abscissa is the mean radiance of the blocks of pixels in an AVHRR 11-μm image. [After Coakley and Bretherton (1982).]

L_{cld}. When averaged over the sector, the instrument noise is negligibly small. Therefore the uncertainty of the mean cloud amount is[11]

$$\Delta\overline{N} = \left\{\left[\frac{\overline{N}\,\Delta L_{\mathrm{cld}}}{L_{\mathrm{cld}} - L_{\mathrm{clr}}}\right]^2 + \left[\frac{(1-\overline{N})\,\Delta L_{\mathrm{clr}}}{L_{\mathrm{cld}} - L_{\mathrm{clr}}}\right]^2\right\}^{1/2}. \tag{8.25}$$

The spatial coherence technique only works about half the time because roughly half of the sectors in an image have cloud systems that are too complex for the method to be applied. When it fails, however, the characteristic arch pattern is missing; thus sectors to which the method cannot be applied are readily identifiable, which improves accuracy. Figure 8.7 shows an example of a two-layer cloud system.

Coakley (1983) explored the possibility of using more than one data channel (3.7 and 11 μm, for example) to retrieve the properties of more than one level

[11] This is based on the assmption that the error of a function y of several independent variables x_i is given by $\Delta y = \left\{\sum_i \left[\left(\frac{\partial y}{\partial x_i}\right)^2 \Delta x_i^2\right]\right\}^{1/2}$, where Δx_i represents the error in the independent variables.

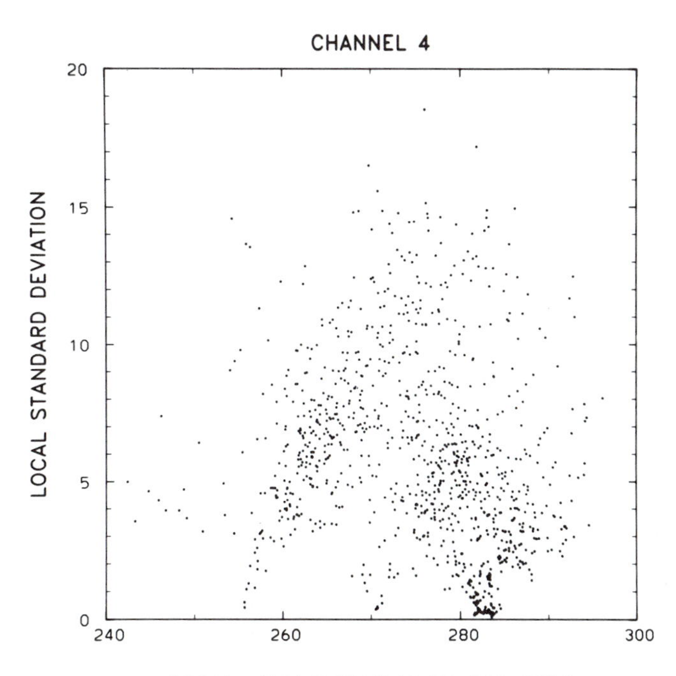

FIGURE 8.7. Spatial coherence diagram for a two-layer cloud system. [After Coakley and Bretherton (1982).]

of cloud. Coakley and Baldwin (1984) discussed an automated method based on the spatial coherence technique for objectively analyzing clouds over the ocean.

It is interesting to note that the spatial coherence technique does not work with visible data. Figure 8.8 shows a spatial coherence diagram for AVHRR channel 1. Although the background brightness is easily identified, clouds are too variable in reflectance to produce a foot. This is thought to be due to small-scale (< 1–4 km) variability in the column liquid water content (Coakley, 1983; Coakley and Baldwin, 1984; Randall *et al.*, 1984; Coakley and Davies, 1986).

8.2.6 Radiative Transfer Techniques

All cloud retrieval techniques can be broken into two steps: (1) cloud detection, and (2) parameter retrieval. Radiative transfer techniques come into play in the second step; after a cloud is determined to exist, radiative transfer calculations are used to determine its properties. The advantage of using radiative transfer calculations is that they allow the retrieval of parameters, such as cloud optical depth and microphysical properties, that are not retrievable with other methods.

Arking and Childs (1985) have used this technique with AVHRR data to retrieve visible optical depth, cloud-top temperature, cloud amount, and an indication of the cloud phase and drop size. They note that given the cloud amount,

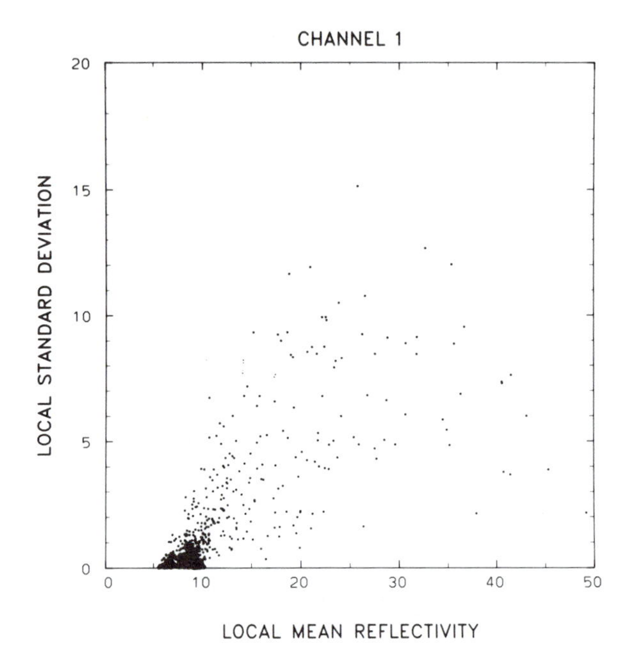

FIGURE 8.8. Visible spatial coherence diagram. [After Coakley and Bretherton (1982).]

visible radiances are affected primarily by the cloud optical depth, 11-μm radiances are affected primarily by the cloud-top temperature, and 3.7-μm radiances (during daylight) are affected primarily by the cloud microphysical parameters (chiefly drop size and drop phase). They chose six sets (models) of microphysical conditions to approximate the range of clouds expected. All of the cloud drops were assumed to be spherical. Three models had ice spheres, and three had liquid water spheres. The ice models had modal drop radii of 4, 8, and 32 μm; the water models had modal drop radii of 4, 8 and 16 μm. Radiative transfer calculations determined the reflectance, transmittance, and absorptance of plane-parallel cloud layers as tabulated functions of optical depth, zenith angle, and azimuth angle at the three wavelengths. Armed with these numbers, the retrieval schemes proceeds as follows.

- A histogram of visible reflectance and 11-μm brightness temperature is constructed within a $2.5° \times 2.5°$ analysis area, and clusters of points are identified. One cluster must represent the surface. Mean reflectance, 11-μm brightness temperature, and 3.7-μm radiance are calculated for the surface cluster and serve as the surface (clear) values.
- In the cloudy cluster, the cloud amount is assumed to be one, and the 3.7-μm radiance is calculated from the surface parameters and the tabulated functions. Comparing the observed and calculated 3.7-μm radiances, one of the six microphysical models can be selected as most representative of

the clouds in the sector.[12] This step is somewhat subjective, but it narrows the range of possible optical depths and cloud-top temperatures.

- Using the clear radiances and the tabulated functions for the microphysical model just selected, the measured visible reflectance and 11-μm brightness temperature of each pixel are transformed into a visible optical depth, a cloud-top temperature, and a cloud amount.

Rossow *et al.* (1989) have developed a different radiative transfer technique. They use radiative transfer models that include the atmosphere and surface as well as clouds to retrieve surface and cloud parameters. Only visible and 11-μm radiances are used. Daily observations from the Scanning Radiometer on board the NOAA 5 satellite were employed. In highly simplified form, their method may be described as follows. First, for each area under consideration, the surface reflectance and skin temperature are found. This is accomplished by examining the time history of pixels in the area. The four largest brightness temperatures and the four minimum reflectances for a month are found. Since using extreme values tends to bias the results, however, the four extreme values are not used directly. Rather, the average of the reflectances paired with the four largest brightness temperatures and the average of the brightness temperatures paired with the four minimum reflectances are used. This transformation reduces problems caused by using extrema. Cloud-free radiative transfer models (see below) are used to convert the mean clear-sky brightness temperature and reflectance into a skin temperature and a surface reflectance. The reflectance is assumed to be constant for the month, but the skin temperature over land is adjusted daily by the difference between the NMC-analyzed shelter temperature and the monthly mean shelter temperature.

Next, cloud properties are retrieved by comparing all pixels with radiative transfer models. The basic idea is that any change from clear-sky radiances is due to clouds. Each pixel is assumed to be 100% clear or 100% cloudy. The infrared model is applied first. Daily NMC temperature and moisture profiles plus the skin temperature retrieved above are used to calculate the radiance seen by the satellite. Clouds, which are assumed to be black, are inserted at mandatory pressure levels. The two levels with calculated radiances which bracket the observed IR radiance are found. The cloud-top temperatures and heights that were used to calculate the matching radiances are interpolated to calculate the retrieved values.

The visible model is more complicated. It consists of five parts: (1) an ozone layer (which absorbs at 0.6 μm), (2) a cloud-free, Rayleigh scattering layer, (3) a cloud layer, (4) a second cloud-free layer, and (5) the surface. The cloud is a homogeneous, plane-parallel layer with an effective mean drop radius of 10 μm. Conservative Mie scattering is calculated. The visible model is used once to calculate a large table of radiance as a function of (1) optical depth (0–100), (2)

[12] The procedure has since been changed to fit a microphysical model to each cluster in the sector (A. Arking, personal communication, 1988). This has the effect of allowing more than one type of cloud (for example cirrus and stratocumulus) to exist in the sector.

viewing and illumination angles, (3) cloud-top height, and (4) surface reflectance. To retrieve the optical depth for a particular pixel, the viewing and illumination angles are obtained from navigation files, the cloud-top height is obtained from the IR model, and the surface reflectance is obtained from the monthly mean. The remaining variable, optical depth, is determined by interpolating between the calculated radiances that best match the observed visible radiance.

Finally, to account for noise (of all kinds), any pixel whose retrieved optical depth is less than 1.2 or whose retrieved cloud-top height is less than 1.4 km is classified as clear. Requiring pixels to pass both tests results in some clouds (e.g., some low stratus and some cirrus) escaping detection. On the other hand, few pixels that are classified as cloudy are actually clear.

8.2.6.1 The ISCCP Cloud Algorithm

The International Satellite Cloud Climatology Program (ISCCP) is the first project of the World Meteorological Organization's World Climate Research Programme (Schiffer and Rossow, 1983). Its goal is to collect global visible and 11-μm satellite data from polar and geostationary satellites and to process them into a cloud climatology. Data collection began in July 1983, and is planned to continue through 1995 (Rossow and Schiffer, 1991).

To keep the amount of data within manageable limits, the original satellite data have been sampled at 30-km and 3-h intervals. The scan spots continue to have the original size of the IR pixels (about 7 km for GOES IR data), and the visible data are averaged to match.[13]

The algorithm for retrieval of these parameters was developed in a series of workshops (Rossow *et al.*, 1985). The process began with an extensive analysis of existing techniques that use visible and IR information: simple visible and infrared thresholds, the threshold technique of Minnis and Harrison (1984), the radiative transfer technique of Rossow *et al.* (1989), and the histogram techniques of Desbois *et al.* (1982) and Simmer *et al.* (1982). Since little is available in the way of "cloud truth," the analysis consisted primarily of comparing retrieved cloud amounts with cloud amounts derived from 1-km-resolution visible data. The methods were also compared with each other and with the mean cloud amount for all methods. The main conclusion was that most of the time all methods worked well. When cloud amount for each method was compared with the average cloud amount for the six methods, the correlation coefficient ranged

[13] It is important to note that this sampling is not equivalent to "throwing away" data. The goal of ISCCP is to obtain information on clouds in grid boxes approximately 280 km on a side. In each box, therefore, there are about 87 pixels. If the data were sampled at 15-km intervals, the number of pixels would increase by a factor of four (300%), which means that the amount of processing would increase by the same 300%. Our knowledge about clouds in the grid cell would certainly increase, but probably not by more than a few percent—not enough to justify a 300% increase in processing.

from 0.83 to 0.93, and the bias ranged from −5.3% to +2.7%. All of the methods, however, had difficulties in "low contrast" situations in which cloud radiance differed little from clear-sky radiance. Among the other conclusions were (1) that since all techniques benefited from better assessment of the clear-sky radiance, the ISCCP algorithm should make clear-sky radiance determination a top priority and (2) that a threshold technique should be adopted for cloud detection because it is the best understood, it is the easiest to interpret, and it seems to be more sensitive to difficult-to-detect clouds than other methods.

The ISCCP algorithm, developed as a result of the intercomparison tests, proceeds in four steps. In the first step a clear-sky radiance map is constructed. This step is the most time consuming part of the algorithm. It rests on the observation that temporal and spatial variations of clear-sky radiances are small compared to the variations caused by clouds. The IR radiance map is constructed as follows. First, pixels that are easily identified as cloudy are eliminated. Subregions of $\sim(500\ \text{km})^2$ over ocean and $\sim(250\ \text{km})^2$ over land are examined. Any pixel that is much colder (~10 K) than the warmest pixel in the subregion is labeled cloudy. These pixels need not be considered further. Next, pixels are compared with the pixel in the same location on the day before and the day after. If the pixels differ by less than a small temperature (~1 K, variable with surface type), the pixel is labeled clear. For pixels that survive this test, the 5-day mean, 5-day maximum, 15-day mean, 15-day maximum, 30-day mean, and 30-day maximum brightness temperatures are calculated. Finally, these numbers, ranked in order of preference, are used in the clear-sky radiance map. Several tests, which depend on surface type, are performed to ensure that the numbers are reasonable. Over land, the 5-day mean is selected unless the tests show it to be cloud contaminated. Testing continues until, if all else fails, the 30-day maximum brightness temperature is used. Over ocean, the same process is used, but it starts with the 15-day mean. A visible clear-sky reflectance map is constructed by a similar, but independent, process.

The second step consists in applying thresholds to detect clouds. The thresholds consist not of absolute values, as in previously mentioned techniques, but of increments above the clear-sky values. The two thresholds are ΔT, an IR brightness temperature difference, and ΔR, a visible reflectance difference. These increments are designed to take into account both instrument noise and uncertainty in the clear-sky values. Each pixel is assumed to be totally clear or totally cloudy. Any pixel whose IR brightness temperature is more than $2\Delta T$ colder than the clear-sky brightness temperature or whose visible reflectance is more than $2\Delta R$ greater than the clear-sky reflectance is labeled cloudy. Any pixel whose IR brightness temperature is more than ΔT but less than $2\Delta T$ colder than the clear-sky brightness temperature or whose visible reflectance is more than ΔR but less than $2\Delta R$ greater than the clear-sky reflectance is labeled marginally cloudy. Warmer and darker pixels are labeled clear. Information about both of these two tests is retained so that users can examine only infrared-detected clouds, or only visible-detected clouds, or any combination. Over land, ΔT and ΔR are 6 K and 0.06, respectively; over ocean they are 3 K and 0.03, respectively. Cloud amounts are determined in the forth step by counting the cloudy pixels in the 2.5° × 2.5° regions.[14]

In the third step, cloud radiative properties for each cloudy or marginally cloudy pixel are determined using the technique of Rossow (1989). By comparing the observed visible reflectance and infrared brightness temperature with radiative transfer calculations, the cloud-top temperature and the cloud optical depth (during daylight) are determined. Using TOVS temperature profiles, the cloud-top pressure is determined from the cloud-top temperature. The philosophy here is that although all pixels are assumed to be either 100% cloud-covered or 0% cloud-covered, variations in cloud IR temperature and visible reflectance are explained in a radiatively consistent way as variations in cloud-top temperature and cloud optical depth.

In the fourth step, the data are histogrammed as functions of optical depth and cloud-top pressure for each 2.5° × 2.5° region (280-km grid cell). Data are also histogrammed in terms of a cloud type, which is determined by the cloud-top pressure and cloud optical thickness (Fig. 8.9). In all, 132 parameters are recorded for each cell, including parameters that reveal how the classifications were made. Table 8.1 lists the retrieved parameters. Monthly summaries of the data are also constructed (Table 8.2). Data tapes of the retrieved ISCCP parameters are available from the Satellite Data Services Division of NESDIS. Figure 8.10 shows the mean cloud amount retrieved by the ISCCP algorithm for July 1983 and January 1984. Figure 8.11 shows the mean zonal cloud amounts for the same months.

8.2.6.2 The Nimbus 7 THIR/TOMS Cloud Climatology

Data from the Temperature Humidity Infrared Radiometer (THIR) and the Total Ozone Mapping Spectrometer (TOMS) on the Nimbus 7 satellite have been used to produce a 6-year cloud climatology (Stowe *et al.*, 1988) in large part for interpreting data from the Earth Radiation Budget experiment (see Section 10.2.2.1). This climatology relies more heavily on thresholds and bispectral techniques than it does on radiative transfer methods, but it is discussed in this section so that it may be compared with the ISCCP algorithm.

The infrared THIR data and the ultraviolet TOMS data were processed separately, using threshold algorithms, then merged (for daylight scenes). The infrared algorithm starts with an estimate of the surface temperature from the U.S. Air Force 3D Nephanalysis (Fye, 1978). An empirical correction for water vapor was made to yield an estimate of what the 11.5-μm clear-sky radiance should be. The uncertainty in this clear-sky radiance was also estimated. If the THIR 11.5-μm radiance was smaller than the clear-sky radiance minus the uncertainty, the scene was classified as cloudy. If the scene was warmer than the clear sky radiance plus the uncertainty, the scene was also classified as cloudy based on the observation

[14] Actually, the boxes are equal area boxes. They extend 2.5° in the latitudinal direction, but approximately 2.5°secΘ (Θ is latitude) in the longitudinal direction so that the boxes have approximately the same area as a 2.5° × 2.5° box at the equator. (See Rossow and Garder, 1984.)

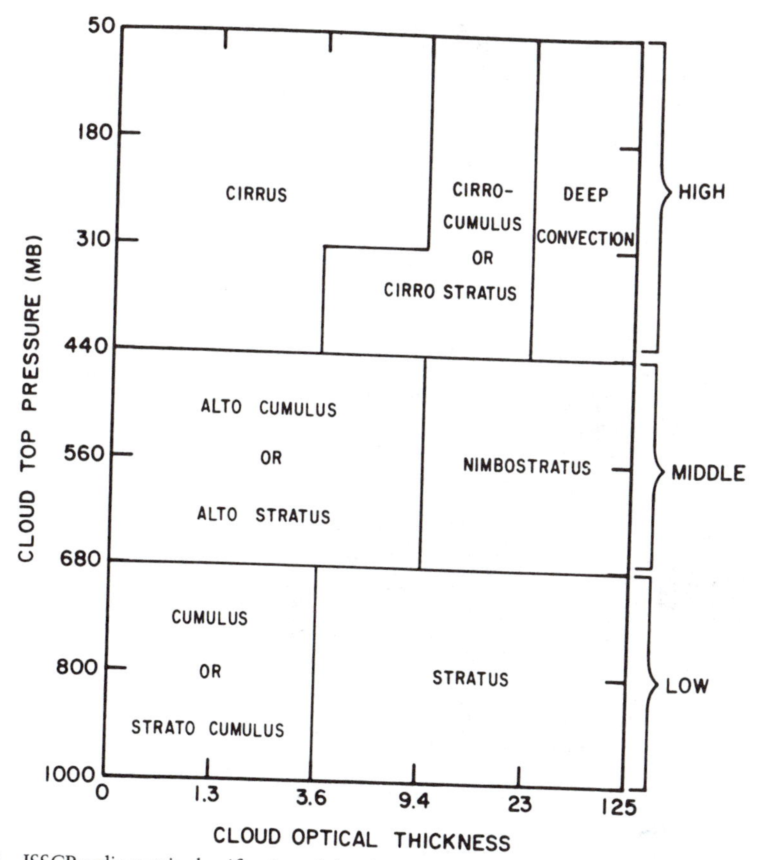

FIGURE 8.9. ISSCP radiometric classification of cloudy pixels by the measured values of optical thickness and cloud-top pressure. At night only cloud-top pressure is determined, so that only the low, middle, and high cloud types are counted. [After Rossow and Schiffer (1991).]

that clouds over locations with significant surface temperature inversions can be detected as dark regions in infrared images. For cold clouds, an attempt was made to classify the cloud height (low, middle, high) based on a climatological lapse rate.

The ultraviolet algorithm is interesting because both ocean and land have low ultraviolet reflectivity, whereas land can have visible reflectivities exceeding 40%. First, a surface reflectivity was estimated from the TOMS-measured reflectivity using a multiple Raleigh scattering model (Dave, 1964). Two wavelengths were used, 0.36 and 0.38 μm, neither of which is absorbed by ozone. The two reflectivities were averaged to produce a "0.37-μm" reflectivity. The cloud fraction for a single TOMS scan spot was modeled as linear with respect to the 0.37-μm reflectivity: 8% reflectivity corresponds to no clouds; 50% reflectivity corresponds to 100% cloud cover.

The two algorithms differ in their treatment of a single scan spot: in the infrared algorithm each scan spot is either cloudy or cloud free; in the ultraviolet algorithm

TABLE 8.1. C1 Data Products of the International Satellite Cloud Climatology Project[a]

Cloud amount and distribution information
Total number of image pixels
Total number of cloudy pixels
Total number of IR-cloudy pixels
Total number of marginal cloudy pixels
Total number of IR-only cloudy pixels
Number of cloudy pixels in 7 cloud-top-pressure classes
Number of cloudy pixels in 35 cloud-top-pressure/optical-depth classes
Average total cloud properties
Cloud-top pressure without visible channel information
Cloud-top pressure with visible channel information
Cloud-top pressure of marginal clouds
Spatial variation of cloud-top pressure
Cloud-top temperature without visible channel information
Cloud-top temperature with visible channel information
Cloud-top temperature of marginal clouds
Spatial variation of cloud-top temperature
Cloud optical thickness
Cloud optical thickness of marginal clouds
Spatial variation of cloud optical thickness
Average surface properties
Surface pressure
Surface temperature
Surface visible reflectance
Snow/ice cover fraction
Topography and land/water flag
Average radiances
IR-cloudy
IR-clear
Spatial variation of IR
VIS-cloudy
VIS-clear
Spatial variation of VIS
Viewing geometry and day/night flag
Satellite identification
Average atmospheric properties
Atmospheric temperatures for nine levels
Surface temperature
Tropopause temperature
Tropopause pressure
Precipitable water amount for five levels
Column ozone abundance
Source of atmospheric data

[a] Global information provided every 3 h for each 280-km grid cell. [After Rossow and Schiffer (1991).]

TABLE 8.2. C2 Data Products of the International Satellite Cloud Climatology Project[a]

Cloud amount information
Monthly average cloud amount
Monthly frequency of cloud occurrence
Monthly average IR-cloud amount
Monthly average marginal cloud amount
Average total cloud properties
Cloud-top pressure (PC)
Average spatial and temporal variations of PC
Cloud-top temperature (TC)
Average spatial and temporal variations of TC
Cloud optical thickness (TAU)
Average spatial and temporal variations of TAU
Cloud water path (PATH)
Average spatial and temporal variations of PATH
Cloud albedo (ALB)
Average spatial and temporal variations of ALB
Average properties (amount, PC, TC, TAU) for cloud types
Low cloud (IR-only)
Middle cloud (IR-only)
High cloud (IR-only)
Cumulus cloud
Stratus cloud
Altocumulus, altostratus cloud
Nimbostratus cloud
Cirrus cloud
Cirrocumulus, cirrostatus cloud
Deep convective cloud
Average surface properties
Surface temperature (TS)
Average temporal variations of TS
Surface visible reflectance (RS)
Snow/ice cover fraction
Average atmospheric properties
Surface pressure (PS)
Surface temperature (TS)
Temperature at 500 hPa (T5)
Tropopause temperature (PT)
Tropopause pressure (PP)
Stratospheric temperature (ST)
Column water amount (PW)
Column ozone amount (O3)

[a] Global, monthly average information provided at eight equally spaced times of day. [After Rossow and Schiffer (1991).]

each spot is partly cloudy. The difference lies in the resolution of the instruments and in the resolution of the desired product. The goal was to estimate cloudiness for target areas about 165 km square. The THIR has 20-km resolution at nadir; thus there are many THIR scan spots per target area. Treating each THIR spot

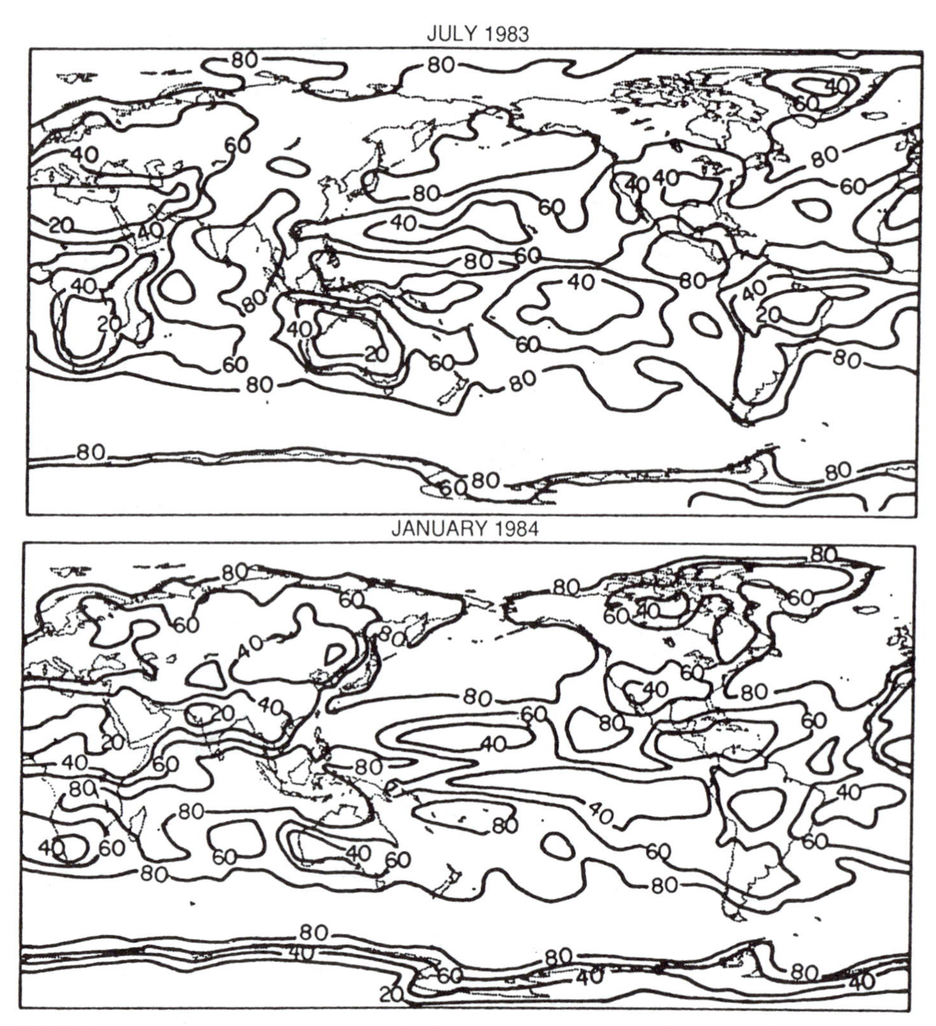

FIGURE 8.10. Mean cloud amounts retrieved by the ISCCP algorithm for July 1983 and January 1984. [Courtesy of G. Garrett Campbell, Cooperative Institute for Research In the Atmosphere, Colorado State University.]

as cloudy or clear results in an estimate of the cloud fraction over the target area. TOMS, on the other hand, has 50-km resolution at nadir and 150 × 200-km resolution at the extreme scan positions. Since there were so few TOMS scan spots per target area, it was desirable to obtain cloud fraction estimates from each one. Fortunately, the uniformly dark background of the snow- and ice-free surface, coupled with the nearly linear dependence of reflectivity on cloud fraction, made single-spot cloud fractions retrievable. The Air Force 3D Nephanalysis snow/ice cover fields were used to determine whether the ultraviolet algorithm should be applied.

For nighttime observations, the THIR cloud data alone were used. During

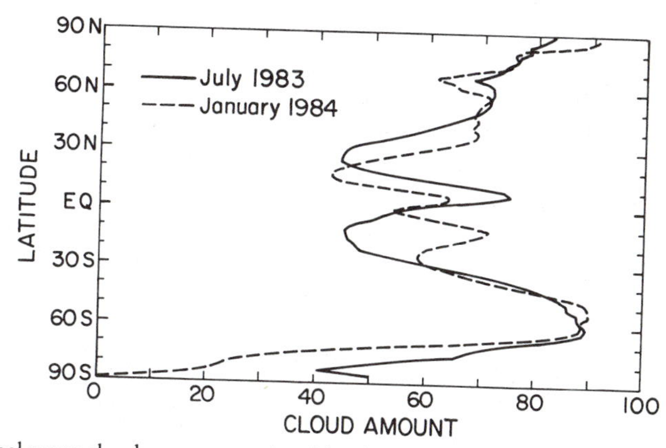

FIGURE 8.11. Zonal mean cloud amounts retrieved by the ISCCP algorithm for July 1983 and January 1984. Note that the global mean cloud amount is greater than 60%, indicating that the 50% figure usually cited may be too low. [Courtesy of G. Garrett Campbell, Cooperative Institute for Research In the Atmosphere, Colorado State University.]

daylight hours, THIR cloud data and TOMS cloud data were merged in a weighted averaging scheme. The TOMS cloud amounts were given greater weight if the THIR estimated no cloud or low cloud because low clouds are difficult to detect at 11.5, μm due to the small difference between the clouds and the surface. Low clouds, however, are good reflectors of visible and ultraviolet light. For cases in which middle and high cloud dominated, the THIR cloud amounts were given more weight than the TOMS cloud amounts.

Finally, the TOMS reflectivities were used to identify cirrus and deep convective clouds. The Nimbus 7 THIR/TOMS cloud climatology is available from the National Space Science Data Center.

8.2.7 Geometric Techniques

A quite different approach to cloud height retrieval involves making geometric measurements with satellite images. A technique that is restricted to low sun angles uses observations of cloud shadows on the Earth's surface (Fujita, 1963; Smith and Reynolds, 1976). Knowing the positions of the sun and the satellite, one can use the distance from a cloud to its shadow to calculate the height of the edge of the cloud.

A more useful technique is *stereoscopy* or simply *stereo*, which involves simultaneously observing a cloud with two satellites (Minzner *et al.*, 1976; Hasler, 1981; Hasler *et al.*, 1983). Just as two eyes allow depth perception, views by two satellites allow determination of the height of clouds. Suppose, for example, that GOES–East and GOES–West view the same cloud. If the GOES–West image is mapped into the perspective of the GOES–East satellite, any feature that is not at sea level will appear displaced, with respect to the GOES–East image, in the mapped GOES–West image (Fig. 8.12). The magnitude of the displacement between the location of the cloud in the two images can be related through

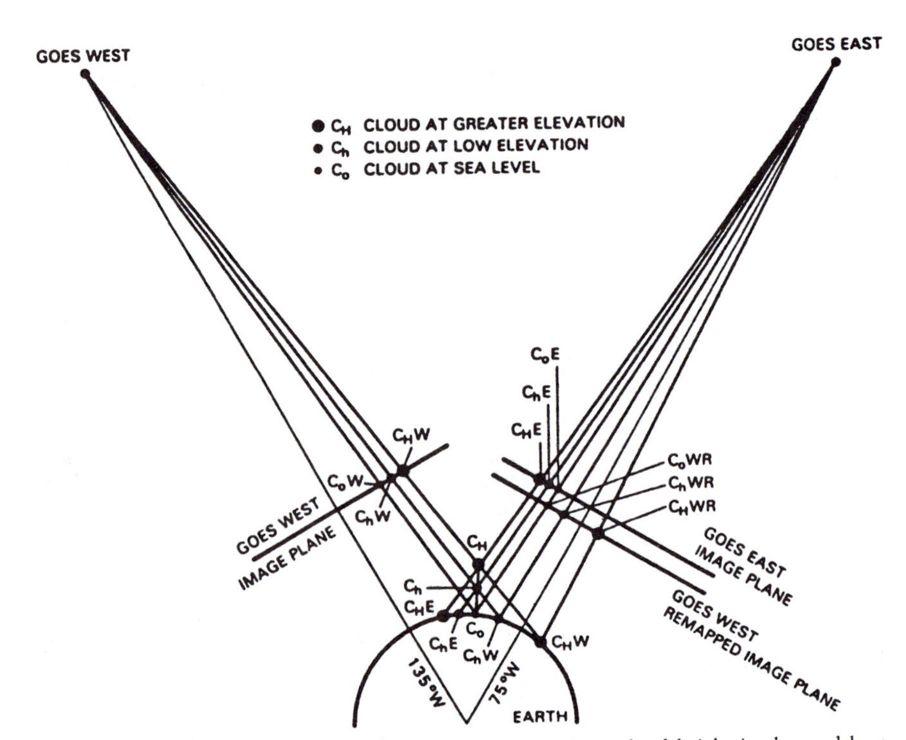

FIGURE 8.12. Simplified two-dimensional diagram illustrating how cloud height is observed by two geostationary satellites. When a GOES–East image is superimposed on a mapped GOES–West image, the separation (parallax) between the location of the same cloud in the two images is proportional to the cloud height. [After Hasler (1981).]

spherical trigonometry to the height of the cloud. Plate 6 shows a perspective view of a thunderstorm cloud top with height contours derived from stereoscopic analysis. Hasler (1981) estimates that with GOES data, stereo heights are accurate to within ± 0.5 km. Stereo images can also be formed from one geostationary image and one low Earth orbiter image.

The great advantage of stereo over other methods of cloud height determination is that stereo depends only on geometry; it does not depend on accurate knowledge of the temperature profile of the atmosphere in which the cloud exists nor on the assumption that the cloud top has the temperature of the air in which it is rising. The disadvantage of stereo is that it requires two satellites to view the same cloud feature at nearly the same time (within ~10 s) so that cloud motion is not confused with cloud height. Although there are large areas of overlapping coverage by two geostationary satellites, no two geostationary satellites are routinely synchronized to make stereo measurements. (GOES–East and GOES–West images are normally scheduled 15 min apart, for example.) With the launch of the next generation of geostationary satellites, such synchronization may become routine.

8.3 CLOUDS FROM MICROWAVE RADIOMETRY

In the microwave portion of the spectrum, the liquid water content of clouds can be measured. This quantity is not measurable in other portions of the spectrum. Recall from Chapter 3 that both water vapor and cloud droplets weakly absorb microwave radiation and that scattering of microwave radiation is negligible for non-precipitating drops. For centimeter wavelengths, water vapor absorption peaks at 22.235 GHz, the center of a weak water vapor rotation line (see Fig. 3.13). Figure 8.13 shows that the absorption by water droplets increases with frequency. The absorption coefficient for each is proportional to the amount of water (liquid or vapor) per unit volume. Over the ocean, which provides a uniform, cold background due to the low emittance of water, atmospheric water vapor and liquid water increase the brightness temperature measured by a satellite-borne radiometer. By exploiting two frequencies, one near 22.235 GHz and one in the atmospheric window near 31 GHz, column-integrated water vapor and liquid water content (both in units of mass per unit area) can be retrieved as follows.

Applying the Rayleigh–Jeans law (Eq. 3.11) to the integrated form of Schwarzchild's equation (Eq. 6.6), the brightness temperature measured by a microwave

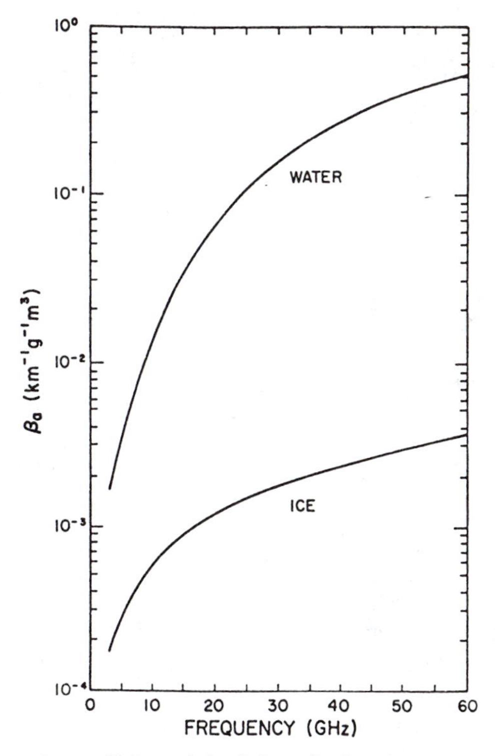

FIGURE 8.13. Mass absorption coefficient of cloud drops in the microwave region. [After Westwater (1972).]

radiometer is

$$T_B(\lambda,\mu) = \varepsilon_o(\mu)T_o\tau_o^{1/\mu} + [1 - \varepsilon_o(\mu)]T_{space}\tau_o^{2/\mu} + \int_{\tau_o}^{1} TW_\lambda(\tau,\mu)\, d\tau, \quad (8.26)$$

where ε_o is the surface emittance (strongly dependent on polarization and μ; weakly dependent on T_o), T_o is the surface temperature, and τ_o is the vertical transmittance from the surface to the satellite.

Since the transmittance τ_o is close to one, an adequate approximation is $T \approx T_o$. Ignoring the small second term, removing T_o from the integral, and integrating the third term with the aid of Eq. 6.8 yields (cf. Grody, 1976)

$$T_B \approx T_o\{1 - [1 - \varepsilon_o(\mu)]\tau_o^{2/\mu}\}. \quad (8.27)$$

Information on surface properties is contained in T_o and ε_o; atmospheric information is contained in τ_o. If U_L and U_V are the column-integrated liquid water content and water vapor content (U_V is also called the precipitable water) and β_L and β_V are the mass absorption coefficients for cloud droplets and water vapor, respectively, then

$$\tau_o^{2/\mu} \approx \exp\left[-\frac{2}{\mu}(\beta_L U_L + \beta_V U_V + \delta_{O_2})\right] \approx \frac{T_o - T_B}{[1 - \varepsilon_o(\mu)]T_o} \quad (8.28)$$

where δ_{O_2} is the small oxygen optical depth. Taking the logarithm,

$$\beta_L U_L + \beta_V U_V \approx -\frac{\mu}{2}\ln\left(\frac{T_o - T_B}{[1 - \varepsilon_o(\mu)]T_o}\right) - \delta_{O_2}. \quad (8.29)$$

Most work (e.g., Grody *et al.*, 1980; Wilheit and Chang, 1980) has approximated T_o as a constant (280 K) and assumed that ε_o is a constant. If this equation is applied at 22 and 31 GHz, then U_L and U_V can be retrieved as linear combinations of μ $\ln[280 - T_B]$:

$$U_L = a_{22}\mu \ln[280 - T_B(22)] + a_{31}\mu \ln[280 - T_B(31)] + a_0, \quad (8.30a)$$

$$U_V = b_{22}\mu \ln[280 - T_B(22)] + b_{31}\mu \ln[280 - T_B(31)] + b_0. \quad (8.30b)$$

The coefficients can be found by using a more accurate radiative model or (for U_V, at least) by comparing with observations. It must be noted that independent measurements of liquid water with which to compare satellite estimates are essentially nonexistent; thus the accuracy of the satellite estimates is not known.

The Soviet satellite Kosmos 243 carried the first microwave radiometer capable of making such measurements (Gurvich and Demin, 1970). U.S. satellites Nimbus 5, 6, 7, Seasat, and two DMSP satellites have carried instruments from which liquid water and precipitable water could be retrieved. The Indian satellites Bhaskara 1 and 2 also carried microwave-sensing instruments.

8.4 STRATOSPHERIC AEROSOLS

Aerosol measurements from satellites rely chiefly on the scattering of sunlight, although a few infrared techniques have been explored. Because somewhat different techniques are involved, we will discuss stratospheric aerosols and tropospheric aerosols separately.

Limb scanning techniques (McCormick and Lovill, 1983), with their large optical path, are well suited to the measurement of stratospheric aerosols, which may have number densities of only 1–10 cm^{-3}. Tropospheric aerosols are perhaps 1000 times more abundant. The most frequently used technique is solar occultation, in which the transmittance of solar radiation is measured during spacecraft sunrise or sunset. A great advantage of solar occultation is that it is nearly self-calibrating; unattenuated solar radiation is measured before each sunset or after each sunrise. Only the ratio of the attenuated to the unattenuated radiation (the transmittance) is used in the retrieval process. Variations in solar output and in instrument gain are eliminated.

McCormick *et al.* (1979) and McCormick (1983) review solar occultation experiments. The first such experiment was performed aboard Apollo—Soyuz in 1975. The Stratospheric Aerosol Measurement (SAM) experiment consisted of a hand-held, one-channel sun photometer with which astronauts made measurements during two sunrises and two sunsets. The measurements were centered at 0.83 μm. The results were sufficiently promising to prompt the development of two complementary instruments to fly on unmanned satellites.

The Stratospheric Aerosol Measurement II (SAM II) experiment was launched aboard the Nimbus 7 satellite in October 1978. It also was a one-channel sun photometer, but its spectral range was centered at 1 μm, where absorption due to atmospheric gases is negligible. Only Rayleigh scattering and scattering by aerosols are important. Since Nimbus 7 had equator crossing times near local noon and midnight, all of the sunrises and sunsets observed by SAM II were in polar regions, 64°–80° north latitude and 64°–80° south latitude.

The Stratospheric Aerosol and Gas Experiment I (SAGE I) was launched aboard a dedicated Applications Explorer Mission satellite (AEM 2) in February 1979. Its orbit had an inclination of 55°, which caused its orbital plane to precess with respect to the sun. It made sunrise and sunset measurements between 79°N and 79°S (depending on season), and it sampled all longitudes in about a month. Thus the SAGE I coverage complemented that of SAM II. About 15 sunrises and 15 sunsets were observed per day for the first 4 months of operation until a problem with the power supply limited observations to sunsets only. Observations were made until November 1981. SAGE I made measurements in four bands centered at 0.385, 0.45, 0.60, and 1.0 μm. All of these bands are sensitive to Rayleigh and aerosol scattering. In addition, the 0.6-μm band is sensitive to ozone (O_3), and the 0.385- and 0.45-μm bands are sensitive to nitrogen dioxide (NO_2). SAGE I had about a 0.5-km field of view at the Earth's limb.

SAGE II was launched in October 1984 on the Earth Radiation Budget Satellite (ERBS). Like AEM 2, ERBS is in a nonsunsynchronous orbit. SAGE II is similar to SAGE I except that it has seven channels centered at 1.02, 0.936, 0.600, 0.525, 0.452, 0.448, and 0.385 μm. The 0.936-μm channel provides the ability to retrieve stratospheric water vapor.

The basic measurement in solar occultation is transmittance (τ_λ) through the limb of the atmosphere. Transmittance is measured as a function of tangent height (h) defined for any ray as the minimum height above the surface (Fig. 4.26). A small correction to the tangent height for the bending of the light ray by refraction is sometimes made. Transmittance is related to the integral along the ray's path of the extinction coefficient:

$$-\ln[\tau_\lambda(h)] = \delta_\lambda(h) = \int_{-\infty}^{\infty} \sigma_e(\lambda)\, dx, \tag{8.31}$$

where $\delta_\lambda(h)$ is the optical depth along the ray path. If we assume that σ_e varies spatially only with z, the height above the Earth's surface, then $\sigma_e(\lambda, z)$ can be retrieved as a solution of Abel's equation (Avaste and Keevallik, 1983; Arfken, 1970, p. 734):

$$\sigma_e(\lambda, z) = \frac{1}{\pi}\int_z^{\infty} \frac{d\ln\tau_\lambda}{dh}[(h + r_e)^2 - (z + r_e)^2]^{-1/2}\, dh, \tag{8.32}$$

where r_e is the radius of the Earth. Since the function $[(h + r_e)^2 - (z + r_e)^2]^{-1/2}$ is strongly peaked at $h = z$, solar occultation has the same advantage as limb sounding; most of the information comes from near the tangent height.

The extinction coefficient is composed of several terms:

$$\sigma_e = \sigma_{\text{Rayleigh}} + \sigma_{\text{aerosol}} + \sigma_{O_3} + \sigma_{NO_2}, \tag{8.33}$$

where σ_{Rayleigh} is the extinction due to Rayleigh scattering by air molecules, σ_{aerosol} is the extinction due to scattering by aerosols, and σ_{O_3} and σ_{NO_2} are the absorption due to ozone and nitrogen dioxide, respectively. Extinction profiles for each of the components could be retrieved from σ_e profiles calculated using Eq. 8.32. However, Eq. 8.32 is usually abandoned in favor of the iterative approach of Chu and McCormick (1979) in which the component extinction profiles are retrieved directly.

Chu and McCormick, who developed their method to invert SAGE I data, approximate Eq. 8.31 as

$$\delta_\lambda(h_i) = \sum_j \sigma_e(\lambda, z_j)\, \Delta x_{ij}, \tag{8.34}$$

where Δx_{ij} is the length traversed by the ray whose tangent height is h_i in the layer represented by z_j. They begin by assuming initial profiles of σ_{Rayleigh}, σ_{aerosol}, σ_{O_3}, and σ_{NO_2}. Each constituent is retrieved using measurements in the channel that is most sensitive to that constituent: σ_{aerosol} at 1.0 μm, σ_{O_3} at 0.60 μm, σ_{NO_2} at 0.45 μm, and σ_{Rayleigh} at 0.385 μm. The procedure begins at 1.0 μm. The optical depth due solely to aerosols (the dominant constituent at this wavelength) is approximated by subtracting the other constituents:

$$\delta_{\text{aerosol}}^{(k)}(h_i) = \tilde{\delta}_\lambda(h_i) - \sum_j (\sigma_{\text{Rayleigh}}^{(k)}(z_j) + \sigma_{O_3}^{(k)}(z_j) + \sigma_{NO_2}^{(k)}(z_j))\, \Delta x_{ij}, \tag{8.35}$$

where the superscript (k) notes the kth iteration. Note that the $\tilde{\delta}_\lambda(h_i)$, being directly related to transmittance, are treated as measured quantities. If the procedure has converged, then it should be true that for every tangent height h_i:

$$\xi_i^{(k)} \equiv \frac{\delta_{\text{aerosol}}^{(k)}(h_i)}{\sum_j \sigma_{\text{aerosol}}^{(k)}(z_j)\,\Delta x_{ij}} = 1. \tag{8.36}$$

If not, then each $\sigma_{\text{aerosol}}(z_j)$ is iterated using the Twomey *et al.* (1977) modification of Chahine's (1970) relaxation method:[15]

$$\sigma_{\text{aerosol}}^{(k+1)}(z_j) = \sigma_{\text{aerosol}}^{(k)}(z_j) \prod_i \left[1 + (\xi_i^{(k)} - 1)\frac{\Delta x_{ij}}{\Delta x_{jj}}\right], \tag{8.37}$$

where the product is taken over all tangent heights, and for a given j, Δx_{jj} is the maximum of Δx_{ij} for all values of i. The process is repeated in turn at 0.6, 0.45, and 0.385 μm[16] to retrieve σ_{O_3}, σ_{NO_2}, and σ_{Rayleigh}, respectively. A convergence criterion based on the uncertainty in the measured transmittance is applied to determine whether another iteration is necessary. This scheme was modified slightly for SAGE II data (Chu *et al.*, 1989) to retrieve water vapor and to retrieve aerosol extinction at four wavelengths, which will aid in determining the size distribution.

An extensive series of experiments were conducted to compare SAGE I and SAM II aerosol extinction profiles with each other and with other instruments (e.g., Russell *et al.*, 1981, 1984; Yue *et al.*, 1984). Figure 8.14 shows a comparison between aerosol extinction profiles measured by SAGE I and SAM II over Alaska. Since the instruments scan up and down across the disk of the sun, and since the instruments have angular fields of view (0.15 mrad) that are considerably less than the angle subtended by the sun (9.3 mrad), several transmittance measurements are made at each tangent height. The standard deviation of the transmittance can be converted into an estimate of the error in the extinction profile at each altitude (Chu and McCormick, 1979). Roughly, aerosol extinction can be retrieved to within about 10% between the tropopause and about 25 km, which includes nearly all of the stratospheric aerosols. When compared with dustsondes, lidar, and aircraft-mounted samplers, satellite measurements of aerosol extinction are within experimental uncertainties (Russell *et al.*, 1984).

[15] The Twomey *et al.* modification uses information at all levels, not just at the peak of the weighting function. Note that $\delta x_{ij}/\Delta x_{jj} = 1$ at the peak of the weighting function (Δx_{ij}) but is nonzero at other tangent heights. Since Δx_{ij} is sharply peaked at the tangent height, Chu and McCormick perform the calculation only within 4 km of the tangent height.

[16] The extinction coefficients are wavelength-dependent. To use the aerosol extinction coefficient, for example, at wavelengths other than 1.0 μm where it is calculated, the wavelength dependence must be known. At the other wavelengths, however, the aerosol extinction contributes only a minor amount to the total extinction; thus the wavelength dependence does not have to be precisely known. Chu and McCormick use a log-normal size distribution of 75% sulfuric acid droplets at a number density of 10 cm^{-3} to approximate the aerosol distribution and derive the wavelength dependence.

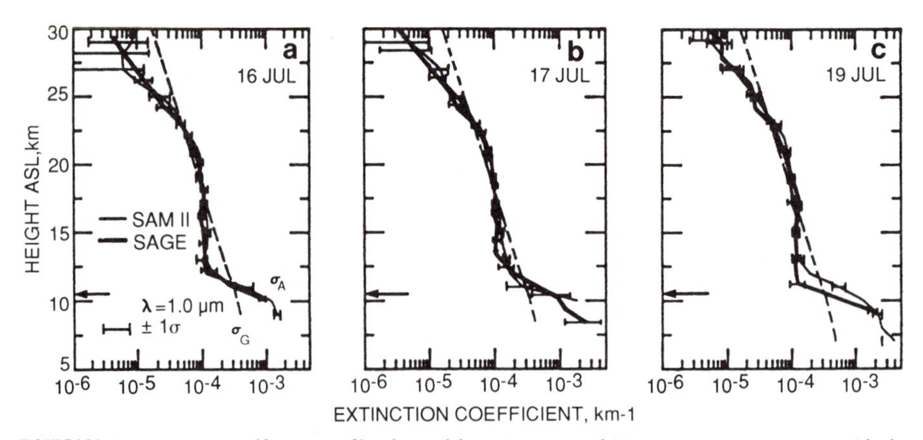

FIGURE 8.14. Extinction coefficient profiles derived from SAGE I and SAM II measurements over Alaska in 1979. The dashed line is extinction due to atmospheric gases (σ_G); the solid lines are extinction due to aerosols (σ_A). The arrow indicates the height of the tropopause. [After Russell *et al.* (1984).]

SAGE II aerosol data have also been compared to SAM II data (Yue *et al.*, 1989) and to other correlative data sets,[17] in both Europe and the United States, with results similar to those for SAGE I, but at four wavelengths. Figure 8.15 shows typical extinction profiles for the four SAGE II wavelengths. Attempts have been made to use the four SAGE II wavelengths to retrieve information about the composition and size distribution of stratospheric aerosols (Livingston and Russell, 1989; Wang *et al.*, 1989).

Satellite observations have been used to study the injection and dispersal of volcanic aerosols (Kent and McCormick, 1984) and stratospheric clouds that form in the winter near the poles (McCormick *et al.*, 1982). In addition, although tropospheric clouds often block measurements below the tropopause, a surprising number of retrievals can be made in the middle to upper troposphere (Kent *et al.*, 1988).

The former Soviet Union experimented with another method, called the Earth's *aureole method*, for measuring extinction coefficients in the stratosphere. Avaste and Keevallik (1983) reported that a four-channel (1.35, 1.9, 2.2, and 2.7 μm) radiometer was flown on the Salyut 4 and Salyut 6 space stations in 1975 and 1978, respectively. The radiometer measured scattered radiance as a function of tangent height in the Earth's limb. Assuming only single scattering, it can be shown that the measured radiance is given by

$$L_\lambda(h) = C[1 - \tau_\lambda(h)], \tag{8.38}$$

where $\tau_\lambda(h)$ is given by Eq. 8.31. C is a constant:

$$C = \frac{1}{4\pi}\tilde{\omega}_o E_{sun}(\lambda)[p(\psi_{sun}) + 2A\mu_{sun}]. \tag{8.39}$$

[17] See the special SAGE II section of the *Journal of Geophysical Research*, Vol. 94 (1989), No. D6.

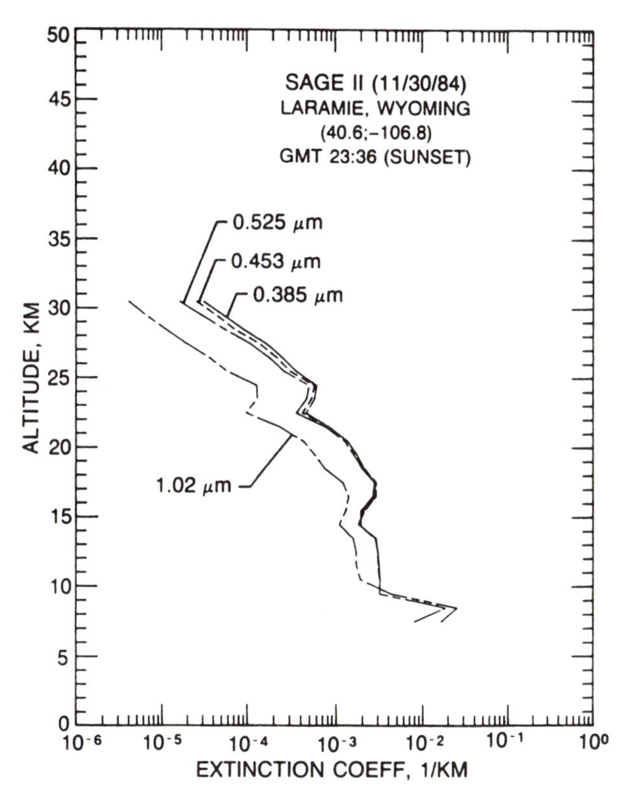

FIGURE 8.15. Typical vertical profiles of aerosol extinction at the four SAGE II wavelengths on November 30, 1984. [After Wang *et al.*, 1989).]

The first term of C is the radiance due to single scattering of directly illuminated aerosols: $\tilde{\omega}_o$ is the single scatter albedo, $E_{sun}(\lambda)$ is the solar irradiance, and $p(\psi_{sun})$ is the scattering phase function for the angle between the sun, tangent point, and satellite. The second term of C is the radiance due to single scattering from aerosols illuminated by radiation reflected from the surface or clouds: A is the albedo below the tangent point, and μ_{sun} is the cosine of the solar zenith angle at the tangent point. The 2 comes from approximating the surface as a Lambertian reflector and integrating over all scattering angles below the tangent point. From Eq. 8.38 we have

$$\frac{d \ln \tau_\lambda}{dh} = \left(\frac{1}{L - C}\right)\frac{dL}{dh}. \tag{8.40}$$

The extinction profile can be obtained by substituting this expression into Eq. 8.32.

The primary advantage of the Earth's aureole method over solar occultation is that measurements are not limited to sunrise and sunset events; they can be made nearly continuously during the sunlit portion of each orbit. They have essentially the same horizontal resolution as limb soundings. The disadvantage is that one must have knowledge of the single scatter albedo and the surface or cloud albedo near the tangent point. A secondary advantage is that aureole measurements can apparently be made at greater altitudes than can occultation measurements. Avaste and Keevallik report making measurements of noctilucent clouds at 80–85 km.

8.5 TROPOSPHERIC AEROSOLS

Studies of tropospheric aerosols also rely on observations of scattered radiation. Instead of the dark, uniform background of space, however, these techniques must deal with the Earth's surface as background. Often, observations are made over the ocean.

The satellite-observed radiance in visible and near-infrared wavelengths can be written (e.g., Durkee *et al.*, 1986)

$$\tilde{L} = L_o \tau_o + L_{\text{Rayleigh}} + L_{\text{aerosol}}, \tag{8.41}$$

where L_o is the radiance leaving the surface, τ_o is the transmittance from the surface to the satellite, L_{Rayleigh} is the radiance due to Rayleigh scattering by air molecules, and L_{aerosol} is the radiance scattered by the aerosols.

Over the ocean, L_o includes sun glint and reflectance by suspended particles as well as Fresnel reflection. Sun glint is caused by reflection of solar radiation from facets of a rough ocean surface (Cox and Munk, 1954). It occurs near the point where the sun's specular reflection is expected. The area of sun glint increases with roughness and is related to wind speed. Because the sun is so bright, reflectance from aerosols is quickly lost in the large sun glint signal. Similarly, areas with suspended particles (including plankton) must be avoided. With these exceptions, however, the ocean makes a good background. Its reflection in the red portion of the spectrum is less than 0.5%, and at wavelengths longer than 0.7 μm the ocean's reflectance is near zero (Ramsey, 1968). In areas free of sun glint and suspended particles, L_o can be approximated as zero at red and near-infrared wavelengths.

L_{Rayleigh} is always present and must be subtracted from $\tilde{L}$ to obtain the radiance due to aerosol scattering. Kaufman (1979) has developed a simple model, based on the two-stream approximation of Coakley and Chylek (1975), to calculate the radiance due to Rayleigh scattering.

Assume that the satellite-observed radiance is free of radiation reflected from the ocean and that it has been corrected for Rayleigh scattering. Assume further that only single scattering need be considered. Then, by Eq. 3.45, the radiative transfer equation may be written:

$$\mu \frac{dL_{\text{aerosol}}}{d\delta_\lambda} = -L_{\text{aerosol}} + \frac{\tilde{\omega}_o}{4\pi} E_{\text{sun}} p(\psi_{\text{sun}}) \exp\left(-\frac{\delta_{\text{aerosol}} - \delta_\lambda}{\mu_{\text{sun}}}\right), \tag{8.42}$$

where δ_λ is the vertical optical depth, δ_{aerosol} is the vertical aerosol optical depth of the entire atmosphere, and μ is the cosine of the satellite viewing angle. The solution of this first order, linear, ordinary differential equation is

$$L_{\text{aerosol}} = \frac{\tilde{\omega}_o}{4\pi} E_{\text{sun}} p(\psi_{\text{sun}}) \left(\frac{\mu_{\text{sun}}}{\mu_{\text{sun}} + \mu}\right)\left\{1 - \exp\left[-\delta_{\text{aerosol}}\left(\frac{\mu_{\text{sun}} + \mu}{\mu\mu_{\text{sun}}}\right)\right]\right\}. \quad (8.43)$$

Since δ_{aerosol} is often much less than one, the satellite-observed radiance can be further approximated as

$$L_{\text{aerosol}} = \frac{\tilde{\omega}_o}{4\pi} E_{\text{sun}} p(\psi_{\text{sun}}) \frac{\delta_{\text{aerosol}}}{\mu}. \quad (8.44)$$

Aerosol optical depth can be retrieved from observations of L_{aerosol}, if the other elements of Eq. 8.44 are known. Three of the necessary parameters are easily obtained. E_{sun} is a function of time of year, the solar spectrum, and the spectral response function of the satellite radiometer. The solar zenith angle is a function of time and the location of the scan spot. The satellite viewing angle is a function of the locations of the satellite and the scan spot. The remaining two parameters, $\tilde{\omega}_o$ and $p(\psi_{\text{sun}})$, depend on the index of refraction and size distribution of the aerosols. Thus in contrast with solar occultation, some knowledge of the properties of the aerosols must be known before optical depth can be retrieved.

Two basic types of aerosols have been studied using satellite data: marine aerosols (haze) and Saharan dust. Griggs (1975) compared surface-based measurements of marine aerosol optical depth (using a Volz sun photometer) with Landsat observations at 0.55, 0.65, and 0.75 μm and found that optical depth could be retrieved within $\pm 10\%$. Griggs (1979) found similar results with GOES 1 and NOAA 5 Scanning Radiometer data. Fett and Isaacs (1979) and Isaacs (1980) studied the effects of marine aerosols on DMSP visible imagery. Hindman *et al.* (1984) and Durkee *et al.* (1986) have used AVHRR data and Nimbus 7 Coastal Zone Color Scanner data to study marine aerosols off the California coast. These last three studies used the marine aerosol models of Shettle and Fenn (1979). Durkee *et al.* (1991) have used AVHRR channels 1 and 2 to retrieve aerosol optical depth. The combination of two channels lessens the uncertainty caused by assuming a single scatter albedo and a scattering phase function.

Satellite data were first used to study Saharan dust by Fraser (1976). He used a radiative transfer model to convert Landsat radiances to vertically integrated aerosol mass. Carlson and Wendling (1977) and Carlson (1979) were able to map dust optical depth over the Atlantic using NOAA 3 VHRR data. Norton *et al.* (1980) studied Saharan dust using GOES data. They were able to estimate the dust optical depth to ± 0.1 accuracy for optical depths up to 0.4. Recently Deuze *et al.* (1988) studied Saharan aerosols over the Mediterranean Sea using Meteosat and AVHRR data.

In January 1989, NESDIS began producing weekly maps of aerosol optical depth based on AVHRR channel 1 data (Rao *et al.*, 1988). The maps have a 100-km grid and include all the world's oceans. The retrieval scheme relies on a lookup table calculated with a radiative transfer model. The model includes Rayleigh scattering and ozone absorption as well as aerosol scattering. The table is calculated as a function of solar zenith angle, satellite zenith angle, and the relative azimuth angle. Given these three parameters, aerosol optical depth can be interpolated on the basis of the observed radiance. To avoid sun glint, retrievals are made only in the antisolar half of each scan line. Clouds are filtered in two ways. First, techniques used in sea surface temperature retrievals are applied (McClain *et al.*, 1985). Second, the difference between the triple-window SST (Eq. 6.73c) and the split-window SST (Eq. 6.73b) is calculated. Rao *et al.* have observed that due to the reflectance of clouds at 3.7 μm these SST estimates differ by at least 5 K in cloudy areas. Thus they use a 5-K threshold to detect and eliminate cloudy scenes. Figure 8.16 shows an example of aerosol optical depth for the Atlantic Ocean. Based on comparison with surface measurements (Griggs and Stowe, 1984), it is estimated that the one-channel scheme is capable of retrieving aerosol optical depth within ± 0.03 to ± 0.05. If the global background optical depth is roughly 0.1, the method's accuracy is in the 30–50% range. In high-signal dust outbreaks such as that shown in Fig. 8.16, however, the relative accuracy is much better.

Aerosol measurements over land are also possible if one knows the surface reflectance. Fraser *et al.* (1984) have used GOES visible data to estimate summertime aerosol optical depth over the eastern United States. They then related the optical depth to the column-integrated mass density of sulfate particles.

Measurements outside the visible to near infrared portion of the spectrum also appear to be useful. Ackerman (1989) reports that the difference between the brightness temperatures at 3.7 and 11 μm is sensitive to aerosol optical depth.

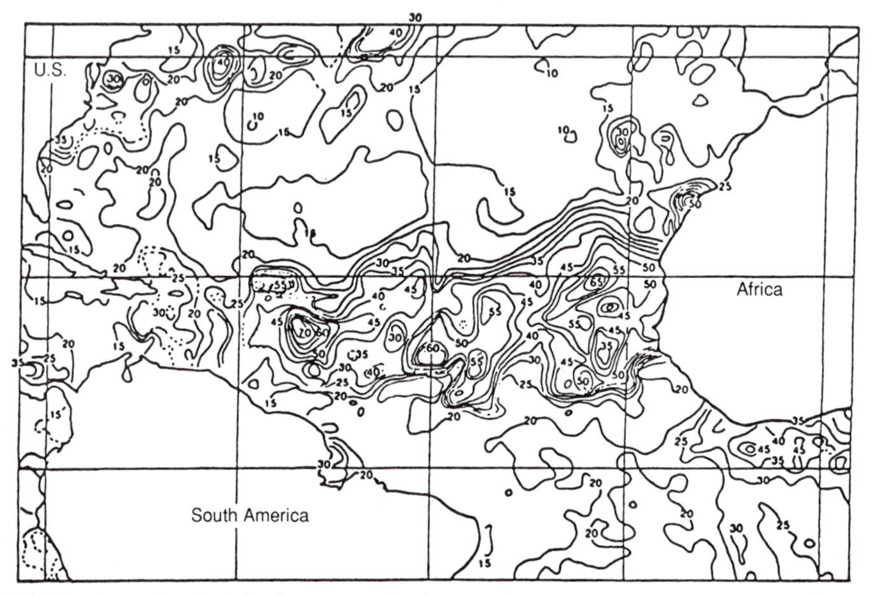

FIGURE 8.16. Aerosol optical depths ($\times 100$) for the period 18–25 June 1987. [Courtesy of Larry Stowe, NOAA.]

He used this information to plot aerosol optical depth over the Arabian Peninsula. Also, Lee *et al.* (1986) report that the difference between brightness temperatures at 11 and 12 μm is sensitive to aerosol optical depth.

Bibliography

Ackerman, S. A. (1989). Using the radiative temperature difference at 3.7 μm and 11 μm to track dust outbreaks. *Remote Sens. Environ.*, **27**, 129–133.

Arfken, G. (1970). *Mathematical Methods for Physicists*, 2nd ed., Academic Press, New York.

Arking, A. (1964). Latitudinal distribution of cloud cover from Tiros III photographs. *Science*, **143**, 569–572.

Arking, A., and J. D. Childs (1985). Retrieval of cloud cover parameters from multispectral satellite images. *J. Climate Appl. Meteor.*, **24**, 322–333.

Avaste, O. A., and S. H. Keevallik (1983). Remote monitoring of aerosols from space. *Adv. Space Res.*, **2**, 87–93.

Bunting, J. T., and R. P. d'Entremont (1982). *Improved Cloud Detection Utilizing Defense Meteorological Satellite Program Near Infrared Measurements.* Air Force Geophysics Lab. Tech. Rep. AFGL-TR-82-0027, Hanscom AFB, MA.

Carlson, T. N. (1979). Atmospheric turbidity in Saharan dust outbreaks as determined by analyses of satellite brightness data. *Mon. Wea. Rev.*, **107**, 322–335.

Carlson, T. N., and P. Wendling (1977). Reflected radiance measured by NOAA 3 VHRR as a function of optical depth for Saharan dust. *J. Appl. Meteor.*, **16**, 1368–1371.

Chahine, M. T. (1970). Inverse problems in radiative transfer: Determination of atmospheric parameters. *J. Atmos. Sci.*, **27**, 960–967.

Chahine, M. T. (1974). Remote sounding of cloudy atmospheres. I. The single cloud layer. *J. Atmos. Sci.*, **31**, 233–243.

Chahine, M. T. (1975). An analytical transformation for remote sensing of clear-column atmospheric temperature profiles. *J. Atmos. Sci.*, **32**, 1946–1952.

Chahine, M. T. (1977). Remote sounding of cloudy atmospheres. II. Multiple cloud formations. *J. Atmos. Sci.*, **34**, 744–757.

Chahine, M. T. (1982). Remote sensing of cloud parameters. *J. Atmos. Sci.*, **39**, 159–170.

Chu, W. P., M. P. McCormick, J. Lenoble, C. Brogniez, and P. Pruvost (1989). SAGE II inversion algorithm. *J. Geophys. Res.*, **94**, 8339–8351.

Chu, W. P. and M. P. McCormick (1979). Inversion of stratospheric aerosol and gaseous constituents from spacecraft solar extinction data in the 0.38–1.0-μm wavelength region. *Appl. Optics*, **18**, 1404–1413.

Coakley, J. A., Jr., and P. Chylek (1975). The two-stream approximation in radiative transfer: Including the angle of the incident radiation. *J. Atmos. Sci.*, **32**, 409–418.

Coakley, J. A., Jr. (1983). Properties of multilayered cloud systems from satellite imagery. *J. Geophys. Res.*, **88**, 10818–10828.

Coakley, J. A., Jr., and D. G. Baldwin (1984). Towards the objective analysis of clouds from satellite imagery data. *J. Climate Appl. Meteor.*, **23**, 1065–1099.

Coakley, J. A., Jr., and F. P. Bretherton (1982). Cloud cover from high-resolution scanner data: Detecting and allowing for partially filled fields of view. *J. Geophys. Res.*, **87**, 4917–4932.

Coakley, J. A., Jr., and R. Davies (1986). The effect of cloud sides on reflected solar radiation as deduced from satellite observations. *J. Atmos. Sci.*, **43**, 1025–1035.

Coakley, J. A., Jr., R. L. Bernstein, and P. A. Durkee (1987). Effect of ship-stack effluents on cloud reflectivity. *Science*, **238**, 1020–1022.

Cox, C. C., and W. Munk (1954). Measurements of the roughness of the sea surface from photographs of the sun's glitter. *J. Optical Soc. Am.*, **44**, 838–850.

Dave, J. V. (1964). Meaning of successive iteration of the auxiliary equation in the theory of radiative transfer. *Astrophys. J.*, **140**, 1292–1303.

Desbois, M., G. Seze, and G. Szejwach (1982). Automatic classification of clouds on METEOSAT imagery: Application to high-level clouds. *J. Appl. Meteor.*, **21**, 401–412.

Deuze, J. L., C. Devaux, M. Herman, R. Santer, and D. Tanre (1988). Saharan aerosols over the south of France: Characterization derived from satellite data and ground based measurements. *J. Appl. Meteor.*, **27**, 680–686.

Duda, R. O., and P. E. Hart (1973). *Pattern Classification and Scene Analysis*. Wiley, New York.

Durkee, P. A., D. R. Jensen, E. E. Hindman, and T. H. Vonder Haar (1986). The relationship between marine aerosol particles and satellite-derived radiance. *J. Geophys. Res.*, **91**, 4063–4072.

Durkee, P. A., F. Pfeil, E. Frost, and R. Shema (1991). Global analysis of aerosols and their effects on climate. *Atmos. Environ.*, **25A**, 2457.

Ebert, E. (1987). A pattern recognition technique for distinguishing surface and cloud types in the polar region. *J. Climate Appl. Meteor.*, **26**, 1412–1427.

Fett, R. W., and R. G. Isaacs (1979). Concerning causes of "anomalous gray shades" in DMSP visible imagery. *J. Appl. Meteor.*, **18**, 1340–1351.

Fraser, R. S. (1976). Satellite measurement of mass of Sahara dust in the atmosphere. *Appl. Optics*, **15**, 2471–2479.

Fraser, R. S., Y. J. Kaufman, and R. L. Mahoney (1984). Satellite measurements of aerosol mass and transport. *Atmos. Environ.*, **18**, 2577–2584.

Fujita, T. T. (1963). Use of TIROS pictures for studies of the internal structure of tropical storms. *Mesometeorology Project Research Paper No. 25.*, The University of Chicago.

Fye, F. K. (1978). *The AFGWC Automated Cloud Analysis Model.* AFGWC Tech. Memo. 78-002, U.S. Air Force Global Weather Central, Offutt Air Force Base, Nebraska.

Griggs, M. (1975). Measurements of atmospheric aerosol optical thickness over water using ERTS–1 data. *J. Air Pollution Control*, **25**, 622–626.

Griggs, M. (1979). Satellite observations of atmospheric aerosols during the EOMET cruise. *J. Atmos Sci.*, **36**, 695–698.

Griggs, M., and L. L. Stowe (1984). Measurements of aerosol optical parameters from satellites. In G. Fiocco (ed.), *IRS '84: Current Problems in Atmospheric Radiation.* A. Deepak Publishing, Hampton, VA, pp. 42–45.

Grody, N. C. (1976). Remote sensing of atmospheric water content from satellites using microwave radiometry. *IEEE Trans. Antennas Propag.*, **AP-24**, 155–162.

Grody, N. C., A. Gruber, and W. C. Shen (1980). Atmospheric water content over the tropical Pacific derived from the Nimbus–6 scanning microwave spectrometer. *J. Appl. Meteor.*, **19**, 986–996.

Gurvich, A. S., and V. V. Demin (1970). Determination of the total moisture content in the atmosphere from measurements on the Cosmos 243 satellite. *Atmos. Ocean Phys.*, **6**, 453–457.

Harris, R., and E. C. Barrett (1978). Toward an objective nephanalysis. *J. Appl. Meteor.*, **17**, 1258–1266.

Harshvardhan (1982). The effect of brokenness on cloud-climate sensitivity. *J. Atmos. Sci.*, **39**, 1853–1861.

Harshvardhan and J. A. Weinman (1982). Infrared radiative transfer through a regular array of cuboidal clouds. *J. Atmos. Sci.*, **39**, 431–439.

Hasler, A. F. (1981). Stereographic observations from geosynchronous satellites: An important new tool for the atmospheric sciences. *Bull. Am. Meteor. Soc.*, **62**, 194–212.

Hasler, A. F., R. Mack, and A. Negri (1983). Stereoscopic observations from meteorological satellites. *Adv. Space Res.*, **2**, 105–113.

Henderson-Sellers, A. (ed.) (1984). *Satellite Sensing of a Cloudy Atmosphere: Observing the Third Planet.* Taylor & Francis, London.

Hindman, E. E., P. A. Durkee, P. C. Sinclair, and T. H. Vonder Haar (1984). Detection of marine aerosol particles in coastal zones using satellite imagery. *Int. J. Remote Sensing*, **5**, 577–586.

Isaacs, R. G. (1980). Investigation of the effect of low level maritime haze in DMSP VHR and LF imagery. *Naval Environmental Prediction Research Facility Contractor Report CR 80-06*, Monterey, CA.

Jacobowitz, H., W. L. Smith, and A. J. Drummond (1972). *Satellite Measurements of Aerosol Backscat-*

tered Radiation from the Nimbus F Earth Radiation Budget Experiment. NOAA Tech. Rep. NESS 60, Washington, DC.

Kaufman, Y. J. (1979). Effect of the Earth's atmosphere on contrast for zenith observations. *J. Geophys. Res.*, **84**, 3165–3172.

Kent, G. S., and M. P. McCormick (1984). SAGE and SAM II measurements of global stratospheric aerosol optical depth and mass loading. *J. Geophys. Res.*, **89**, 5303–5314.

Kent, G. S., U. O. Farrukh, P. H. Wang, and A. Deepak (1988). SAGE I and SAM II measurements of 1 μm aerosol extinction in the free troposphere. *J. Appl. Meteor.*, **27**, 269–279.

Koffler, R., A. G. DeCotiis, and P. K. Rao (1973). A procedure for estimating cloud amount and height from satellite infrared radiation data. *Mon. Wea. Rev.*, **101**, 240–243.

Lee, T. Y., R. S. Fraser, and Y. Kaufman (1986). Satellite measurements of physical properties of Saharan dust. *Preprints: Second Conference on Satellite Meteorology/Remote Sensing and Applications*, American Meteorological Society, Boston, pp. 294–297.

Livingston, J. M., and P. B. Russell (1989). Retrieval of aerosol size distribution moments from multiwavelength particulate extinction measurements. *J. Geophys. Res.*, **94**, 8425–8433.

Mack, R. A., A. F. Hasler, and R. F. Adler (1983). Thunderstorm cloud top observations using satellite stereoscopy. *Mon. Wea. Rev.*, **111**, 1949–1964.

McClain, E. P., W. G. Pichel, and C. C. Walton (1985). Comparative performance of AVHRR-based multichannel sea surface temperatures. *J. Geophys. Res.*, **90**, 11587–11601.

McCleese, D. J., and L. S. Wilson (1976). Cloud top heights from temperature sounding instruments. *Quart. J. Roy. Meteor. Soc.*, **102**, 781–790.

McCormick M. P., and J. E. Lovill (eds.) (1983). Space observations of aerosols and ozone. *Adv. Space Res.*, **2**, 213 pp.

McCormick, M. P. (1983). Aerosol measurements from Earth orbiting spacecraft. *Adv. Space Res.*, **2**, 73–86.

McCormick, M. P., H. M. Steele, P. Hamill, W. P. Chu, and T. J. Swissler (1982). Polar stratospheric cloud sightings by SAM II. *J. Atmos. Sci.*, **29**, 1387–1397.

McCormick, M. P., P. Hamill, T. J. Pepin, W. P. Chu, T. J. Swissler, and L. R. McMaster (1979). Satellite studies of stratospheric aerosol. *Bull. Am. Meteor. Soc.*, **60**, 1038–1046.

McKee, T., and S. K. Cox (1974). Scattering of visible radiation by finite clouds. *J. Atmos. Sci.*, **31**, 1885–1892.

McKee, T., and S. K. Cox (1976). Simulated radiance patterns for finite cubic clouds. *J. Atmos. Sci.*, **33**, 2014–2020.

McMillin, L. M., and C. Dean (1982). Evaluation of a new operational technique for producing clear radiances. *J. Appl. Meteor.*, **21**, 1005–1014.

Menzel, W. P., W. L. Smith, and T. R. Stewart (1983). Improved cloud motion wind vector and altitude assignment using VAS. *J. Climate Appl. Meteor.*, **22**, 377–384.

Minnis, P., and E. F. Harrison (1984). Diurnal variability of regional cloud and clear-sky radiative parameters derived from GOES data. Part I: Analysis method. *J. Climate Appl. Meteor.*, **23**, 993–1011.

Minzner, R. A., W. E. Shenk, J. Steranka, and R. D. Teagle (1976). Cloud heights determined stereographically from imagery recorded simultaneously by two synchronous meteorological satellites, SMS–1 and SMS–2. *EOS*, **57**, 593.

Norton, C. C., F. R. Mosher, B. Hinton, D. W. Martin, D. Santek, and W. Kuhlow (1980). A model for calculating desert aerosol turbidity over the oceans from geostationary satellite data. *J. Appl. Meteor.*, **19**, 633–644.

Phulpin, T., M. Derrien, and A. Brard (1983). A two-dimensional histogram procedure to analyze cloud cover from NOAA satellite high-resolution imagery. *J. Climate Appl. Meteor.*, **22**, 1332–1345.

Platt, C. M. R. (1979). Remote sounding of high clouds: I. Calculation of visible and infrared optical properties from lidar and radiometer measurements. *J. Appl. Meteor.*, **18**, 1130–1143.

Platt, C. M. R. (1983). On the bispectral method for cloud parameter determination from satellite VISSR data: Separating broken cloud and semitransparent cloud. *J. Climate Appl. Meteor.*, **12**, 429–439.

Platt, C. M. R., and A. C. Dilley (1979). Remote sounding of high clouds: II. Emissivity of cirrostratus. *J. Appl. Meteor.*, **18**, 1144–1150.

Prabhakara, C., I. Wang, A. T. C. Chang, and P. Gloersen (1983). A statistical examination of Nimbus–7 SMMR data and remote sensing of sea surface temperature, liquid water content in the atmosphere and surface wind speed. *J. Climate Appl. Meteor.*, **22**, 2023–2037.

Ramsey, R. C. (1968). *Study of the Remote Measurement of Ocean Color.* Final Report: NASW-1658. TRW, Redondo Beach, California.

Randall, D. A., J. A. Coakley, Jr., C. W. Fairall, R. A. Kropfli, and D. H. Lenschow (1984). Outlook for research on subtropical marine stratiform clouds. *Bull. Am. Meteor. Soc.*, **65**, 1290–1301.

Rao, C. R. N., L. L. Stowe, E. P. McClain, and J. Sapper (1988). Development and application of aerosol remote sensing with AVHRR data from the NOAA satellites. In P. V. Hobbs and M. P. McCormick (eds.), *Aerosols and Climate*, A. Deepak Publishing, Hampton, VA.

Reynolds, D. W., and T. H. Vonder Haar (1977). A bispectral method for cloud parameter determination. *Mon. Wea. Rev.*, **105**, 446–457.

Reynolds, D. W., T. B. McKee, and K. S. Danielson (1978). Effects of cloud size and cloud particles on satellite-observed reflected brightness. *J. Atmos. Sci.*, **35**, 160–164.

Rossow, W. B., and R. A. Schiffer (1991). ISSCP cloud data products. *Bull. Am. Meteor. Soc.*, **72**, 2–20.

Rossow, W. B. (1989). Measuring cloud properties from space: A review. *J. Climate*, **2**, 201–213.

Rossow, W. B., and L. Garder (1984). Selection of a map grid for data analysis and archival. *J. Climate Appl. Meteor.*, **23**, 1253–1257.

Rossow, W. B., F. Mosher, E. Kinsella, A. Arking, M. Desbois, E. Harrison, P. Minnis, E. Ruprecht, G. Seze, C. Simmer, and E. Smith (1985). ISCCP cloud algorithm intercomparison. *J. Climate Appl. Meteor.*, **24**, 877–903.

Rossow, W. B., L. C. Garder, and A. A. Lacis (1989). Global, seasonal cloud variations from satellite radiance measurements Part I: Sensitivity of analysis. *J. Climate*, **2**, 419–458.

Russell, P. B., M. P. McCormick, T. J. Swissler, J. M. Rosen, D. J. Hofmann, and L. R. McMaster (1984). Satellite and correlative measurements of the stratospheric aerosol. III: Comparison of measurements by SAM II, SAGE, dustsondes, filters, impactors and lidar. *J. Atmos. Sci.*, **41**, 1791–1800.

Russell, P. B., M. P. McCormick, T. J. Swissler, W. P. Chu, J. M. Livingston, W. H. Fuller, J. M. Rosen, D. J. Hofmann, L. R. McMaster D. C. Woods, and T. J. Pepin (1981). Satellite and correlative measurements of the stratospheric aerosol. II: Comparison of measurements by SAM II, dustsondes, and airborne lidar. *J. Atmos. Sci.*, **38**, 1295–1302.

Saunders, R. W. (1985). Monthly mean cloudiness observed from METEOSAT–2. *J. Climate Appl. Meteor.*, **24**, 114–127.

Saunders, R. W., and G. E. Hunt (1983). Some radiation budget and cloud measurements derived from Meteosat 1 data. *Tellus*, **35B**, 177–188.

Schiffer, R. A., and W. B. Rossow (1983). The International Satellite Cloud Climatology Project (ISCCP): The first project of the World Climate Research Programme. *Bull. Am. Meteor. Soc.*, **64**, 779–784.

Shenk, W. E., and R. J. Curran (1973). A multi-spectral method for estimating cirrus cloud top heights. *J. Appl. Meteor.*, **12**, 1213–1216.

Shenk, W. E., and V. V. Salomonson (1972). A simulation study exploring the effects of sensor spatial resolution on estimates of cloud cover from satellites. *J. Appl. Meteor.*, **11**, 214–220.

Shettle, E. P., and R. W. Fenn (1979). *Models for the Aerosols of the Lower Atmosphere and the Effects of Humidity Variations on Their Optical Properties.* Air Force Geophysics Laboratory Tech. Rep. AFGL-TR-79-0214, Hanscom Air Force Base, MA.

Simmer, C., E. Raschke, and E. Ruprecht (1982). A method for determination of cloud properties from two-dimensional histograms. *Ann. Meteor.*, **18**, 130–132.

Smith, E. A., and D. W. Reynolds (1976). Comparison of cloud top height determinations from three independent sources: Satellite IR measurements: Satellite viewed cloud shadows: Radar. *Proceedings of the Symposium on Meteorological Observations from Space: Their contribution to the First GARP Global Experiment.* COSPAR, Boulder, CO.

Smith, W. L. (1968). An improved method for calculating tropospheric temperature and moisture from satellite radiometer measurements. *Mon. Wea. Rev.*, **96**, 387–396.
Smith, W. L., and C. M. R. Platt (1978). Comparisons of satellite-deduced cloud heights with indications from radiosonde and ground-based laser measurements. *J. Appl. Meteor.*, **17**, 1796–1802.
Smith, W. L., and H. M. Woolf (1976). The use of eigenvectors of statistical covariance matrices for interpreting satellite sounding radiometer data. *J. Atmos. Sci.*, **33**, 1127–1140.
Smith, W. L., H. M. Woolf, and W. J. Jacob (1970). A regression method for obtaining real-time temperature and geopotential height profiles from satellite spectrometer measurements and its application to Nimbus 3 "SIRS" observations. *Mon. Wea. Rev.*, **98**, 582–603.
Smith, W. L., H. M. Woolf, P. G. Abel, C. M. Hayden, M. Chalfant, and N. Grody (1974). *Nimbus–5 Sounder Data Processing System.* NOAA Tech. Memo. NESS 57, Washington, DC.
Stowe, L. L., C. G. Wellemeyer, T. F. Eck, H. Y. M. Yeh, and the Nimbus–7 Cloud Data Processing Team (1988). Nimbus–7 global cloud climatology. Part I: Algorithms and validation. *J. Climate*, **1**, 445–470.
Susskind, J., J. Rosenfield, D. Reuter, and M. T. Chahine (1984). Remote sensing of weather and climate parameters from HIRS2/MSU on TIROS–N. *J. Geophys. Res.*, **89**, 4677–4697.
Szejwach, G. (1982). Determination of semi-transparent cirrus cloud temperature from infrared radiances: Application to METEOSAT. *J. Appl. Meteor.*, **21**, 384–393.
Twomey, S., B. Herman, and R. Rabinoff (1977). An extension to the Chahine method of inverting the radiative transfer equation. *J. Atmos. Sci.*, **34**, 1085–1090.
Wang, P.-H., M. P. McCormick, T. J. Swissler, M. T. Osborn, W. H. Fuller, and G. K. Yue (1989). Inference of stratospheric aerosol composition and size distribution from SAGE II satellite measurements. *J. Geophys. Res.*, **94**, 8435–8446.
Weinman, J. A., and Harshvardhan (1982). Solar reflection from a regular array of horizontally finite clouds. *Appl. Optics*, **21**, 2940–2944.
Westwater, E. R. (1972). *Microwave Emission from Clouds.* NOAA Tech. Rep. 219-WPL 18. Boulder, CO.
Wielicki, B. A., and J. A. Coakley Jr. (1981). Cloud retrieval using infrared sounder data: Error analysis. *J. Appl. Meteor.*, **20**, 157–169.
Wilheit, T. T. and A. T. C. Chang (1980). An algorithm for retrieval of ocean surface and atmosphere parameters from the observations of the scanning multichannel microwave radiometer. *Radio Sci.*, **15**, 525–544.
Yue, G. K., M. P. McCormick, and W. P. Chu (1984). A comparative study of aerosol extinction measurements made by the SAM II and SAGE satellite experiments. *J. Geophys. Res.*, **89**, 5321–5327.
Yue, G. K., M. P. McCormick, W. P. Chu, P. Wang, and M. T. Osborn (1989). Comparative studies of aerosol extinction measurements made by the SAM II and SAGE II satellite experiments. *J. Geophys. Res.*, **94**, 8412–8424.

9

Precipitation

AT FIRST, ONE might think that if the clouds have been located in a satellite image (as in Chapter 8) precipitation estimation should be a simple task. Unfortunately, early researchers found that only a small fraction of clouds produce rain at any one time (Nagle and Serebreny, 1962). Much of the research into techniques for estimating precipitation with satellite data involves separating raining from nonraining clouds. An excellent book covering this subject is *The Use of Satellite Data in Rainfall Monitoring* by E. C. Barrett and D. W. Martin (1981).

Precipitation-estimation techniques may be divided into three categories: those that use visible or infrared data, those that use passive microwave data, and future techniques that will use radar. The first three sections of this chapter discuss these techniques. The final section discusses the related topic of severe weather detection and monitoring.

9.1 VISIBLE AND INFRARED TECHNIQUES

Visible and infrared techniques are grouped together because they share a common characteristic: The radiation does not penetrate through the cloud. Visible and infrared techniques estimate precipitation falling from the bottom of the cloud based on radiation coming from the top and/or the side of the cloud, depending on viewing geometry. All visible and infrared precipitation-estimation schemes are necessarily indirect; a cloud's brightness or equivalent blackbody temperature may be related to the rain falling from it, but the raindrops themselves are not directly sensed.

Because of the indirect nature of the relationship between satellite-measured visible or infrared radiance and precipitation, the techniques discussed below are not universally applicable. Techniques developed for the tropics may not perform well in the extratropics. Similarly, techniques developed to estimate monthly rainfall may not be useful when estimating hourly rainfall. When selecting a visible or infrared technique, one should carefully compare one's application with the application for which the technique was developed. In fact, Barrett (1988) argues that a hierarchy of techniques is necessary to estimate rainfall from satellite data under the wide range of meteorological conditions and user needs which are encountered in practice.

A further note is in order here, although it applies just as well to microwave and radar techniques as it does to visible and infrared techniques. There are two ways of verifying precipitation estimates from satellite data; one may compare either with radar or with rain-gauge estimates. Radar probably provides the more reasonable comparison because it samples volumes of the atmosphere which are comparable with the size of satellite pixels. However, radar meteorologists generally do not claim better than a factor-of-2 accuracy for their precipitation estimates. Rain gauges make very accurate measurements, but a standard rain gauge samples an area only about 10^{-1} m^2. Satellite estimates, on the other hand, sample an area roughly 10^6–10^8 m^2. Since rainfall is highly variable spatially, comparison of satellite estimates and rain-gauge estimates would be poor even if the satellite estimates were perfect. We will attempt to give accuracy figures for the techniques discussed in this chapter, but the reader should be aware that verification of satellite precipitation estimates is difficult at best. Another point to remember is that averaging reduces noise. Thus satellite precipitation estimates averaged over large areas or long time periods will compare better with "truth" than will estimates for smaller areas or shorter times. With these caveats, we are ready to begin.

Barrett and Martin (1981) divide visible and infrared techniques into four categories: cloud indexing, life history, bispectral, and cloud model techniques. We will follow their classification system.

9.1.1 Cloud Indexing

The oldest (and still useful) precipitation-estimation technique is known as *cloud indexing*. It rests on the observation that it is fairly easy to identify cloud types in satellite images. In its most fundamental form, cloud indexing assigns a rain rate to each cloud type. The rain at a particular location (or in a particular area) can then be written

$$R = \sum_i r_i f_i, \tag{9.1}$$

where r_i is the rain rate assigned to cloud type i, and f_i is the fraction of time that the point is covered with (or fraction of the area covered by) cloud type i.

9.1.1.1 Barrett

Cloud indexing was pioneered by Barrett (1970), who wanted to estimate precipitation over Australia and the "Maritime Continent." He used once-daily nephanalyses produced by NOAA's predecessor, the Environmental Science Service Administration (ESSA), to estimate monthly precipitation with a slightly more complex scheme[1] than that represented by Eq. 9.1 (Fig. 9.1). Barrett's cloud types were cumulonimbus, stratiform, cumuliform, stratocumuliform, and cirriform. He found that a cubic polynomial function of the satellite-estimated rain depth could account for 90% of the variance in rain-gauge-observed precipitation (Fig. 9.2).

Barrett's cloud indexing technique has been profitably applied to a large number of locations, most of which have either few or no rain gauges (see Barrett and Martin, 1981). In a modification of the technique, satellite images are used to interpolate precipitation in an area sparsely covered by rain gauges (Barrett, 1980).

Barrett (1973) attempted to forecast daily precipitation in another modification of the cloud indexing technique. He used the 800-hPa wind to construct a quadrant from which the air passing over a station in the next day could be expected to come (Fig. 9.3). He then estimated the 24-h precipitation in the quadrant by cloud indexing and forecast that amount of precipitation at the station. Comparisons with rain-gauge data showed that the forecasting scheme was able to correctly classify rain into one of seven categories (including no rain) about 75% of the time.

In collaboration with NESDIS, this technique has been implemented as an interactive scheme on a digital image processing system (Moses and Barrett, 1986). Called BIAS (Bristol/NOAA InterActive Scheme), the scheme allows an analyst to combine climatology and surface synoptic station data with satellite rain estimates.

9.1.1.2 Follansbee

Whereas Barrett (1970) wanted high-spatial-resolution precipitation estimates averaged over a month, Follansbee (1973) wanted daily precipitation estimates averaged over a state-sized area. Using once-daily, afternoon, visible satellite images, he visually estimated the fraction of a state covered by cumulonimbus

[1] Barrett's cloud indexing equation also included a mean monthly cloudiness and the probability of rainfall from each cloud type as well as the rain rate. Since these factors are multiplicative, however, Eq. 9.1 still applies, but r_i must be interpreted as an effective rain rate.

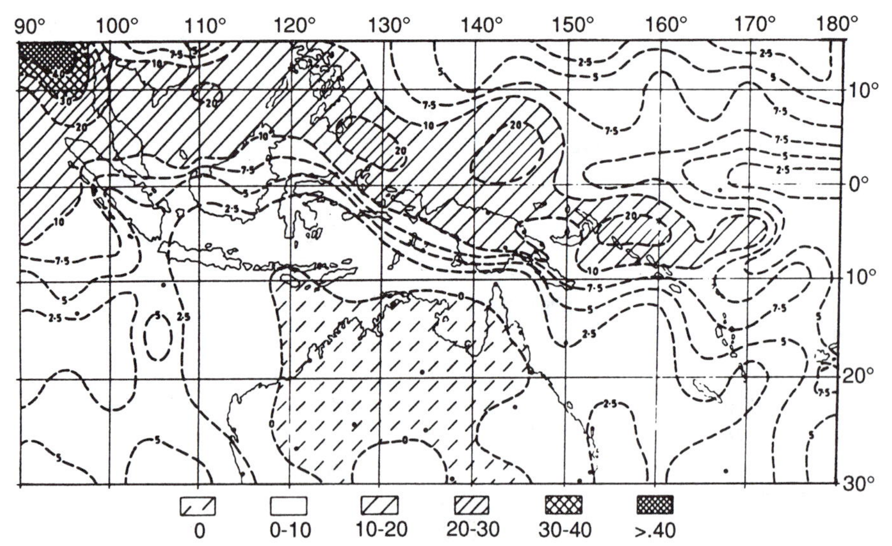

FIGURE 9.1. Precipitation (inches) for the month of July 1966 estimated by the cloud indexing method of Barrett (1970).

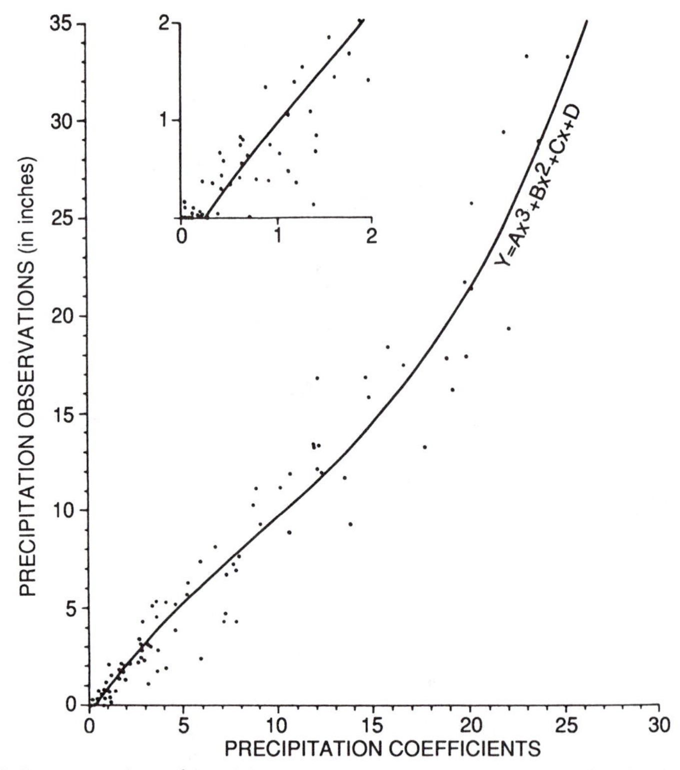

FIGURE 9.2. Rain-gauge-estimated precipitation versus precipitation estimated by cloud indexing for the area in Fig. 9.1 during the months of March–June 1966. [After Barrett (1970).]

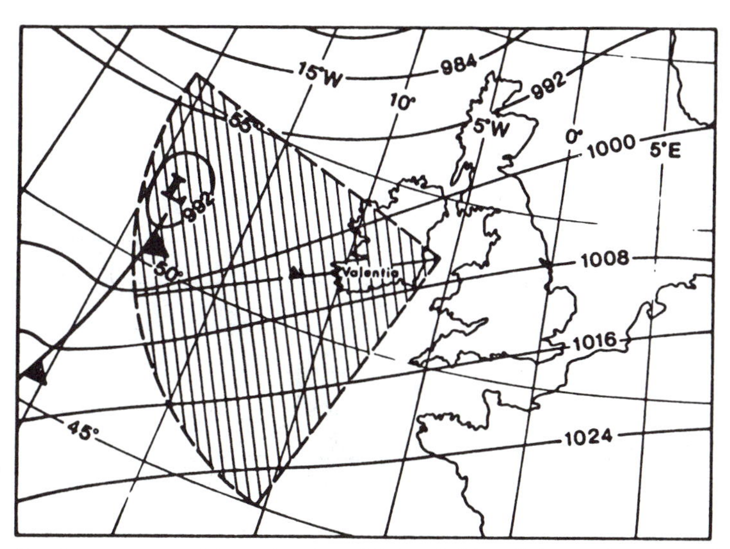

FIGURE 9.3. Quadrant constructed for estimation of daily rainfall at Valentia Island, Ireland. The 800 hPa wind is used to construct the quadrant. [After Barrett (1973).]

(r = 1 inch/day), nimbostratus (r = 0.25 inch/day), or cumulus congestus (r = 0.02 inch/day) and applied Eq. 9.1.

Figure 9.4 shows Follansbee's results for Arkansas precipitation in July and August, 1971. These results are typical of precipitation estimation schemes. Peaks in the rain-gauge estimates are almost always associated with peaks in the satellite estimates because rain falls from clouds, which satellites detect well. However, while the satellite estimates sometimes closely followed the rain-gauge estimates (August), they sometimes did not (July). This is because a simple indexing scheme cannot take into account precipitation efficiencies and other factors that affect the rain rate from a particular cloud type on a particular day. Also, once-daily observations can miss precipitation, particularly convective precipitation.

For verification purposes, Follansbee divided precipitation into classes (in inches): 0, 0.01–0.10, 0.11–0.20, 0.21–0.30, 0.31–0.50, 0.51–1.00, 1.01–2.00, and 2.01–5.00. He found that 79% of the time, the satellite estimate was within one category of the rain-gauge-estimated average daily precipitation in a state.

The above schemes are designed for convective precipitation (as are most satellite precipitation-estimation schemes). Follansbee and Oliver (1975) wanted to estimate cool-season precipitation in Asia chiefly for agricultural purposes. Cool-season precipitation in the extratropics is dominated by traveling, synoptic-scale systems which have a variable rain rate. Follansbee and Oliver developed an interpolation scheme for use with twice-daily satellite images. Figure 9.5 illustrates their technique. On each of a pair of satellite images separated by approximately 12 h, an analyst outlines the area thought to be precipitating, and then interpolates the movement of the system in order to estimate at each point the number of

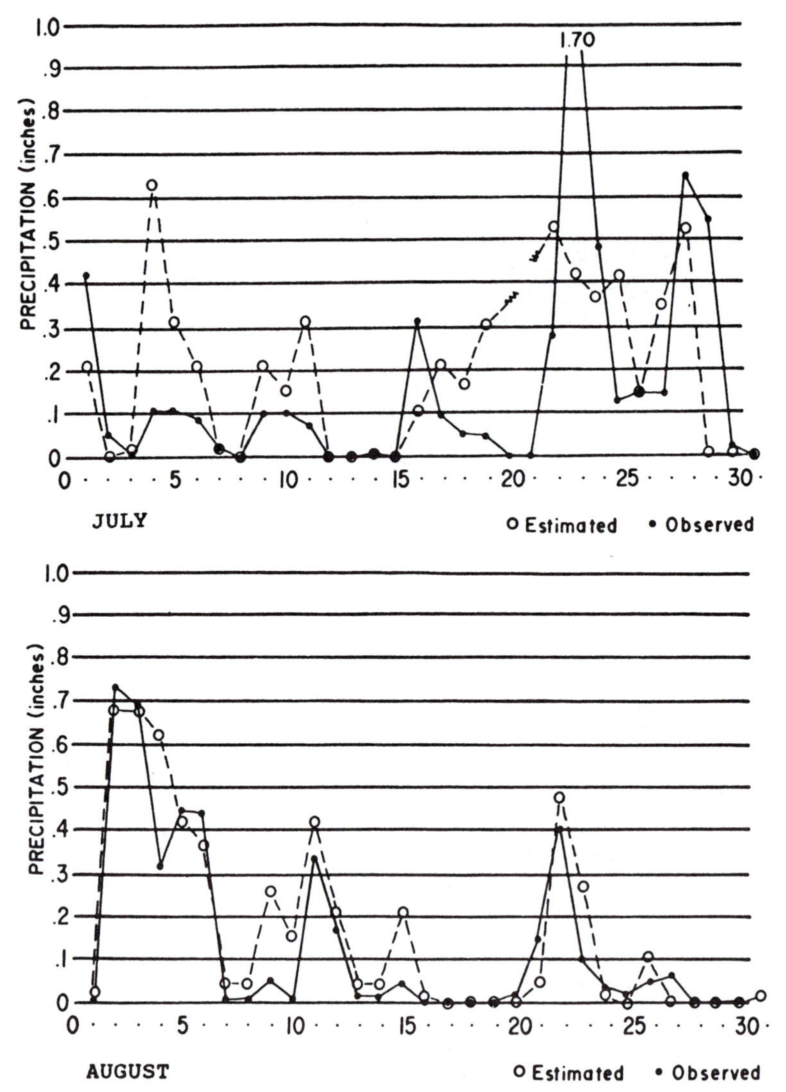

FIGURE 9.4. Arkansas precipitation for the months of July and August 1971 estimated by Follansbee (1973).

hours during which precipitation fell. At any point, the 12-h precipitation is based on the duration (D, in hours) of precipitation, on the normal monthly precipitation (P_N, which helps correct for the variable precipitation rate in synoptic-scale systems), and on an empirical constant (0.0075). The equation they used is

$$P_{12} = 0.0075\, D\, P_N. \tag{9.2}$$

For a 4-week period, the technique's bias was nearly zero, and its mean absolute difference was 55% of the mean when compared to Soviet rain gauges. For daily

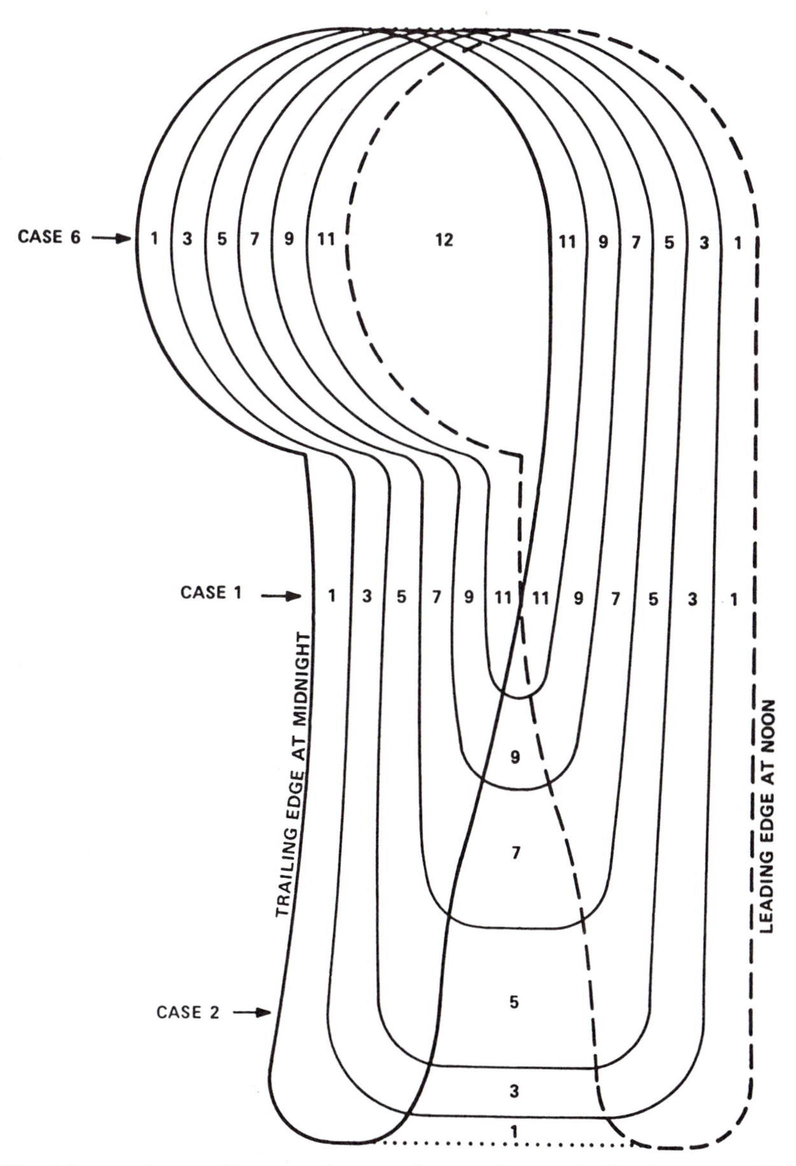

FIGURE 9.5. Schematic diagram illustrating the interpolation technique of Follansbee (1976). The heavy lines outline the raining area of an extratropical cyclone with a trailing front. The solid heavy line is the position at midnight, the dashed heavy line outlines the position at noon. The thin lines (and the numbers) indicate the number of hours that a point was under a raining area.

precipitation, the mean absolute difference was approximately 100% of the mean. Although the 100% figure seems large, it is for a one-day period at a single station; in the above schemes, precipitation was averaged over a large area or over a long time. When averaged over a given area, the mean absolute difference would fall. It must also be kept in mind that *depending on one's application*, even a 100% mean absolute difference may be adequate.

9.1.1.3 Kilonsky and Ramage

Kilonsky and Ramage (1976) developed a technique that is still frequently employed to estimate precipitation over the tropical ocean. They used once-daily visible satellite mosaics computer-mapped into a Mercator projection by NESDIS. In these mosaics, they located "highly reflective clouds" (HRCs). Since tropical oceanic rainfall is dominated by deep clouds, the frequency with which a location is covered by HRCs is related to precipitation. They found that when the monthly precipitation on small coral islands in the Pacific was correlated with the number of days that the island was covered by HRCs, the correlation coefficient was .75 significant at the 1% level. Garcia (1981) found that in the tropical Atlantic (in the GATE area) the simple Kilonsky–Ramage technique performed better than expected (Fig. 9.6) in comparison to the more complex Griffith–Woodley tech-

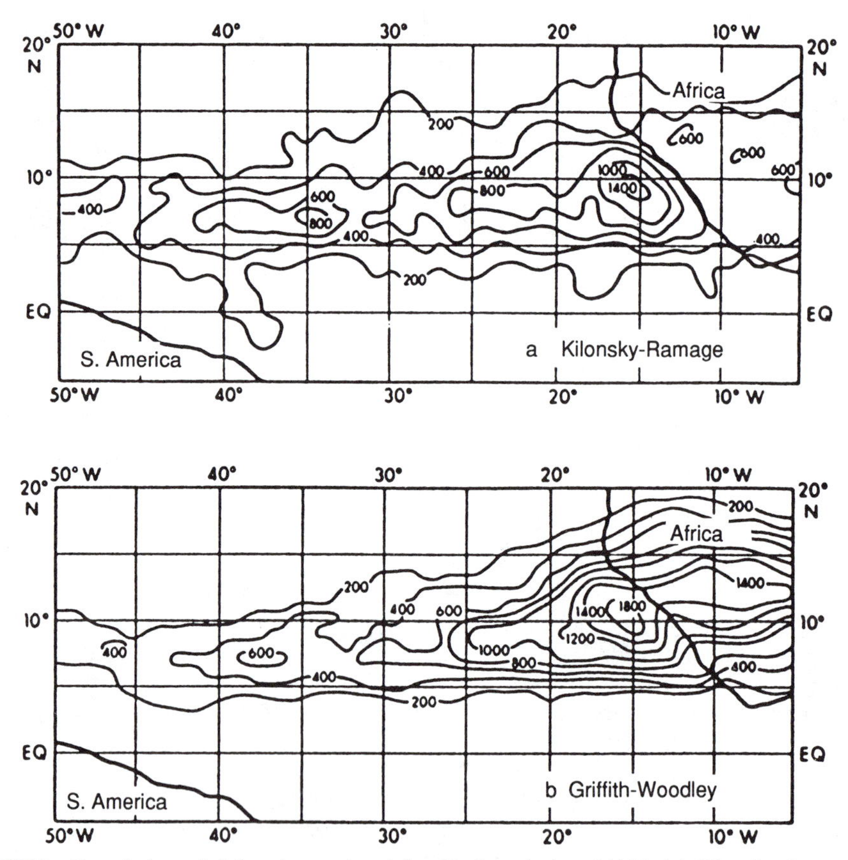

FIGURE 9.6. Cumulative rainfall estimates (mm) for 80 days during GATE: (a) Kilonsky–Ramage technique, (b) Griffith–Woodley technique. [After Garcia (1981).]

nique (Section 9.1.3.2). Of interest is the regression equation used by Garcia (1981):

$$R = 62.6 + 37.4\,N_{\mathrm{D}}, \tag{9.3}$$

where R is the monthly rainfall in millimeters, and N_{D} is the number of days during the month that the location was covered by HRCs.

9.1.1.4 Arkin

The final technique to be discussed in this section was developed by Arkin (1979) to estimate tropical precipitation for climatological purposes.[2] It is similar to the Kilonsky–Ramage technique, but it uses infrared GOES data. Arkin found that for the GATE area, radar-estimated precipitation was highly correlated (correlation coefficient .86) with the fraction of the area covered by pixels colder than 235 K. Other thresholds were tried but yielded lower correlations. Of course, the correlation coefficient depends on the area and time over which the precipitation is estimated. Richards and Arkin (1981) tested averaging areas between 0.5° × 0.5° and 2.5° × 2.5° latitude and averaging times from 1 to 24 h. They found that the correlations increase with averaging area and with averaging time.

Arkin and Meisner (1987) call their precipitation estimate the GOES Precipitation Index (GPI). They use a 235 K threshold and a constant rain rate[3] of 3 mm h^{-1}, which are appropriate values for estimating tropical precipitation in areas approximately 2.5° × 2.5° of latitude. The precise equation is

$$\mathrm{GPI} = 3\,f\,\Delta t, \tag{9.4}$$

where GPI is an estimate of the mean rain depth (millimeters) in the area, f is the fraction of area colder than the threshold, and Δt is the time (hours) for which f applies (e.g., if the images are collected each 3 h, then $\Delta t = 3$).

NMC's Climate Analysis Center (CAC) has an important archive of reduced-volume GOES data, called the GOES Histogram Data Set, primarily for the study of climate. Every 3 h, images from each GOES satellite are blocked into 2.5° × 2.5° latitude–longitude boxes between 50° north and 50° south and between 175° east and 25° west. The infrared counts in each box are histogrammed (binned) in 16 classes (Table 9.1). Each box is thus described by 16 numbers in comparison with roughly 2000 infrared pixels. The data are further compressed by summing each time period for one half month (i.e., eight time periods per day are kept, but each one is summed for a half-month period). The final data set is reduced in volume by a factor of approximately 2000 over the full resolution data. With this archive, the GPI (and other parameters) can easily be calculated for any time

[2] Actually this is a threshold technique rather than a true cloud indexing technique, but it fits best in this section.

[3] This represents the slope of the regression between areally averaged rain rate and fractional area with equivalent blackbody temperatures less than the threshold. The small intercept in this regression is ignored.

TABLE 9.1. The GOES Histogram Data Set

Bin	*Temperature* (K)	*Count*
1	>270	<119
2	266–270	119–128
3	261–265	129–138
4	256–260	139–148
5	251–255	149–158
6	246–250	159–168
7	241–245	169–177
8	236–240	178–182
9	231–235	183–187
10	226–230	188–192
11	221–225	193–197
12	216–220	198–202
13	211–215	203–207
14	201–210	208–217
15	191–200	218–227
16	<191	>227

period after the archive's start in December 1981.[4] Using the full-resolution GOES images, such calculations would be prohibitively expensive.

Recently, the World Climate Research Programme has begun a Global Precipitation Climatology Project, which aims at estimating precipitation in 2.5° × 2.5° boxes over the entire globe for periods ranging from five days to one month. The data from all five geosynchronous satellites are being binned as in the GOES Histogram Data Set, but they are being averaged in 5-day instead of 15-day periods. In addition, data from the NOAA polar orbiting satellites are being histogrammed for the first time. Microwave data from the DMSP SSM/I will also be used. This challenging project will contribute to both precipitation estimation algorithms and to methods that merge precipitation estimates from different sources (World Climate Research Programme, 1986).

9.1.2 Bispectral Techniques

Clouds that are bright in visible images are more likely to precipitate than dark clouds because brightness is related to optical depth and thus to cloud thickness. Clouds that are cold in infrared images are more likely to precipitate than warm clouds because cold clouds have higher tops than warm clouds. There are exceptions to these rules, however. Stratus clouds are bright, but do not rain as much, nor as often, as cumulonimbus clouds. Cirrus clouds are cold, but do not produce as much precipitation as some warmer clouds. Bispectral methods

[4] Phenomena with periods between 1 and 15 days cannot be studied because of the half-month averaging.

attempt to combine these rules by saying that clouds which have the best chance of raining are both cold and bright (Lethbridge, 1967). Lesser amounts (lower probabilities) of precipitation can be expected from cold-but-dark clouds (cirrus) and bright-but-warm clouds (stratus).

9.1.2.1 Dittberner and Vonder Haar

Among the first to use a bispectral technique to estimate precipitation were Dittberner and Vonder Haar (1973), who studied the Indian Summer Monsoon. They developed a relationship of the form

$$P = c_1 E + c_2 A + P_o, \tag{9.5}$$

where P is percent of normal seasonal precipitation, E is the seasonal mean infrared radiant exitance, A is the seasonal mean visible albedo, and the remaining parameters are regression coefficients. The technique successfully separated weak from intense monsoons, whereas visible or infrared data alone could not.

9.1.2.2 Lovejoy and Austin

Lovejoy and Austin (1979a) compared SMS/GOES visible and infrared data with radar data in GATE and around Montreal. They used brightness and temperature observations together to determine whether it was raining. They constructed two 2D (two-dimensional) histograms: a raining histogram and a nonraining histogram. Each satellite pixel was assigned to one of these histograms based on radar observations. The histogram axes were visible count (x axis) and infrared count (y axis). To normalize the visible images for varying solar zenith angle, the visible count was scaled between the minimum (0) and the maximum (1) for that image. Twenty-five bins were used for each axis. Figure 9.7 shows these two histograms for GATE day 248 at 1300 UTC. Notice that the raining pixels are clustered near the cold, bright portion of the histogram.

The next step was to calculate precipitation probabilities for each bin as the ratio of the number of raining pixels to total pixels in each bin (Fig. 9.8). These numbers are useful in themselves, for example, to map precipitation probabilities beyond the range of the radar (Bellon *et al.*, 1980).

In the final step, Lovejoy and Austin (1979a) determined a probability threshold to delineate raining pixels from nonraining pixels by minimizing a loss function (the fraction of incorrectly classified pixels). The contour in Fig. 9.8 separates raining from nonraining pixels.

Accuracy of the Lovejoy and Austin scheme is somewhat difficult to assess. One measure of success is the fraction of pixels that are correctly classified. In the GATE area, Lovejoy and Austin correctly classified between 82% and 98%. Around Montreal, the percentage ranged from 80% to 94%. This statistic can be somewhat misleading, however. Except in unusual cases, rain will not cover the entire area. For the cases examined in GATE, the maximum fraction of area covered by rain was 45.1%; at Montreal it was 24.9%. These could be considered the maximum possible error because a scheme in which all pixels are classified as nonraining would have this error. The 98% success rate in GATE occurred at a

NO RAIN

0	0	0	0	0	0	0	0	0	0	0	0	0	0	0	0	0	0	0	0	0	0	0	0	0
0	9	5	1	0	1	0	0	0	0	0	0	0	0	0	0	0	0	0	0	0	0	0	0	0
0	37	23	10	1	2	0	3	0	0	0	0	0	0	0	0	0	0	0	0	0	0	0	0	0
0	40	32	13	16	12	10	9	8	5	3	1	1	1	0	0	0	0	0	0	0	0	0	0	0
1	38	77	23	15	24	9	7	17	4	3	5	1	0	0	0	0	0	0	0	0	0	0	0	0
0	15	54	20	19	12	10	11	13	7	10	4	6	1	2	1	2	0	1	0	0	0	0	0	0
0	6	42	14	18	16	16	14	7	6	6	4	1	3	2	0	0	0	0	0	0	0	0	0	0
0	2	33	14	10	15	14	11	11	7	1	0	0	6	2	1	1	0	0	0	0	0	0	0	0
0	4	28	22	12	10	9	8	9	5	9	7	2	4	3	2	1	0	1	0	0	0	0	0	0
0	1	27	20	17	7	6	8	9	5	8	2	1	2	2	0	1	2	1	0	0	1	0	0	0
0	2	12	24	20	6	7	8	8	6	6	2	2	1	0	2	0	0	0	0	0	0	0	0	0
0	1	11	30	20	6	4	9	5	6	7	4	2	0	1	2	2	0	0	1	0	0	0	0	0
0	0	13	19	27	10	15	9	13	6	7	3	1	0	3	3	0	0	0	0	0	0	0	0	0
0	0	3	25	53	34	10	16	8	10	7	5	1	2	3	2	0	0	0	0	0	0	0	0	0
0	1	0	17	47	40	16	13	17	13	14	8	1	2	5	0	1	1	0	0	0	0	0	0	0
0	0	2	2	18	44	29	15	16	9	15	13	4	0	6	5	1	0	0	0	0	0	0	0	0
0	0	0	4	10	27	34	20	30	13	34	26	20	11	17	12	7	3	0	0	0	0	0	0	0
0	0	0	0	2	3	11	9	24	30	35	33	29	35	36	26	20	7	1	0	1	0	0	0	0
0	0	0	0	0	0	0	2	2	4	26	31	34	44	45	39	45	33	21	12	2	0	0	0	0
0	0	0	0	0	0	0	0	0	0	0	8	17	29	33	56	47	42	28	21	7	2	0	0	0
0	0	0	0	0	0	0	0	0	0	0	0	0	5	4	7	13	6	23	13	16	7	0	0	0
0	0	0	0	0	0	0	0	0	0	0	0	0	0	1	1	2	10	10	15	19	16	6	0	0
0	0	0	0	0	0	0	0	0	0	0	0	0	0	0	0	0	0	0	0	0	3	0	0	0
0	0	0	0	0	0	0	0	0	0	0	0	0	0	0	0	0	0	0	0	0	0	0	0	0
0	0	0	0	0	0	0	0	0	0	0	0	0	0	0	0	0	0	0	0	0	0	0	0	0

IR (vertical axis, arrow upward); VIS (horizontal axis, arrow rightward)

b

RAIN

0	0	0	0	0	0	0	0	0	0	0	0	0	0	0	0	0	0	0	0	0	0	0	0	0
0	0	0	0	0	0	0	0	0	0	0	0	0	0	0	0	0	0	0	0	0	0	0	0	0
0	0	1	1	0	0	1	0	0	0	0	0	0	0	0	0	0	0	0	0	0	0	0	0	0
0	0	1	0	2	2	0	0	1	0	1	1	0	0	0	0	0	0	0	0	0	0	0	0	0
0	0	0	0	1	0	1	3	1	1	0	3	0	2	0	0	0	0	0	0	0	0	0	0	0
0	1	1	2	2	0	1	1	2	3	2	5	2	2	0	1	1	0	0	0	0	0	0	0	0
0	0	2	0	0	1	0	2	3	1	1	7	2	2	6	4	1	0	0	0	0	0	0	0	0
0	0	0	2	0	0	0	0	4	2	4	1	0	6	4	2	3	1	0	0	0	0	0	0	0
0	0	1	1	1	0	1	1	1	3	2	5	2	7	8	9	6	0	0	0	0	0	0	0	0
0	0	0	0	0	0	1	2	1	3	2	2	0	4	4	4	3	2	1	1	0	0	0	0	0
0	0	0	0	0	0	0	0	1	1	2	2	1	3	4	6	2	3	0	0	0	0	0	0	0
0	0	0	0	0	0	1	1	0	1	4	3	0	3	3	1	8	7	0	2	0	0	0	0	0
0	0	0	0	0	0	0	0	2	2	2	3	0	0	1	4	4	0	0	2	0	0	0	0	0
0	0	0	0	0	0	0	0	1	4	6	7	2	1	6	4	2	1	1	0	0	0	0	0	0
0	0	0	0	0	0	0	0	1	0	0	3	3	2	3	4	3	3	1	0	1	1	0	0	0
0	0	0	0	0	0	0	0	0	0	2	3	5	5	6	4	4	3	2	3	2	4	1	0	0
0	0	0	0	0	0	0	1	1	3	6	8	6	9	9	12	22	17	8	10	5	2	2	0	0
0	0	0	0	0	0	0	0	2	4	3	7	17	12	20	22	21	31	26	66	35	23	1	0	0
0	0	0	0	0	0	0	0	0	0	1	5	11	28	32	37	20	51	50	66	51	47	6	0	0
0	0	0	0	0	0	0	0	0	0	0	1	3	6	16	29	35	62	84	168	144	126	37	1	0
0	0	0	0	0	0	0	0	0	0	0	0	0	0	4	3	11	41	50	65	105	184	81	16	3
0	0	0	0	0	0	0	0	0	0	0	0	0	0	0	0	6	25	19	26	43	108	85	15	3
0	0	0	0	0	0	0	0	0	0	0	0	0	0	0	0	0	0	1	3	3	27	17	3	0
0	0	0	0	0	0	0	0	0	0	0	0	0	0	0	0	0	0	0	0	0	0	0	0	0
0	0	0	0	0	0	0	0	0	0	0	0	0	0	0	0	0	0	0	0	0	0	0	0	0

IR (vertical axis, arrow upward); VIS (horizontal axis, arrow rightward)

FIGURE 9.7. Two-dimensional histograms of GOES data for (a) nonraining and (b) raining pixels as determined by radar. The data are for GATE area on 5 September 1974 at 1300 UTC. (Infrared counts increase downward, infrared radiance increases upward.) [After Lovejoy and Austin (1979a).]

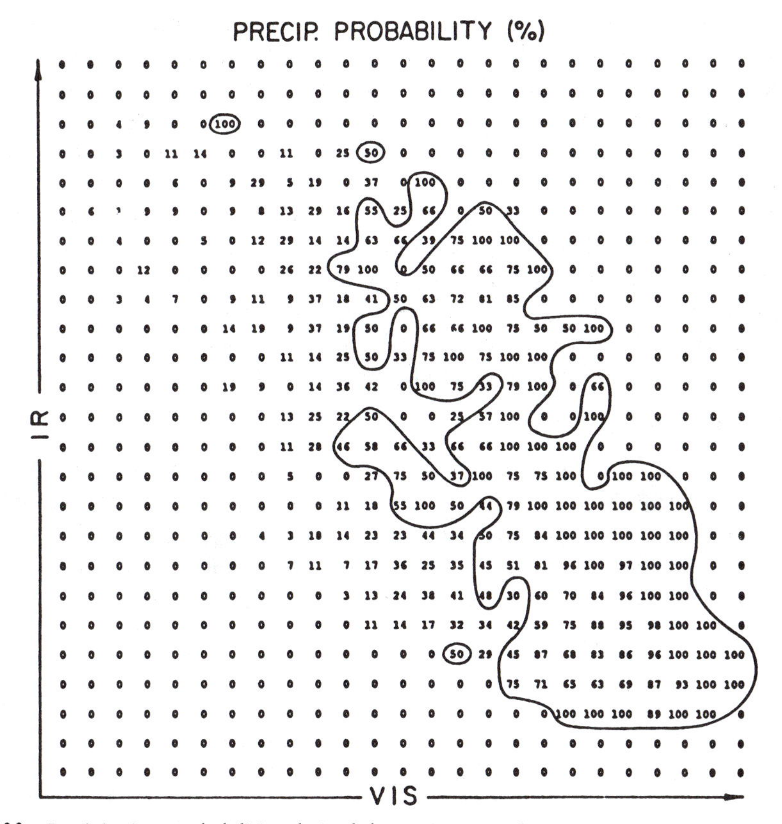

FIGURE 9.8. Precipitation probabilities derived from Fig. 9.7. The contour separates raining from nonraining pixels based on an optimum probability. [After Lovejoy and Austin (1979a).]

a time when only 2.5% of the area was covered with rain; thus the minimum success rate was 97.5%. Another way to measure success is to ask the question, of pixels that are classified as raining, what fraction are not raining? This measure is called the *false-alarm ratio* (FAR). The FAR averaged 35% for GATE and about 45% for Montreal. Lovejoy and Austin (1979a) offer other measures of success that are probably better than the two mentioned here but are also less intuitive; the reader is referred to their paper.

Lovejoy and Austin studied only one stratiform rain case. Their technique worked less well in the stratiform case than in the convective cases. The major difficulty was that the histogram for the raining pixels was very similar to that for the nonraining pixels, with the raining pixels only slightly colder and brighter that the nonraining pixels and with considerable overlap in the distributions. In other words, visible and infrared data contain little information that can be used to distinguish raining from nonraining stratiform clouds.

Lovejoy and Austin compared their bispectral technique to monospectral threshold techniques. The bispectral technique always performed better than either visible or infrared thresholds.

Bellon *et al.* (1980) proposed a short-term forecasting technique based on the Lovejoy—Austin scheme. First, precipitation probabilities were mapped using bispectral satellite data. Second, a cross-correlation procedure (see Section 7.1.2) determined the motion of precipitating regions. Third, the precipitation was advected ahead in time to yield a short-term forecast. This method (known as RAINSAT) is used operationally in Canada by the Atmospheric Environment Service. The scheme was modified slightly to produce three classes of precipitation (light, moderate, heavy), rather than precipitation probabilities, and to use only infrared data at night. Three to six hour forecasts are made (King, 1984).

Bellon and Austin (1986) found that satellite-derived rainfall accumulations during daylight hours over the less populated areas of Canada can be more accurate than those derived from existing rain gauge networks.

9.1.2.3 Tsonis and Isaac

Tsonis and Isaac (1985) have modified the Lovejoy–Austin method by using a clustering technique similar to those used for cloud detection (see Section 8.2.2). They delineate raining areas by classifying pixels in clusters (Fig. 9.9). The raining cluster is determined from radar data. For warm season rain in Canada, Tsonis and Isaac achieved a *probability of detection* (POD, the ratio of pixels classified

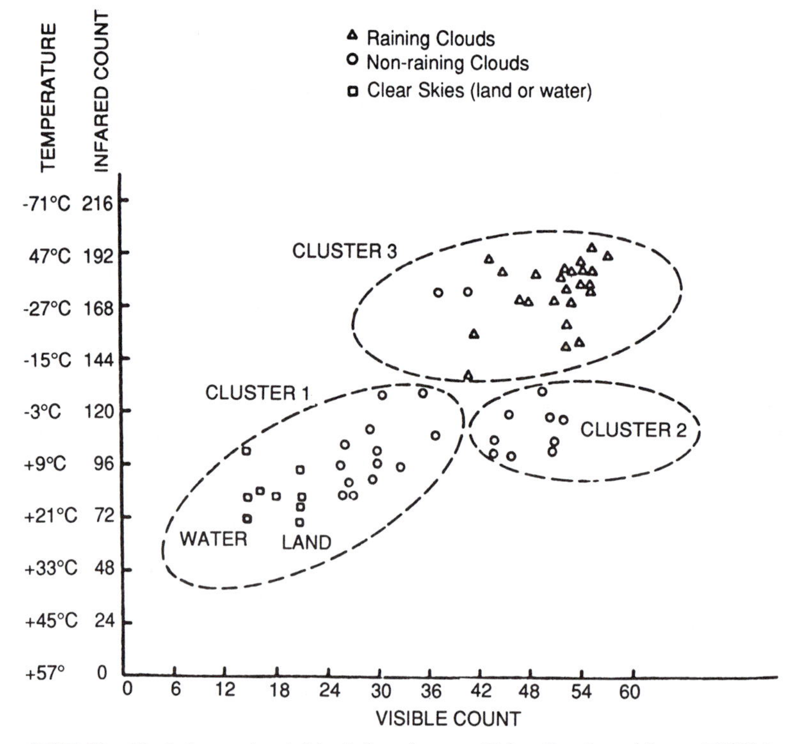

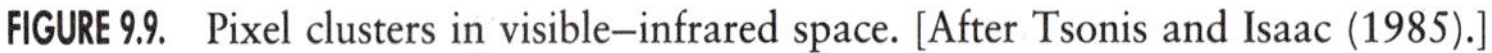

FIGURE 9.9. Pixel clusters in visible–infrared space. [After Tsonis and Isaac (1985).]

as raining to total number of radar-classified raining pixels) of 66% and a FAR of 37%. Eighty percent of the pixels were correctly classified.

Tsonis and Isaac also found that their technique performs better for convective than for nonconvective cases. In nonconvective cases, the POD was higher, but the FAR was also higher.

Note that Tsonis and Isaac normalize the GOES visible counts in what has become a standard correction. Since the visible count is proportional to the square root of the radiance, they divide the count by $[\cos(\zeta)]^{1/2}$, where ζ is the solar zenith angle. This approximates the count value which would be measured if the sun were directly overhead.

9.1.3 Life-History Techniques

The rain rate of a cloud, particularly a convective cloud, is a function of the stage in its life cycle. Life-history techniques take into account a cloud's life cycle. Geostationary satellite data are required for these techniques, and more than one image is necessary for the algorithms.

As with most satellite techniques, life-history techniques go back to the earliest days when the necessary data became available. Sikdar (1972), using ATS 3 data, found a relationship between the time rate of change of the area of a cirrus anvil and the rain rate of the thunderstorm. Since anvil growth rate is better related to storm severity, it will be discussed in Section 9.4.

9.1.3.1 Stout, Martin, and Sikdar

A technique that illustrates the life history method is that of Stout, Martin, and Sikdar (1979). They examined the relationship between radar-estimated rain rate and satellite-measured area of cloud for an isolated thunderstorm (Fig. 9.10). The essential point is that precipitation peaks while the cloud area is rapidly growing; precipitation is much reduced at the time of maximum cloud area. Stout *et al.* approximated this characteristic by adding a term to the rain-rate equation:

$$R = a_0 A + a_1 \frac{dA}{dt}, \tag{9.6}$$

where A is the cloud area, dA/dt is the time rate of change of the cloud area, and a_0 and a_1 are empirically determined coefficients. Because a_1 is positive, this equation ensures that the rain rate will be larger in the growing stage than in the decaying stage of the cloud.

Using half-hourly GOES data in the GATE area, Stout *et al.* adopted a threshold count of 172 to define clouds in the visible channel, and 160 (250 K) in the infrared The satellite areas were regressed against radar-estimated rain rates. Table 9.2 gives the regression coefficients a_0 and a_1 which best fit radar and satellite data. Note that the resulting rain rate is in units of $m^3\ s^{-1}$, which is called the *volumetric rain rate*. It is the volume of water falling from the cloud per unit time.

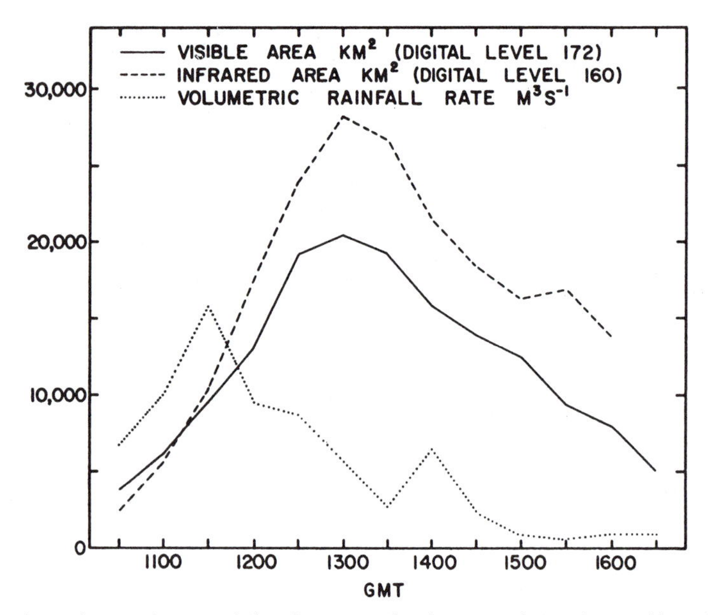

FIGURE 9.10. The evolution of a typical thunderstorm. Cloud area as observed in visible and infrared GOES images and radar-estimated volumetric rain rate are plotted against time. [After Stout *et al.* (1979).]

Comparing rainfall from a single thunderstorm for a 30-min period, the standard error of estimate was 76% of the mean rain rate for the infrared estimate and 62% for the visible estimate. These are very good estimates in comparison with some of those above. The improvement comes about because (1) Stout *et al.* were dealing with a single cloud type in a single area, and (2) their rain-rate equation better approximates the relationship between satellite-observed cloud area and precipitation. Comparing hourly rainfall over the GATE area (10^5 km^2) estimated by radar and by the Stout *et al.* technique produced a .84 correlation coefficient and a 0.25-mm standard error of estimate.

9.1.3.2 Griffith and Woodley

A widely applied precipitation-estimation technique is known as the *Griffith–Woodley technique* (Griffith *et al.*, 1976; Woodley *et al.*, 1980). Originally developed to estimate extra-area effects of cloud seeding in Florida, it has been

TABLE 9.2. Stout *et al.* Regression Coefficients

Band	a_0 (m s^{-1})	a_1 (m)
Infrared	5.4×10^{-7}	2.8×10^{-3}
Visible	5.2×10^{-7}	2.6×10^{-3}

applied to a wide range of tropical and midlatitude convective precipitation. The Griffith–Woodley scheme has evolved over time; different papers use slightly different parameterizations. A concise summary of the scheme is given in Negri *et al.* (1984).

Griffith and Woodley began by comparing satellite images (first visible, now infrared) with rain-gauge-calibrated Miami radar data. The scheme rests on an empirical attempt to estimate from the satellite images what the associated radar echo for each cloud would be. The scheme is entirely automated; one supplies a series of digital images to the computer code.

To estimate the precipitation from a single cloud, the cloud—defined as anything colder than 253 K—is first followed for its entire lifetime to determine its maximum areal extent (A_m). Clouds that merge or split are terminated, and the resulting clouds are treated as new clouds. Figure 9.11a (Griffith *et al.*, 1978) shows the empirical curves used to determine the radar echo area from the satellite-estimated area of the cloud (A_c). The echo area (A_e) is estimated as a fraction of the maximum cloud area depending on the ratio A_c/A_m and the sign of the time rate of change of A_c. A_m itself determines which curve to use in Fig. 9.11a. The rain rate is estimated using Fig. 9.11b (Woodley *et al.*, 1980; Griffith *et al.*, 1980) and a knowledge of the ratio of the echo area to the maximum echo area. The rain volume falling from the cloud is then the product of (1) the rain rate, (2) the echo area, (3) the time interval between successive satellite images, and (4) an empirical factor (Griffith *et al.*, 1978) that starts at 1.00 and increases to a maximum of 3.24 essentially as the mean temperature of the cloud top decreases. Finally the total rain volume is apportioned within the cloud; one half of the rain falls uniformly below the coldest 10% of the cloud top, the remaining half falls below the next warmest 40% of the cloud top. No rain falls in the warmest half of the cloud.

The Griffith–Woodley technique has been applied to many areas of the globe: Florida, the U.S. High Plains, Venezuela, and the tropical Atlantic. In each case the scheme was modified slightly, either in the way precipitation is apportioned below the cloud or in the adjustment factor. In GATE, for example, the rain was uniformly distributed below the cloud (Woodley *et al.*, 1980). In the U.S. High Plains, the precipitation was adjusted by a factor derived from the ratio of High Plains to Florida rain-gauge data or derived from the ratio of cloud model precipitation in the High Plains and in Florida (Griffith *et al.*, 1981). The basic (Florida) relationships between cloud area and precipitation embodied in Fig. 9.11, however, have remained unchanged.

The accuracy of the Griffith–Woodley technique is perhaps best portrayed by Griffith (1987a,b). She used GOES data to calculate precipitation over a 3.6×10^6-km^2 area in the central United States during August 1979. Statistics were accumulated for hourly and daily time periods and for the entire area as well as for specific rain-gauge locations (points). A portion of her results are shown in Table 9.3. Like other techniques, the Griffith–Woodley technique performs best for longer time periods and larger areas. The technique could not be considered useful for estimating how much rain falls in a single rain gauge, except, perhaps,

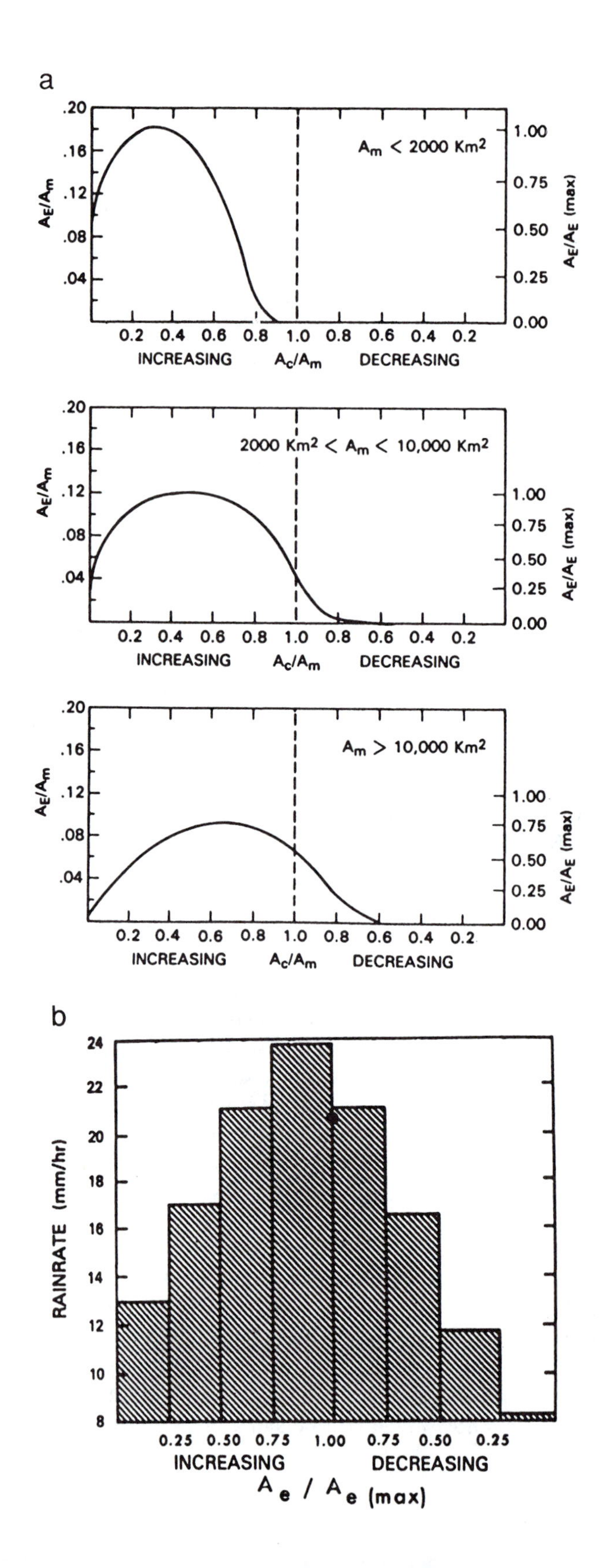
a
A_m < 2000 Km²
2000 Km² < A_m < 10,000 Km²
A_m > 10,000 Km²
A_E/A_m
A_E/A_E (max)
INCREASING
A_c/A_m
DECREASING
b
RAINRATE (mm/hr)
INCREASING
DECREASING
A_e / A_e (max)

TABLE 9.3. Error Statistics for the Griffith–Woodley Technique

	Area		*Point*	
Statistic	*Daily*	*Hourly*	*Daily*	*Hourly*
Rms error (mm)	1	0.1	14	5
Correlation coefficient	0.72	0.60	0.22	−0.05
Rms error ÷ mean	0.43	0.78	1.24	1.35

for periods much longer than a day. Daily rainfall over a large area, however, is well represented by Griffith–Woodley, and hourly precipitation is acceptable (depending, of course, on one's requirements).

Recently, Negri *et al.* (1984) and Griffith (1987a,b) have critically examined the Griffith–Woodley technique. The weakest part of the technique seems to be the process of tracking clouds throughout their lifetimes before assigning precipitation. While convective clouds do go through a life cycle, it is often not observable with visible or infrared satellite data because the cirrus anvils of neighboring clouds interact. In an area such as southern Florida, although there can be dozens of active thunderstorms on a summer afternoon, late in the day there may be only one cloud, as defined by the 253-K isotherm. This explains in part why most visible and infrared precipitation schemes tend to underestimate convective precipitation early in the day and to overestimate it later. Negri *et al.* (1984) and Griffith (1987a,b) showed independently that a simplified precipitation scheme, which does not track clouds (each one is treated as if it existed in only one image), performs no worse than the full Griffith–Woodley scheme, and it takes only 10% of the computer time.

9.1.3.3 About Convective Clouds

Research into precipitation estimation techniques has revealed some basic properties of convective clouds. The most important of these are

- There is a high correlation between a cloud's area (as measured in visible or infrared wavelengths) and the total volume of rain per unit time falling from it (Lovejoy and Austin, 1979b).
- There is little correlation between the visible brightness or infrared temperature of a point and the rain rate beneath that point (Lovejoy and Austin, 1979b; Weiss and Smith, 1987).

FIGURE 9.11. (a) Echo area/cloud area relationships for infrared data in the Griffith–Woodley technique, normalized by maximum cloud area. Curves are the actual Griffith–Woodley approximations to empirical data. [Adapted from Griffith *et al.* (1978) by Negri *et al.* (1984).] (b) The Griffith–Woodley relationship between rain rate and echo area. The singular point at 20.7 mm h^{-1} is for clouds at their maximum echo area. [Adapted from Griffith *et al.* (1980) by Negri *et al.* (1984).]

The first property has been confirmed with radar data. Doneaud *et al.* (1984) and Lopez *et al.* (1983) have shown that the volume of rain that falls from a convective cloud through its lifetime is very nearly proportional to the area-time integral (ATI):

$$\mathrm{ATI} \equiv \int_{t_T} A(t)\,dt, \tag{9.7}$$

where $A(t)$ is the raining area of a cloud (defined by a reflectivity threshold; say, 20 dBZ) at time t, and t_T is the cloud lifetime. This means that the total rain which falls from a convective cloud is insensitive to the details of precipitation structure; only the raining area is important. It also means that if satellite data can define the raining area of a convective cloud, the volumetric rainfall can be estimated (Doneaud *et al.*, 1987). This is the reason for the success of most life history techniques.

The second property is a result of the fact that there is no universal relationship between rain and what is observable on the outside surface of the cloud, which is what visible and IR instruments sense.

9.1.4 Cloud Model Techniques

To improve precipitation estimation techniques based on visible and infrared satellite data, we believe that it is necessary to build the physics of the cloud into the retrieval process. One way to do this is through the use of cloud models. Several investigators have attempted to use cloud models to relate satellite observations to precipitation.

9.1.4.1 Gruber

The earliest such attempt was by Gruber (1973a), who noted that Kuo's (1965) parameterization of convection could be used to relate fractional cloud cover to rain rate. In extremely simplified form, the scheme is as follows. Consider a grid box into which moisture is flowing at the rate I. If Q is the amount of moisture necessary to completely saturate the box, then the rate at which the box is filling with cloud could be estimated as I/Q. However, convective clouds do not live forever. Suppose that all of the clouds which form are convective and have identical lifetimes t_L. Then the rate at which clouds are dying is approximately c/t_L, where c is the fractional cloud cover. If one makes the further assumption that the rate of influx of moisture I is exactly the same as the rain rate R, then one arrives at the equilibrium relationship $R = Qc/t_L$. Gruber developed his scheme using Florida Area Cumulus Experiment (FACE) data from June and July 1970. He calculated Q from a nearby sounding, c from infrared satellite data, and R from radar data. The resulting t_L correlated well with observed cloud lifetimes and averaged 30 min. He then tested this scheme on a squall line that moved through Illinois and Indiana on 3 July 1970 (Gruber, 1973b). Assuming t_L = 30 min, the mean R was estimated to be 3.8 mm h^{-1}, whereas gauge measurements averaged 2.5 mm h^{-1}.

9.1.4.2 Wylie

Another use of cloud models is in adjusting calibration coefficients. Most of the above techniques were developed in a particular location. The changes necessary to apply them elsewhere are not obvious. Wylie (1979) attempted to use the one-dimensional cloud model of Simpson and Wiggert (1969) to adjust the precipitation estimates. He used the method of Stout *et al.* (1979) to estimate precipitation in GATE and around Montreal. Wylie adjusted the satellite rain estimates for the six cases around Montreal by the ratio of precipitation estimates made using the cloud model. Substantially improved results were obtained in five of the six cases. Griffith *et al.* (1981) used the Wylie technique to apply the Florida-calibrated Griffith–Woodley technique to the U.S. High Plains.

9.1.4.3 Convective—Stratiform Technique

Adler and Mack (1984) studied the ability of a 1D (one-dimensional) cloud model to explain differences in cloud top temperature-rain rate relationships in Florida and Oklahoma. In general they found that the 1D model was able to explain differences in the curves.

Adler and Negri (1988) applied the results of Adler and Mack (1984) in a tropical precipitation estimation scheme that they call the *convective–stratiform technique* (CST). In simplified form the technique can be described as follows.

First, a 1D cloud model is run using a representative sounding as input. The outputs are (1) a relationship between cloud top temperature and rain rate and (2) a relationship between cloud top temperature and raining area.[5]

Second, infrared satellite data are analyzed. Local minima in the IR temperatures are found and screened to eliminate thin, nonprecipitating cirrus. The remaining minima are assumed to be convective elements protruding from the top of cirrus anvils. Around each convective element the modal temperature in an area approximately 80 km on a side is calculated. This temperature is assumed to represent the anvil temperature. The average of all anvil temperatures is used as a threshold for stratiform precipitation.

Third, precipitation is assigned to the convective elements. The rain rate and raining area are determined from the cloud-top temperature using the output of the 1D cloud model. To map the rain, the calculated rain rate is assigned to pixels in a spiral fashion. starting at the center of the temperature minimum, and continuing until the raining area is reached.

Finally, to every point that is colder than the stratiform threshold and that did not receive any convective precipitation, a 2-mm-h^{-1} stratiform rain rate is assigned.

The CST is still experimental, however when tested with Florida data, it outperformed all other visible/infrared techniques. In particular, because it assigns convective precipitation only to temperature minima, the CST showed much less

[5] This step is more difficult than it sounds because 1D cloud models require significant experience to be applied meaningfully.

tendency to overestimate precipitation late in the day when cirrus debris is widespread but convection has decreased. It remains to be seen whether convective elements embedded in the anvil, and thus undetectable in infrared data, contribute substantially to convective precipitation. Perhaps this question could be answered with microwave imagery.

9.1.4.4 Scofield and Oliver

The final visible–infrared scheme to be discussed is that developed by Scofield and Oliver (1977) for the purpose of operationally estimating convective precipitation, particularly heavy precipitation. The operational scheme is explained by Scofield (1987); however, the scheme is continually being improved. The Scofield–Oliver technique is in use around the clock at the Synoptic Analysis Branch of NESDIS, which issues statements on the location and intensity of heavy precipitation. The technique can be and is used in many other locations that have access to geostationary images.

Barrett and Martin (1981) classified Scofield–Oliver as a life-history technique. Although it does follow the life history of clouds, we believe that its success is due to the fact that it is firmly based on a conceptual cloud model; thus we discuss it in the cloud model section.

The Scofield–Oliver technique is quite different from other visible—infrared techniques in that it is not automated (though parts of it could be). It relies on the judgment of a satellite meteorologist to locate precipitation-associated signatures in satellite images and to assign an appropriate rain rate to points affected by these features. The technique is thus subjective, but it can take into account phenomena that automated techniques cannot.

The Scofield–Oliver technique is designed to use (primarily) enhanced, half-hourly GOES infrared images (enhanced using the MB curve; see Chapter 5), which are widely available in forecast offices (Fig. 9.12). A decision tree is used to assign a rain rate to a particular point at the time of the satellite image. A simplified version of this decision tree is presented in Fig. 9.13 and discussed here.

The first step is for the analyst to determine whether the cloud is convective; the technique only applies to convective clouds.

If the analyst is using a video display device rather than hard-copy photographs, the second step should be to determine the likely temperature of the top of the cirrus anvil. This can be estimated from a sounding by calculating the equilibrium level (the temperature at which the moist adiabat through the lifting condensation level crosses the environmental temperature sounding). If the equilibrium level is significantly warmer than −62°C, the MB enhancement curve should be adjusted to make the boundary between black and repeat gray fall on the equilibrium-level temperature.

The main loop starts with the analyst outlining the area of active convection. Clues to the location of this area include (Scofield, 1987):

- The active area is near the portion of the anvil with the tightest gradient of IR temperature or near the center of an anvil which has a uniform IR temperature gradient.

- Overshooting tops are in the active area.
- The anvil is brighter and/or more textured over the active area.
- Comparing the last two pictures, the active area is under the half of the anvil bounded by the edge that moves least.
- The active area is near the upper level (between 500 and 200 hPa) upwind end of anvil.
- The active area is near low-level inflow.
- Any area under radar echo is an active area.

This step is the most subjective, but it can be performed reliably by a trained analyst.

In the next step a preliminary rainfall rate is assigned; this is the most complicated step. In some situations, such as convection initiated by boundary intersections or that embedded in a moist environment, the rain "bursts," that is, its onset is rapid and its rain rate is intense. If rain appears to be bursting, the analyst assigns a rain rate of 25–50 mm per half hour in the first (and possibly the second) image of a cloud's lifetime. Otherwise, the analyst assigns a rain rate between trace and 50 mm per half hour depending on the following parameters: the enhancement level of the cloud top, the rate of change of size (usually length) of the anvil, the sign of the time rate of change of the temperature of the coldest tops, the strength of the divergence aloft, and the strength of the low-level inflow. In this step the rain rates come from a lookup table. More rain is assigned to colder clouds, to clouds that are growing in size or becoming colder, and to clouds that have strong divergence aloft or low-level inflow. Relatively less rain is assigned to decaying clouds, to those that are warming, and to those decreasing in area.

The initial rain estimate is augmented by 8 mm per half hour in the area of any overshooting tops (most easily located in visible imagery), by 13 mm per half hour in the region of cloud mergers, and by 5–13 mm per half hour if the environment is likely to be saturated (upwind edge of storm stationary for an hour or more). The base rain rate is the rate after this step.

Two multiplicative correction factors are applied to the base rain rate. The first is called the *speed-of-storm factor*. Rapidly moving thunderstorms produce less rain at a point than more slowly moving storms because they stay over a point for less time. The speed-of-storm factor is between 1.00 (for stationary storms) and 0.25 for rapidly moving storms. The second multiplicative correction is the *moisture-correction factor*. It is numerically equal to the precipitable water (in inches) times the mean surface–500 hPa relative humidity (0–1). This factor takes into account the fact that storms in moister environments have higher rain rates. The final half-hourly rain rate is the base rain rate times the speed-of-storm factor times the moisture-correction factor.

If the cloud remains active, the analyst goes to the next half hour period and repeats the main loop.

The gross features of heavy convective precipitation are almost always well described by the Scofield–Oliver technique, which makes it quite useful for flash flood forecasting (Fig. 9.14). Because of extreme spatial variations of convective precipitation (Huff, 1967; Vogel, 1981), comparison with rain-gauge observations

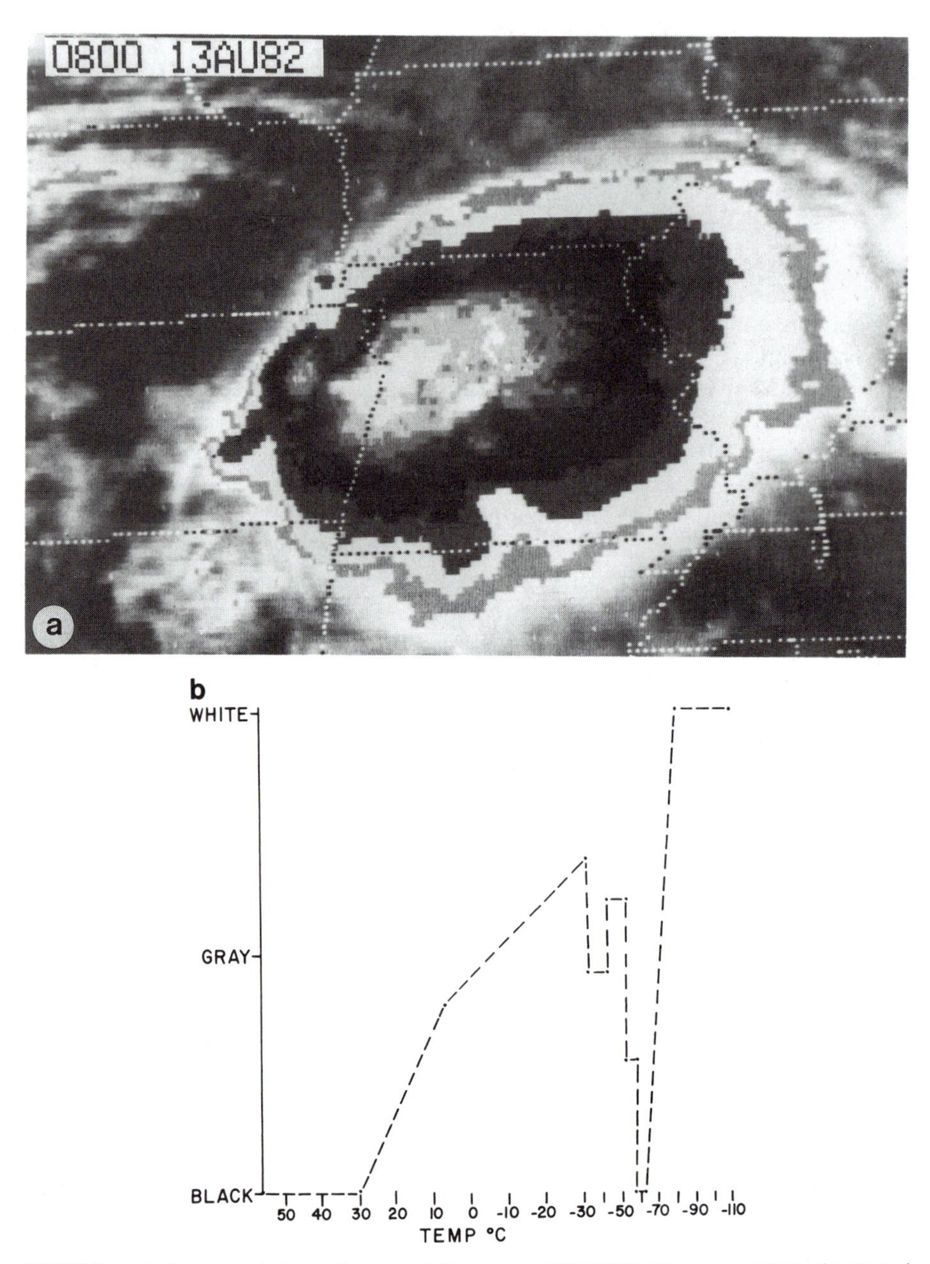

FIGURE 9.12. (a) Enhanced infrared imagery (MB curve), 0730 UTC 13 August 1982. (b) Digital enhancement curve (MB). (c) Radar summary chart, 0735 UTC 13 August 1982. [After Scofield (1987).]

will always be noisy. However, for storm total precipitation, Scofield (1987) found that the Scofield–Oliver technique agrees with rain gauges within about 30%.

A precipitation-estimation technique similar to that of Scofield and Oliver has

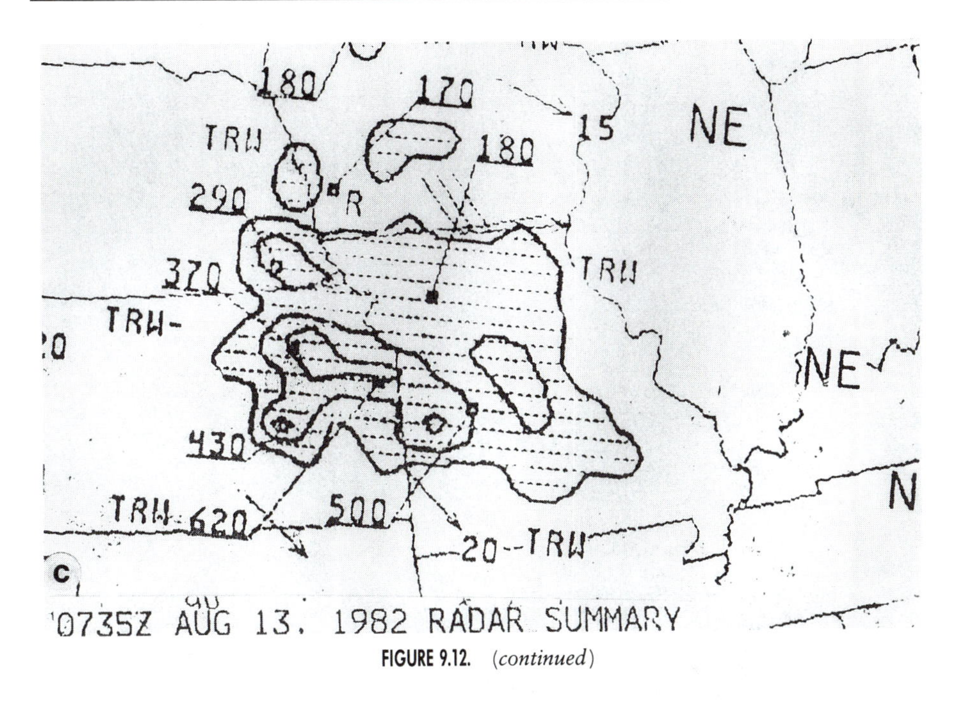

FIGURE 9.12. *(continued)*

been developed for use with tropical cyclones (Spayd and Scofield, 1984). A related technique has been developed for extratropical cyclones (Scofield and Spayd, 1984). Both of these are in operational use by NESDIS.

Martin and Howland (1986) have developed a precipitation-estimation scheme called "grid history," which can be viewed as a highly streamlined, semiautomated Scofield–Oliver-type technique. A grid is placed over the area of interest. At each grid point, IR data are examined. If the brightness temperature is warmer than a threshold, the automatic decision is no rain. If the brightness temperature is colder than a second threshold, heavy rain is automatically assigned. At intermediate temperatures, an analyst, who views the image (and the accompanying visible image) is asked to assign a rain category (light, moderate, or heavy). By using empirically determined rainfall rate coefficients and summing the rainfall through a day, the daily rainfall can be determined within about a factor of 2.

9.2 PASSIVE MICROWAVE TECHNIQUES

At visible and infrared wavelengths, clouds are opaque; precipitation must be estimated from measurements made at the tops of clouds. At microwave wavelengths, cloud droplets only weakly interact with the radiation. The advantage of the microwave portion of the spectrum is that microwave radiation penetrates clouds. Precipitation-size drops interact strongly with microwave radiation, which allows their detection by microwave radiometers. The disadvantage of

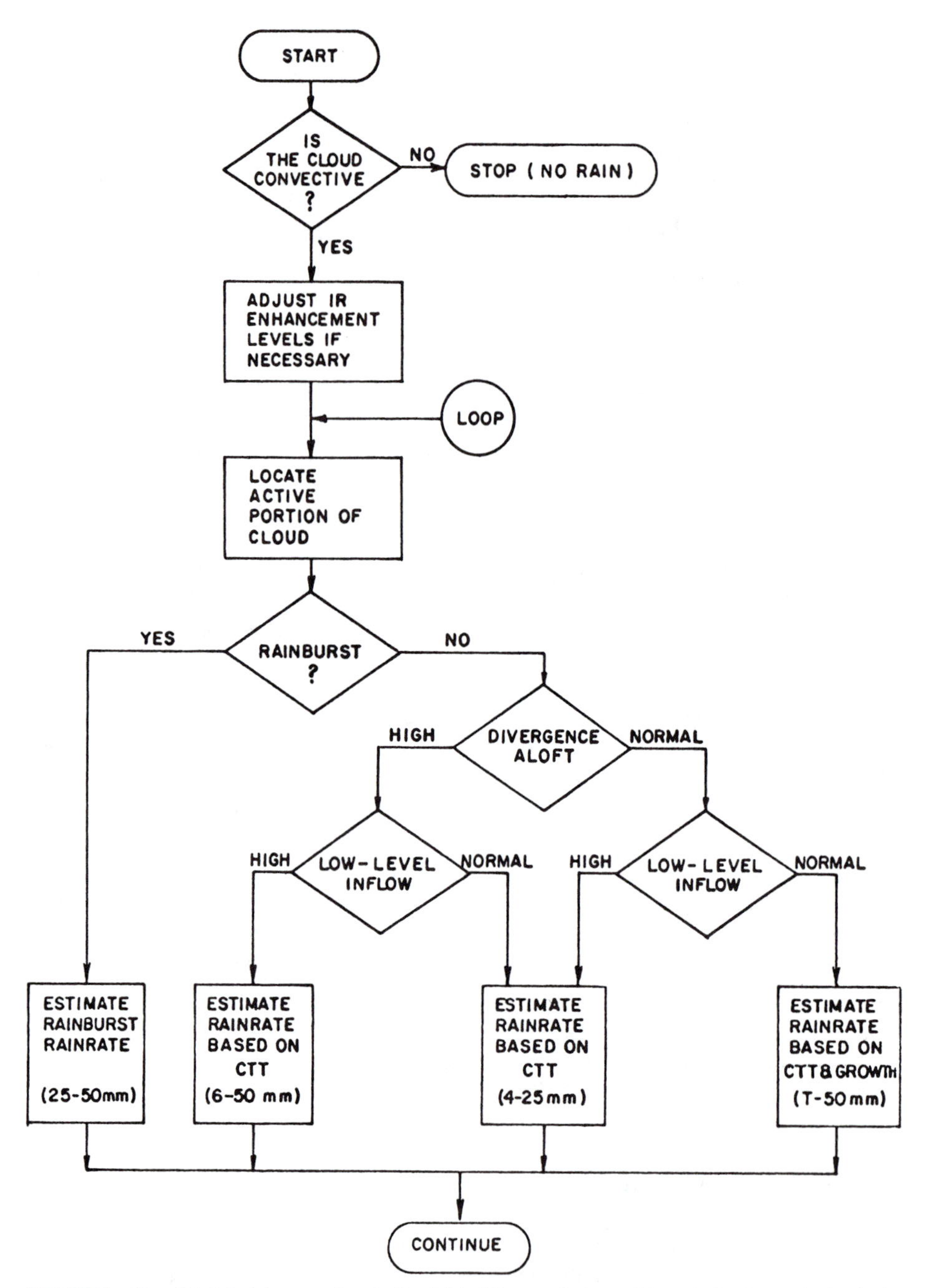

FIGURE 9.13. Simplified decision tree for the Scofield–Oliver technique. (Consult the Synoptic Analysis Branch of NESDIS for the currently operational technique.)

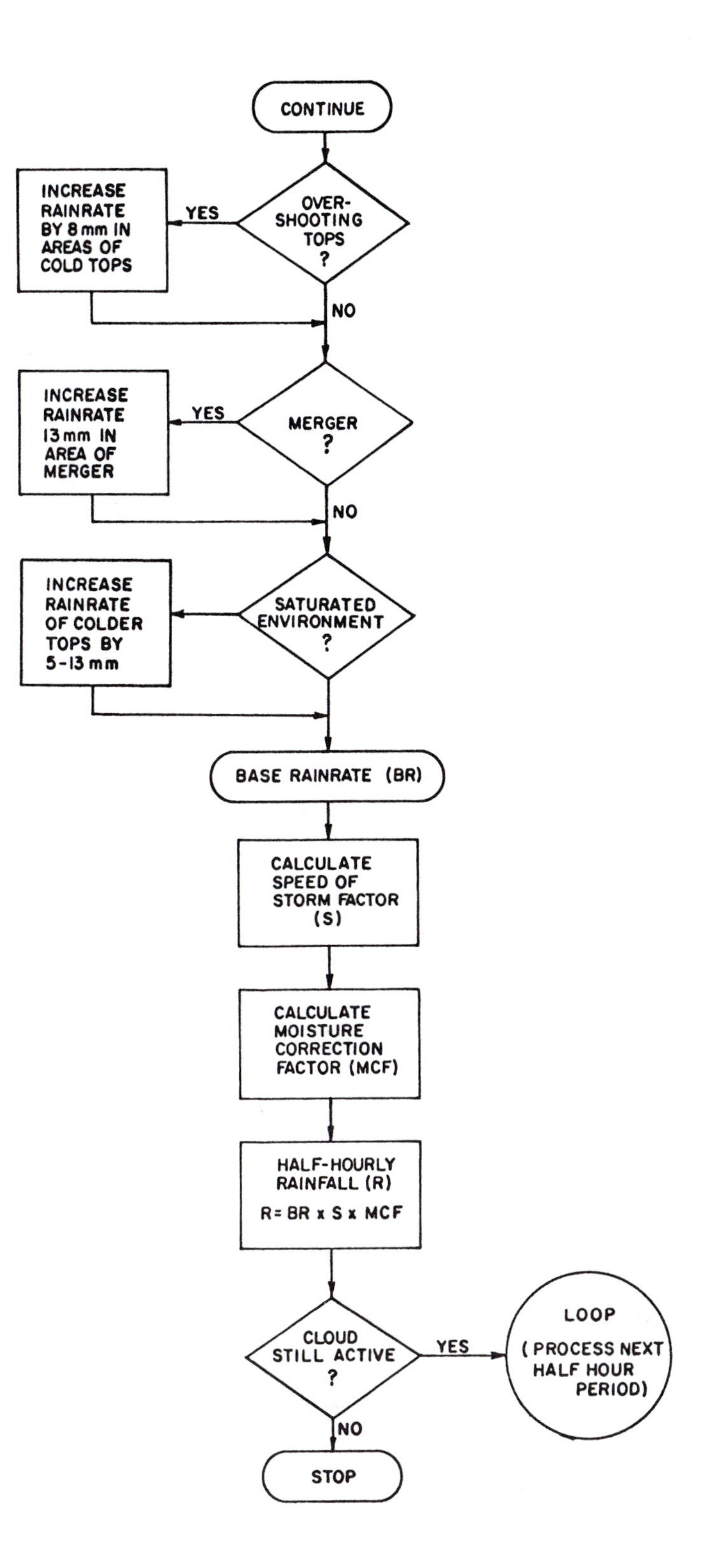
CONTINUE
OVER-
SHOOTING
TOPS
?
YES
INCREASE
RAINRATE
BY 8 mm IN
AREAS OF
COLD TOPS
NO
MERGER
?
YES
INCREASE
RAINRATE
13 mm IN
AREA OF
MERGER
NO
SATURATED
ENVIRONMENT
?
INCREASE
RAINRATE
OF COLDER
TOPS BY
5-13 mm
BASE RAINRATE (BR)
CALCULATE
SPEED OF
STORM FACTOR
(S)
CALCULATE
MOISTURE
CORRECTION
FACTOR (MCF)
HALF-HOURLY
RAINFALL (R)
R= BR x S x MCF
CLOUD
STILL ACTIVE
?
YES
LOOP
(PROCESS NEXT
HALF HOUR
PERIOD)
NO
STOP

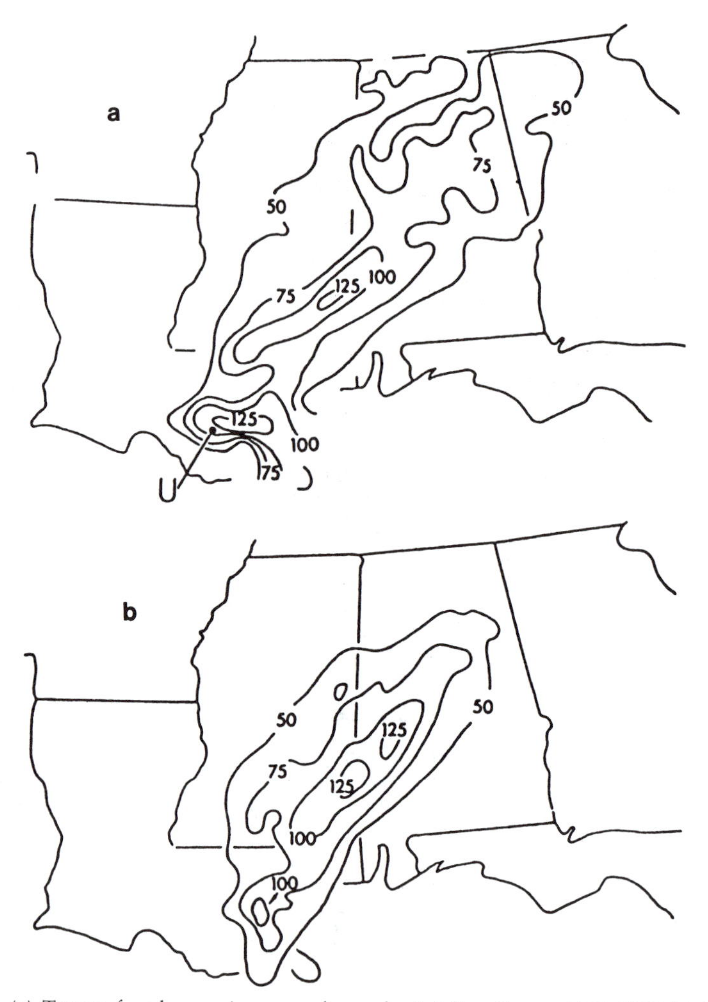

FIGURE 9.14. (a) Twenty-four-hour rain-gauge-observed rainfall ending at 1200 UTC 28 December 1983. (b) Scofield–Oliver estimated rainfall between 1900 UTC 27 December and 1000 UTC 28 December 1983. [After Scofield (1987).]

microwave precipitation-estimation techniques is that the radiometers have had poor spatial and temporal resolution, a point to which we will return.

The physical basis of microwave precipitation schemes is somewhat more complicated than for visible/infrared techniques. Spencer *et al.* (1989) have calculated the scattering and absorption properties of rain[6] for the three main wavelengths that have been used to measure precipitation (Fig. 9.15). The following important properties can be seen:

- Ice essentially does not absorb microwave radiation; it only scatters.
- Liquid drops both absorb and scatter, but absorption dominates.

[6] Assuming a Marshall–Palmer raindrop-size distribution. The reader should be aware that the rain-rate scale on the figure is dependent on this assumption.

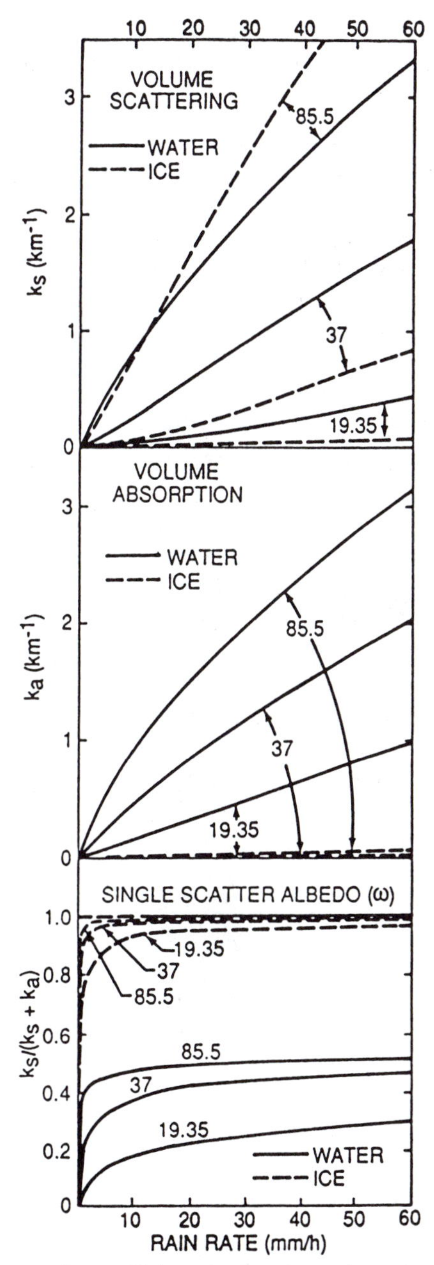

FIGURE 9.15. Mie volume scattering coefficients (top), volume absorption coefficients (middle), and single scattering albedos (bottom) for a Marshall–Palmer precipitation size distribution of water and ice spheres at three frequencies (GHz). (Note that Spencer *et al.* use k to symbolize volume scattering coefficient, whereas we use σ.) [After Spencer *et al.* (1989).]

- Scattering and absorption both increase with frequency and with rain rate. However, scattering by ice increases much more rapidly with frequency than scattering by liquid.

Two general conclusions can be drawn. First, the microwave spectrum can be divided roughly into three parts. Below about 22 GHz, absorption is the primary mechanism affecting the transfer of microwave radiation. Scattering does occur, but it is of secondary importance. Above about 60 GHz, scattering dominates absorption. Between 22 and 60 GHz, both scattering and absorption are important. Second, at different frequencies, microwave radiometers observe different parts of the rain structure. Below 22 GHz, any ice above the rain is nearly transparent; microwave radiometers respond directly to the rain layer. Above 60 GHz, however, ice scattering is the dominant process; microwave radiometers sense only the ice and cannot see the rain below. Thus precipitation estimates made at higher frequencies are necessarily more indirect than those made at lower frequencies.

It must also be noted that cloud droplets, water vapor, and oxygen all absorb (but do not scatter) microwave radiation and thus have the potential to confuse precipitation estimates based on absorption.

If a model of precipitation and its atmospheric environment are assumed, the radiative transfer equation may be used to calculate microwave brightness temperature as a function of rainfall rate. While such calculations depend strongly on the assumptions made about the atmosphere (e.g., on the amount of cloud water and on the structure of the rain layer), they show the general behavior of microwave brightness temperatures in the presence of rain. One set of calculations, for a nadir-viewing instrument, is shown in Fig. 9.16.

The behavior of the brightness temperature—rain-rate curves can be understood qualitatively by examining the radiative transfer equation. Using the Rayleigh–Jeans approximation and assuming for simplicity that we are looking straight down (cf. Eq. 3.25),

$$\frac{dT_\mathrm{B}}{dz} = \sigma_\mathrm{a}(T - T_B) + \sigma_\mathrm{s}(\langle T'_B \rangle - T_B), \tag{9.8}$$

where $\langle T'_B \rangle$ is the directionally weighted mean brightness temperature (cf. Eq. 3.23)

$$\langle T'_B \rangle = \frac{1}{4\pi}\int_{4\pi} T'_B\, p(\psi_s) d\Omega'. \tag{9.9}$$

If we ignore scattering for the moment, the equation becomes

$$\frac{dT_\mathrm{B}}{dz} = \sigma_\mathrm{a}(T - T_B). \tag{9.10}$$

Let's assume that T is nearly constant in the rain layer ($T = T_\mathrm{A}$) and note that the volume absorption coefficient is nearly zero except in the rain layer. Integrating this simplified equation gives

$$T_\mathrm{B} \approx T_\mathrm{BS}\tau + T_\mathrm{A}(1 - \tau), \tag{9.11}$$

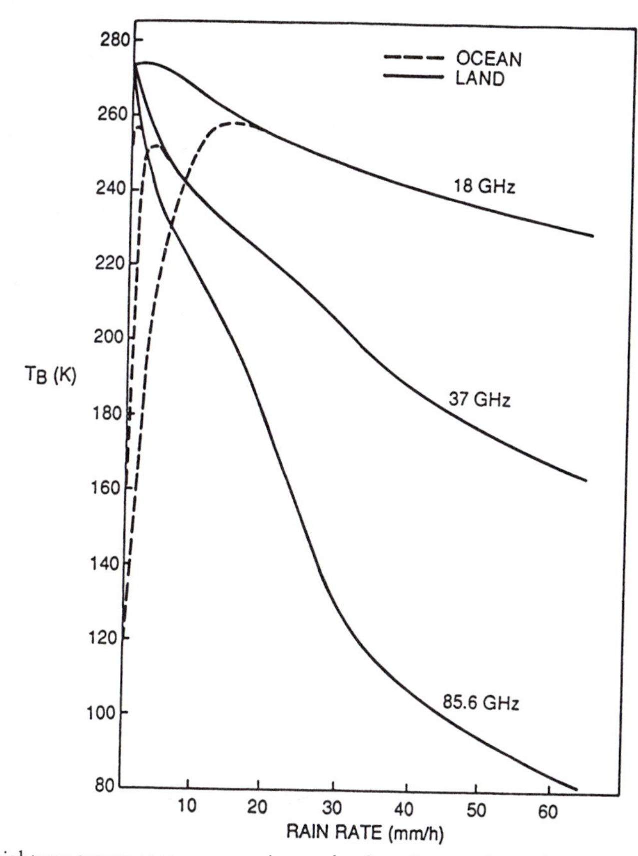

FIGURE 9.16. Brightness temperature versus rain rate for three frequencies. [After Spencer *et al.* (1989).]

where τ is the transmittance of the rain layer, and T_{BS} is the brightness temperature of the surface, that is the sum of surface emission and reflected sky brightness temperature:

$$T_{BS} \approx \varepsilon T_S + (1 - \varepsilon)[T_A(1 - \tau) + T_{space}\tau]. \tag{9.12}$$

Combining:

$$T_B \approx \varepsilon\tau T_S + (1 - \varepsilon)\tau^2 T_{space} + [1 - \varepsilon\tau - (1 - \varepsilon)\tau^2]T_A. \tag{9.13}$$

Ignoring the T_{space} term, Eq. 9.13 can be written

$$T_B \approx T_A\left[1 + \varepsilon\left(\frac{T_S}{T_A} - 1\right)\tau - (1 - \varepsilon)\tau^2\right]. \tag{9.14}$$

The transmittance of the rain layer is given approximately by

$$\tau \approx \exp(-\sigma_a D), \tag{9.15}$$

where D is the depth of the rain layer. These equations best represent the low frequencies (<22 GHz) where ice above the rain layer is nearly transparent.

Examination of Eq. 9.14 shows that for $\tau = 1$ (no rain), $T_B \approx \varepsilon T_S$. As τ decreases (rain increases), T_B converges to T_A. This is as expected because as τ approaches zero, the rain layer will obscure the surface while becoming increasingly visible itself. For water surfaces, T_B increases dramatically with rain rate because ε is small (Fig. 9.17), thus the water background is cold ($\varepsilon T_S \approx 150$ K). Raining areas are easily detected because they contrast well with the cold background. Over dry land, where $\varepsilon_o \approx 0.9$, the change in T_B with increasing rain rate is small (or absent) and not useful for rainfall estimation.

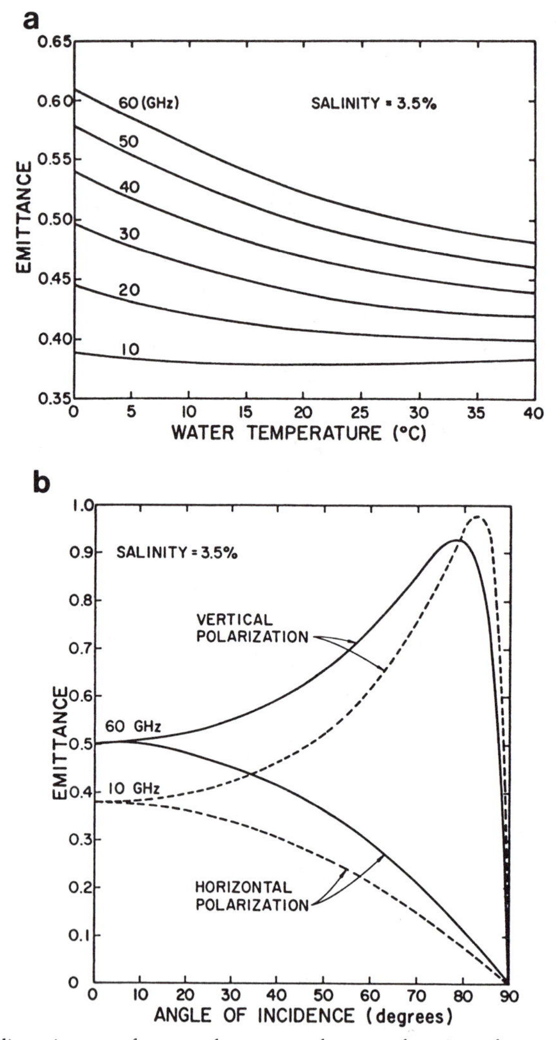

FIGURE 9.17. (a) Nadir emittance of a smooth ocean surface as a function of sea surface temperature. (b) Emittance of a smooth ocean surface as a function of zenith angle. [After Kidder (1979).]

Two difficulties in retrieving rain rate from brightness temperature measurements with this technique should be noted. First, rain rate is related to σ_a, but satellite-measured brightness temperature is related to the product $\sigma_a D$. To retrieve rain rate, the depth of the rain layer must be known. Second, cloud droplets and water vapor contribute an unknown amount to σ_a.

Increasing the rain rate or the microwave frequency causes scattering to become more important. The second term in Eq. 9.8 reveals the effect that scattering has on brightness temperatures (see Section 3.3). Because the scattering phase function has the property that

$$\frac{1}{4\pi}\int_{4\pi} p(\psi_s)d\Omega' \equiv 1. \tag{9.16}$$

$\langle T'_B \rangle$ is simply a weighted average of the brightness temperature from all directions. The scattering phase function divided by 4π is the weight. If $\langle T'_B \rangle$ is greater than T_B at a point, the upwelling radiation stream is augmented; if it is less, the upwelling radiation is diminished. Since space has an extremely low brightness temperature ($T_{space} \approx 2.7$ K), scattering tends to lower the observed brightness temperature by "averaging in" T_{space} and thus decreasing $\langle T'_B \rangle$. This explains why over both land and ocean brightness temperature decreases with increasing rain rate (above a threshold rain rate where scattering becomes significant). Ice hydrometeors are particularly good at lowering brightness temperature because they scatter but have little compensating emission. At the higher frequencies, which are more ice-sensitive, brightness temperature decreases more rapidly with increasing rain rate than at the lower frequencies.

Attempts to use microwave radiometry to estimate precipitation can be divided into two categories: absorption schemes and scattering schemes. Schemes that are absorption-based have the longest history and will be discussed first.

9.2.1 Absorption Schemes

The launch of the Electrically Scanning Microwave Radiometer on the Nimbus 5 satellite (ESMR–5) in December 1972 provided the first opportunity to estimate precipitation with microwave techniques. Wilheit *et al.* (1977) attempted to use these data to estimate precipitation over the ocean. Employing a radiative transfer model, they calculated the 19.35-GHz brightness temperature as a function of rain rate (Fig. 9.18a). Since theirs is an absorption technique, the thickness of the rain layer must be considered: a thicker rain layer will have a higher optical depth and a higher brightness temperature (over the ocean). Wilheit *et al.* assumed that the rain layer extended from the freezing level to the surface. Thus the height of the freezing level, which is indicated in the figure, is the thickness of the rain layer.

Wilheit *et al.* attempted to verify their calculations by comparing ESMR–5 observations over the ocean with Miami radar estimates of precipitation. Because the radar had considerably better spatial resolution than did ESMR–5, they averaged the radar estimates inside each ESMR–5 scan spot. Their results are shown in Fig. 9.18b. In general, the ESMR–5 estimates are within a factor of two of the radar estimates.

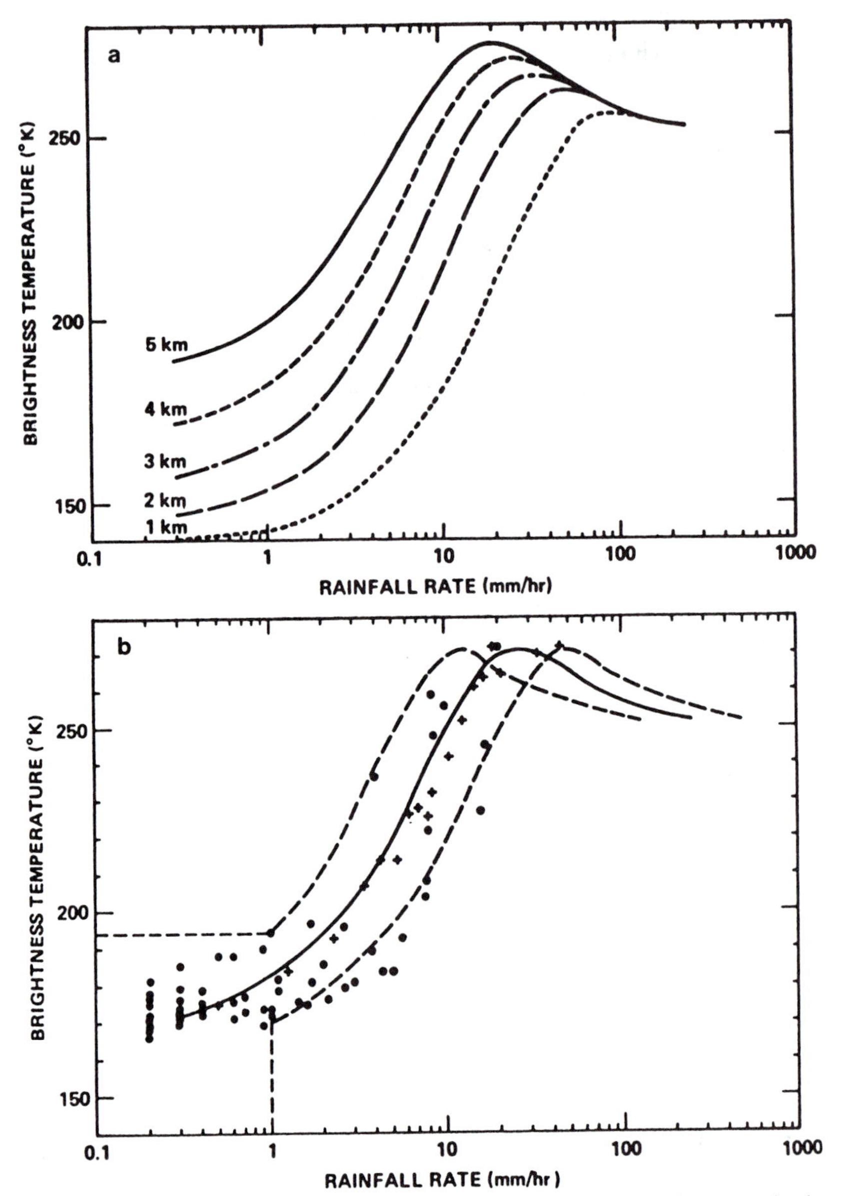

FIGURE 9.18. (a) Calculated 19.35-GHz brightness temperature as a function of rain rate for freezing levels of 1–5 km. (b) Brightness temperature as a function of rain rate: Nimbus 5 ESMR vs. WSR–57 radar (dots) and inferred from ground-based measurements of brightness temperature and direct measurements of rain rate (crosses). The solid line is the calculated brightness temperature for a 4-km freezing level. The dashed lines represent departure of 1 mm h^{-1} or a factor of 2 in rain rate (whichever is greater) from the calculated curve. [After Wilheit *et al.* (1977).]

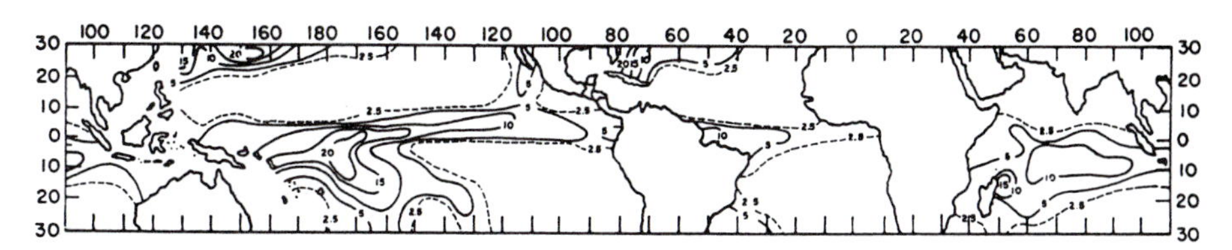

FIGURE 9.19. Precipitation frequencies (percentage of days with rain) over the tropical oceans for the period December 1972 to February 1973 determined from Nimbus 5 ESMR measurements. [After Kidder and Vonder Haar (1977).]

The Wilheit *et al.* technique, and those related to it, rely on the increase of brightness temperature with rain rate over the ocean. They cannot determine rain rates greater than a saturation rain rate, which decreases with increasing microwave frequency. For this reason, lower frequencies are preferred for oceanic precipitation estimation.

Several groups attempted to use microwave observations to map precipitation. Kidder and Vonder Haar (1977), for example, used brightness temperature thresholds to discriminate raining from nonraining pixels. They calculated precipitation frequencies over the tropical oceans (Fig. 9.19). Rao *et al.* (1976) processed all of the ESMR–5 data to produce an atlas of instantaneous rain rates over the oceans.

The Nimbus 6 satellite carried a second ESMR (ESMR–6) that had several important changes. First, it scanned in a cone in front of the spacecraft rather than in a plane through nadir. The advantage of this configuration is that the ground resolution and the viewing angle are the same for all scan spots, which lessens problems with angle-dependent parameters such as surface emittance. Second, it made measurements in two polarizations: horizontal and vertical[7] (ESMR–5 made only horizontal polarization measurements). Third, it measured 37-GHz radiation. These modifications were made in part to better estimate precipitation over land. The higher frequency is more sensitive to precipitation over land, and the polarization measurements were designed to help eliminate lakes and wet soil (Weinman and Guetter, 1997).

Rodgers *et al.* (1979) compared radar measurements with ESMR–6 measurements over the southeastern United States. They found that using polarization measurements they could discriminate precipitation from dry ground and from wet ground. The basis for the discrimination is that dry ground is warmer than precipitation, and wet ground is more polarized (vertical > horizontal) than precipitation (Fig. 9.20). Unfortunately, Rodgers *et al.* found that the technique does not work for surface temperatures less than 15°C. Even more unfortunate, the technique retrieves no information on rain rate.

[7] Horizontal polarization means that the electric vector oscillates perpendicular to the line joining the satellite to the scan spot and parallel to the Earth's surface at the scan spot. Vertical polarization means the electric vector is perpendicular both to the line joining the satellite to the scan spot and to the horizontal polarization vector.

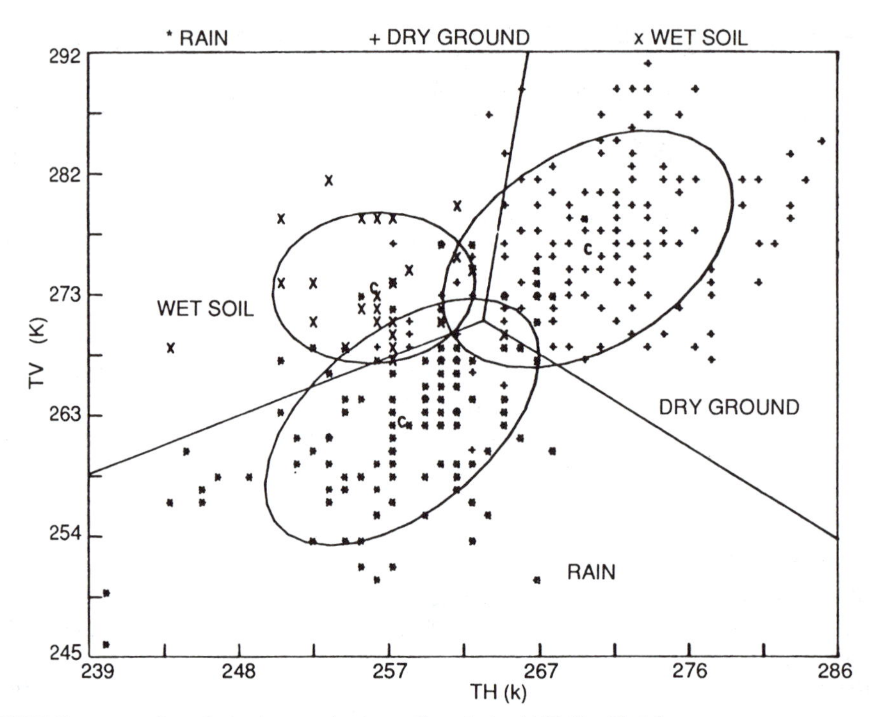

FIGURE 9.20. Vertically polarized versus horizontally polarized ESMR–6 brightness temperatures over the southeastern United States. [After Rodgers *et al.* (1979).]

The Nimbus 7 satellite carried the Scanning Multichannel Microwave Radiometer (SMMR) mentioned in previous chapters. This five-wavelength, dual-polarization instrument has been used to estimate precipitation by Spencer *et al.* (1983c) and Spencer (1984). They used a stepwise multiple regression approach to relate the 10 SMMR brightness temperatures to rain rate as determined from manually digitized microfilm of NWS radar PPI displays. Precipitation rate is being operationally retrieved from SSM/I data by the U.S. Air Force using a similar scheme.

There are two problems with the absorption approach to estimation of rain rate using passive microwave radiometry. First, as pointed out by Lovejoy and Austin (1980), cloud water and rain water are difficult to separate, especially using a single wavelength. Perhaps the multiple wavelength approach of Spencer *et al.* offers hope because the SMMR channels contain some independent information on liquid water.

The second problem is more serious and is, in fact, a combination of two problems: a beam-filling problem and a nonlinearity problem. Because of the inverse relationship between wavelength and antenna size required to achieve a desired ground resolution, microwave radiometers have had large footprints, 500 km^2 at least. These areas are too large to be filled with a uniform rain rate. The radiometer, of course, averages the brightness temperature over the footprint, but because the brightness temperature is highly nonlinear in rain rate (Fig. 9.18b), the average brightness temperature does not produce a good estimate of the mean rain rate in the footprint. In fact, the rain rate estimated by using the radiometer-measured brightness temperature will always underestimate the footprint-mean rain rate. This underestimation can be quite serious (Lovejoy and Austin, 1980). Again, a multiple wavelength approach may offer some relief because of different sensitivities to rain at different wavelengths, but to make substantial improvements, larger antennas are required.

9.2.2 Scattering

Precipitation is the only atmospheric constituent that scatters microwave radiation; the others merely absorb. Thus if scattering can be detected and quantified, an unambiguous precipitation signal will have been found. Over water backgrounds, polarization measurements appear to provide a means to measure scattering.

Consider a conically scanning instrument (ESMR–6, SMMR, SSM/I) that views each scan spot with an angle of incidence near 50°. The emission from a water surface will be highly polarized (Fig. 9.17), and Eq. 9.14 applies independently to each polarization:

$$T_B^H \approx T_A \left[1 + \varepsilon^H \left(\frac{T_S}{T_A} - 1 \right) \tau - (1 - \varepsilon^H)\tau^2 \right], \qquad (9.17a)$$

$$T_B^V \approx T_A \left[1 + \varepsilon^V \left(\frac{T_S}{T_A} - 1 \right) \tau - (1 - \varepsilon^V)\tau^2 \right], \qquad (9.17b)$$

where the superscripts H and V refer to horizontal and vertical polarization, respectively. Neither τ nor T_A has a superscript because the atmosphere absorbs but does not scatter; it has the same effect on all polarizations. The solid line in Fig. 9.21 shows the result of plotting T_B^H versus T_B^V for all values of τ ($0 \le \tau \le 1$). Although Eqs. 9.15 are nonlinear, T_B^H versus T_B^V is very nearly a straight line.

Spencer (1986) noted that if nonscattering material is added to the atmosphere above a nonraining, oceanic scan spot, the observed brightness temperatures will move along the no-scattering line. If scattering particles (precipitation) are introduced, the point moves off the no-scattering line. Since scattering lowers brightness temperature, observations in which precipitation occurs will fall between the no-scattering line and the no-polarization line of Fig. 9.21.

Spencer (1986) then asserted that the precipitation rate is linearly related to the distance from the no-scattering line. This assertion needs to be verified by theoretical and observational work, but if true, it will help solve a problem caused by the large footprint of microwave instruments: the linearity of the relationship means that the satellite measurements are linearly related to the footprint-averaged rain rate.

Spencer *et al.* (1989) introduced a quantity called *polarization-corrected brightness temperature* (PCT), which is a measure of the distance from the no-scattering line. To obtain the PCT for a particular observation (T_B^V, T_B^H), one draws a line through this point parallel[8] to the no-scattering line. The parallel direction is

[8] The slope of this line is $\frac{dT_B^H}{dT_B^V} = \frac{T_A - \varepsilon^H T_s}{T_A - \varepsilon^V T_s}$.

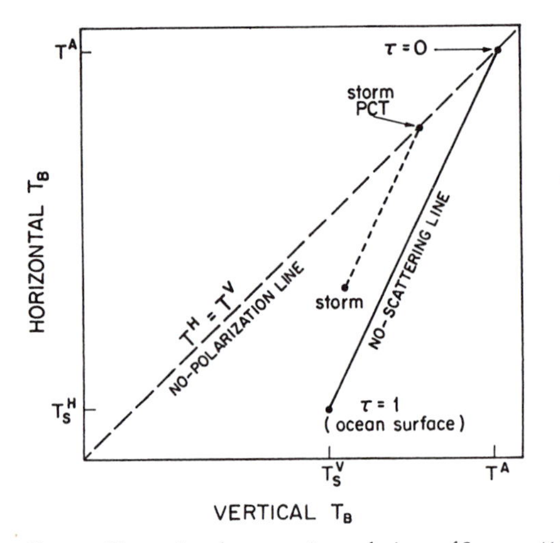

FIGURE 9.21. Polarization diagram illustrating the scattering technique of Spencer (1986). [After Spencer *et al.* (1989).]

chosen because changing the amount of absorbing material in the path, such as cloud droplets and water vapor, will cause the observed point to move parallel to the no-scattering line. The PCT is the temperature at the point that the line intersects the no-polarization line (see Fig. 9.21). Spencer *et al.* (1989) have begun studies of 85.5 GHz data from the SSM/I instrument on DMSP. They calculated PCT $= 1.818T_B^V - 0.818T_B^H$, and found that the 255-K isotherm corresponds well with the outline of precipitating regions as determined by radar. Work continues on ways to relate PCT to rain rate.

Scattering-based methods to estimate precipitation are too new for a realistic assessment of accuracy, but to us they appear promising.

9.2.3 Cloud Model Techniques

As in the case of visible and infrared techniques, microwave techniques can benefit from the use of a cloud model, Mugnai and Smith (1984) were the first to couple a two-dimensional, time-dependent cloud model with a microwave radiative transfer model to produce estimates of microwave brightness temperature at satellite altitude from convective clouds. These studies continue (Mugnai and Smith, 1988; Smith and Mugnai, 1988, 1989; Mugnai *et al.*, 1990; Adler *et al.*, 1991) and have revealed or reinforced the importance of cloud water, ice crystals, and the vertical micophysical structure of the cloud on the upwelling microwave radiation. By considering the vertical structure of the cloud, improved rainfall estimation schemes may be found, and it may be possible to use multifrequency passive microwave observations to estimate the vertical profiles of microphysical quantities such as rain drops, cloud drops, ice particles, and graupel (Smith *et al.*, 1992).

9.2.4 Soil Moisture and API

Soil moisture is another parameter that may be estimated using passive microwave observations. As mentioned above, dry soil has a high emittance; water surfaces have low emittances. If one adds water to the soil, the emittance falls and becomes polarized. Grody (1983) has used channels 1 and 2 (50.30 and 53.74 GHz) of the Microwave Sounding Unit to retrieve surface emittance. In essence, channel 2, which peaks in the lower troposphere, is used to correct the surface-sensing channel 1 for atmospheric effects. Emittance is retrieved as a linear combination of the two brightness temperatures. A more complicated scheme has been investigated by Jones (1989), who combined SSM/I data, GOES data, and soundings to retrieve soil emittance.

With a knowledge of "normal" emittance at a particular location (which depends on soil type and vegetation), microwave observations can be used to detect changes in emittance and therefore of soil moisture. Since soil moisture is changed by precipitation, these emittance changes may serve as an antecedent precipitation index (API), which is related to the precipitation that has fallen since the last pass of the satellite. If this technique works, it will help solve one of the major problems of microwave sensors to date: they only fly on polar-orbiting satellites; thus observations of instantaneous rain rate are available a maximum of four times per day, assuming two polar-orbiting satellites.

9.2.5 Temporal Resolution

We cannot leave the subject of passive microwave sensing of precipitation without noting that precipitation is extremely variable in time and space. Only a geostationary microwave instrument is capable of the temporal resolution necessary to study precipitating systems. Because large antennas (on the order of several meters) are necessary to provide useful ground resolution, placing a microwave instrument in geostationary orbit presents technological challenges. We believe, however, that a passive microwave imager/sounder in geostationary orbit will be extremely valuable in the forecasting and research of weather systems. We urge that the technological problems be pursued with all speed.

9.3 RADAR

The dream of placing a precipitation-sensing radar in Earth orbit predates the launch of the first satellite (Wexler, 1954). That no such radars have been launched indicates the complexity of the task. In this section we review the physics of radar detection of precipitation, we outline some of the problems associated with satellite-borne radars, and we mention an instrument that will likely be the first precipitation-sensing radar in space. This discussion is necessarily brief. The reader is referred to Chapter 14 of Barrett and Martin (1981) and to Bandeen and Katz (1975) for more detail.

9.3.1 Radar Fundamentals

Radar works by transmitting a pulse of microwave radiation and making measurements of the radiation returned from precipitation (or other objects). The power, frequency, and phase of the returned signal can be measured.

The power returned to the radar can be calculated from the radiative transfer equation. Suppose that the radar transmits a pulse of radiation of duration Δt and power P_t. (For National Weather Service WSR–57 radars, $\Delta t = 4\ \mu s$ and $P_t = 450$ kW.) The pulse travels outward as a spherical shell of thickness $c\Delta t$ (1200 m for WSR–57s), where c is the speed of light (Fig. 9.22a). The power is not radiated in all directions, however, but in a very narrow cone (Fig. 9.22b). For a circular paraboloid antenna, the beam width (i.e., the full width at half maximum of the main lobe) is given by

$$\theta = \frac{\alpha\lambda}{d}, \tag{9.18}$$

where θ is the beam width in radians, λ is the wavelength of the transmitted signal, and d is the antenna diameter (10 cm and 3.36 m, respectively, for WSR–57s). For a perfect circular paraboloid antenna, $\alpha \approx 1.02$, and for a practical antenna, $\alpha \approx 1.26$ (Battan, 1973, p. 165). The beam width of a WSR–57 is thus about 2°.

For the sake of argument, suppose that all of the radar's power is transmitted uniformly into solid angle $\Omega_t = 2\pi[1 - \cos(\theta/2)]$. The transmitted radiance is then

$$L_t = \frac{P_t}{\Omega_t A}, \tag{9.19}$$

where A is the area of the antenna. Most weather radars are operated in so-called nonattenuating wavelengths where absorption by atmospheric gases is negligible and where absorption and scattering out of the transmitted beam by clouds and precipitation are small. Thus the transmitted radiance is very nearly the radiance received by the sampled volume (shaded in Fig. 9.22a).

The sampled volume is that portion of the spherical shell from which radiation is being received at a particular time. If the pulse is transmitted from time $t = -\frac{1}{2}\Delta t$ to time $t = +\frac{1}{2}\Delta t$, then, due to the finite pulse length, at time t radiation will be returned to the antenna from ranges between $r = \frac{1}{2}c(t - \frac{1}{2}\Delta t)$ and

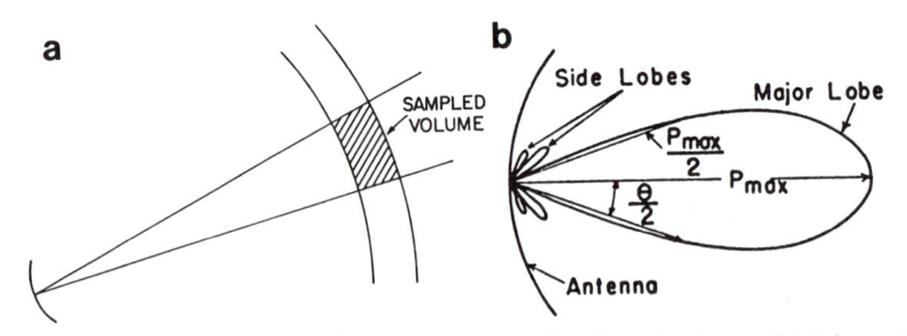

FIGURE 9.22. (a) Schematic diagram of the volume sampled by the radar beam. (b) Schematic cross section of a radar beam from a parabolic antenna. [After Battan, L. J., *Radar Observation of the Atmosphere*. Univ. Chicago Press, © 1973 by the University of Chicago.]

$r = \frac{1}{2}c(t + \frac{1}{2}\Delta t)$. That is, at time t the sampled volume is centered at range $r = \frac{1}{2}ct$, and has depth $\Delta r = \frac{1}{2}c\Delta t$. Because of the necessity of the two-way trip radiation must make from the antenna to the location of interest and back to the antenna, the sampled volume moves at half the speed and (to conserve energy) has half the depth of the transmitted pulse. The sampled volume for a WSR–57 has a depth of about 600 m and, at a range of 100 km, has a diameter of about 3.8 km.

Assuming that the sampled volume is completely filled with particles with volume scattering coefficient σ_s and scattering phase function $p(\psi_s)$, the radiance backscattered toward the antenna is given by[9]

$$L_r = \Delta r \frac{\sigma_s}{4\pi} \int_{4\pi} L' p(\psi_s) d\Omega' \approx \Delta r \frac{\sigma_s}{4\pi} L_t p(180°)\Omega_r, \tag{9.20}$$

where $\Omega_r = A/r^2$ is the solid angle of the antenna subtended at the sampled volume. Finally, the returned power is simply

$$P_r = L_r A \Omega_t = \frac{P_t A \Delta r}{4\pi r^2} \sigma_s p(180°), \tag{9.21}$$

which is one form of the radar equation. [See Battan (1973) for other forms and for corrections for nonuniform radiance in the beam.] Note that the returned power from a single drop would be proportional to r^{-4}, but that the returned power from a volume is proportional to r^{-2}. This is explained by the fact that the volume sampled—and thus the number of drops scattering radiation back to the radar—increases as r^2.

The product $\sigma_s p(180°)$ is the backscattering cross section per unit volume. It is numerically equal to

$$X_B \equiv \sigma_s p(180°) = \sum_{\text{vol}} \frac{\pi D^2}{4} Q_s p(180°), \tag{9.22}$$

where Q_s is the scattering efficiency, D is the drop diameter, and the sum is taken over all drops in a unit volume. For drops in the Rayleigh size range $p(180°) = 1.5$, and (from Eq. 3.52),

$$Q_s = \frac{8}{3}\left(\frac{\pi D}{\lambda}\right)^4 |K|^2, \tag{9.23}$$

where $K = (m^2 - 1)/(m^2 + 2)$ is a weak function of temperature and wavelength, but a strong function of phase (ice or water) Thus, for Rayleigh drops, we obtain

$$X_B = \frac{\pi^5}{\lambda^4} |K|^2 \sum_{\text{vol}} D^6 \tag{9.24}$$

and

$$P_r = \frac{C |K|^2 Z}{r^2}, \tag{9.25}$$

[9] Radar transmits highly polarized radiation. Since we are only considering scattering in the backward direction, however, this nonpolarized radiative transfer equation applies.

where C is a constant that depends on properties of a particular radar (antenna area, wavelength, transmitted power, pulse duration), and

$$Z \equiv \sum_{\mathrm{vol}} D^6 \tag{9.26}$$

is called the *radar reflectivity factor*. Z is usually measured in units of millimeters to the sixth power per cubic meter ($\mathrm{mm}^6\ \mathrm{m}^{-3}$).

The power returned to a radar is measured in decibels:

$$\mathrm{dB} \equiv 10 \log_{10}\left(\frac{P_{\mathrm{r}}}{P_{\mathrm{ref}}}\right). \tag{9.27}$$

The reference power P_{ref} is often taken to be that power which would be returned if each cubic meter of atmosphere contained one drop with a diameter of 1 mm ($Z = 1\ \mathrm{mm}^6\ \mathrm{m}^{-3}$). When referenced in this way, the returned power is indicated by the symbol dBZ.

If the droplets are not in the Rayleigh region, the Mie scattering equations can be used to calculate the power returned to the radar. In this case, an *effective radar reflectivity factor* Z_{e} is calculated.

If one has an estimate of the drop size distribution and knowledge of the velocity at which drops of a particular diameter fall, one can use the radar reflectivity factor to calculate the precipitation rate. These relationships often take the form

$$Z = aR^{\mathrm{b}}, \tag{9.28}$$

which is known as a *Z–R relationship*. The rainfall rate R is measured in units of millimeters per hour ($\mathrm{mm\ h}^{-1}$). Z–R relationships are somewhat variable; Battan (1973) states that fairly typical relationships are as follows: for stratiform rain, $Z = 200R^{1.6}$; for orographic rain, $Z = 31R^{1.71}$; for thunderstorm rain, $Z = 486R^{1.37}$. Snowflakes reflect microwave radiation much less well than do rain drops because $|K|^2$ is much less for ice than for water. Snowflakes are also not spherical, and both the fall speed and the "flake" size distribution is poorly known. Empirical observations, however, indicate that an appropriate Z–R relationship for snow may be $Z = 2000R^2$ (Battan, 1973).

Radiation from a radar antenna is polarized. Most weather radars transmit (and receive) linearly polarized radiation. On radars equipped to do so, it is possible to measure reflectivities with different polarizations. Polarizations studied include horizontal linear, vertical linear, and circular. In addition, one can transmit with one polarization and receive with a different polarization. Measurements of reflectivity at different polarizations can give information on drop shape, which is related to rainfall rate.

Although nonattenuating wavelengths are chosen for most ground-based weather radars, measurements of attenuation can provide information on rain rate. At attenuating wavelengths (those less than 10 cm), the returned power is reduced by a factor

$$\exp\left[-2\int(\sigma_{\mathrm{a}} + \sigma_{\mathrm{s}})dr\right].$$

If measurements at a single wavelength are used, the extinction coefficient ($\sigma_e = \sigma_s + \sigma_a$) can be derived from Z with the aid of an assumed drop size distribution. If measurements are made at two wavelengths, both the radar reflectivity factor and the extinction coefficient can be retrieved. Rain rate can be estimated from σ_e as well as from Z; R is nearly proportional to σ_e.

Horizontal motions of drops in the sampled volume cause frequency shifts in the returned signal due to the Doppler effect. There are two types of frequency shifts. Random motion of the drops causes a spread, but no shift, in the spectrum of returned radiation. Systematic frequency shifts are caused by nonrandom translations of the drops. If the frequency of the returned radiation is measured with a "Doppler radar," the speed of translation of the drops in the radial direction (toward or away from the radar) can be calculated:

$$V \equiv \frac{dr}{dt} = \frac{1}{2}\lambda\Delta\nu, \tag{9.29}$$

where $\Delta\nu$ is the frequency shift. If the wind speed is 50 m s^{-1}, the frequency shift on a 10-cm (3-GHz) radar is only 1000 Hz. Yet even these small shifts can be accurately measured to yield radial drop speed and, thus, an estimate of the radial wind speed. NEXRAD (next-generation radar), which is replacing the WSR–57 radars, is a Doppler radar.

Finally, the phase of the returned radiation can be measured. By comparing the phases of horizontally and vertically polarized measurements, information on drop shape can be obtained. In particular, since large drops associated with higher rainfall rates tend to flatten as they fall, the difference in phases between the two polarizations can be used to estimate rain rate.[10]

9.3.2 Practical Problems

Designing a precipitation-sensing radar to be used on a satellite platform is not a simple undertaking. Dennis (1963) and Bandeen and Katz (1975) discuss the problems. Some of the major ones are the following:

- Radar is an active system, which means that the power consumption is large. A WSR–57 has a transmitted power of 450 kW, a pulse duration of 4 μs, and a pulse repetition frequency of 164 Hz. This means that for the transmitter alone, an average power of 295 W is transmitted. Typical satellite instruments consume an order of magnitude less power.
- Satellite-borne instruments in low Earth orbit move rapidly, roughly 7 km s^{-1}. Accurate radar measurements usually require that several observations ($\sim$10) of the each volume be averaged to increase the signal-to-noise (S/N) ratio. This is difficult from a rapidly moving satellite, but it can be dealt with by increasing the pulse repetition frequency. Such an increase, however, compounds the next problem.

[10] R. J. Doviak and D. S. Zrnic, National Severe Storms Laboratory, Norman, Oklahoma, personal communication, 1989.

- The rapid motion of satellites means that the radar antenna may not be looking in the right direction when the signal returns from the target precipitation.
- Narrow beam widths are required for two reasons. First, the long distance from the target requires narrow beam widths to achieve acceptable ground resolution. Second, unlike ground-based radar in which the beam rises gently in the atmosphere, every beam from a satellite-borne radar intercepts the surface. Narrow beam widths are required to reduce surface clutter for range bins near the surface (Fig. 9.23). A narrow beam can be accomplished only by means of a large antenna or operating at higher (attenuating) frequencies or both.
- A low-orbiting satellite will see most points on Earth a maximum of twice per day. Since precipitation changes on a time scale considerably shorter than 12 h, satellite radar observations are limited either to instantaneous or to climatological time scales.

9.3.3 A Planned Satellite-Borne Radar

The Tropical Rainfall Measuring Mission (TRMM), planned to fly in the late-1990s, will carry the first space-borne, precipitation-sensing radar. It has been designed to solve some of the above problems. TRMM is described in Chapter 11.

9.4 SEVERE THUNDERSTORMS

An important application of satellite data is in the detection and forecasting of severe weather. Severe weather takes many forms, but in this section we discuss

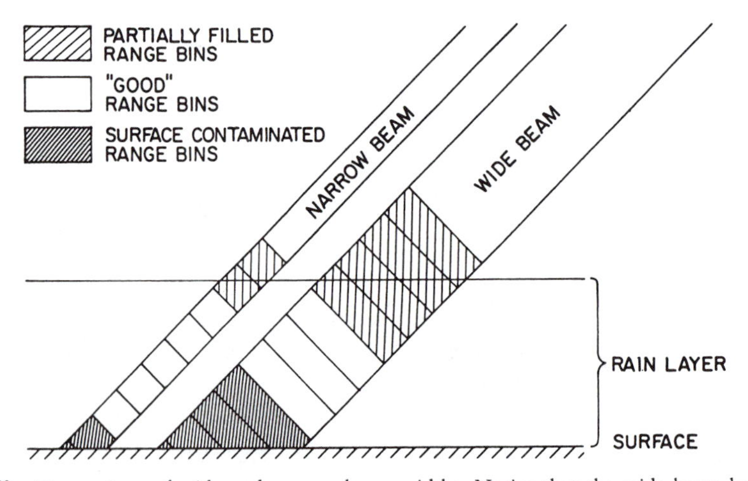

FIGURE 9.23. Comparison of wide and narrow beam widths. Notice that the wide beam has fewer "good" range bins and more partially filled and surface-contaminated range bins than does the narrow beam.

only severe weather produced by thunderstorms because the techniques involved are closely related to precipitation estimation. Maddox and Vonder Haar (1988) provide a summary of this work.

Much of the work done in severe thunderstorm detection and forecasting falls in the realm of image interpretation; satellite images or loops of these images are examined qualitatively for indications of severe weather. This approach to severe weather is discussed in Chapter 5. In this section, we concentrate on objective techniques that use digital data.

9.4.1 Severe Storm Observations

How can satellite data be used to distinguish ordinary thunderstorms from severe thunderstorms? This question is very important from a public safety point of view.[11] Severe thunderstorms are known to have stronger updrafts, to be taller, and to process more water vapor than less intense storms. Most of the work on severe thunderstorm detection is based on these properties.

The large vertical mass flux in a severe thunderstorm causes large divergence in the outflow (anvil) region, which is related to the areal growth rate of the anvil:

$$\frac{1}{A}\frac{dA}{dt} = \nabla \cdot V, \tag{9.30}$$

where $\nabla \cdot V$ is the divergence of the wind velocity. The areal growth rate is approximated by counting the number of pixels, N, colder than a temperature threshold:

$$\frac{1}{A}\frac{dA}{dt} \approx \frac{1}{N}\frac{dN}{dt}. \tag{9.31}$$

In a modification of the work of Sikdar *et al.* (1970) and Sikdar (1972), Adler and Fenn (1979a) chose not to examine the entire anvil area, which often contains many thunderstorms in various stages of development or decay. Rather, they used very cold GOES brightness temperature thresholds ($\leq$226 K) that apply to storm tops which penetrate the cirrus anvil. In fact, they chose the temperature for which $6 \leq N \leq 30$. This area is large enough to overcome some of the sensor resolution and response problems, but small enough to represent a single storm. Figure 9.24 shows a thunderstorm growth rate diagram for a storm that occurred in Nebraska and Iowa on 6 May 1975. The ordinate is the logarithm of the number of pixels colder than each temperature threshold; the slope of each line is $(1/N)dN/dt$. Severe weather reports tend to occur during or just after periods of rapid growth. Adler and Fenn found that on this day a $(1/N)dN/dt$ threshold of $3.5 \times 10^{-3}\,\text{s}^{-1}$, separated severe from nonsevere thunderstorms with a probability of detection (POD) of 0.67 and a false-alarm ratio (FAR) of 0.41.

[11] An equally important question from this viewpoint is, How can satellite data be used to detect flash flooding? This question is answered by the precipitation-estimation techniques discussed above, some of which were designed in large part to detect flash floods.

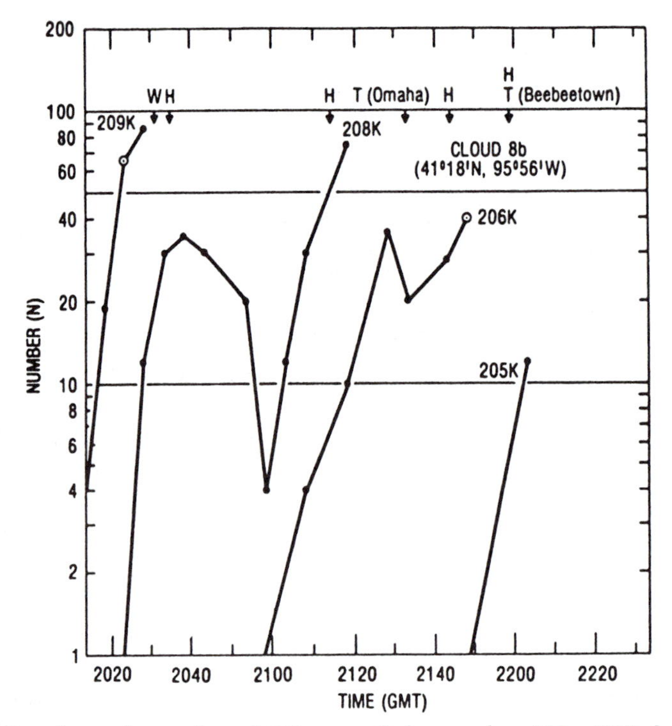

FIGURE 9.24. Growth rate diagram for a thunderstorm that occurred on 6 May 1975. Severe weather reports are indicated at the top: W = high winds, H = hail, T = tornado. [After Adler and Fenn (1979a).]

Colder storms are taller and thus more likely to be severe. Adler and Fenn (1979a) also examined the minimum brightness temperature achieved by a storm (T_{min}). They found that on their case study day a T_{min} threshold of 212.8 K separated severe from nonsevere storms with a POD of 0.93 and a FAR of 0.50. Combining T_{min} and $(1/N)dN/dt$ in a linear discriminant analysis, they found that severe storms could be separated from nonsevere storms with a POD of 0.73 and a FAR of 0.31. In related work, Adler and Fenn (1981) examined 11 tornadic storms and found that all were associated with rapid decreases in T_{min} (Fig. 9.25). In eight cases, the rapid T_{min} decrease preceded tornado touchdown (by 30–45 min). In the other three cases (Fig. 9.25f—h), the tornado occurred during the decrease. In three observations (Fig. 9.25i—k), Doppler radar showed that the rapid T_{min} decrease was associated with the formation of a mesocyclone.

Lead time, or the time between classification of the storm as severe and the first report of the occurrence of severe weather, is important for severe weather warnings. Adler and Fenn (1979a) found that the median lead time was 24 min using the T_{min} threshold, 27.5 min using the $(1/N)dN/dt$ threshold, and 7 min using the combined threshold. Thus these indices appear to be useful for severe weather detection and warnings *if rapid-scan (~5 min interval) satellite data are available.*

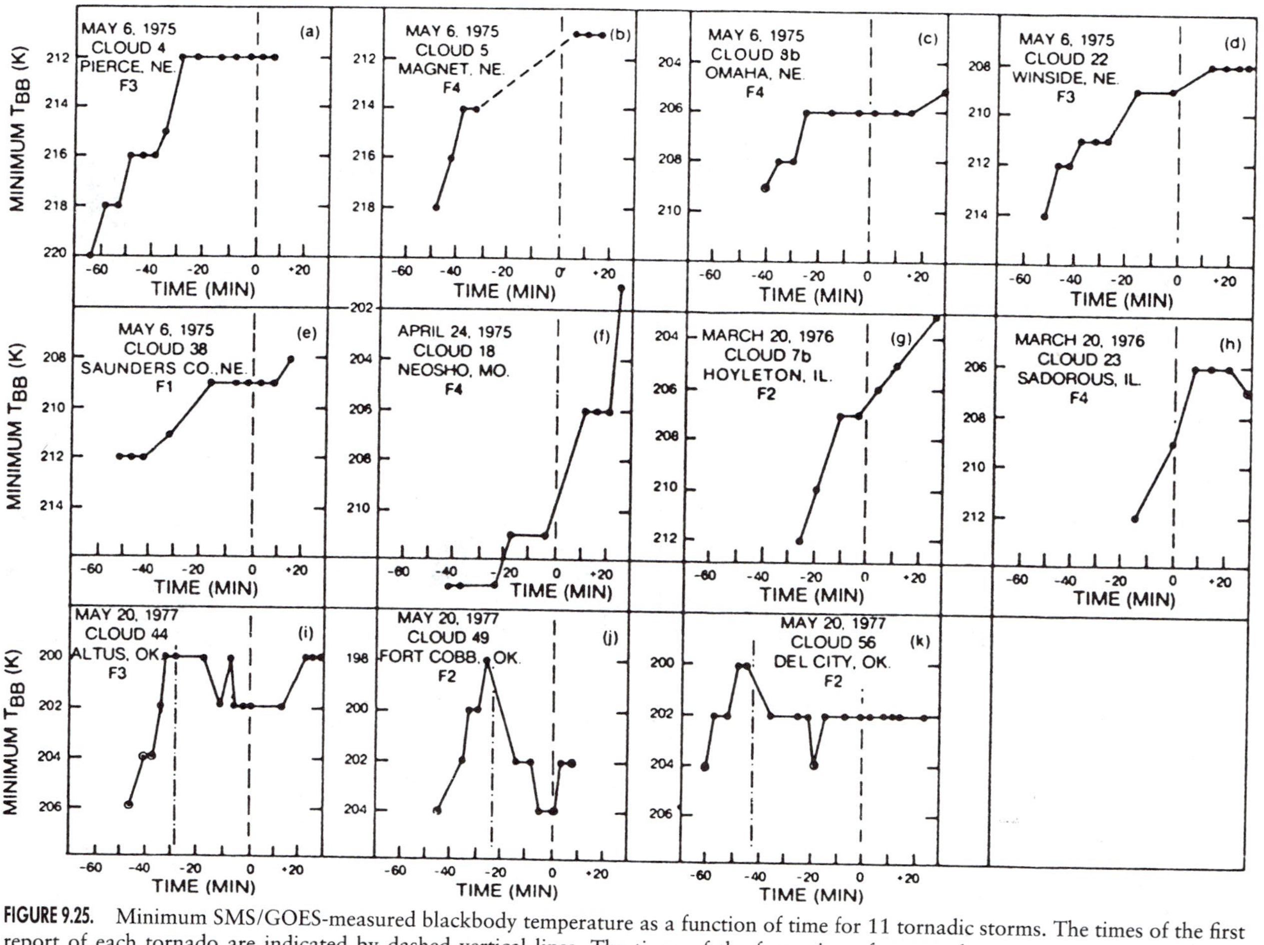

FIGURE 9.25. Minimum SMS/GOES-measured blackbody temperature as a function of time for 11 tornadic storms. The times of the first report of each tornado are indicated by dashed vertical lines. The times of the formation of mesocyclones in the three cases where Doppler radar observations were available (i, j, k) are indicated by dash–dot vertical lines. [After Adler and Fenn (1981).]

A variation on the T_{min} method was introduced by Reynolds (1980). He suggested that the minimum brightness temperature be compared with the tropopause temperature as a way to normalize storms occurring in different environments. Adler *et al.* (1985) found that indeed $T_{min}-T_{trop}$ is a good indicator of severe thunderstorms. However, they also found that the modal brightness temperature of the cirrus anvil, which is obtained from the satellite data rather from rawinsonde data, is a slightly better indicator of severity. Using 5 K as a threshold (T_{min} at least 5 K colder than T_{mode}) Adler *et al.* found a POD of 0.79 and a FAR of 0.37.

The ascent rate of the top of a thunderstorm is an indicator of the updraft strength. Again using rapid-scan GOES IR data, Adler and Fenn (1979b) calculated the time rate of change of equivalent blackbody temperature (dT_{BB}/dt). They estimated the vertical velocity of the cloud top using an environmental lapse rate:

$$w = \frac{\dfrac{dT_{BB}}{dt}}{\left(\dfrac{\partial T}{\partial z}\right)_{env}}. \tag{9.32}$$

In a small sample of storms Adler and Fenn found that when w exceeded 4 m s^{-1}, the storm was likely to be severe (POD = 0.67, FAR = 0.14).

Adler *et al.* (1985) combined $(1/N)dN/dt$ and dT^{BB}/dt into a Thunderstorm Index (TI). Using a threshold TI to estimate severity, they found that severe storms could be discriminated from nonsevere storms with a POD of 0.84 and a FAR of 0.39. The median lead time was 14 min for hail or tornadoes, and 25 min for tornadoes only. Severe thunderstorms can display a cold V pattern (Negri, 1982; McCann, 1983; Heymsfield and Blackmer, 1988). Adler *et al.* found that when the V pattern occurs, severe weather is likely to be imminent, but many severe storms do not display the pattern.

The GOES IR sensor is not ideally suited for severe thunderstorm detection. The brightness temperature of cloud elements smaller than the approximately 7-km-diameter scan spots cannot be adequately measured; for cold tops, too warm a temperature will be measured. A second problem is the response time of the GOES IR sensor (see Section 5.4.7.6), which causes small cloud elements to be shifted down scan and to appear warmer than they are. Warm biases cause the heights of the clouds to be underestimated. A solution to this problem is to make stereoscopic height measurements using the 1-km-resolution GOES visible data.

Mack *et al.* (1983) used 3-min-interval stereoscopic cloud height measurements to study thunderstorms. As expected, the stereoscopic heights were higher than the IR heights by 1.4–6.8 km, with the larger value occurring early in the cloud's life, and the smaller value occurring later. Interestingly, stereo ascent rates compared well with IR ascent rates. Mack *et al.* estimate that the stereoscopic ascent rates were accurate within about 1.3 m s^{-1}. Stereoscopic measurements would be extremely valuable in severe thunderstorm monitoring. However, since stereo requires the synchronization of two satellites, it is rarely available.

9.4.2 Severe Storm Forecasts

In addition to monitoring severe storms once they form, it is important to monitor the prestorm environment for signs that severe weather might develop. Again, much of this work falls in the realm of image interpretation.

Peslen (1980) used rapid-scan SMS/GOES data on 6 May 1975 to track low-level clouds in a prestorm situation. She found that the severe storms appeared in an area of convergence ahead of a dry line. The maximum convergence was between 10^{-5} and 10^{-4} s^{-1}. Negri and Vonder Haar (1980) also traced low-level clouds to yield winds, but they added surface mixing ratio observations. They calculated the low-level moisture convergence in the prestorm environment and found that centers of moisture convergence generally indicated the areas where severe storms later formed. The maximum moisture convergence was 2.2×10^{-3} g kg^{-1} s^{-1}.

The high resolution of sounding instruments is rarely exploited. Hillger and Vonder Haar (1981) have retrieved temperature, moisture, and stability parameters from Nimbus 6 HIRS/1 data at full resolution (~30 km). They found that convection tended to form in areas detectable with the satellite data but not with coarse-resolution conventional data. Similarly, Chesters *et al.* (1983) found that VAS "split window" channels could retrieve mesoscale, low-level water vapor fields, which are useful for convection forecasting.

Many examples could be given of mesoscale products derived from satellite data. We believe, however, that the full usefulness of satellite data in severe storm forecasting will become apparent only when satellite observations are used to initialize mesoscale numerical weather prediction models.

Bibliography

Adler, R. F., and E. B. Rodgers (1977). Satellite-observed latent heat release in a tropical cyclone. *Mon. Wea. Rev.*, **105**, 956–963.

Adler, R. F., and D. D. Fenn (1979a). Thunderstorm intensity as determined from satellite data. *J. Appl. Meteor.*, **18**, 502–517.

Adler, R. F., and D. D. Fenn (1979b). Thunderstorm vertical velocities estimated from satellite data. *J. Atmos. Sci.*, **36**, 1747–1754.

Adler, R. F., and D. D. Fenn (1981). Satellite-observed cloud-top height changes in tornadic thunderstorms. *J. Appl. Meteor.*, **20**, 1369–1375.

Adler, R. F., and R. A. Mack (1984). Thunderstorm cloud height-rainfall rate relations for use with satellite rainfall estimation techniques. *J. Climate Appl. Meteor.*, **23**, 280–296.

Adler, R. F., M. J. Markus, and D. D. Fenn (1985). Detection of severe Midwest thunderstorms using geosynchronous satellite data. *Mon. Wea. Rev.*, **113**, 769–781.

Adler, R. F., and A. J. Negri (1988). A satellite infrared technique to estimate tropical convective and stratiform rainfall. *J. Appl. Meteor.*, **27**, 30–51.

Adler, R. F., H.-Y. M. Yeh, N. Prasad, W.-K. Tao, and J. Simpson (1991). Microwave simulations of a tropical rainfall system with a three-dimensional cloud model. *J. Appl. Meteor.*, **30**, 924–953.

Allison, L., E. Rodgers, T. Wilheit, and R. Fett (1974). Tropical cyclone rainfall as measured by the Nimbus 5 electrically scanning microwave radiometer. *Bull. Am. Meteor. Soc.*, **55**, 1074–1089.

Arkin, P. A. (1979). The relationship between fractional coverage of high cloud and rainfall accumulations during GATE over the B-scale array. *Mon. Wea. Rev.*, **107**, 1382–1387.

Arkin, P. A., and B. Meisner (1987). The relationship between large-scale convective rainfall and cold cloud over the Western Hemisphere during 1982–1984. *Mon. Wea. Rev.*, **115**, 51–74.

Atlas, D. (1982). Adaptively pointing spaceborne radar for precipitation measurements. *J. Appl. Meteor.*, **21**, 429–431.

Augustine, J. A., C. G. Griffith, W. L. Woodley, and J. G. Meitín (1981). Insights into errors of SMS-inferred GATE convective rainfall. *J. Appl. Meteor.*, **20**, 509–520.

Bandeen, W. R., and I. Katz (1975). Active microwave sensing of the atmosphere. In R. E. Matthews (ed.), *Active Microwave Workshop Report*, NASA SP-376, Washington, DC, pp. 287–367.

Barrett, E. C. (1970). The estimation of monthly rainfall from satellite data. *Mon. Wea. Rev.*, **98**, 322–327.

Barrett, E. C. (1973). Forecasting daily rainfall from satellite data. *Mon. Wea. Rev.*, **101**, 215–222.

Barrett, E. C. (1980). The use of satellite imagery in operational rainfall monitoring in developing countries. In V. V. Salomonson and P. D. Bhavsar (eds.), *The Contribution of Space Observations to Water Resource Management*. Pergamon Press, Oxford and New York, pp. 163–178.

Barrett E. C., and D. W. Martin (1981). *The Use of Satellite Data in Rainfall Monitoring*. Academic Press, New York, 340 pp.

Barrett, E. C. (1988). Hierarichal approaches to operational satellite rainfall monitoring. *International Workshop on Satellite Techniques for Estimating Precipitation*, USAID, Washington, DC.

Battan, L. J. (1973). *Radar Observation of the Atmosphere*. Univ. Chicago Press, Chicago, 324 pp.

Bellon, A., and G. L. Austin (1986). On the relative accuracy of satellite and rain-gauge rainfall measurements over middle latitudes during daylight hours. *J. Climate Appl. Meteor.*, **25**, 1712–1724.

Bellon, A., S. Lovejoy, and G. L. Austin (1980). Combining satellite and radar data for the short range forecasting of precipitation. *Mon. Wea. Rev.*, **108**, 1554–1566.

Biondini, R. (1976). Cloud motion and rainfall statistics. *J. Appl. Meteor.*, **15**, 1100–1111.

Blackmer, R. H., Jr., and S. M. Serebreny (1968). Analysis of maritime precipitation using radar data and satellite cloud photographs. *J. Appl. Meteor.*, **7**, 122–131.

Bristor, C. L., and M. A. Ruzecki (1960). TIROS 1 photographs of the Midwest storm of April 1, 1960. *Mon. Wea. Rev.*, **88**, 315–326.

Browning, K. A. (1979). The FRONTIERS plan: A strategy for using radar and satellite imagery for very-short-range precipitation forecasting. *Meteorol. Mag.*, **108**, 161–184.

Cherna, E., A. Bellon, G. L. Austin, and A. Kilambi (1985). An objective technique for the delineation and extrapolation of thunderstorms from GOES data. *J. Geophys. Res.*, **90**, 6203–6210.

Chesters, D., L. W. Uccellini, and W. D. Robinson (1983). Low-level water vapor fields from the VISSR Atmospheric Sounder (VAS) "split window" channels. *J. Climate Appl. Meteor.*, **22**, 725–743.

Del Beato, R., and S. L. Barrell (1985). Rain estimation in extratropical cyclones using GMS imagery. *Mon. Wea. Rev.*, **113**, 747–755.

Dennis, A. S. (1963). *Rainfall Determination by Meteorological Satellite Radar*. Final Report to NASA, Stanford Research Institute, Menlo Park, CA, 105 pp.

Dittberner, G. J., and T. H. Vonder Haar (1973). Large scale precipitation estimates using satellite data; application to the Indian Monsoon. *Arch. Met. Geoph. Biokl.* Ser. B, **21**, 317–334.

Doneaud, A. A., J. R. Miller, Jr., L. R. Johnson, T. H. Vonder Haar, and P. Laybe (1987). The area–time integral technique to estimate convective rain volumes over areas applied to satellite data–a preliminary investigation. *J. Climate Appl. Meteor.*, **26**, 156–169.

Doneaud, A. A., S. Ionescu-Niscov, D. L. Priegnitz, and P. L. Smith (1984). The area–time integral as an indicator for convective rain volumes. *J. Climate Appl. Meteor.*, **23**, 555–561.

Doviak, R. J., and D. S. Zrnic (1984). *Doppler Radar and Weather Observations*. Academic Press, Orlando.

Follansbee, W. A. (1973). *Estimation of Average Daily Rainfall from Satellite Cloud Photographs*. NOAA Tech. Memo. NESS 44, Washington, DC, 39 pp.

Follansbee, W. A., and V. J. Oliver (1975). *A Comparison of Infrared Imagery and Video Pictures in the Estimation of Daily Rainfall from Satellite Data*. NOAA Tech. Memo. NESS 62, Washington, DC, 14 pp.

Follansbee, W. A. (1976). *Estimation of Daily Precipitation over China and the USSR Using Satellite Imagery*. NOAA Tech. Memo. NESS 81, Washington, DC, 30 pp.

Fujita, M., K. Okamoto, H. Masuko, and N. Fugono (1985a). Quantitative measurements of path-integrated rain rate by an airborne microwave radiometer over the ocean. *J. Atmos. Ocean. Tech.*, **2**, 285–292.

Fujita, M., K. Okamoto, S. Yoshikado, and K. Nakamura (1985b). Inference of rain rate profile and path-integrated rain rate by an airborne microwave rain scatterometer. *Radio Sci.*, **20**, 631–642.

Garcia, O. (1981). A comparison of two satellite rainfall estimates for GATE. *J. Appl. Meteor.*, **20**, 430–438.

Griffith, C. G., and W. L. Woodley (1973). On the variation with height of the top brightness of precipitating convective clouds. *J. Appl. Meteor.*, **12**, 1086–1089.

Griffith, C. G., W. L. Woodley, S. Browner, J. Teijeiro, M. Maier, D. W. Martin, J. Stout, and D. N. Sikdar (1976). *Rainfall Estimation from Geosynchronous Satellite Imagery during Daylight Hours*. NOAA Tech. Rep. ERL 356-WMPO 7, Boulder, CO, 106 pp.

Griffith, C. G., W. L. Woodley, P. G. Grube, D. W. Martin, J. Stout, and D. N. Sikdar (1978). Rain estimation from geosynchronous satellite imagery–visible and infrared studies. *Mon. Wea. Rev.*, **106**, 1153–1171.

Griffith, C. G., W. L. Woodley, J. S. Griffin, and S. C. Stromatt (1980). *Satellite-Derived Precipitation Atlas for the GARP Atlantic Tropical Experiment*. Division of Public Documents, U.S. Government Printing Office, Washington, DC, 284 pp.

Griffith, C. G., J. A. Augustine, and W. L. Woodley (1981). Satellite rain estimation in the U.S. High Plains. *J. Appl. Meteor.*, **20**, 53–56.

Griffith, C. G. (1987a). *The Estimation from Satellite Imagery of Summertime Rainfall over Varied Space and Time Scales*. NOAA Tech. Memo. ERL ESG-25, Boulder, CO, 102 pp.

Griffith, C. G. (1987b). Comparisons of gauge and satellite rain estimates for the central United States during August 1979. *J. Geophys. Res.*, **92**, 9551–9566.

Grody, N. C. (1983). Severe storm observations using the Microwave Sounding Unit. *J. Climate Appl. Meteor.*, **22**, 609–625.

Gruber, A. (1973a). *An Examination of Tropical Cloud Clusters Using Simultaneously Observed Brightness and High Resolution Infrared Data from Satellites*. NOAA Tech. Memo. NESS 50, Washington, DC, 22 pp.

Gruber, A. (1973b). Estimating rainfall in regions of active convection. *J. Appl. Meteor.*, **12**, 110–118.

Hakkarinen, I. M., and R. F. Adler (1988). Observations of convective precipitation at 92 and 183 GHz: Aircraft results. *Meteor. Atmos. Phys.*, 164–182.

Heymsfield, G. M., and R. H. Blackmer, Jr. (1988). Satellite-observed characteristics of Midwest severe thunderstorm anvils. *Mon. Wea. Rev.*, **116**, 2200–2224.

Hillger, D. W., and T. H. Vonder Haar (1981). Retrieval and use of high-resolution moisture and stability fields from Nimbus 6 HIRS radiances in pre-convective situations. *Mon. Wea. Rev.*, **109**, 1788–1806.

Huff, F. A. (1967). Rainfall gradients in warm season rainfall. *J. Appl. Meteor.*, **6**, 435–437.

Jones, A. S. (1989). *Microwave Remote Sensing of Cloud Liquid Water and Surface Emittance over Land Regions*. M.S. thesis, Dept. Atmospheric Science, Colorado State University, Ft. Collins, 145 pp.

Kidder, S. Q., and T. H. Vonder Haar (1977). Seasonal oceanic precipitation frequencies from Nimbus 5 microwave data. *J. Geophys. Res.*, **82**, 2083–2086.

Kidder, S. Q. (1979). *Determination of Tropical Cyclone Surface Pressure and Winds from Satellite Microwave Data*. Atmospheric Science Paper No. 307, Colorado State University, Fort Collins, 87 pp.

Kilonsky, B. J., and C. S. Ramage (1976). A technique for estimating tropical open-ocean rainfall from satellite observations. *J. Appl. Meteor.*, **15**, 972–975.

King, P. (1984). Test results from RAINSAT. *Nowcasting II: Mesoscale Observations and Very-Short-Range Weather Forecasting*. Proceedings of 2nd International Symposium on Nowcasting, ESA SP-208, ESA Scientific & Technical Publications Branch, Noordwijk, Netherlands, pp. 139–144.

Kuo, H. L. (1965). On formation and intensification of tropical cyclones through latent heat release by cumulus convection. *J. Atmos. Sci.*, **22**, 40–63.

Lethbridge, M. (1967). Precipitation probability and satellite radiation data. *Mon. Wea. Rev.*, **95**, 487–490.

Lopez, R. E., J. Thomas, D. O. Blanchard, and R. L. Holle (1983). Estimation of rainfall over an extended region using only measurements of the area covered by radar echoes. *Preprints: 21st Conf. Radar Meteor.*, American Meteorological Society, Boston, pp. 681–686.

Lovejoy, S., and G. L. Austin (1979a). The delineation of rain areas from visible and IR satellite data for GATE and mid-latitudes. *Atmosphere-Ocean*, **17**, 77–92.

Lovejoy, S., and G. L. Austin (1979b). The sources of error in rain amount estimating schemes for GOES visible and IR satellite data. *Mon. Wea. Rev.*, **107**, 1048–1054.

Lovejoy, S., and G. L. Austin (1980). The estimation of rain from satellite-borne microwave radiometers. *Q. J. Roy. Meteor. Soc.*, **106**, 255–276.

Mack, R. A., A. F. Hasler, and R. F. Adler (1983). Thunderstorm cloud top observations using satellite stereoscopy. *Mon. Wea. Rev.*, **111**, 1949–1964.

Maddox, R. A., and T. H. Vonder Haar (1988). The use of satellite observations. Chapter 12 of E. Kessler (ed.), *Instruments and Techniques for Thunderstorm Observations and Analysis*. Univ. Oklahoma Press, Norman, 191–214.

Martin, D. W., and M. R. Howland (1986). Grid history: A geostationary satellite technique for estimating daily rainfall in the tropics. *J. Climate Appl. Meteor.*, **25**, 184–195.

McCann, D. W. (1983). The enhanced-V: A satellite observable severe storm signature. *Mon. Wea. Rev.*, **111**, 887–894.

Meneghini, R., J. Eckerman, and D. Atlas (1983). Determination of rain rate from a spaceborne radar using measurements of total attenuation. *IEEE Trans Geosci. Remote Sens.* **GE-21**, 34–43.

Meneghini, R., and D. Atlas (1986). Simultaneous ocean cross section and rainfall measurements from space with a nadir-looking radar. *J. Atmos. Ocean. Tech.*, **3**, 400–413.

Moses, J. F., and E. C. Barrett (1986). Interactive procedures for estimating precipitation from satellite imagery. *Proceedings of the Workshop on Hydrologic Applications of Space Technology* held in Cocoa Beach, Florida, August 1985. IAHS Publ. No. 160, pp. 25–39.

Mugnai, A., and E. A. Smith (1984). Passive microwave radiation transfer in an evolving cloud medium. In G. Fiocco (ed.), *IRS '84: Current Problems in Atmospheric Radiation*, A. Deepak Publishing, Hampton, VA, pp. 297–300.

Mugnai, A., and E. A. Smith (1988). Radiative transfer to space through a precipitating cloud at multiple microwave frequencies. Part I: Model description. *J. Appl. Meteor.*, **27**, 1055–1073.

Mugnai, A., H. J. Cooper, E. A. Smith, and G. J. Tripoli (1990). Simulation of microwave brightness temperatures of an evolving hailstorm at SSM/I frequencies. *Bull. Am. Meteor. Soc.*, **71**, 2–13.

Nagle, R. E., and S. M. Serebreny (1962). Radar precipitation echo and satellite cloud observations of a maritime cyclone. *J. Appl. Meteor.*, **1**, 279–295.

NASA (1988). *Preliminary Project Plan for the Tropical Rainfall Measuring Mission (TRMM). Phase A, Part I: Technical Plan.* NASA/Goddard Space Flight Center, Greenbelt, MD.

Negri, A. J., and T. H. Vonder Haar (1980). Moisture convergence using satellite-derived wind fields: A severe local storm case study. *Mon. Wea. Rev.*, **108**, 1170–1182.

Negri, A. J., and R. F. Adler (1981). Relation of satellite-based thunderstorm intensity to radar-estimated rainfall. *J. Appl. Meteor.*, **20**, 288–300.

Negri, A. J. (1982). Cloud-top structure of tornadic storms on 10 April 1979 from rapid scan and stereo satellite observations. *Bull. Am. Meteor. Soc.*, **63**, 1151–1159.

Negri, A. J., R. F. Adler, and P. J. Wetzel (1984). Rain estimation from satellites: An examination of the Griffith–Woodley Technique. *J. Climate Appl. Meteor.*, **23**, 102–116.

Negri, A. J., and R. F. Adler (1987a). Visible and infrared rain estimation. I: A grid cell approach. *J. Climate Appl. Meteor.*, 1553–1564.

Negri, A. J., and R. F. Adler (1987b). Visible and infrared rain estimation. II: A cloud definition approach. *J. Climate Appl. Meteor.*, 1565–1576.

Peslen, C. A. (1980). Short-interval SMS wind vector determinations for a severe local storms area. *Mon. Wea. Rev.*, **108**, 1407–1418.

Prabhakara, C., G. Dalu, G. L. Liberti, J. J. Nucciarone, and R. Suhasini (1992). Rainfall estimation over oceans from SMMR and SMM/I microwave data. *J. Appl. Meteor.*, **31**, 532–552.

Rao, M. S. V., and J. S. Theon (1977). New features of global climatology revealed by satellite-derived oceanic rainfall maps. *Bull. Am. Meteor. Soc.*, **58**, 1285–1288.

Rao, M. S. V., W. V. Abbott, and J. S. Theon (1976). *Satellite-Derived Global Oceanic Rainfall Atlas (1973 and 1974)*. NASA SP-410, Washington, DC, 31 pp. plus appendixes.

Reynolds, D. W., and E. A. Smith (1979). Detailed analysis of composited digital radar and satellite data. *Bull. Am. Meteor. Soc.*, **60**, 1024–1037.

Reynolds, D. W. (1980). Observations of damaging hailstorms from geosynchronous satellite digital data. *Mon. Wea. Rev.*, **108**, 337–348.

Richards, F., and P. Arkin (1981). On the relationship between satellite-observed cloud cover and precipitation. *Mon. Wea. Rev.*, **109**, 1081–1093.

Rodgers, E. B., H. Siddalingaiah, A. T. C. Chang, and T. T. Wilheit (1979). A statistical technique for determining rainfall over land employing Nimbus 6 ESMR measurements. *J. Appl. Meteor.*, **18**, 978–991.

Savage, R. C., and J. A. Weinman (1975). Preliminary calculations of the upwelling radiance from rainclouds at 37.0 and 19.35 GHz. *Bull. Am. Meteor. Soc.*, **56**, 1272–1274.

Savage, R. C. (1978). The radiative properties of hydrometeors at microwave frequencies. *J. Appl. Meteor.*, **17**, 904–911.

Scofield, R. A., and V. J. Oliver (1977). *A Scheme for Estimating Convective Rainfall from Satellite Imagery*. NOAA Tech. Memo. NESS 86, Washington, DC, 47 pp.

Scofield, R. A. and L. E. Spayd Jr. (1984). *A Technique That Uses Satellite, Radar, and Conventional Data for Analyzing and Short-Range Forecasting of Precipitation from Extratropical Cyclones*. NOAA Tech. Memo. NESDIS 8, Washington, DC, 51 pp.

Scofield, R. A. (1987). The NESDIS operational convective precipitation estimation technique. *Mon. Wea. Rev.*, **115**, 1773–1792.

Shifrin, K. S., and M. M. Chernyak (1968). Microwave absorption and scattering by precipitation. In *Transfer of Microwave Radiation in the Atmosphere*, Israel Program for Scientific Translations, pp. 69–78.

Sikdar, D. N., V. E. Suomi, and C. E. Anderson (1970). Convective transport of mass and energy in severe storms over the United States–an estimate from a geostationary altitude. *Tellus*, **22**, 521–532.

Sikdar, D. N. (1972). ATS–3 observed cloud brightness field related to a meso-to-subsynoptic scale rainfall pattern. *Tellus*, **24**, 400–413.

Simpson, J., and V. Wiggert (1969). Models of precipitating cumulus towers. *Mon. Wea. Rev.*, **97**, 471–489.

Simpson, J., R. F. Adler, and G. R. North (1988). A proposed tropical rainfall measuring mission (TRMM) satellite. *Bull. Am. Meteor. Soc.*, **69**, 278–295.

Smith, E. A., and A. Mugnai (1988). Radiative transfer to space through a precipitating cloud at multiple microwave frequencies. Part II: Results and analysis. *J. Appl. Meteor.*, **27**, 1074–1091.

Smith, E. A., and A. Mugnai (1989). Radiative transfer to space through a precipitating cloud at multiple microwave frequencies. Part III: Influence of large ice particles. *J. Meteor. Soc. Japan*, **67**, 739–755.

Smith, E. A., A. Mugnai, H. J. Cooper, G. J. Tripoli, and X. Xiang (1992). Foundations for statistical-physical precipitation retrieval from passive microwave satellite measurements. Part I: Brightness-temperature properties of a time-dependent cloud radiation model. *J. Appl. Meteor.*, **31**, 506–531.

Spayd, L. E., Jr. and R. A. Scofield (1984). *A Tropical Cyclone Precipitation Estimation Technique Using Geostationary Satellite Data*. NOAA Tech. Memo. NESDIS 5, Washington, DC, 36 pp.

Spencer, R. W., W. S. Olson, R. Wu, D. W. Martin, J. A. Weinman, and D. A. Santek (1983a).

Heavy thunderstorms observed by the Nimbus–7 scanning multi-channel microwave radiometer. *J. Climate Appl. Meteor.*, **22**, 1041–1046.

Spencer, R. W., B. B. Hinton, and W. S. Olson (1983b). Nimbus–7 37 GHz radiances correlated with radar rain rates over the Gulf of Mexico. *J. Climate Appl. Meteor.*, **22**, 2095–2099.

Spencer, R. W., D. W. Martin, B. Hinton, and J. A. Weinman (1983c). Satellite microwave radiances correlated with radar rain rates over land. *Nature*, **304**, 141–143.

Spencer, R. W. (1984). Satellite passive microwave rain rate measurement over croplands during spring, summer, and fall. *J. Climate Appl. Meteor.*, **23**, 1553–1562.

Spencer, R. W. (1986). A satellite passive 37-GHz scattering-based method for measuring oceanic rain rates. *J. Climate Appl. Meteor.*, **25**, 754–766.

Spencer, R. W., H. M. Goodman, and R. E. Hood (1989). Precipitation retrieval over land and ocean with the SSM/I: Identification and characteristics of the scattering signal. *J. Atmos. Ocean. Tech.*, **6**, 254–273.

Stout, J. E., D. W. Martin, and D. N. Sikdar (1979). Estimating GATE rainfall with geosynchronous satellite images. *Mon. Wea. Rev.*, **107**, 585–598.

Theon, J. S. (1973). A multispectral view of the Gulf of Mexico from Nimbus 5. *Bull. Am. Meteor. Soc.*, **54**, 934–937.

Tsang, L., J. A. Kong, E. Njoku, D. H. Staelin, and J. W. Waters (1977). Theory for microwave thermal emission from a layer of cloud or rain. *IEEE Trans. Antennas Propag.*, **AP-25**, 650–657.

Tsonis, A. A. (1984). On the separability of various classes from the GOES visible and infrared data. *J. Climate Appl. Meteor.*, **23**, 1393–1410.

Tsonis, A. A., and G. A. Isaac (1985). On a new approach for instantaneous rain area delineation in the midlatitudes using GOES data. *J. Climate Appl. Meteor.*, **24**, 1208–1218.

Vogel, J. L. (1981). Raingauge network sampling statistics. In D. Atlas and O. Thiele (eds.), *Precipitation Measurements from Space Workshop Report*, Goddard Space Flight Center, Greenbelt, MD.

Vonder Haar, T. H. (1969). Meteorological applications of reflected radiance measurements from ATS 1 and ATS 3. *J. Geophys. Res.*, **74**, 5404–5412.

Weinman, J. A., and P. J. Guetter (1977). Determination of rainfall distributions from microwave radiation measured by the NIMBUS 6 ESMR. *J. Appl. Meteor.*, **16**, 437–442.

Weinman, J. A., and R. Davies (1978). Thermal microwave radiances from horizontally finite clouds of hydrometeors. *J. Geophys. Res.*, **83**, 3099–3107.

Weiss, M., and E. A. Smith (1987). Precipitation discrimination from satellite infrared temperatures over the CCOPE mesonet region. *J. Climate Appl. Meteor.*, **26**, 687–697.

Wexler, H. (1954). Observing the weather from a satellite vehicle. *J. Br. Interplanetary Sci.*, **13**, 269–276.

Wexler, H. (1957). The satellite and meteorology. *J. Astronaut*, **4**, 1–6.

Wilheit, T. T., J. S. Theon, W. E. Shenk, L. Allison, and E. B. Rodgers (1976). Meteorological interpretations of the images from the Nimbus 5 electrically scanned microwave radiometer. *J. Appl. Meteor.*, **15**, 166–172.

Wilheit, T. T., A. T. C. Chang, M. S. V. Rao, E. B. Rodgers, and J. S. Theon (1977). A satellite technique for quantitatively mapping rainfall rates over the oceans. *J. Appl. Meteor.*, **16**, 551–560.

Wilheit, T. T., J. L. King, E. B. Rodgers, R. A. Nieman, B. M. Krupp, A. S. Milman, J. S. Stratigos, and H. Siddalingaiah (1982). Microwave radiometric observations near 19.35, 35, 92 and 183 GHz of precipitation in tropical storm Cora. *J. Appl. Meteor.*, **21**, 1137–1145.

Wilheit, T. T. (1986). Some comments on passive microwave measurement of rain. *Bull. Am. Meteor. Soc.*, **67**, 1226–1232.

Wilheit, T. T., A. T. C. Chang, and L. S. Chiu (1991). Retrieval of monthly rainfall indices from microwave radiometric measurements using probability distribution functions. *J. Atmos. Ocean. Tech.*, **8**, 118–136.

Woodley, W. L., and B. Sancho (1971). A first step toward rainfall estimation from satellite cloud photographs. *Weather*, 279–289.

Woodley, W. L., C. G. Griffith, J. S. Griffin, and S. C. Stromatt (1980). The inference of GATE convective rainfall from SMS–1 imagery. *J. Appl. Meteor.*, **19**, 388–408.

World Climate Research Programme (1986). *Global Large-Scale Precipitation Data Sets for the WCRP*. WCP-111, WMO/TD No. 94, Geneva, Switzerland.

Wu, R., and J. A. Weinman (1984). Microwave radiances from precipitating clouds containing aspherical ice, combined phase, and liquid hydrometeors. *J. Geophys. Res.*, **89**, 7170–7178.

Wu, R., J. A. Weinman, and R. T. Chin (1985). Determination of rainfall rates from GOES satellite images by a pattern recognition technique. *J. Atmos. Ocean. Tech.*, **2**, 314–330.

Wylie, D. P. (1979). An application of a geostationary satellite rain estimation technique to an extratropical area. *J. Appl. Meteor.*, **18**, 1640–1648.

10

Earth Radiation Budget

ABSORPTION OF SOLAR radiation and emission of terrestrial radiation drive the general circulation of the atmosphere and are largely responsible for the Earth's weather and climate. In this chapter we discuss satellite measurements of the constituent radiative quantities which comprise the radiation budget of Earth. The very first successful meteorological measurements from space were of the Earth's radiation budget. They were made by a Suomi radiometer from the Explorer 7 satellite launched 13 October 1959 (see Chapter 1). The long history of radiation budget measurements is reviewed by House *et al.* (1986). In this chapter we concentrate on recent satellite experiments. We begin by discussing measurements of the solar constant, then we proceed to the radiation budget at the top of the atmosphere and at the Earth's surface.

10.1 THE SOLAR CONSTANT

The solar energy reaching the Earth is traditionally quantified as the solar constant (see Section 3.7). To be precise, the solar constant is the annual average solar irradiance received outside the Earth's atmosphere (at the top of the atmosphere) on a surface normal to the incident radiation and at the Earth's mean distance from the sun. The actual solar irradiance varies by ±3.4% from the solar constant during the year due to the eccentricity of the Earth's orbit about the sun, but it varies as a result of other causes also. Ground-based measurements of the solar constant date back to 1913. The problem with ground-based measurements is that they must be corrected for the effects of the atmosphere. Space-based measurements require no such correction; they date from 1975.

Both the Nimbus 6 and 7 satellites carried an Earth Radiation Budget (ERB) experiment (Smith *et al.*, 1975; Jacobowitz *et al.*, 1978). Part of the instrument was designed to monitor the sun's output. Once each orbit, a 10-channel radiometer viewed the sun for approximately 3 min. The 10 channels were designed to measure the solar irradiance in several spectral intervals. On the Nimbus 7 ERB, channel 10 was replaced with channel 10C, an electrically self-calibrating cavity radiometer, which measures the total solar irradiance in the spectral range from less than 0.2 μm to greater than 50 μm (Hickey *et al.*, 1988a). The Nimbus 7 10C data began on 16 November 1978 and are still being collected as of this writing.

The Solar Maximum Mission satellite carried an active cavity radiometer [ACRIM; Willson (1979, 1984), Willson and Hudson (1988)]. Launched 16 February 1980, SMM made solar irradiance measurements until December 1989.

The Earth Radiation Budget Experiment (see Chapter 4) has flown on the Earth Radiation Budget Satellite (launched 5 October 1984) as well as the NOAA 9 and 10 satellites (launched 12 December 1984 and 17 September 1986, respectively).

Figure 10.1 shows the annually averaged Nimbus 7 ERB and Solar Max ACRIM solar constant measurements. Shown for comparison is the annually averaged sunspot number calculated at the Zurich Observatory. The solar constant seems to follow the sunspot cycle. The mean Nimbus 7 solar constant, measured over an 11-year solar cycle, is 1372.0 W m^{-2} with a standard deviation of 0.6 W m^{-2}. The mean Solar Max solar constant over a period shorter than a solar cycle is 1367.5 W m^{-2} with a standard deviation of 0.4 W m^{-2}. The ERBE measurements are approximately 1365 W m^{-2}.

The difference between the measures of the solar constant is attributable to the great difficulty in calibrating in the laboratory an instrument to measure the solar constant in space. The radiometric standard used for calibration is thought to be accurate to within about ±0.5% or about ±7 W m^{-2}. It must be noted, however, that the measurements by the instruments are highly correlated. Thus *changes* in the annually averaged solar constant can be determined quite accurately, perhaps within a tenth of a watt per square meter, even though the absolute magnitude can be determined only within a few watts per square meter.

The reader should note, however, that the above numbers must be characterized as preliminary for several reasons.

- Work continues on possible errors in the irradiance observations.
- While the year-to-year change in the annually averaged solar constant is small, on the order of 0.2 W m^{-2}, the annual range of daily observations

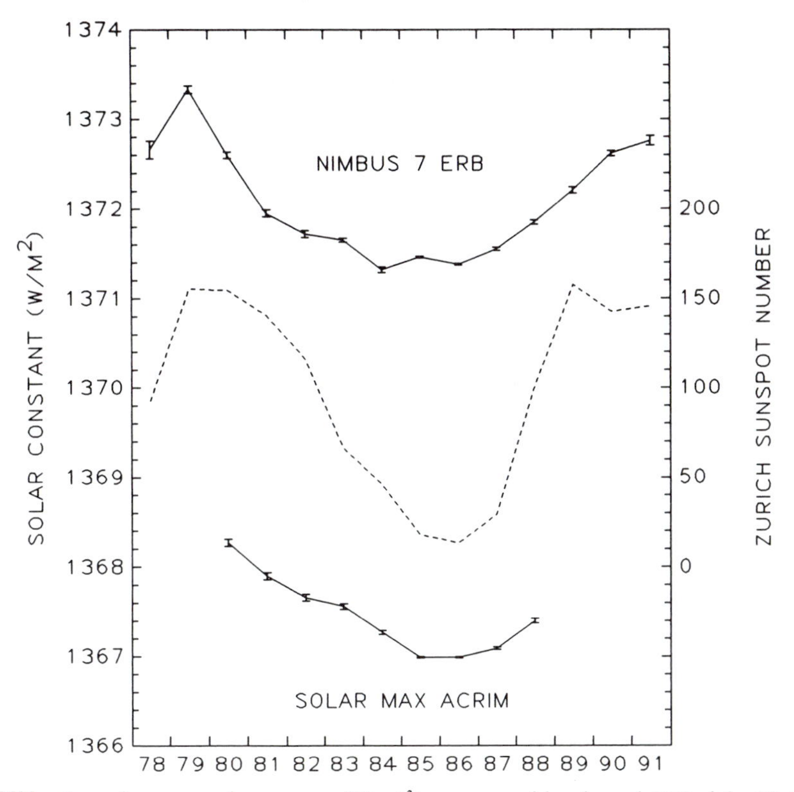

FIGURE 10.1. Annual average solar constant (W m^{-2}) as measured by channel 10C of the Nimbus 7 ERB and by the Solar Maximum Mission ACRIM. The dashed line is the annual average Zurich sunspot number. The ERB and ACRIM data are from Table 4 of Hoyt *et al.* (1992). The sunspot data are courtesy of the NOAA/NESDIS/National Geophysical Data Center, Boulder, Colorado.

is large, about 3 W m^{-2}. Plots of daily irradiance values appear quite noisy, although some of the fluctuation is correlated with solar storms.

- Observations over several solar sunspot cycles will be required to characterize the sun's behavior.

Continued observations of the sun are important both to resolve the differences between the various instruments and to determine the cause and effects of the variable solar output. Solar constant measurements continue to be made on Nimbus 7, the Earth Radiation Budget Satellite (ERBS), and NOAA 9 and 10. As of this writing, solar constant measurements are being made on the Upper Atmosphere Research Satellite (UARS), and they will probably be made on the Earth Observing System, scheduled to be launched in the late 1990s (see Chapter 11).

10.2 TOP OF THE ATMOSPHERE RADIATION BUDGET

The goal of radiation budget studies is to measure the incoming and outgoing radiation as a function of time and space. Incoming radiation is irradiance and is given the symbol $E(t,\Theta,\Psi,h)$, where t is time, Θ is latitude, Ψ is longitude, and h is height above the Earth's surface. Outgoing radiation is radiant exitance and is given the symbol $M(t,\Theta,\Psi,h)$. Usually we are not interested in the wavelength dependence of these quantities except for the division between shortwave and longwave components. Shortwave radiation, usually defined as radiation with wavelengths less than 5 μm, is primarily reflected solar radiation. Longwave radiation (> 5 μm) is primarily emitted terrestrial radiation.

Satellite methods to measure radiation budget quantities depend on the altitude in which one is interested. The two most common altitudes are the top of the atmosphere (TOA, often defined for radiation budget purposes as 30 km) and the surface. These two altitudes will be discussed separately.

At the top of the atmosphere, the quantities we want to measure are the shortwave radiant exitance (M_{SW}), and the longwave radiant exitance (M_{LW}), which is also called the *outgoing longwave radiation* (OLR). With these quantities we can calculate the albedo

$$A(t,\Theta,\Psi) = \frac{M_{SW}(t,\Theta,\Psi)}{E_{sun}\mu_{sun}}, \tag{10.1}$$

the absorbed solar radiation

$$E_{abs}(t,\Theta,\Psi) = (1 - A)E_{sun}\mu_{sun}, \tag{10.2}$$

and the net radiation

$$E_{net}(t,\Theta,\Psi) = (1 - A)E_{sun}\mu_{sun} - M_{LW} = E_{sun}\mu_{sun} - M_{SW} - M_{LW}. \tag{10.3}$$

In these equations, $\mu_{sun} = \mu_{sun}(t,\Theta,\Psi)$ is the cosine of the solar zenith angle, and $E_{sun} = E_{sun}(t)$ is the solar constant S_{sun} adjusted for distance from the sun,

$$E_{sun} = S_{sun}\left(\frac{\bar{d}_{sun}}{d_{sun}(t)}\right)^2, \tag{10.4}$$

where $d_{sun}(t)$ is the Earth–sun distance at time t, and $\bar{d}_{sun}$ is the mean Earth–sun distance (1 *astronomical unit*, AU).

Instruments used for radiation budget studies can be divided into two spectral categories. *Broadband sensors* attempt to measure spectrally integrated radiation budget quantities. *Narrowband sensors* measure only narrow portions of the spectrum; extrapolation to other portions of the spectrum is necessary for radiation budget calculations. The AVHRR on the NOAA satellites is a narrowband sensor (Fig. 4.5); the ERBE sensors are broadband (Fig. 4.19).

Radiation budget instruments may also be divided into two field-of-view (FOV) categories: *wide-field-of-view* (WFOV) sensors and *narrow-field-of-view* (NFOV) sensors. WFOV sensors measure radiation from horizon to horizon. They are often called *flat-plate sensors* because they typically consist of a flat element that measures the irradiance received at the satellite. They are said to have a *cosine response function* because in the equation that relates irradiance to radiance (Eq. 3.3), the incoming radiance is weighted by the cosine of the sensor zenith angle.

Since angular resolution for a sensor is stated in terms of the half-power angles, flat-plate sensors have a 120° angular resolution. The ERBE and Nimbus 7 ERB both have flat-plate sensors. An instrument which employs flat-plate sensors is also called a *nonscanner* because nadir-pointing, flat-plate sensors detect radiation from horizon to horizon without scanning.

NFOV sensors measure radiation in a narrow cone. AVHRR is a NFOV instrument (Fig. 4.6), and the scanning instruments on the Nimbus 6 and 7 ERB and on the ERBE are also NFOV sensors, but they are not as narrow as the AVHRR sensor. An instrument that employs a NFOV sensor is often called a *scanner* because to monitor the Earth disk, the sensor must scan.

Flat-plate sensors have few moving parts; thus they can operate for a decade or longer, which makes them useful for climate studies. NFOV scanners typically wear out in a few years. On the other hand, scanners make high-spatial-resolution measurements, which are useful for regional radiation budget studies. WFOV sensors make measurements which have been integrated over large areas, typically a circle of radius 2500–3500 km at the surface. WFOV measurements are useful for global or large-scale studies.

A polar-orbiting satellite is a very good platform for measuring the radiation budget. A satellite-mounted instrument makes two observations daily, separated by about 12 h, of every point on Earth. These are sufficient to calculate the monthly average radiation budget, which is usually the time scale of interest. There are some problems, however:

- The measurements are made at the satellite height, not at the top of the atmosphere (the *inverse problem*).
- The twice-daily observations may not adequately sample the diurnal variation, particularly of shortwave radiation. A diurnal model may be necessary (the *diurnal problem*).
- Radiation budget quantities which have been integrated over wavelength, shortwave or longwave, are necessary. Satellite sensors, however, do not measure exactly these quantities (the *spectral correction problem* or *unfiltering problem*).
- To measure the radiant exitance from a point, one must observe the point from all possible directions. Satellites, however, observe a single point from a single direction (the *angular dependence problem*).

In summary, to retrieve accurate radiation budget quantities, we must know the radiance in all directions, at all times, in all wavelengths, and at all heights. Satellites can observe a larger fraction of the necessary data than other methods, but not all. We must use modeling to fill the gaps in the measurements. The three radiation budget data sets described below differ in their handling of these problems.

10.2.1 The NOAA Data Set

Since 1974, with one gap, NOAA/NESDIS has derived radiation budget quantities from its polar-orbiting satellites. The data set has been constructed using data from several satellites and two different instruments: (1) the Very High Resolution Radiometer (VHRR), which flew on NOAA satellites prior to NOAA 6; and (2) the AVHRR, since the launch of TIROS N. Retrieval of VHRR-based radiation budget products between June 1974, and March 1978, is described by Gruber (1978). Retrieval of AVHRR-based radiation budget products since January 1979, is similar and is described by Gruber *et al.* (1983). Here we summarize the retrieval scheme of Gruber *et al.* (1983).

Afternoon satellites (see Table 4.2) are used so as to have the highest solar elevation angle and to avoid, to the extent possible, missing data at high latitudes in the winter hemisphere. The five-channel, 1-km-resolution AVHRR data are too voluminous to all be processed, especially since the final radiation budget products have roughly 200-km resolution. Two of the five channels are processed: channel 1 (0.58–0.68 μm) represents solar wavelengths, and channel 5 (11.5–12.5 μm) represents infrared wavelengths. Channel 5 was chosen because it is more sensitive to water vapor than channel 4; thus it is thought to better represent the outgoing longwave radiation. However, some satellites do not have channel 5; channel 4 (10.5–11.5 μm) is used for these satellites.

Next, to reduce data processing, *global area coverage* (GAC) data are used instead of the full-resolution *local area coverage* (LAC) data. The scheme to produce GAC data from the 1-km-resolution data is as follows. Along a scan line, four contiguous scan spots are averaged, and a fifth one is skipped. Two scan spots at each end of the 2048-element scan line are ignored. This results in a 409-element scan line. The next two scan lines are skipped entirely, and the fourth scan line is averaged as above. This results in a data set which at nadir has about 4-km resolution in the cross-track direction and 3.3-km resolution in the along-track direction. To further reduce processing, the GAC data are processed in 11×11 blocks. These blocks are called *GAC targets*. At nadir, they are about 36 km along-track by 44 km cross-track. If 60 or more of the 121 GAC scan spots are present, the counts for all scan spots in a GAC target are averaged, and the average counts in the two channels represent one radiation budget observation. If fewer than 60 scan spots are present (due to data loss), the entire block is treated as missing.

With the average count for a GAC target calculated, the visible count is converted into an unnormalized albedo using a linear equation determined in the prelaunch calibration. The infrared count is converted into a radiance using on-board blackbody and space observations (Lauritson *et al.*, 1979; see also Chapter 4). At this point, the data consist of calibrated, spatially averaged radiometric quantities at the altitude of the satellite.

The next step converts the data into radiation budget estimates. The visible albedo is normalized by dividing by the cosine of the solar zenith angle at the center of the GAC target. The assumptions in this calculation are that (1) all reflected radiation is isotropic (and thus that one viewing angle is sufficient to estimate the total reflected radiation in all directions), (2) the albedo in the entire solar spectrum is the same as the albedo measured in the narrow spectral band of the AVHRR channel 1, and (3) the measured albedo is the daily average albedo.

These assumptions cause errors in the retrieved albedo, but they produce useful albedo estimates, and the data processing is much simpler than in more accurate schemes discussed below. Ascending and descending portions of the orbit are processed separately. Albedo is only calculated for the daylight half orbit. In addition, the available solar radiation ($\mu_{\text{sun}} E_{\text{sun}}$, averaged over the entire day) is calculated as

$$\bar{E}_{\text{sun}} = \frac{S_{\text{sun}}}{t_{\text{D}}} \int_{\text{sunrise}}^{\text{sunset}} \left(\frac{\bar{d}_{\text{sun}}}{d_{\text{sun}}(t)} \right)^2 \mu_{\text{sun}}(t)\, dt, \tag{10.5}$$

where t is the time in seconds, t_{D} is the day length (86,400 s), and the solar constant S_{sun} is assumed to be 1353 W m^{-2}. The daily averaged absorbed solar radiation is then calculated as

$$\bar{E}_{\text{abs}} = (1 - A)\bar{E}_{\text{sun}}. \tag{10.6}$$

The infrared radiance is first corrected for viewing angle. An empirical limb darkening correction (Abel and Gruber, 1979) converts the radiance viewed at zenith angle ζ to zenith angle zero:

$$L(0) = L(\zeta) + [c_1 + c_2\, L(\zeta)](\sec\zeta - 1) + [c_3 + c_4\, L(\zeta)](\sec\zeta - 1)^2, \tag{10.7}$$

where the constants c_1 through c_4 are given in Gruber *et al.* (1983). Next a spectral correction is applied to convert narrowband observations to broadband radiant exitance. Abel and Gruber (1979) showed, based on radiative transfer modeling, that flux temperature $[T_{\text{F}} \equiv (M_{\text{LW}}/\sigma)^{1/4}]$ is statistically related to a simple function of the equivalent blackbody temperature in the 10–12-μm window. Accordingly, the radiance $L(0)$ is converted to equivalent blackbody temperature (T_{BB}) using the Planck function. Then the flux temperature is calculated as

$$T_{\text{F}} = T_{\text{BB}}(a + b\, T_{\text{BB}}), \tag{10.8}$$

where, again, the coefficients a and b are given by Gruber *et al.* (1983). Finally, the outgoing longwave radiation is calculated:

$$M_{\text{LW}} = \sigma\, T_{\text{F}}^4 = \text{OLR}. \tag{10.9}$$

OLR values are calculated and stored separately for ascending and descending half orbits; thus two OLR observations of any point are made each day. In short, both angular and spectral corrections are made to the OLR, and two observations per day are available to estimate the daily average radiant exitance.

The final step in the procedure is to map the data. They are mapped onto two 125 × 125 polar stereographic maps (arrays), one Northern Hemisphere and one Southern Hemisphere, which have about 100-km resolution at the equator and 200-km resolution at the poles. Each GAC target is assigned to the box in which its center falls, and all values in the box are averaged for each day. From the polar stereographic arrays, a 2.5° × 2.5° latitude–longitude map is constructed using bilinear interpolation. The four parameters mapped and archived are OLR daytime, OLR nighttime, absorbed solar radiation, and available solar radiation

($\bar{E}_{sun}$, which is not archived in 2.5° × 2.5° format). The data are available from the Satellite Data Services Division of NESDIS.[1] The OLR data have appeared in atlases (Janowiak *et al.*, 1985; Gruber *et al.*, 1986).

The accuracy of radiation budget estimates from satellite instruments is difficult to assess because of the lack of independent observations. The primary means for verification has been comparison with other satellite measurements. Gruber and Jacobowitz (1985) compared NOAA OLR estimates from the TIROS N satellite with OLR estimates made from Nimbus 7 ERB measurements. They found that the monthly, globally averaged OLR estimates agreed to within about 1–2 W m^{-2} rms. Comparison of the monthly zonal averages showed rms differences of up to 9 W m^{-2}. Unfortunately, the algorithms for both the NOAA and the Nimbus 7 ERB radiation budget parameters have changed since 1985; thus these accuracy figures may no longer be applicable.

The shortwave parameters (albedo, absorbed solar radiation, and thus net radiation) are considered to be less accurate than the OLR. The main reason for this conclusion is that the OLR estimates include angular and spectral corrections while the shortwave parameters do not. We know of no published accuracy estimates for the shortwave parameters.

The primary advantage to the NOAA radiation budget data set is that it is a long-term data set which will continue for many years. The OLR data have been used for many purposes and are a very valuable tool for climate studies.

10.2.2 Nimbus 7 Data Set

The Earth Radiation Budget (ERB) instrument, flown on the Nimbus 6 and 7 satellites, was the first radiation budget instrument to combine nonscanning, WFOV instruments with scanning, NFOV instruments on the same satellite. This combination has become standard[2] due to the complimentary properties the two designs. The ERB is not the most accurate radiation budget instrument to have flown (that honor belongs to the Earth Radiation Budget Experiment to be discussed in the next section), but it produced a long-term data set including the best angular dependence models assembled to date. Its history is interesting because it demonstrates the difficulty and challenge of making laboratory-quality measurements from space. Kyle *et al.* (1993) review the 20-year history of the ERB instrument.

The Earth Radiation Budget (ERB) instrument was designed to be flown on the Nimbus 6 satellite. It was launched 12 June 1975. A second ERB was flown on the Nimbus 7 satellite, launched 24 October 1978. (See the *Nimbus* 6 and *Nimbus 7 User's Guides*.) Each ERB is a 22-channel instrument. Channels 1–10 directly observe the sun to measure the solar flux density in several bands (see Section 10.1). Channels 11–14 are nonscanning, WFOV (flat-plate) channels. Channels 11 and 12 measure total flux density (0.2 to >50 μm). Channel 12 is

[1] Kidwell (1991) describes the positioning of the maps and the format of the data tapes.

[2] The EOS may not have WFOV sensors.

the operational total channel; channel 11 is normally shuttered to slow degradation due to exposure to sunlight. It is unshuttered occasionally to compare with channel 12. Channel 13 is the shortwave channel. It is essentially the same as channels 11 and 12, but it is covered with two Suprasil W domes to measure radiation in the range 0.2–3.8 μm. Channel 14 is a near-IR channel; it has a filter (0.695–2.8 μm) between two Suprasil W domes. Channels 15–22 scan and have a narrow field of view (0.25° × 5.12°). Channels 15–18 are shortwave channels (Suprasil W filter, 0.2–4.8 μm), and channels 19–22 are longwave channels (coated diamond interference filters, 4.5–50 μm). Channels 19–22 are calibrated in space by observing an on-board blackbody and space. None of the other channels are directly calibrated, although the satellite is occasionally tipped so that the WFOV channels directly view the sun. The NFOV scanner operated for 20 months; the WFOV and solar channels continue to operate.

Neither the Nimbus 6 nor the Nimbus 7 ERB performed in space the way they did in prelaunch calibrations. A variety of problems were encountered (Maschhoff *et al.*, 1984). Degradation of sensors due to exposure to sunlight is always a problem in space. The channel 13 and 14 Suprasil W domes degraded asymmetrically because of differential exposure to sunlight and collisions with exospheric gases (Ardanuy and Rea, 1984). There were also changes in offset during the normal 3-days-on, 1-day-off schedule of the instrument. Heating of the filter domes caused errors in the WFOV channels. Internal reflections from the baffles in channel 12 caused a negative bias due to radiation escaping to cold space. Heating of the Earth-facing side of the instrument at sunrise caused changes in the calibration. As a result of these and other problems, the Nimbus 6 ERB data were never fully processed. Only an OLR data set was produced (Bess and Smith, 1987a; Bess *et al.*, 1989).

A very large effort was made to understand and eliminate problems in the Nimbus 7 ERB data. Ground measurements on nearly identical instruments (Maschhoff *et al.*, 1984) and studies of in-flight data (Kyle *et al.*, 1984) led to new calibrations and corrections. In fact, all of the Nimbus 7 ERB data were reprocessed to reflect these corrections (Kyle *et al.*, 1993). Readers are encouraged to read these papers because they illuminate the techniques and complexities of processing radiation budget data.

The algorithms for retrieval of parameters from the Nimbus 7 ERB are given in Jacobowitz *et al.* (1984a). Here we outline the principles that govern the retrieval.

10.2.2.1 Angular Dependence Models

Perhaps the most important products produced from the Nimbus 7 ERB data are the angular dependence models [ADMs; Suttles *et al.* (1988, 1989)]. *Angular dependence model* is a generic term used to refer to models of both reflection of shortwave radiation and longwave limb darkening. The purposes of the ADM are to estimate the radiant exitance from a location based on a radiance observation from a single direction and to estimate the diurnal variation of the radiant exitance. These models are used to retrieve radiation budget parameters from the ERBE Scanner, and they were used to reprocess the Nimbus 7 ERB data (Kyle *et al.*, 1993). The technique was originally developed by Raschke *et al.* (1973) to process Nimbus 3 data, and for shortwave radiation it may be derived from basic principles as follows.

Recall from Eq. 3.63 that the radiance reflected from a surface or scene is given by

$$L_r(\theta_r,\phi_r) = \int_0^{2\pi}\int_0^{\pi/2} L_i(\theta_i,\phi_i)\gamma_r(\theta_r,\phi_r;\theta_i,\phi_i)\cos\theta_i\sin\theta_i d\theta_i d\phi_i, \quad (10.10)$$

where subscript *i* refers to incident radiance, subscript *r* refers to reflected radiance, and γ_r is the bidirectional reflectance. For solar radiation at the top of the atmosphere, L_i all comes from essentially one direction $(\theta_{sun},\phi_{sun})$, so that

$$\begin{aligned} L_r(\theta_r,\phi_r) &= \mu_{sun}L_{sun}\Omega_{sun}\gamma_r(\theta_r,\phi_r;\theta_{sun},\phi_{sun}) \\ &= \mu_{sun}E_{sun}\gamma_r(\theta_r,\phi_r;\theta_{sun},\phi_{sun}), \end{aligned} \quad (10.11)$$

where L_{sun} is the solar radiance, Ω_{sun} is the solar solid angle, and $E_{sun} = L_{sun}\Omega_{sun}$ is the distance-corrected solar constant. The reflected radiant exitance[3] is

$$M_r(\theta_{sun}) = \int_0^{2\pi}\int_0^{\pi/2} L_r(\theta_r,\phi_r)\cos\theta_r\sin\theta_r d\theta_r d\phi_r \quad (10.12)$$

or

$$M_r(\theta_{sun}) = \mu_{sun}E_{sun}\int_0^{2\pi}\int_0^{\pi/2} \gamma_r(\theta_r,\phi_r;\theta_{sun},\phi_{sun})\cos\theta_r\sin\theta_r d\theta_r d\phi_r. \quad (10.13)$$

Since the solar radiation incident on the scene is $\mu_{sun}E_{sun}$, the albedo $(A = M_r/E_i)$ must be

$$A(\theta_{sun}) = \int_0^{2\pi}\int_0^{\pi/2} \gamma_r(\theta_r,\phi_r;\theta_{sun},\phi_{sun})\cos\theta_r\sin\theta_r d\theta_r d\phi_r. \quad (10.14)$$

Dividing the radiant exitance (Eq. 10.13) by the observed radiance (Eq. 10.11) and substituting Eq. 10.14 gives

$$M_r = \frac{L_r}{\gamma_r/A} = \frac{\pi L_r}{\pi\gamma_r/A} = \frac{\pi L_r}{\xi_r}, \quad (10.15)$$

where ξ_r is the anisotropic reflectance factor (Eq. 3.67). Thus, with a knowledge of ξ_r, one can in theory convert a radiance measurement into an estimate of the radiant exitance from a scene.

To calculate the required anisotropic reflectance factor, Suttles *et al.* (1988) divided the Earth into five surface types: ocean, land, snow, desert, and mixed

[3] In theory, M_r depends on ϕ_{sun} as well as θ_{sun}. One would expect a different radiant exitance if the sun were shining parallel to long ocean waves rather than perpendicular to them, for example. However, azimuth angle is important for relatively few surfaces, and for those few, the orientation of the surface is probably not known with any certainty. Therefore, the azimuthal dependence of M_r and of albedo is simply dropped.

land-ocean (coastal areas). They also chose four cloud categories: clear (0–5% cloud cover), partly cloudy (5–50%), mostly cloudy (50–95%), and overcast (95–100%). There would be 20 scene types if all types of clouds were allowed over all surfaces. However, Suttles *et al.* allowed only one overcast scene (regardless of surface type), they eliminated the difficult-to-discriminate partly-cloudy-over-snow and mostly-cloudy-over-snow categories, and they combined land and desert for partly and mostly cloudy situations. Thus they arrived at 12 scene types: clear over land, ocean, snow, desert, and land–ocean mix; partly cloudy over ocean, land or desert, and land–ocean mix; mostly cloudy over ocean, land or desert, and land–ocean mix; and overcast. Nimbus 7 ERB NFOV observations between 1 April 1979 and 22 June 1980 were binned in one of these scene types using the Nimbus 7 THIR/TOMS cloud algorithm (see Chapter 8) to determine cloudiness.

The Nimbus 7 ERB NFOV channels could scan in many directions, including parallel to the satellite subtrack (see Jacobowitz *et al.*, 1978). This made it possible for the ERB to view a given location from several viewing angles as it approached, passed over, and receded from the location. For each of the scene types, the NFOV data were further binned into 10 solar zenith angle bins, into seven viewing zenith angle bins, and into eight relative azimuth angle[4] bins (for a total of 5880 bins, when all azimuth angles were pooled for the vertical viewing angle). The azimuth angles, ϕ_R, were relative to the sun's azimuth and were assumed to be symmetric on each side of the sun.

To calculate $\xi_r(\theta_r, \theta_{sun}, \phi_R)$, all radiances were normalized to the mean Earth–sun distance and to overhead sun [i.e., the radiances were multiplied by $(d_{sun}/\bar{d}_{sun})^2 \sec\theta_{sun}$]. The average radiance, $\bar{L}$, was then calculated in each bin. For each surface type and for each solar zenith angle, the radiant exitance was calculated by integrating over viewing zenith angle and relative azimuth angle using Eq. 10.12. The anisotropic factor is then given by

$$\xi_r(\theta_r, \theta_{sun}, \phi_R) = \frac{\pi \bar{L}(\theta_r, \theta_{sun}, \phi_R)}{M_r(\theta_{sun})}. \tag{10.16}$$

The albedo for each scene type and each solar zenith angle was calculated by

$$A(\theta_{sun}) = \frac{M_r(\theta_{sun})}{S_{sun}}. \tag{10.17}$$

(The solar constant was used because the radiances were normalized to overhead sun and to the mean Earth–sun distance.)

Since Nimbus 7 is a sunsynchronous satellite with approximately a noon equator crossing time, not all solar zenith angles were observed for every surface type. Deserts, for example, occur mainly in the subtropics, but Nimbus 7 can observe high solar zenith angles only in the high latitudes. Empty angular bins were filled using several techniques. The most important is the Helmholtz Reciprocity

[4] Defined as the difference between the sun's azimuth and the viewing azimuth such that reflection away from the sun has a zero relative azimuth, and reflection toward the sun has a 180° relative azimuth.

Principle, which states that the bidirectional reflectance is invariant when the incoming and outgoing angles are interchanged:

$$\gamma_r(\theta_1,\phi_1;\theta_2,\phi_2) = \gamma_r(\theta_2,\phi_2;\theta_1,\phi_1). \tag{10.18}$$

When an entire solar zenith angle bin was empty, the bidirectional reflectances for the other bins were calculated, and Eq. 10.18 was used to calculate the bidirectional reflectances of the missing bin. Then Eq. 10.14 was used to recover the albedo of the missing bin, and finally, the anisotropic reflectance function was calculated from Eq. 10.16. Judicious use of interpolation, extrapolation, and smoothing was also necessary to construct the ADMs.[5] Finally, to conserve energy, it must be true that

$$\frac{1}{\pi}\int_0^{2\pi}\int_0^{\pi/2} \xi_r(\theta_r,\phi_r;\theta_{sun},\phi_{sun})\cos\theta_r\sin\theta_r d\theta_r d\phi_r \equiv 1. \tag{10.19}$$

All anisotropic reflectance factors were adjusted to satisfy Eq. 10.19.

Since GOES data samples the diurnal variation of albedo better than Nimbus 7, GOES-measured albedos were combined with the Nimbus 7 albedos to yield a hybrid estimate of albedo as a function of solar zenith angle (directional model) for each surface type (see Section 10.2.4).

The anisotropic reflectance factors and the albedos are tabulated in Suttles *et al.* (1988), and the data for a few scenes are plotted in Fig. 3.22. The anisotropic factors range from 0.41 to 12.76 (both over clear ocean). This range is an indication of the error in assuming that reflectance is isotropic, which is the same as assuming that the anisotropic reflectance factor is always one (see Green *et al.*, 1990a).

Suttles *et al.* (1989) calculated a longwave angular dependence model from Nimbus 7 ERB data and THIR/TOMS cloud data. The process for longwave calculations was similar to that for the shortwave. The goal is to produce an anisotropic factor that relates the observed longwave radiance to an estimate of the longwave radiant exitance. The same 12 scene types used in the shortwave calculations were used in the longwave calculations. The longwave data were also binned into seven viewing zenith angle bins. However, the longwave observations were assumed to be independent of solar zenith angle and relative azimuth angle. Instead, ten latitude bins and four season bins were used (for a total of 3360 bins). Day and night values were calculated separately, but they were averaged for the final ADM.

The longwave anisotropic factor is tabulated in Suttles *et al.* (1989). The longwave ADM is much less variable than the shortwave ADM. The longwave anisotropic factor for all surfaces is similar; it starts at around 1.05 for overhead viewing, and decreases to approximately 0.90 for oblique viewing. The longwave anisotropic factor ranges from 0.84 to 1.07.

[5] Note that although reciprocity was used to fill empty bins, the final models do not satisfy reciprocity in general. (See Green *et al.* (1990a).)

Finally, Suttles *et al.* (1988, 1989) calculated parameters required to determine cloudiness: standard deviation of shortwave and longwave radiance and the correlation between shortwave and longwave radiance in each bin.

10.2.2.2 Scanner Products

The original ERB algorithms (Jacobowitz *et al.*, 1984a) resulted in NFOV albedos that were about 10% too high and OLR values that were about 3 W m^{-2} too low compared with the WFOV data (Arking and Vemury, 1984; Kyle *et al.*, 1985). Also, improved calibrations had been found, and improved algorithms had been developed for the ERBE instrument. Therefore, the Nimbus 7 ERB data were reprocessed with the ERBE algorithms (Kyle *et al.*, 1993).

The NFOV data (May 1979–May 1980) were processed in two ways to yield albedo, OLR, and net radiation. The first method is the traditional method developed by Raschke *et al.* (1973) and refined into the ERBE algorithm. The corrected NFOV radiances were binned daily into 18,630 boxes each approximately (166 $km)^2$ and covering the globe. Each of these boxes is called a *subtarget area*. Each of the radiances was transformed into a radiant exitance ($M_r = \pi L/\xi_r$) using the ADMs discussed above.

The heart of this algorithm is the method of selecting the scene type (one of the 12 discussed above) and thus which anisotropic factor to use. The surface type is selected from a geographic database. Then the cloud cover is determined using a *maximum-likelihood cloud estimation* (MLCE) method (Wielicki and Green, 1989). MLCE is a bispectral technique that computes the probability that the observed radiance pair (shortwave and longwave) belongs to one of the four cloud cover categories. This probability is computed using the radiance means, standard deviations, and shortwave–longwave correlations[6] discussed above assuming that the radiances for each cloud-cover category are described by a bivariate normal frequency distribution. The cloud cover category with the highest probability is selected. Diekmann and Smith (1989) found that the cloud-cover category is correctly identified about two thirds of the time, and incorrectly identified by one cloud category about one third of the time.

During processing, it was discovered that at large satellite viewing zenith angles the retrieved albedos were too high and the longwave radiant exitances were too low. To alleviate this problem, observations from satellite viewing zenith angles greater than 75° (70° for ERBE) were rejected.

At this point in the processing, the instantaneous values of OLR and albedo have been calculated. The next step produces daily averaged values. The daily averaged OLR ($\bar{M}_{LW}$) is simply the mean of the noon (ascending) and the midnight (descending) observations. The mean albedo is more complicated and is discussed by Brooks *et al.* (1986). The technique uses the albedo versus solar zenith angle data calculated by Suttles *et al.* (1988). Suppose that at time t_{obs} the satellite

[6] The ERBE team decided that the shortwave–longwave correlations were not important and they do not use them for ERBE processing.

observes a target area and that the albedo calculated using the anisotropic reflectance factor is $A^*[\theta_{sun}(t_{obs})]$ Let $A'[\theta_{sun}(t)]$ be the albedo of the scene as a function of θ_{sun} as calculated by Suttles *et al.* (1988). Then an estimate of albedo at another time during the day is

$$A[\theta_{sun}(t)] = A^*[\theta_{sun}(t_{obs})] \left(\frac{A'[\theta_{sun}(t)]}{A'[\theta_{sun}(t_{obs})]} \right). \tag{10.20}$$

In other words, the albedo estimates of Suttles *et al.*, coupled with a knowledge of the sun's elevation angle, supply the time variation of albedo, and the observed albedo supplies the magnitude. The daily mean reflected shortwave radiation can then be estimated as

$$\bar{M}_{SW} = \frac{1}{t_D} \int_{sunrise}^{sunset} \mu_{sun} E_{sun} A[\theta_{sun}(t)]\, dt. \tag{10.21}$$

The average incident solar radiation is, of course,

$$\bar{E}_{sun} = \frac{1}{t_D} \int_{sunrise}^{sunset} \mu_{sun} E_{sun}\, dt. \tag{10.22}$$

The average daily albedo, then, is

$$\bar{A} = \frac{\bar{M}_{SW}}{\bar{E}_{sun}}, \tag{10.23}$$

and the daily net radiation is

$$\bar{E}_{net} = (1 - \bar{A})\bar{E}_{sun} - \bar{M}_{LW}. \tag{10.24}$$

Note that the solar constant was assumed to be 1371.5 W m^{-2} in all of the Nimbus 7 ERB processing.

$\bar{A}$, $\bar{M}_{LW}$, and $\bar{E}_{net}$ in the subtarget areas are archived at the National Space Science Data Center for the days that the instrument was turned on between May 1979 and May 1980.

The second way in which the NFOV data were processed was with the *sorting into angular bins* (SAB) *algorithm* (Kyle *et al.*, 1993). Developed by Arking and Vemury (1986) as a means of checking the predecessor to the MLCE method, the SAB method does not use angular dependence models. Instead, observations of an area are accumulated until the area has been observed from sufficiently many angles that the radiant exitance may be evaluated by directly integrating Eq. 10.12. Clearly the SAB method cannot produce high temporal resolution. The NSSDC archive contains zonal mean quantities calculated daily, but the regional quantities are only available as monthly averages. Also, the retrievals were not in the $(166 \text{ km})^2$ subtarget areas but in $(500\text{-km})^2$ areas called *target areas*. (Each target area is composed of nine subtarget areas; the globe is covered by 2070 target areas.) The SAB method is still primarily useful for checking the MLCE algorithm, and the two methods compare well, according to Kyle *et al.* (1993).

10.2.2.3 WFOV Products

The Nimbus 7 ERB makes WFOV measurements of shortwave, near-IR, and total radiation. However, the irradiance measured by the satellite (E) is not the radiant exitance of the Earth (M). King and Curran (1980) discuss the relationship between E and M. The WFOV data have been processed by two different algorithms. The first is simple; it basically corrects for the height of the satellite. The second algorithm attempts to desmooth the data so that regional variations can be studied.

In the first algorithm, the shortwave and near-IR data are converted to albedos as follows. If the positions of the satellite and the sun are known, a calculation can be made to determine the maximum irradiance which the satellite is likely to measure.[7] This is done by assuming that the Earth is a perfectly reflecting ($A = 1$) Lambertian surface ($\xi = 1$), for which the bidirectional reflectance is a constant π^{-1}. The radiance[8] reflected from a point with latitude Θ and longitude Ψ is

$$L_{\mathrm{r}}^{*}(\Theta,\Psi;\,\Theta_{\mathrm{sun}},\Psi_{\mathrm{sun}}) = \gamma_{\mathrm{r}}\mu_{\mathrm{sun}}E_{\mathrm{sun}} = \frac{1}{\pi}\mu_{\mathrm{sun}}E_{\mathrm{sun}}, \tag{10.25}$$

where $\mu_{\mathrm{sun}} = \mu_{\mathrm{sun}}(\Theta,\Psi;\,\Theta_{\mathrm{sun}},\Psi_{\mathrm{sun}})$ is the cosine of the solar zenith angle at the point (Θ,Ψ). The solar constant is assumed to be 1371.5 W m^{-2}. The maximum shortwave irradiance that a satellite above this Lambertian Earth would measure is then

$$E_{\mathrm{SW}}^{*}(r_{\mathrm{s}},\Theta_{\mathrm{s}},\Psi_{\mathrm{s}}) = \int_{0}^{2\pi}\int_{0}^{\pi/2} L_{\mathrm{r}}^{*}(\Theta,\Psi;\,\Theta_{\mathrm{sun}},\Psi_{\mathrm{sun}})\cos\theta\,\sin\theta\,d\theta\,d\phi \tag{10.26}$$

or

$$E_{\mathrm{SW}}^{*}(r_{\mathrm{s}},\Theta_{\mathrm{s}},\Psi_{\mathrm{s}}) = \frac{1}{\pi}E_{\mathrm{sun}}\int_{0}^{2\pi}\int_{0}^{\pi/2} \mu_{\mathrm{sun}}\cos\theta\,\sin\theta\,d\theta\,d\phi, \tag{10.27}$$

where $(r_{\mathrm{s}},\Theta_{\mathrm{s}},\Psi_{\mathrm{s}})$ is the position of the satellite, and (θ,ϕ) are directions from the flat-plate sensor. In this calculation, the point (Θ,Ψ) is the location where the ray from the satellite intersects the top of the atmosphere; i.e., both Θ and Ψ are functions of r_{s}, Θ_{s}, Ψ_{s}, θ, and ϕ. (Note that $\cos\theta$ is the response of the flat-plate sensor.) If $E_{\mathrm{SW}}(r_{\mathrm{s}},\Theta_{\mathrm{s}},\Psi_{\mathrm{s}})$ is the measured shortwave irradiance, then the albedo is

$$A(r_{\mathrm{s}},\Theta_{\mathrm{s}},\Psi_{\mathrm{s}}) = \frac{E_{\mathrm{SW}}(r_{\mathrm{s}},\Theta_{\mathrm{s}},\Psi_{\mathrm{s}})}{E_{\mathrm{SW}}^{*}(r_{\mathrm{s}},\Theta_{\mathrm{s}},\Psi_{\mathrm{s}})}. \tag{10.28}$$

This estimate assumes that the albedo is constant throughout the field of view of the flat-plate sensor and that the reflection is isotropic. The albedo for the near-

[7] We will call this the maximum irradiance even though greater irradiances could be measured if reflectance is not isotropic.

[8] Radiance is usually specified as a function of zenith and azimuth angles. Here, to emphasize the Earth–sun–satellite geometry, we specify radiance as a function of the latitude and longitude of the viewed location and of the sun.

IR channel is similarly calculated, but the wavelength response of the filter is taken into account in calculating E_{sun}. The data are binned into target areas. The mean albedo for the target area is the sum of the measured irradiances divided by the sum of the maximum irradiances.

In the first algorithm, the longwave irradiance is calculated by subtracting the shortwave irradiance from the total irradiance ($E_{LW} = E_{TOT} - E_{SW}$). Then the irradiance is converted into a radiant exitance at the top of the atmosphere. The historical way to do this (Smith *et al.*, 1977; Jacobowitz *et al.*, 1979) is to assume that the radiation from Earth is isotropic, which means that the radiation can be thought of as emanating from a point source at the center of the Earth and decreasing in intensity as the square of the distance from the center of the Earth. If the measured longwave irradiance is E_{LW}, then the TOA radiant exitance must be

$$M_{LW} = E_{LW}\left(\frac{r_e + h_s}{r_e + h_{TOA}}\right)^2, \tag{10.29}$$

where r_e is the radius of the Earth, h_s is the height of the satellite, and h_{TOA} is the height of the assumed top of the atmosphere (15 km for Nimbus 7 ERB retrievals). This is called a *shape factor retrieval* (Green, 1981) or *inverse square law retrieval*. The shape factor is

$$\left(\frac{r_e + h_{TOA}}{r_e + h_s}\right)^2$$

and is constant for a circular orbit. M_{LW} is assigned to the target area which contains the subsatellite point, and all observations within a target area are averaged.

The net radiation is calculated for each target area as

$$\bar{E}_{net} = (1 - kA)\bar{E}_{sun} - \bar{M}_{LW}, \tag{10.30}$$

where $\bar{M}_{LW}$ is the average of the noon and midnight OLR observations, $\bar{E}_{sun}$ (Eq. 10.22) is the mean daily solar insolation on the target area, and k is a factor to transform A into a mean daily value. The factor k is based on a hybrid directional reflectance model (90% cloud/ocean, 10% land) derived from Raschke *et al.* (1973); the model is applied globally.

Nine years (November, 1978–October, 1987) of albedo, OLR, and net radiation computed from Nimbus 7 WFOV observations are archived at the National Space Science Data Center for daily and monthly time periods. Figure 10.2 shows the zonally averaged albedo, OLR, and net radiation for the period November 1978 to October 1986. The Earth's radiation budget is quite consistent from year to year, but important interannual variations occur, for example, the 1982–1983 El Niño event.

The WFOV instrument is still operational, and radiation budget products will be produced past October, 1987, as funds become available.

From an altitude of 955 km, a Nimbus 7 WFOV sensor simultaneously views over 130 (500 km)^2 target areas. Each observation, however, is assigned to the

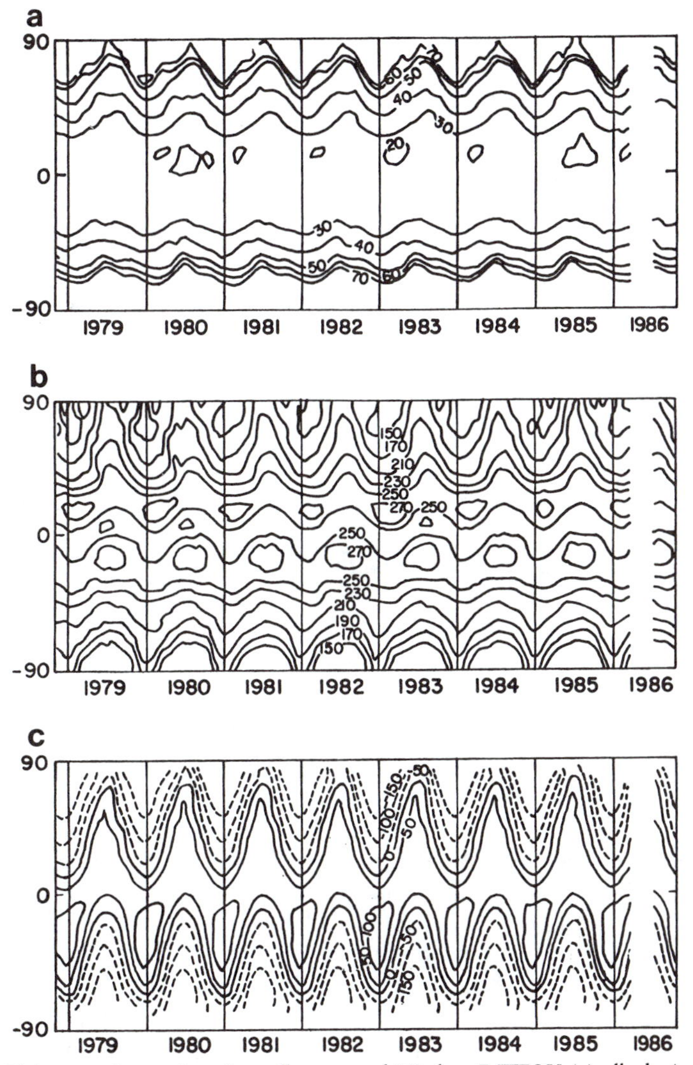

FIGURE 10.2. Eight-year time series of zonally averaged Nimbus 7 WFOV (a) albedo (percent), (b) longwave radiant exitance (W m^{-2}), and (c) net radiation (W m^{-2}). [Courtesy of G. Garrett Campbell, Cooperative Institute for Research in the Atmosphere, Colorado State University.]

single target area which is directly below the satellite. The resulting radiation budget fields therefore appear to be very smooth in comparison with NFOV fields. A large effort has gone into desmoothing or deconvoluting the Nimbus 7 ERB data. This is the second algorithm with which the WFOV data have been processed.

The longwave data can be desmoothed in a straightforward manner. Green (1981) explains two ways to desmooth the longwave data. One way involves a spherical harmonic expansion of the longwave fields. The satellite-measured irradiance is given by

$$E_{LW}(\Theta_s, \Psi_s) = \frac{1}{\pi}\int_0^{2\pi}\int_0^{\pi/2} M_{LW}(\Theta, \Psi)\, \xi(\Theta, \Psi, \zeta) \cos\theta \sin\theta \, d\theta \, d\phi, \quad (10.31)$$

where $L_{LW}(\Theta, \Psi, \theta, \phi) = \pi^{-1} M_{LW}(\Theta, \Psi)\, \xi(\Theta, \Psi, \zeta)$ and ζ is the zenith angle to the satellite from Θ, Ψ. Suppose that both $E_{LW}(\Theta_s, \Psi_s)$ and $M_{LW}(\Theta, \Psi)$ are expanded in spherical harmonics (Y_n^m):

$$E_{LW}(\Theta_s, \Psi_s) = \sum_{n=0}^{N}\sum_{m=-n}^{n} a_n^m \, Y_n^m(\Theta_s, \Psi_s) \quad (10.32a)$$

$$M_{LW}(\Theta, \Psi) = \sum_{n=0}^{N}\sum_{m=-n}^{n} b_n^m \, Y_n^m(\Theta, \Psi). \quad (10.32b)$$

The a_n^m are calculated from observations obtained over a sufficiently long period of time that all the "boxes" (perhaps target areas) are filled. This takes a minimum of several days. The b_n^m are the spherical harmonic expansion coefficients of the sought solution for M_{LW}. Smith and Green (1981) showed that if the anisotropic factor ξ is a function of viewing zenith angle ζ only (not latitude or longitude) then spherical harmonics are eigenfunctions of the measurement equation 10.31:

$$\frac{1}{\pi}\int_0^{2\pi}\int_0^{\pi/2} Y_n^m(\Theta, \Psi)\, \xi(\zeta) \cos\theta \sin\theta \, d\theta \, d\phi = \lambda_n Y_n^m(\Theta, \Psi), \quad (10.33)$$

where the λ_n are easily calculated eigenvalues of Eq. 10.33. Therefore, the a_n^m are very simply related to the b_n^m:

$$b_n^m = \lambda_n^{-1} a_n^m. \quad (10.34)$$

This technique is called *deconvolution*.[9] Since $\lambda_0 = [(r_e + h_{TOA})/(r_e + h_s)]^2$, deconvolution starts at the same point as the shape factor retrieval. The λ_n decrease with increasing n, so deconvolution amplifies higher spatial harmonics.

A second way to desmooth the longwave data is called *parameter estimation* by Green (1981) and is discussed in detail by Twomey (1977). Here the Earth is divided into regions each with a constant radiant exitance M_k and known anisotropic reflectance factor $\xi_k(\zeta)$. Suppose that there are N such regions. Each satellite irradiance observation is given by

$$E_i = \sum_{k=1}^{N} B_{ik} M_k, \quad (10.35)$$

where

$$B_{ik} = \frac{1}{\pi}\xi_k(\zeta_{ik}) \cos\theta_{ik} \Omega_{ik}, \quad (10.36)$$

[9] Care must be taken to choose the optimal number of spherical harmonics. Too small a number will not desmooth the data as much as they could be. Too large a number will introduce artifacts into the data.

and Ω_{ik} is the solid angle of area k subtended at the ith position of the satellite. A large number of observations ($>N$) can be used to find the least squares best solution[10] for the M_k.

The shortwave data are more difficult to desmooth than the longwave data because the shortwave radiant exitance depends strongly on the position of the sun and because the shortwave anisotropic factor is more complicated than the longwave anisotropic factor. In particular, the deconvolution technique does not work because spherical harmonics are not eigenfunctions of the measurement equation. Parameter estimation can work, but retrieval of M_{SW} is not really appropriate because it depends so strongly on the position of the sun. A more appropriate parameter to retrieve is the albedo of an area, which depends only on surface and atmospheric properties. Hucek *et al.* (1987) have combined the above two techniques to retrieve albedo. They start with the measurement equation:

$$E_{SW}(\Theta_s, \Psi_s) = \frac{1}{\pi} \int_0^{2\pi} \int_0^{\pi/2} M_{SW}(\Theta, \Psi)\, \xi(\zeta) \cos\theta \sin\theta \, d\theta \, d\phi, \qquad (10.37)$$

and they expand M_{SW} as $\mu_{sun} E_{sun} A$. They further expand the albedo as a truncated spherical harmonic series. For each shortwave WFOV observation, Eq. 10.37 is a linear equation in the unknown spherical harmonic expansion coefficients. Proceeding as in parameter estimation, a large number of E_{SW} measurements are used to calculate the spherical harmonic coefficients. Hucek *et al.* use 289 coefficients ($N = 16$), which they estimate to be the number of independent pieces of information that are retrievable from the WFOV data. The albedo of each target area is then calculated using the coefficients.

Figure 10.3 compares the smooth and desmoothed WFOV zonal averages. Desmoothing results in a field in which the highs are higher and the lows are lower; thus they are more like scanner observations. The field mean does not change in the desmoothing process. Desmoothing makes the WFOV data more useful for regional radiation budget studies. If one only wants to study the global mean radiation budget, desmoothing is not necessary.

One year (July 1983 to June 1984) of Nimbus 7 ERB WFOV data have been desmoothed and are archived at the National Space Science Data Center (Kyle *et al.*, 1993). The shortwave observations were desmoothed by the method of Hucek *et al.* (1987), and the longwave data were processed by the deconvolution method. Ardanuy *et al.* (1987) detail the specific process for retrieval of the longwave radiant exitance and the net radiation.

The best estimates of the accuracy of the Nimbus 7 ERB parameters are determined by comparing ERB data with ERBE data. We therefore defer until the end of the next section a discussion of ERB accuracy.

[10] Again, care must be taken when choosing N. Too large a number creates a system of equations which is difficult to solve; too small a number results in less than optimal desmoothing. Also, there can be stability problems in finding the solution (see Twomey, 1977).

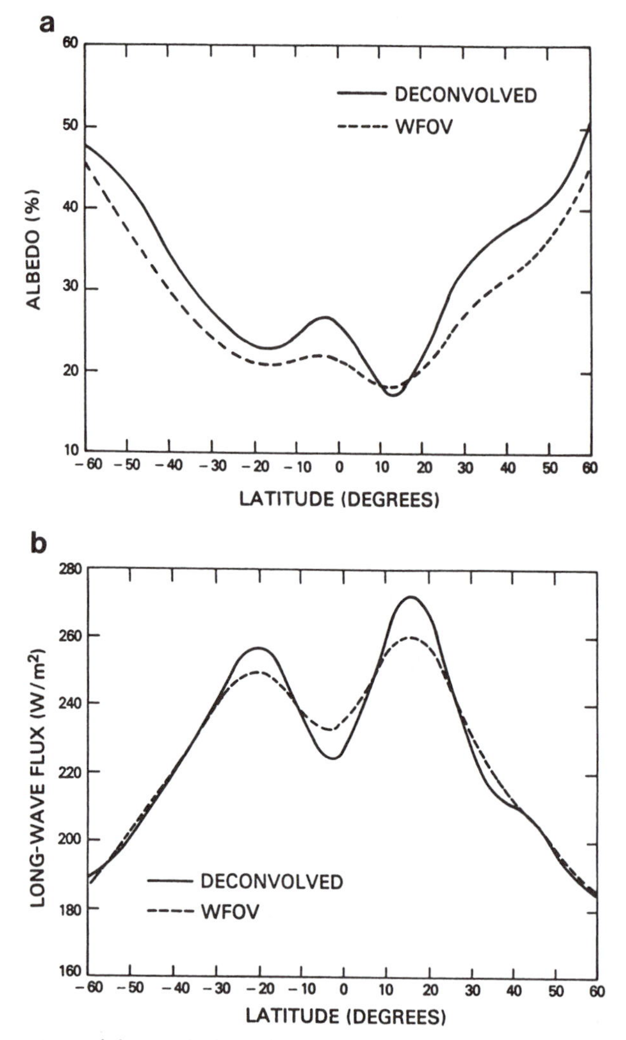

FIGURE 10.3. Comparison of desmoothed Nimbus 7 WFOV data with standard (smooth) data: (a) albedo (percent) and (b) longwave radiant exitance (W m^{-2}). [After Ardanuy *et al.* (1987).]

10.2.3 ERBE Data Set

The Earth Radiation Budget Experiment (ERBE) is similar to the Nimbus 6 and 7 ERB. Each instrument consists of a scanning NFOV instrument, a non-scanning WFOV instrument, and a solar monitor. (The ERBE instruments are described in section 4.1.6). However, the ERBE was designed to improve upon the ERB, and thus is different in many important ways. Among the improvements are the following:

- Identical ERBE instruments have been launched on three satellites: ERBS,

NOAA 9, and NOAA 10. The ERBS is in a non-sunsynchronous orbit (see section 2.4.3); in about 37 days the orbit precesses through all local times. Thus the ERBS samples the diurnal variation of radiation budget quantities. In addition, three satellites yield more frequent observations, thus reducing uncertainty in the time- and space-averaged products.
- All sensors are calibrated in space. The sensors internally view either a blackbody or a tungsten lamp. The blackbody is monitored with a thermistor, and the tungsten lamps are monitored with photodiodes. Externally, the sensors view space.
- The instruments have better pre-launch characterization. Mathematical models of each sensor were constructed and verified before launch so that problems which occur in space can be better understood and more easily corrected.
- The ERBE scanner has longwave, shortwave, and total sensors, whereas the Nimbus 7 scanner had only shortwave and total sensors. The third sensor provides a redundancy check and makes possible improved corrections for non-flat spectral response. The three channels are used to estimate broadband longwave and shortwave radiation in a linear conversion in which the coefficients are dependent on scene type, viewing zenith angle, solar zenith angle, and relative azimuth angle.
- The non-scanner has a medium field-of-view (MFOV) sensor which serves as a calibration check between the WFOV and NFOV sensors.

The retrieval algorithms used for ERBE and ERB are also similar because many of the algorithms developed for ERBE (Smith *et al.*, 1986; Brooks *et al.*, 1986) were applied when the ERB data were reprocessed (Kyle *et al.*, 1993). Again, however, there are important improvements in the ERBE algorithms.

First, ERBE has better scene identification. The twelve scene types developed from Nimbus 7 ERB data and used in the reprocessing of Nimbus 7 NFOV data were refined slightly for ERBE processing, for example by separating the single desert scene into vegetated and non-vegetated desert scenes.[11]

Second, while the scanner was operational, NFOV scene types were used to improve retrieval of WFOV albedo. The ERB algorithm assumed that reflected shortwave radiation was isotropic (Eq. 10.27). When both the ERBE scanner and the ERBE non-scanner are operating, the scanner sweeps out the area viewed by the non-scanner. The scene types identified with the scanner can be used to include an anisotropic factor into Eq. 10.27:

$$E^*_{SW}(r_s, \Theta_s, \Psi_s) = \frac{1}{\pi} E_{sun} \int_0^{2\pi} \int_0^{\pi/2} \mu_{sun}\, \xi \cos\theta \sin\theta \, d\theta \, d\phi. \tag{10.38}$$

The albedo is then calculated by dividing the observed irradiance by E^*_{SW}. In this calculation, the albedo is still assumed constant over the field of view, but the

[11] The vegetated desert was included in the land scene, and the non-vegetated desert became the desert model (R. N. Green, personal communication, 1990).

assumption of isotropy is no longer necessary. When the scanner data is not available, a climatological anisotropic factor is used.

Third, a numerical filter is used to desmooth non-scanner data. Developed by House and Jafolla (1980) and refined by Green (1983), the numerical filter relies on the fact that the WFOV instruments oversample the radiation field. That is, in the time between samples (0.8 s, for ERBE), the satellite moves only a fraction of a field of view. Thus an estimate of what a nadir-pointing, higher-resolution instrument would see can be derived by a weighted average of observations along the subsatellite track:

$$M_i = \sum_{j=-N}^{N} w_j E_{i+j}. \tag{10.39}$$

To conserve energy, the weights must sum to the reciprocal of the shape factor. The advantage of this technique over deconvolution is that it does not smear observations across local times because observations separated by several minutes are used instead of observations over several days. The disadvantage is that the numerical filter technique only enhances the resolution along the ground track and does not improve the across-track resolution. Deconvolution enhances the radiation field in both directions.

Fourth, better spatial and temporal averaging procedures are used (Brooks *et al.*, 1986). A table of observations during a month is kept for each location. The table is formed of day and local time bins. The observations from multiple satellites are included in the table, and averages are carefully performed to determine monthly mean quantities. This averaging includes, for example, fitting a half sine curve to the daytime longwave measurements over land, which Brooks and Minnis (1984) showed is an improvement over simple averaging. The scanner data are mapped onto a 2.5° × 2.5° latitude–longitude grid system, which is finer resolution than the $(500\ \text{km})^2$ grid used for the Nimbus 7 ERB data. The enhanced resolution (numerically filtered) WFOV and MFOV data are mapped onto a 5° × 5° grid, and the shape factor WFOV and MFOV data are mapped onto a 10° × 10° grid. The reader is referred to Brooks *et al.* (1986) for further details on the complicated task of producing accurate monthly mean radiation budget quantities from instantaneous satellite-measured irradiances.

An important aspect of the ERBE data processing is that clear scenes are evaluated by themselves in addition to their use in the averages; ERBE produces an estimate of average clear-sky radiation for each month. The difference between clear-sky and mean radiation is caused mainly by clouds. This difference can be used to investigate cloud forcing, the effect of clouds on the radiation budget of Earth. Ramanathan *et al.* (1989) showed using ERBE data that in April, 1985, averaged over the globe, clouds caused an extra 44.5 W m^{-2} in shortwave energy to be reflected to space. They also caused a reduction in longwave loss to space of 31.3 W m^{-2}. Clouds, therefore, appear to cool the climate on average. In April, 1985, clouds lowered the globally averaged net radiation by 13.2 W m^{-2}. The effects of clouds on radiation budget and climate remain an active research topic which would be impossible to investigate without satellite radiation budget measurements. This concept is now being used by many researchers as a critical test of general circulation models.

Note that ERBE uses two parameters which are different from those used in Nimbus 7 ERB processing. ERBE algorithms use a solar constant of 1365 W m^{-2} compared to the ERB value of 1371.5 W m^{-2}. The ERBE algorithms locate the top of the atmosphere at an altitude of 30 km versus 15 km in the ERB algorithms.

ERBE data from all three satellites and the many retrieved products are archived at the National Space Science Data Center. The archive includes raw data (counts); solar irradiances; instantaneous scanner and non-scanner measurements; regional, zonal, and global averages of longwave, shortwave, and albedo on 2.5° and larger scales; and monthly average products. Barkstrom *et al.* (1989) describe the archived products. As an example of these many products, Fig. 10.4 shows the mean and clear albedo and longwave radiant exitance for June 1985. These maps

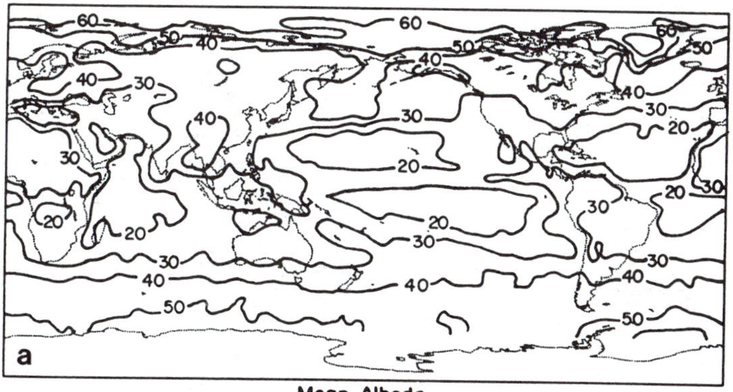

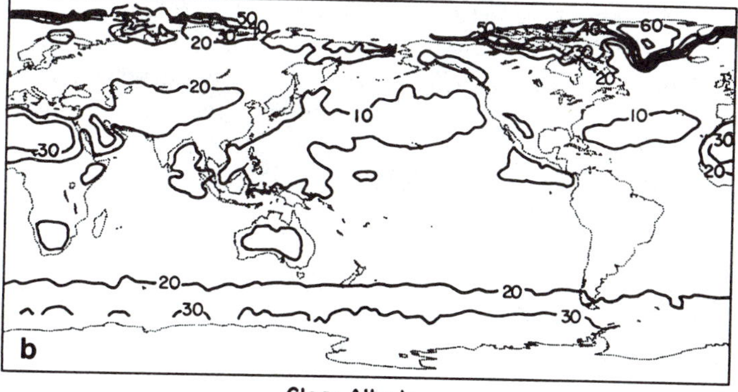

FIGURE 10.4. Combined ERBS and NOAA 9 scanner data for June 1985: (a) mean albedo (percent), (b) clear albedo (percent), (c) mean longwave radiant exitance (W m^{-2}), (d) clear longwave radiant exitance (W m^{-2}). [Courtesy of G. Garrett Campbell, Cooperative Institute for Research in the Atmosphere, Colorado State University.]

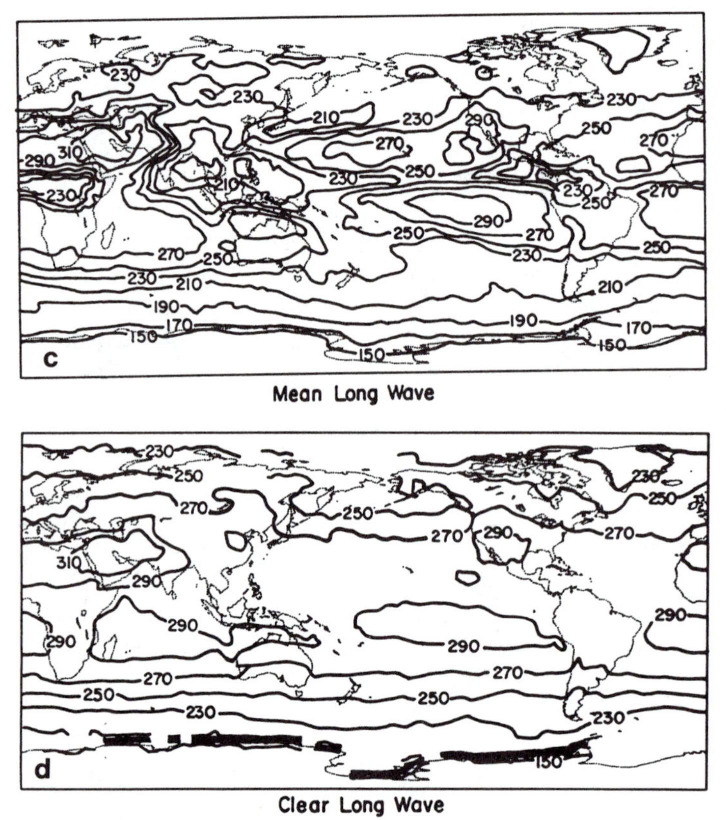

FIGURE 10.4. (*continued*)

were constructed from combined ERBS and NOAA 9 scanner data. The zonal averages for the same time period are shown in Fig. 10.5.

There are basically three ways to estimate the accuracy of satellite-based radiation budget measurements: (1) compare the measurements made by different satellites, (2) compare measurements made by the same satellite, and (3) estimate the accuracy by estimating the uncertainty in each aspect of the measurement and/or retrieval and compounding the uncertainties. All three types of analysis have been performed.

Kyle *et al.* (1989b) have compared Nimbus 7 ERB WFOV measurements with ERBE measurements. They found that "the Nimbus 7 WFOV shortwave sensor reads about 2.5% higher than the ERBS sensor, while the Nimbus 7 total channel reads about 1% below the ERBS channel near midnight and 1% above it at noon." The global average of albedo agreed with combined ERBS/NOAA 9 NFOV albedo within 0.03%, and the OLR agreed within 0.16%. They conclude that the agreement between the ERB and ERBE data sets is good enough to allow the sets to be combined for long-term studies of the radiation budget. They warn, however, that the characteristics of each instrument must be kept in mind.

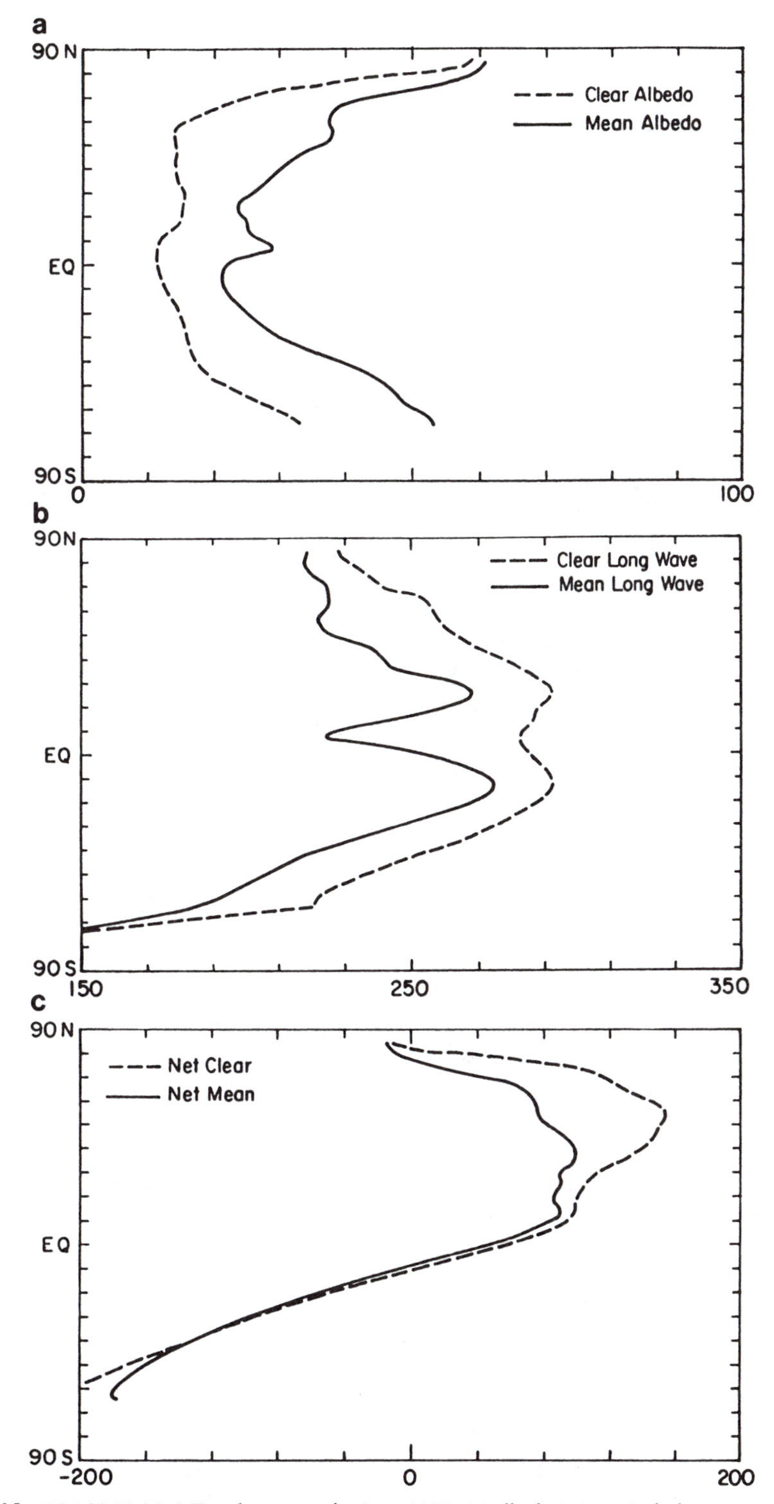

FIGURE 10.5. ERBS/NOAA 9 Zonal averages for June 1985: (a) albedo (percent), (b) longwave radiant exitance (W m^{-2}), (c) net radiation (W m^{-2}). [Courtesy of G. Garrett Campbell, Cooperative Institute for Research in the Atmosphere, Colorado State University.]

Green *et al.* (1990b) have compared the ERBE sensors with themselves. The scanner has longwave, shortwave, and total channels. The total channel was assumed to be correct. At night, the longwave channel should give the same reading as the total channel. They found that this was true within about 1% for both ERBS and NOAA 9. During the day, the shortwave channel should be the difference between the total channel and the longwave channel. They found that for ERBS the agreement was within 0.5%, but the NOAA 9 shortwave channel appears to be about 2% too high. Using the scanner measurements, the non-scanner measurements can be simulated. Green *et al.* found that "Generally, the MFOV measurements are within 2% to 3% of the scanner measurements. The WFOV TOT channel is very good, and the WFOV SW channel is about 3% high for ERBS and excellent for NOAA 9."

Barkstrom *et al.* (1989) list ten validation criteria and many researchers who are pursuing them using various data for comparison. They note that "ERBE data products are complex assemblages of data and models. Thus, their uncertainties are difficult to compute." They expect uncertainties in instantaneous scanner radiances to be about 1% for longwave observations and 2–3% for shortwave observations. For instantaneous radiant exitances from $2.5° \times 2.5°$ boxes they expect ± 5 W m^{-2} for longwave observations and ± 15 W m^{-2} for shortwave observations. For monthly averaged radiant exitances on a regional scale, they expect uncertainties of ± 5 W m^{-2} for both longwave and shortwave measurements. Finally, they estimate that the uncertainty in ERBE measurements of the annual, global average of the net radiation is about ± 5 W m^{-2}.

Unfortunately, these uncertainty estimates do not give the complete picture of the accuracy of radiation budget measurements, nor are they universally applicable. The primary sources of uncertainty are two: inadequate sampling of the diurnal variation, and inaccurate knowledge of the angular dependence of the shortwave radiation. Instrument uncertainty is of secondary importance. Uncertainty analysis of ERBE data is an ongoing task. More and better uncertainty estimates will appear in the literature. Potential users of the data need to carefully investigate the uncertainties in the particular products which they want to use.

One very important aspect of these uncertainties is that they prevent us from measuring subtle trends in climate such as those expected from increased CO_2 concentration. There are trends in the radiation budget data, but the uncertainties prevent them from being statistically significant. We hope that this problem will be resolved with next-generation radiation budget instruments.

10.2.4 Geostationary Products

Geostationary satellites have been used to retrieve radiation budget quantities (Saunders and Hunt, 1983; Minnis and Harrison, 1984a,b,c). Since geostationary satellites were not designed to make radiometric measurements, their data must be carefully processed to be used for this purpose. Three corrections are necessary. First, the data must be converted from counts, which are used to make images, to radiances. This is especially crucial for the visible data because they are not calibrated in space. Second, the narrowband radiances must be converted to broadband radiances. Finally, the radiances must be converted to radiant exitances with the use of angular dependence models.

Saunders and Hunt used Meteosat data. To calibrate the visible channel, they used scene-dependent calibration factors derived from aircraft underflights (Kriebel, 1981). The resulting narrowband radiances were converted to broadband radiant exitances by assuming that the narrowband albedo is equal to the broadband albedo and by applying a Nimbus–7-derived anisotropic reflectance model (Stowe *et al.*, 1980). To convert the infrared data, they used calibration factors derived from sea surface temperature measurements by Morgan (1980) and Koepke (1980). Narrowband radiances were converted to broadband radiant exitances by using regression-based limb darkening and spectral models (Gube, 1980, 1982).

Minnis and Harrison used GOES data. To calibrate the GOES measurements, they collected Nimbus 7 ERB scanner measurements (during November 1978) which were coincident not only in space and time (solar zenith angle) but were also nearly coincident in viewing zenith and relative azimuth angles as well. Quadratic regression equations were developed to relate squared visible counts and infrared radiances to Nimbus 7 radiances. Separate equations were developed for land and ocean backgrounds. The rms errors were about 14% for shortwave radiances and 5–6% for longwave radiances. The regression equations effectively produced calibrated, broadband radiances from the GOES data. To calculate radiant exitances, Minnis and Harrison constructed and used angular dependence models. The longwave model was based on the work of Raschke *et al.* (1973); the shortwave model was based on a combination of GOES and aircraft (Kriebel, 1978) data.

Geostationary satellites have one great advantage over low Earth orbiters: they can observe the diurnal cycle. Figure 10.6 shows an example of diurnal variation of radiation budget parameters calculated by Minnis and Harrison (1984c) for an area over South America during November 1978.

The disadvantage of geostationary satellites is that they observe each point from only one angle, thus the accuracy of radiation budget quantities is directly related to the accuracy of the angular dependence models used in their derivation. This is just the opposite of the case for low-orbiters, which observe from many angles (thus averaging and reducing the effects of angular dependence models) but only once or twice per day. The combination of low-orbiter data and geostationary data should improve the accuracy of radiation budget estimates.

10.3 SURFACE RADIATION BUDGET

For some meteorological applications, particularly agricultural applications, the radiation budget at the Earth's surface is required. The observations of the surface radiation budget are not as direct as observations of the radiation budget at the top of the atmosphere; corrections for the atmosphere must be made. Since

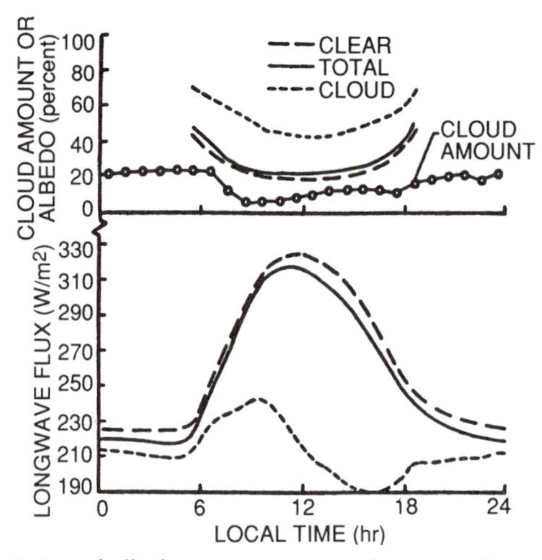

FIGURE 10.6. Diurnal variation of albedo (top, percent) and outgoing longwave radiation (bottom, W m^{-2}) for an area over South America centered at 25.9° South and 68.7° West during November 1978. [After Minnis and Harrison (1984c).]

the surface radiation budget is dominated by clouds, however, satellite data are well suited for this purpose, and they can provide world-wide estimates.

The surface radiation budget may be divided into five quantities: downward shortwave radiation (called insolation), upward shortwave (reflected) radiation, downward longwave radiation, upward longwave radiation, and net radiation.

10.3.1 Downward Shortwave Radiation

Due to interest in the use of solar energy, downward shortwave radiation at the surface has been studied for some years. Following Fritz *et al.* (1964) and Hanson *et al.* (1967), the solar radiation incident at the top of the atmosphere suffers one of three fates: it may be reflected to space, it may be absorbed by the atmosphere, or it may be absorbed at the ground. The governing equation is

$$\mu_{\text{sun}} E_{\text{sun}} = A \mu_{\text{sun}} E_{\text{sun}} + E_{\text{atm}} + (1 - A_{\text{sfc}}) E_{\text{sfc}}, \tag{10.40}$$

where $\mu_{\text{sun}} E_{\text{sun}}$ is the TOA solar insolation as described above, A is the TOA albedo, E_{atm} is the energy absorbed in the atmosphere, E_{sfc} is the downward solar irradiance at the surface, and A_{sfc} is the surface albedo. This equation implicitly includes scattering and multiple reflections between the surface and clouds.

Fritz *et al.* correlated A, measured by TIROS 3, with atmospheric transmittance ($\tau_{\text{atm}} \equiv E_{\text{sfc}} / \mu_{\text{sun}} E_{\text{sun}}$), measured by ground-based instruments. A correlation coefficient of -0.9 was found, which confirms the hypothesis that when clouds (with their high albedo) cover an area, the fraction of solar radiation reaching the surface decreases. Fritz *et al.* went on to explore the fraction of incident radiation which is absorbed in the atmosphere. Hanson *et al.* used more accurate TIROS

4 data from a longer period to show that the solar energy absorbed in the atmosphere over the U.S. during the period March through May 1962 ranged from 9% to 26%, and that absorbed at the ground ranged from 34% to 58%. Non-satellite estimates of the surface albedo were necessary for these studies. Vonder Haar and Ellis (1975) estimated surface albedo from satellite data and surface pyranometer measurements. All of these studies used data from low Earth orbiters, which means that only one observation was made per day during daylight.

Tarpley (1979) developed and Justus *et al.* (1986) improved a statistical technique to use GOES data to estimate E_{sfc} over the U.S. Solving Eq. 10.40 for E_{sfc} we have

$$E_{sfc} = \frac{\mu_{sun}E_{sun} - A\mu_{sun}E_{sun} - E_{atm}}{1 - A_{sfc}}. \tag{10.41}$$

Assuming isotropic reflection,

$$A\mu_{sun}E_{sun} = \pi L, \tag{10.42}$$

where L is the satellite-measured radiance. Eq. 10.41 then becomes

$$E_{sfc} = \frac{\mu_{sun}E_{sun} - \pi L - E_{atm}}{1 - A_{sfc}}. \tag{10.43}$$

The two most important terms in the Eq. 10.43 are the solar insolation ($\mu_{sun}E_{sun}$) and the reflected radiance (πL). It follows that a simple regression equation using these quantities as predictors ought to estimate the surface insolation well. To account for geographic variations in surface albedo and in atmospheric absorption, Tarpley and Justus *et al.* used deviations from the expected minimum radiance rather than radiance itself. Their final regression equation (Justus *et al.*, 1986) is

$$E_{sfc} = \mu_{sun}E_{sun}(a + b\mu_{sun} + c\mu_{sun}^2) + d(\bar{C}^2 - C_{min}^2), \tag{10.44}$$

where $\bar{C}$ and C_{min} are the mean observed and minimum (respectively) GOES visible brightness count (approximately proportional to $L^{-1/2}$). The mean is calculated over a 30–50-km square, which means that the results have this resolution. The most difficult part of this analysis is in keeping an up-to-date map of minimum counts for each satellite viewing time. Justus *et al.* used a 1° latitude-longitude grid for this purpose.

Using this method, Justus *et al.* were able to reproduce hourly surface pyranometer measurements within about 16% rms of the mean hourly insolation and daily surface pyranometer measurements within about 9.5% rms of the daily insolation or about 1.4 MJ m^{-2}. Figure 10.7 shows the mean daily surface insolation over the United States for July and December 1983.

Gautier *et al.* (1980) and Diak and Gautier (1983) developed a physical retrieval scheme based on a simple radiative transfer model. Their model atmosphere consists of an ozone layer; an atmospheric layer, which absorbs and scatters solar radiation; a cloud layer; a subcloud atmospheric layer; and the surface. The absorption and scattering properties of the atmosphere are taken from the litera-

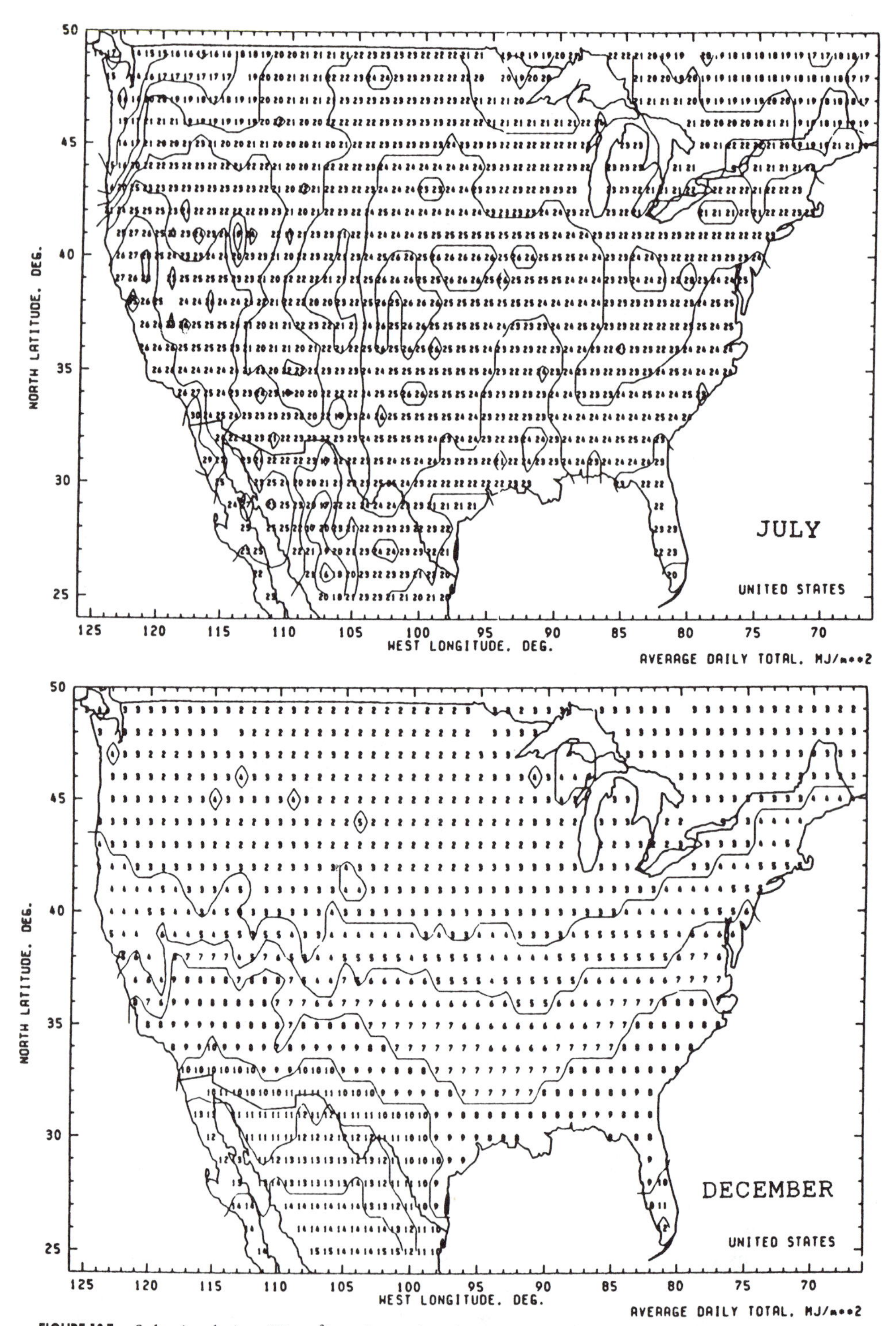

FIGURE 10.7. Solar insolation (W m^{-2}) at the surface for (a) July and (b) December 1983. [Reprinted by permission of the publisher from "Satellite Measured Insolation in the United States, Mexico, and South America" by C. G. Justus, M. V. Paris, and J. D. Tarpley, *Remote Sensing of Environment*, Vol. 20, pp. 57–83. Copyright 1986 by Elsevier Science, Inc.]

ture, and the surface dewpoint (from an analysis) provides an estimate of precipitable water, which is the chief atmospheric absorber of sunlight. A satellite measurement of reflected radiation yields an albedo. If the albedo is lower than a threshold, clear sky is assumed, and the solar insolation at the surface is calculated with a clear-sky model. If the albedo is higher than the threshold, an overcast sky is assumed, and surface solar insolation is calculated from the cloudy model. The technique is simple and computationally easy. Its advantage is that it can be used at every satellite pixel, which provides an estimate of horizontal variability of solar radiation.

In comparison with pyranometers, the method of Diak and Gautier had a standard error of 5.3% of the mean daily insolation, which is roughly the accuracy cited for a well-maintained pyranometer.

Dedieu *et al.* (1987) developed a method similar to that of Diak and Gautier for use with Meteosat data.

Möser and Raschke (1984) also used Meteosat data, but they employed a sophisticated radiative transfer model to convert radiances to surface insolation. Basically, Möser and Raschke assume that clouds perturb the clear-sky insolation $E_{\mathrm{sfc}}^{\mathrm{clr}}$. They find the minimum and maximum observed radiance over an area during a month's time for each observing time. The observed radiances L_{obs} are then normalized by

$$L' = \frac{L_{\mathrm{obs}} - L_{\mathrm{min}}}{L_{\mathrm{max}} - L_{\mathrm{min}}}, \tag{10.45}$$

and the surface insolation is calculated according to

$$E_{\mathrm{sfc}}(\zeta_{\mathrm{sun}}) = [1 - f(L', \zeta_{\mathrm{sun}})E_{\mathrm{sfc}}^{\mathrm{clr}}(\zeta_{\mathrm{sun}})], \tag{10.46}$$

where $f(L', \zeta_{\mathrm{sun}})$ is a tabulated function determined from the radiative transfer model. Möser and Raschke report daily rms errors in Europe of 10–14% (1.8–2.2 MJ m^{-2}) and mean monthly rms errors of 3–7% (0.5–1.3 MJ m^{-2} day-1).

10.3.2 Reflected Shortwave Radiation

The reflected solar radiation is the product of the surface albedo and the downward solar radiation. Thus a knowledge of the surface albedo is the key element in estimating reflected solar radiation.

The basic approach to estimating surface albedo has historically been through the minimum albedo technique (e.g., Raschke and Preuss, 1975). The region of interest is monitored over a period of time, say one month, and the minimum albedo during that time is noted. Since few locations are likely to be cloud-covered for an entire month, the minimum albedo is likely to represent the clear-sky planetary albedo. Corrections for atmospheric absorption and scattering may then be applied to retrieve the albedo of the surface itself (e.g., Preuss and Geleyn, 1980).

Surface albedo is also produced as a byproduct of solar insolation calculations (Gautier et. al., 1980). Dedieu *et al.* (1987) have mapped surface albedos over Europe using Meteosat data.

Pinker and Ewing (1986) note that the surface type is important when using narrowband measurements to infer surface albedo. The albedo of snow, for example is high for wavelengths less than 1 μm, but drops rapidly for longer wavelengths. The albedo of vegetated surfaces, on the other hand, is low in the portion of the spectrum sampled by visible detectors (such as AVHRR channel 1), but rises in the near infrared. Simply assuming that the broadband surface albedo is the same as the narrowband albedo can cause errors if surface type is ignored.

Because Eq. 10.40 is linear in A and A_{sfc}, it is reasonable to assume that an equation of the form

$$A_{sfc} = a + bA \tag{10.47}$$

can convert the satellite-measured planetary albedo A to an estimate of surface albedo. Chen and Ohring (1984) and Koepke and Kriebel (1987) calculate broadband surface albedos from narrowband observations. Koepke (1989) calculates narrowband surface albedos from narrowband AVHRR observations. The coefficients a and b must be functions of solar zenith angle. Koepke (1989) offers tables of a and b for various values of aerosol, ozone, and water vapor concentration.

The accuracy of surface albedo estimates are difficult to determine and are essentially unknown.

10.3.3 Downward Longwave Radiation

Downward longwave radiation at the surface depends primarily on the temperature and moisture structure of the atmosphere and on the character of the clouds. Satellite soundings, therefore, are suitable to estimate downward longwave radiation.

Darnell *et al.* (1983, 1986) have used operationally retrieved TOVS soundings to estimate downward longwave radiation at the surface. The NESDIS operational retrievals (see chapter 6) were processed to yield temperature and moisture information at 50 hPa vertical resolution, then a radiative transfer model was applied to calculate the clear-sky downward radiation. Cloud amount and cloud top height information in the TOVS retrievals were used to calculate the effects of clouds. The clouds were assumed to be 50 hPa thick, thus the cloud base pressure was assumed to be 50 hPa greater than the cloud top pressure. The radiative transfer model was then run to calculate the downward longwave radiation under a plane-parallel cloud. Finally, the clear-sky and overcast results were averaged using the cloud amount as a weight.

Darnell *et al.* (1983) compared essentially instantaneous satellite estimates with ground-based pyrgeometer data and found a 0.87 correlation coefficient and a standard error of estimate of 20 W m^{-2}, which was 6.5% of the mean downward longwave radiation. The largest problem occurred when one site was foggy. The TOVS cloud amount was only 8%, which caused the satellite estimate to be too low by about 80 W m^{-2}.

Darnell *et al.* (1986) improved their data handling and corrected the pyrgeometer data for shortwave heating of the dome. For monthly averaged data, they found a correlation coefficient of 0.98 and a standard error of 10 W m^{-2}, or about 3% of the mean monthly irradiance.

Frouin *et al.* (1988) also used TOVS soundings but they experimented with different ways to estimate the effects of clouds. One method used GOES visible data to estimate cloud amount and cloud optical depth and used GOES infrared data to estimate cloud top height. The optical depth was converted into a cloud depth and finally a cloud base. Other methods used GOES cloud fractions but simpler treatment for cloud base estimation. Over the ocean off the California coast, they found standard errors of 21 to 27 W m^{-2} for half-hourly irradiance estimates and 16 to 22 W m^{-2} for daily estimates. The most sophisticated cloud treatment produced the best results, which indicates the importance of clouds in downward longwave irradiance estimation.

10.3.4 Upward Longwave Radiation

Little longwave radiation is reflected by the surface; nearly all upward longwave radiation is emitted by the surface. The problem of estimating upward longwave radiation, therefore, boils down to estimating the surface temperature. This problem was discussed in section 6.5.2, to which the reader is referred.

10.3.5 Net Radiation

It is possible to combine the above four components into an estimate of net radiation at the surface. Since the errors in each of the components accumulate, however, this is not the approach which has been taken. Rather, research has focused on retrieving net radiation directly from satellite measurements.

Pinker and Corio (1984) began this work by extending the work of Tarpley (1979). They used (1) Tarpley's pyranometer measurements of solar radiation, (2) surface albedo estimates made primarily from aircraft observations (Kung *et al.*, 1964), and (3) surface cloud and temperature observations to construct the daily surface net radiation over the U.S. Great Plains for a period of three summer months. NOAA-retrieved radiation budget parameters from the NOAA 5 satellite (see section 10.2.1) were compared with the ground-based data. Pinker and Corio found a correlation coefficient of 0.69 between TOA net radiation and surface net radiation. The correlation was slightly improved (to 0.76) if both net radiation and daytime OLR were used as predictors.

The NOAA satellites view an area only twice per day, which limits the possible accuracy. Pinker *et al.* (1985) used GOES data to make net radiation estimates. Eleven visible and five infrared observations per day were used to capture the diurnal variation. The satellite net radiation budget estimates were correlated with surface net radiation measurements at five Canadian sites for a two-year period. The daily mean surface net radiation estimated from the TOA net radiation had a standard error of estimate of between 5 and 20% of the mean surface net radiation, or between 6 and 21 W m^{-2}, depending on month and location. The ten-day average net radiation had a standard error of estimate of 5–9% or 5–11 W m^{-2}.

Pinker and Tarpley (1988) found that GOES brightness count alone has nearly as high a correlation with surface net radiation as does TOA net radiation, which supports the argument that clouds are the chief moderator of surface net radiation.

Bibliography

Abel, P., and A. Gruber (1979). *An Improved Model for the Calculation of Longwave Flux at 11 μm.* NOAA Tech. Memo. NESS 106, Washington, D.C.

Ardanuy, P. E., and H. Jacobowitz (1984). A calibration adjustment technique combining ERB parameters from different remote sensing platforms into a long-term data set. *J. Geophys. Res.*, **89**, 5011–5019.

Ardanuy, P. E., H. L. Kyle, R. R. Hucek, and B. S. Groveman (1987). Nimbus 7 Earth Radiation Budget wide field of view climate data set improvement. 2. Deconvolution of Earth radiation budget products and consideration of 1982–1983 El Niño event. *J. Geophys. Res.*, **92**, 4125–4143.

Ardanuy, P., and J. Rea (1984). Degradation asymmetries and recovery of the NIMBUS 7 Earth Radiation Budget shortwave radiometer. *J. Geophys. Res.*, **89**, 5039–5048.

Arking, A., and J. S. Levine (1967). Earth albedo measurements: July 1963 to June 1964. *J. Atmos. Sci.*, **24**, 721–724.

Arking, A., and S. Vemury (1984). The Nimbus 7 ERB data set: A critical analysis. *J. Geophys. Res.*, **89**, 5089–5097.

Atwater, M. A., and J. T. Ball (1981). A surface solar radiation model for cloudy atmospheres. *Mon. Wea. Rev.*, **109**, 878–888.

Bandeen, W. R., R. A. Hanel, J. Lickt, R. A. Stampl, and W. G. Stroud (1961). Infrared and solar radiation measurements from the TIROS II meteorological satellite. *J. Geophys. Res.*, **66**, 3169–3185.

Barkstrom, B. R. (1984). The Earth Radiation Budget Experiment (ERBE). *Bull. Amer. Meteor. Soc.*, **65**, 1170–1185.

Barkstrom, B. R., and G. L. Smith (1986). The Earth Radiation Budget Experiment: Science and implementation. *Rev. Geophys.*, **24**, 379–390.

Barkstrom, B. R., E. Harrison, G. Smith, R. Green, J. Kibler, R. Cess, and the ERBE Science Team (1989). Earth Radiation Budget Experiment (ERBE) archival and April 1985 results. *Bull. Amer. Meteor. Soc.*, 70, 1254–1262.

Bess, T. D., and G. L. Smith (1987a). *Atlas of Wide-Field-of-View Outgoing Longwave Radiation Derived from Nimbus 6 Earth Radiation Budget Data Set–July 1976 to June 1978.* NASA Ref. Pub. RP-1185, Langley Research Center, Hampton, VA.

Bess, T. D., and G. L. Smith (1987b). *Atlas of Wide-Field-of-View Outgoing Longwave Radiation Derived from Nimbus 7 Earth Radiation Budget Data Set–November 1978 to October 1985.* NASA Ref. Pub. RP-1186, Langley Research Center, Hampton, VA.

Bess, T. D., G. L. Smith, and T. P. Charlock (1989). A ten-year monthly data set of outgoing longwave radiation from Nimbus–6 and Nimbus–7 satellites. *Bull. Amer. Meteor. Soc.*, **70**, 480–489.

Bess, T. D., R. N. Green, and G. L. Smith (1980). *Deconvolution and Analysis of Wide-Angle Longwave Radiation Data from Nimbus 6 Earth Radiation Budget Experiment for the First Year.* NASA Tech. Paper 1746, Langley Research Center, Hampton, VA.

Bess, T. D., R. N. Green, and G. L. Smith (1981). Deconvolution of wide field-of-view radiometer measurements of Earth-emitted radiation. Part II: Analysis of first year of Nimbus 6 ERB data. *J. Atmos. Sci.*, **38**, 474–488.

Brooks, D. R., and P. Minnis (1984). Comparison of longwave diurnal models applied to simulations of the Earth Radiation Budget Experiment. *J. Climate Appl. Meteor.*, **23**, 155–160.

Brooks, D. R., E. F. Harrison, P. Minnis, J. T. Suttles, and R. S. Kandel (1986). Development of algorithms for understanding the temporal and spatial variability of the Earth's radiation balance. *Rev. Geophys.*, **24**, 422–438.

Chen, T. S., and G. Ohring (1984). On the relationship between clear-sky planetary and surface albedos. *J. Atmos. Sci.*, **41**, 156–158.

Darnell, W. L., S. K. Gupta, and W. F. Staylor (1983). Downward longwave radiation at the surface from satellite measurements. *J. Climate Appl. Meteor.*, **22**, 1956–1960.

Darnell, W. L., S. K. Gupta, and W. F. Staylor (1986). Downward longwave radiation from sun-synchronous satellite data. *J. Climate Appl. Meteor.*, **25**, 1012–1021.

Dedieu, G., P. Y. Deschamps, and Y. H. Kerr (1987). Satellite estimation of solar irradiance at the surface of the Earth and of surface albedo using a physical model applied to Meteosat data. *J. Climate Appl. Meteor.*, **26**, 79–87.

Diak, G., and C. Gautier (1983). Improvements to a simple physical model for estimating insolation from GOES data. *J. Climate Appl. Meteor.*, **22**, 505–508.

Diekmann, F. J., and G. L. Smith (1989). Investigation of scene identification algorithms for radiation budget measurements. *J. Geophys. Res.*, **94**, 3395–3412.

Duvel, J. P., and R. S. Kandel (1985). Regional-scale diurnal variations of outgoing radiation observed by Meteosat. *J. Climate Appl. Meteor.*, **24**, 335–349.

England, C. F., and G. E. Hunt (1984). A study of the errors due to temporal sampling of Earth's radiation budget. *Tellus*, **36B**, 303–316.

ERBE Science Team (1986). First data from the Earth Radiation Budget Experiment (ERBE). *Bull. Amer. Meteor. Soc.*, **67**, 818–824.

Fritz, S., P. K. Rao, and M. Weinstein (1964). Satellite measurements of reflected solar energy and the energy received at the ground. *J. Atmos. Sci.*, **21**, 141–151.

Frouin, R. D. Gautier, and J. Morcrette (1988). Downward longwave irradiance at the ocean surface from satellite data: Methodology and in situ validation. *J. Geophys. Res.*, **93**, 597–619.

Gautier, C., G. Diak, and S. Masse (1980). A simple physical model to estimate incident solar radiation at the surface from GOES satellite data. *J. Appl. Meteor.*, **19**, 1005–1012.

Green, R. N. (1980). The effect of directional radiation models on the interpretation of Earth radiation budget measurements. *J. Atmos. Sci.*, **37**, 2298–2313.

Green, R. N. (1981). The effect of data analysis techniques on the interpretation of wide-angle longwave radiation measurements. *J. Atmos. Sci.*, **38**, 2045–2055.

Green, R. N. (1983). Accuracy and resolution of Earth radiation budget estimates. *J. Atmos. Sci.*, **40**, 977–985.

Green, R. N., and G. L. Smith (1978). *Parameter Estimation Applied to Nimbus 6 Wide-Angle Longwave Radiation Measurements*. NASA Tech. Paper 1307. Langley Research Center, Hampton, VA.

Green, R. N., J. T. Suttles, and B. A. Wielicki (1990a). Angular dependence models for radiance to flux conversion. SPIE's 1990 Technical Symposium on Optical Engineering and Photonics in Aerospace Sensing, Conference on Long-Term Monitoring of the Earth's Radiation Budget, 16–20 April 1990, Orlando, Florida. [Available from the first author, Atmospheric Sciences Division, NASA Langley Research Center, Hampton, VA 23665.]

Green, R. N., F. B. House, P. W. Stackhouse, X. Wu, S. A. Ackerman, W. L. Smith, and M. J. Johnson (1990b). Intercomparison of scanner and nonscanner measurements for the Earth Radiation Budget Experiment. *J. Geophys. Res.*, **95**, 11785–11798.

Green, R. N., G. L. Smith, L. M. Avis, J. T. Suttles, B. A. Wielicki, J. Robbins, L. Harris, J. Johnson, N. Manalo, and S. G. Gupta (1989). Inversion validation for the Earth Radiation Budget Experiment. In J. Lenoble and M. Geleyn (eds), *International Radiation Symposium 1988*, A. Deepak Publishing, Hampton, Virginia.

Gruber, A. (1978). *Determination of the Earth-Atmosphere Radiation Budget from NOAA Satellite Data*, revised edition. NOAA Tech. Rep. NESS 76, Washington, DC.

Gruber, A., and A. F. Krueger (1984). The status of the NOAA outgoing longwave data set. *Bull. Amer. Meteor. Soc.*, **65**, 958–962.

Gruber, A., and H. Jacobowitz (1985). The longwave radiation estimated from NOAA polar orbiting satellites, an update and comparison with Nimbus–7 ERB results. *Adv. Space Res.*, **5**, 111–120.
Gruber, A., and T. S. Chen (1988). Diurnal variation of outgoing longwave radiation. *J. Climatology*, **8**, 1–16.
Gruber, A., I. Ruff, and C. Earnest (1983). *Determination of the Planetary Radiation Budget from TIROS–N Satellites.* NOAA Tech. Rep. NESDIS 3, Washington, DC.
Gruber, A., M. Varnadore, P. A. Arkin, and J. S. Winston (1986). *Monthly and Seasonal Mean Outgoing Longwave Radiation and Anomalies.* NOAA Tech. Rep. NESDIS 26, Washington, DC.
Gube, M. (1980). Outgoing longwave flux computation from Meteosat data. *ESA Journal*, **4**, 381–396.
Gube, M. (1982). Planetary albedo estimates from Meteosat data. *ESA Journal*, **6**, 53–69.
Hall, J. B., Jr. (ed.) (1982). *Earth Radiation Science Seminars.* NASA Conf. Pub. 2239, Langley Research Center, Hampton, VA.
Hanson, K. J, T. H. Vonder Haar, and V. E. Suomi (1967). Reflection of sunlight to space and absorption by the Earth and atmosphere over the United States during spring 1962. *Mon. Wea. Rev.*, **95**, 354–362.
Hanson, K. J. (1971). Studies of cloud and satellite parameterization of solar irradiation at the Earth's surface. *Proceedings: Miami Workshop on Remote Sensing.* U.S. Department of Commerce, 133–148.
Harrison, E. F., D. R. Brooks, P. Minnis, B. A. Wielicki, W. F. Staylor, G. G. Gibson, D. F. Young, F. M. Denn, and the ERBE Science Team (1988). First estimates of the diurnal variation of longwave radiation from the multiple-satellite Earth Radiation Budget Experiment. *Bull. Amer. Meteor. Soc.*, **69**, 1144–1151.
Hartmann, D., V. Ramanathan, A. Berrior, and G. E. Hunt (1986). Earth radiation budget data and climate research. *Rev. Geophys.*, **24**, 439–468.
Hickey, J. R., B. M. Alton, F. J. Griffin, H. Jacobowitz, P. Pelligrino, R. H. Maschhoff, E. A. Smith, and T. H. Vonder Haar (1982). Extraterrestrial solar irradiance variability: Two and one-half years of measurements from Nimbus–7. *Sol. Energy*, **29**, 125–127.
Hickey, J. R., B. M. Alton, H. L. Kyle, and D. Hoyt (1988a). Total solar irradiance measurements by ERB/Nimbus 7: A review of nine years. *Space Sci. Rev.*, **48**, 321–342.
Hickey, J. R., B. M. Alton, H. L. Kyle, and E. R. Major (1988b). Observation of total irradiance variability from Nimbus satellites. *Adv. Space Res.*, **8**(7), 5–10.
Hickey, J. R., L. L. Stowe, H. Jacobowitz, P. Pelligrino, R. H. Maschhoff, F. B. House, and T. H. Vonder Haar (1980). Initial solar irradiance determinations from Nimbus 7 cavity radiometer measurements. *Science*, **208**, 281–283.
House, F. B., A. Gruber, G. E. Hunt, and A. T. Mecherikunnel (1986). History of satellite missions and measurements of the Earth radiation budget (1957–1984). *Rev. Geophys.*, **24**, 357–377.
House, F. B., and J. C. Jafolla (1980). One dimensional technique for enhancing Earth radiation budget observations from Nimbus 7 satellite. 1980 International Radiation Symposium, IAMAP Radiation Commission, Fort Collins, CO, 292–294.
Hoyt, D. V., H. L. Kyle, J. R. Hickey, and R. H. Maschhoff (1992). The Nimbus–7 total solar irradiance: A new algorithm for its derivation. *J. Geophys. Res.*, **97**, 51–63.
Hucek, R. R., H. L. Kyle, and P. E. Ardanuy (1987). Nimbus 7 Earth Radiation Budget wide field of view climate data set improvement. 1. The Earth albedo from deconvolution of shortwave measurements. *J. Geophys. Res.*, **92**, 4107–4123.
Hunt, G. E., R. Kandel, and A. T. Mecherikunnel (1986). A history of presatellite investigations of the Earth's radiation budget. *Rev. Geophys.*, **24**, 351–356.
Hwang, P. H., L. L. Stowe, H. Y. M. Yeh, H. L. Kyle, and the Nimbus 7 Cloud Data Processing Team (1988). The Nimbus–7 global cloud climatology. *Bull. Amer. Meteor. Soc.*, **69**, 743–752.
Jacobowitz, H., H. V. Soule, H. L. Kyle, F. B. House, and the Nimbus 7 ERB Experiment Team (1984a). The Earth Radiation Budget (ERB) experiment: An overview. *J. Geophys. Res.*, **89**, 5021–5038.
Jacobowitz, H., L. L. Stowe, and J. R. Hickey (1978). The Earth Radiation Budget (ERB) experiment. In *The Nimbus 7 Users' Guide*, NASA Goddard Space Flight Center, Greenbelt, MD, 33–69.
Jacobowitz, H., R. J. Tighe, and the Nimbus 7 ERB Experiment Team (1984b). The Earth radiation budget derived from the Nimbus–7 ERB experiment. *J. Geophys. Res.*, **89**, 4997–5010.

Jacobowitz, H., W. L. Smith, H. B. Howell, F. W. Nagle, and J. R. Hickey (1979). The first 18 months of planetary radiation budget measurements from the Nimbus 6 ERB experiment. *J. Atmos. Sci.*, **36**, 501–507.

Janowiak, J. E., A. F. Krueger, P. A. Arkin, and A. Gruber (1985). *Atlas of Outgoing Longwave Radiation Derived from NOAA Satellite Data.* NOAA Atlas No. 6, Washington, DC.

Justus, C. G., M. V. Paris, and J. D. Tarpley (1986). Satellite measured insolation in the United States, Mexico, and South America. *Rem. Sens. Env.*, **20**, 57–83.

Kidwell, K. B. (1991). *NOAA Polar Orbiter Data User's Guide.* NOAA/NESDIS Satellite Data Services Division, Washington, D.C.

King, M. D., and R. J. Curran (1980). The effect of a nonuniform planetary albedo on the interpretation of Earth radiation budget observations. *J. Atmos. Sci.*, **37**, 1262–1278.

Kopia, L. P., Earth Radiation Budget Experiment scanner instrument. *Rev. Geophys.*, **24**, 400–406.

Koepke, P. (1980). Calibration of the Meteosat IR channel by ground measurements. *Contrib. Atmos. Phys.*, **53**, 442–445.

Koepke, P., and K. T. Kriebel (1987). Improvements in the shortwave cloudfree radiation budget accuracy. Part I: Numerical study including surface anisotropy. *J. Climate Appl. Meteor.*, **26**, 374–395.

Koepke, P. (1989). Removal of atmospheric effects from AVHRR albedos. *J. Appl. Meteor.*, **28**, 1341–1348.

Kriebel, K. T. (1978). Measured spectral bidirectional reflection properties of vegetated surfaces. *Appl. Opt.*, **17**, 253–259.

Kriebel, K. T. (1981). Calibration of the Meteosat VIS channel by airborne measurements. *Appl. Opt.*, **20**, 11–12.

Kung, E. C., R. A. Bryson, And D. H. Lenschow (1964). Study of a continental surface albedo on the basis of flight measurements and structure of the earth's surface cover over North America. *Mon. Wea. Rev.*, **92**, 543–564.

Kyle, H. L., A. Mecherikunnel, P. Ardanuy, L. Penn, B. Groveman, G. G. Campbell, and T. H. Vonder Haar (1990). A comparison of two major earth radiation budget data sets. *J. Geophys. Res.*, **95**, 9951–9970.

Kyle, H. L., F. B. House, P. E. Ardanuy, H. Jacobowitz, R. H. Maschhoff, and J. R. Hickey (1984). New in-flight calibration adjustment of the Nimbus 6 and 7 Earth Radiation Budget wide field of view radiometers. *J. Geophys. Res.*, **89**, 5057–5076.

Kyle, H. L., J. R. Hickey, P. E. Ardanuy, H. Jacobowitz, A. Arking, G. G. Campbell, F. B. House, R. Maschhoff, G. L. Smith, L. L. Stowe, and T. Vonder Haar (1993). The Nimbus Earth Radiation Budget (ERB) experiment: 1975 to 1992. *Bull. Amer. Meteor. Soc.*, **74**, 815–830.

Kyle, H. L., P. E. Ardanuy, and E. J. Hurley (1985). The status of the Nimbus–7 Earth radiation budget data set. *Bull. Amer. Meteor. Soc.*, **66**, 1378–1388.

Lauritson, L. G., Nelson, G. J., and F. W. Porto (1979). *Data Extraction and Calibration of TIROS–N/NOAA Radiometers.* NOAA Tech. Memo. NESS 107, Washington, D.C.

London, J., and T. Sasamori (1971). Radiative energy budget of the atmosphere. *Space Res.*, **11**, 639–649.

Luther, M. R., J. E. Cooper, and G. R. Taylor (1984). Earth Radiation Budget Experiment nonscanner instrument. *Rev. Geophys.*, **24**, 391–399.

Maschhoff, R., A. Jalink, J. Hickey, and J. Swedberg (1984). NIMBUS–Earth radiation budget sensor characterization for improved data reduction fidelity. *J. Geophys. Res.*, **89**, 5049–5056.

Minnis, P., and E. F. Harrison (1984a). Diurnal variability of regional cloud and clear-sky radiative parameters derived from GOES data. Part I: Analysis method. *J. Climate Appl. Meteor.*, **23**, 993–1011.

Minnis, P., and E. F. Harrison (1984b). Diurnal variability of regional cloud and clear-sky radiative parameters derived from GOES data. Part II: November 1978 cloud distributions. *J. Climate Appl. Meteor.*, **23**, 1012–1031.

Minnis, P., and E. F. Harrison (1984c). Diurnal variability of regional cloud and clear-sky radiative parameters derived from GOES data. Part III: November 1978 radiative parameters. *J. Climate Appl. Meteor.*, **23**, 1032–1051.

Morcrette, J., and P. Deschamps (1986). Downward longwave radiation at the surface in clear-sky atmospheres: Comparisons of measured, satellite-derived, and calculated fluxes. ISLSCP: Proceedings of an international conference held in Rome, Italy, 1985; ESA SP-248, ESA, Paris, France, 257–262.

Morgan, J. (1980). Meteosat 1 calibration report. Issue 8, ESOC MDMD/MET.

Möser, W., and E. Raschke (1984). Incident solar radiation over Europe from METEOSAT data. *J. Climate Appl. Meteor.*, **23**, 166–170.

NASA (1978). *The Nimbus 7 Users' Guide.* NASA/Goddard Space Flight Center, Greenbelt, MD.

Ohring, G., A. Gruber, and R. G. Ellingson (1984). Satellite determinations of the relationship between total longwave radiation flux and infrared window radiance. *J. Climate Appl. Meteor.*, **23**, 416–425.

Pinker, R. T., and J. D. Tarpley (1988). The relationship between the planetary and surface net radiation: An update. *J. Appl. Meteor.*, **27**, 957–964.

Pinker, R. T., and J. A. Ewing (1985). On the modeling of surface solar radiation: Model formulation and validation. *J. Climate Appl. Meteor.*, **24**, 389–401.

Pinker, R. T., and J. A. Ewing (1986). Effect of surface properties on the narrow to broadband spectral relationship in clear sky satellite observations. *Rem. Sens. Env.*, **20**, 267–282.

Pinker, R. T., and L. A. Corio (1984). Surface radiation budget from satellites. *Mon. Wea. Rev.*, **112**, 209–215.

Pinker, R. T., J. A. Ewing, and J. D. Tarpley (1985). The relationship between the planetary and surface net radiation. *J. Climate Appl. Meteor.*, **24**, 1262–1268.

Preuss, H. J., and J. F. Geleyn (1980). Surface albedos derived from satellite data and their impact on forecast models. *Arch. Meteorol. Geophys. Bioklmatol., Ser. A*, **29**, 345.

Ramanathan, V., R. D. Cess, E. F. Harrison, P. Minnis, B. R.. Barkstrom, E. Ahmad, and D. Hartmann (1989). Cloud-radiative forcing and climate: Results from the Earth Radiation Budget Experiment. *Science*, **243**, 57–63.

Raschke, E., T. H. Vonder Haar, W. R. Bandeen, and M. Pasternak (1973). The annual radiation balance of the Earth-atmosphere system during 1969–1970 from Nimbus 3 measurements. *J. Atmos. Sci.*, **30**, 341–364.

Raschke, E., and H. J. Preuss (1975). The determination of the solar radiation budget of the Earth's surface from satellite measurement. *Meteorol. Rundsch.*, **32**, 18–28.

Rasool, S. I., and J. H. Bolle (1984). International Satellite Land Surface Climatology Project. *Bull. Amer. Meteor. Soc.*, **65**, 141.

Saunders, R. W., and G. E. Hunt (1983). Some radiation budget and cloud measurements derived from Meteosat 1 data. *Tellus*, **35B**, 177–188.

Saunders, R. W., and G. E. Hunt (1980). Meteosat observations of diurnal variation of radiation budget parameters. *Nature*, **283**, 645–647.

Saunders, R. W., L. L. Stowe, G. E. Hunt, and C. F. England (1983). An intercomparison between radiation budget estimates from Meteosat, Nimbus 7 and TIROS–N satellites. *J. Climate Appl. Meteor.*, **22**, 546–559.

Smith, G. L., and R. N. Green (1981). Deconvolution of wide field-of-view radiometer measurements of Earth-emitted radiation. Part I: Theory. *J. Atmos. Sci.*, **38**, 461–473.

Smith, G. L., R. N. Green, E. Raschke, L. M. Avis, J. T. Suttles, B. A. Wielicki, and R. Davies (1986). Inversion methods for satellite studies of the Earth's radiation budget: Development of algorithms for the ERBE mission. *Rev. Geophys.*, **24**, 407–421.

Smith, W. L., D. T. Hilleary, H. Jacobowitz, H. B. Howell, J. R. Hickey, and A. J. Drummond (1975). The Earth Radiation Budget (ERB) experiment. In *The Nimbus 6 User's Guide.* NASA Goddard Space Flight Center, Greenbelt, MD, 109–139.

Smith, W. L., D. T. Hilleary, H. Jacobowitz, H. B. Howell, J. R. Hickey, and A. J. Drummond (1977). Nimbus–6 Earth Radiation Budget Experiment. *Appl. Opt.*, **16**, 306–318.

Stephens, G. L., Campbell, G. G., and T. H. Vonder Haar (1981). Earth radiation budgets. *J. Geophys. Res.*, **86**, 9739–9760.

Stowe, L. L. (ed.) (1988). *Report of the Earth Radiation Budget Requirements Review–1987.* NOAA Tech. Rep. NESDIS 41, Washington, DC.

Stowe, L. L., and M. Fromm (1983). *Nimbus–7 ERB Sub-Target Radiance Tape (STRT) Data Base.* NOAA Tech. Memo. NESDIS 3, Washington, DC.

Stowe, L. L., C. G. Wellemeyer, T. F. Eck, H. Y. M. Yeh, and the Nimbus 7 Cloud Data Processing Team (1988). Nimbus–7 global cloud climatology. Part I: Algorithms and validation. *J. Climate*, **1**, 445–470.

Stowe, L. L., H. Jacobowitz, and V. R. Taylor (1980). Reflectance characteristics of Earth and cloud surfaces as measured by the ERB scanning channels on the Nimbus 7 satellite. *Proceedings: International Radiation Symposium*, Fort Collins, Colorado, 430–432.

Suomi, V. E. (1958). The radiation balance of the Earth from a satellite. *Ann. IGY*, **1**, 331–340.

Suttles, J. T., R. N. Green, G. L. Smith, B. A. Wielicki, I. J. Walker, V. R. Taylor, and L. L. Stowe (1989). *Angular Radiation Models for Earth-Atmosphere System, Volume II: Longwave Radiation.* NASA Ref. Pub. RP 1184 vol. II, Langley Research Center, Hampton, VA.

Suttles, J. T., R. N. Green, P. Minnis, G. L. Smith, W. F. Staylor, B. A. Wielicki, I. J. Walker, D. F. Young, V. R. Taylor, and L. L. Stowe (1988). *Angular Radiation Models for Earth-Atmosphere System, Volume I: Shortwave Radiation.* NASA Ref. Pub. RP 1184 vol. I, Langley Research Center, Hampton, Virginia.

Suttles, J. T., B. A. Wielicki, and S. Vemury (1992). Top-of-atmosphere radiative fluxes: Validation of ERBE scanner inversion algorithm using Nimbus–7 ERB data. *J. Appl. Meteor.*, **31**, 784–796.

Tarpley, D. (1979). Estimating incident solar radiation at the surface from geostationary satellite data. *J. Climate Appl. Meteor.*, **18**, 1172–1181.

Tarpley, J. D. (1979). Estimating incident solar radiation at the surface from geostationary satellite data. *J. Appl. Meteor.*, **18**, 1172–1181.

Taylor, V. R., and L. L. Stowe (1984). Reflectance characteristics of uniform earth and cloud surfaces derived from Nimbus 7 ERB. *J. Geophys. Res.*, **89**, 4987–4996.

Thompson, S. L., and S. G. Warren (1982). Parameterization of outgoing infrared radiation derived from detailed radiative calculations. *J. Atmos. Sci.*, **39**, 2667–2680.

Twomey, S. (1977). *Introduction to the Mathematics of Inversion in Remote Sensing and Indirect Measurements.* Elsevier, New York, 243 pp.

Vonder Haar, T. H., and V. E. Suomi (1971). Measurements of the Earth's radiation budget from satellites during a five-year period. Part 1: Extended time and space means. *J. Atmos. Sci.*, **28**, 305–314.

Vonder Haar, T. H., E. Raschke, M. Pasternak, and W. Bandeen (1972). The radiation budget of the Earth-atmosphere system as measured from the Nimbus 3 satellite (1969–1970). *Space Res.*, **12**, 491–497.

Vonder Haar, T. H., and J. S. Ellis (1975). Solar energy microclimate as determined from satellite observations. *Proc. Soc. Photo-Opt. Instrum. Eng.*, **68**, 18–22.

Wielicki, B. A., and R. N. Green (1989). Cloud identification for ERBE radiative flux retrieval. *J. Appl. Meteor.*, **28**, 1133–1146.

Willson, R. C. (1984). Measurements of solar total irradiance and its variability. *Space Sci. Rev.*, **38**, 203–242.

Willson, R. C., and H. S. Hudson (1988). Solar luminosity variations in solar cycle 21. *Nature*, **322**, 810–812.

Willson, R. C., S. Gulkis, M. Jansson, H. S. Hudson, and G. A. Chapman (1981). Observations of solar irradiance variability. *Science*, **211**, 700–703.

Willson, R. C. (1979). Active cavity radiometer type IV. *Appl. Optics*, 18(2), 179–188.

Winston, J. S., A. Gruber, T. I. Gray, Jr., M. S. Varnadore, C. L. Earnest, and L. P. Mannello (1979). *Earth-Atmosphere Radiation Budget Analyses Derived from NOAA Satellite Data, June 1974–February 1978.* Vol. 1 and 2, NOAA, Washington, DC.

Wydick, J. E., P. A. Davis, and A. Gruber (1987). *Estimation of Broadband Planetary Albedo from Operational Narrowband Satellite Measurements.* NOAA Tech. Rep. NESDIS 27, Washington, DC.

11

The Future

THE COMING DECADES promise to be an exciting era in satellite meteorology. Nearly all of the currently operational satellites will be replaced with new, more capable satellites, research satellites will be launched, and instruments will be designed which will make possible fundamental new measurements. Writing about this era, however, is somewhat challenging because of the rapid rate of change in technology, politics, and economics. We have chosen to divide the chapter into three sections. The first discusses the nearly certain changes that will occur in the NOAA polar orbiting satellites. The second section discusses the broad, but still evolving, plans for Mission to Planet Earth, which constitutes the bulk of the chapter. Finally, in the third section we discuss some potential observations that we think are important, but that do not appear to be included in the plans now taking shape.

11.1 NOAA K, L, M

NOAA and NASA have completed the specifications for the third generation of TIROS N satellites (NASA, 1988a), which are called NOAA K, L, M. The first of this series is scheduled for launch in 1996.

TABLE 11.1. NOAA K, L, M Instruments

Satellite	*AVHRR/3*	*HIRS/3*	*AMSU-A*	*AMSU-B*	*SBUV/2*	*SEM/2*	*DCS/2*	*S&R/2*
NOAA K	X	X	X	X	X	X	X	X
NOAA L	X	X	X	X		X	X	X
NOAA M	X	X	X	X	X	X	X	X

NOAA K, L, M will not be radically different than previous satellites, although most of the instruments will be modified. Table 11.1 shows the instrument complement for these satellites. In the next three sections, we outline the significant changes to the instruments.

11.1.1 AVHRR/3

The AVHRR/3, which will fly on NOAA K, L, M (NASA, 1988b), will have a sixth channel. Known as channel 3A, the new channel will have a center wavelength of 1.6 μm. This channel will be used during daylight to gain information on the phase of clouds (ice or water) and to discriminate between low cloud and snow cover or ice. It is an improved version of the 1.6-μm Special Sensor C, which flew on the DMSP F4 satellite (Bunting and d'Entremont, 1982). The old channel 3 (3.7 μm) will be called channel 3B. The two channel 3s will operate at different times: 3A during daylight, 3B at night. The instrument will be programmed to automatically switch channels at the terminator.

Several other small changes will be made in AVHRR/3 to improve performance or calibration. Otherwise, the AVHRR/3 will be virtually identical to the AVHRR/2. AVHRR/3 is being build by ITT Aerospace/Optical Division.

11.1.2 HIRS/3

The HIRS/3 will also be very similar to its predecessor, the HIRS/2 (NASA, 1988c). The only major difference is that five of the channels will have different wavelengths (Table 11.2). All of the changes are made to better sense tropospheric

TABLE 11.2. HIRS/3 Channels[a]

	Central Wavelength			
Channel	*Old* (μm)	*New* (μm)	*Half-Power Bandwidth* (cm^{-1})	*Specified NEΔL* ($mW\ m^{-2}\ sr^{-1}\ cm$)
10	8.16	12.47	16	0.15
12	6.72	6.52	55	0.20
15	4.46	4.47	23	0.004
16	4.40	4.45	23	0.004
17	4.24	4.13	28	0.002

[a] Differences from HIRS/2; compare with Table 4.5.

TABLE 11.3. Microwave Instrument Comparison

Parameter	*SSM/T*	*MSU*	*SSM/I*	*AMSU–A*	*AMSU–B*	*SSM/T–2*
Satellites	Current DMSP	Current NOAA	Current DMSP	NOAA K, L, M	NOAA K, L, M	Current DMSP
Channels	7	4	7	15	5	5
Frequency Range (GHz)	50.5–59.4	50.3–57.95	19.35–85.5	23.8–89.0	89.0–183.3	91.6–183.3
NEΔT (K)	0.4–0.6	0.3	0.4–1.7	0.25–1.20	0.8	0.5
Beam width	14°	7.5°	0.3°–1.2°	3.3°	1.1°	3.3°–6.0°
Best Ground Resolution (km)	204	110	12.5–50	48	16	48–84
Scan steps	7	11	64–128	30	90	28
Swath width (km)	2053	2347	1394	2179	2179	2053

temperature and water vapor. HIRS/3 is also being build by ITT Aerospace/Optical Division.

11.1.3 AMSU

The major change on NOAA K, L, M is that the Microwave Sounding Unit will be significantly improved and renamed the Advanced Microwave Sounding Unit (AMSU). In fact it will become two instruments:[1] AMSU–A (being built by Aerojet) and AMSU–B (being supplied by the British Met. Office). Table 11.3 compares current and future microwave instruments.

AMSU–A will have twice the horizontal resolution of the MSU, and AMSU–B will have six times the resolution. This resolution increase will substantially improve the scale on which analysis of the data can be performed.

A second major improvement in AMSU is that it will have a total of 20 channels in comparison with only four on the MSU. Table 11.4 lists the frequencies sensed by these instruments. Further, the new channels will allow the retrieval of water vapor profiles, integrated liquid water, precipitation, and such surface parameters as soil moisture and wind speed (over the ocean), in addition to atmospheric temperature.

AMSU–A will replace the infrared Stratospheric Sounding Unit (SSU) on the current NOAA satellites. Channels 10–14 are stratospheric-sensing channels.

Finally, the retrieval process will also change significantly for the NOAA K, L, M sounding system. On the current NOAA satellites, the HIRS/2 is the primary sounding instrument. The MSU is an important part of the retrieval process, but its chief use is in overcast situations when the infrared sounders cannot be used for tropospheric measurements. On NOAA K, L, M, the primary sounding instrument will be the AMSU. HIRS/3 will be used when cloud conditions permit to improve vertical resolution.

[1] Actually three instruments, because AMSU–A will have two instrument packages, AMSU–A1 and AMSU–A2.

TABLE 11.4. Microwave Frequencies[a] (GHz) and Polarizations[b]

Channel	*SSM/T*	*MSU*	*AMSU–A*	*AMSU–B*	*SSM/I*	*SSM/T–2*
1	50.5H	50.30R	23.8R	89.0R	19.35H	183.3 ± 3R
2	53.2H	53.74R	31.4R	157.0R	19.35V	183.3 ± 1R
3	54.35H	54.96R	50.3R	183.3 ± 1R	22.235V	183.3 ± 7R
4	54.9H	57.95R	52.8R	183.3 ± 3R	37.0H	91.7R
5	58.4V		53.6R	183.3 ± 7R	37.0V	150R
6	58.825V		54.4R		85.5H	
7	59.4V		54.9R		85.5V	
8			55.5R			
9			57.2R			
10			57.29 ± .217R			
11			57.29 ± .322 ± .048R			
12			57.29 ± .322 ± .022R			
13			57.29 ± .322 ± .010R			
14			57.29 ± .322 ± .0045R			
15			89.0R			

[a] *Notation*: $x \pm y \pm z$; x is the center frequency. If y appears, the center frequency is not sensed, but two bands, one on either side of the center frequency, are sensed; y is the distance from the center frequency to the center of the two passbands. If z appears, it is the width of the two passbands. This pattern is easily implemented with radiofrequency receivers, and it effectively doubles the signal (two passbands instead of one).

[b] V = vertical, H = horizontal, R = rotates with scan angle.

11.2 MISSION TO PLANET EARTH

In response to the growing concern with global environmental change, and particularly with human-induced change, NASA initiated in the 1980s a broad interdisciplinary program of satellite and in situ observations and research called Mission to Planet Earth (MTPE).[2] Today, MTPE is an international program with contributions from Japan, Europe, and Canada as well as from non-NASA U.S. agencies. The space component of MTPE will include satellites in sunsynchronous orbit, geostationary orbit, and other orbits as required to make the observations necessary to understand the Earth system. The centerpiece of the MTPE is the Earth Observing System (EOS)—"a series of polar-orbiting and low inclination satellites for long-term global observations of the land surface, biosphere, solid Earth, atmosphere, and oceans" (Asrar and Dokken, 1993). MTPE is currently envisioned as having three phases: a pre-EOS era, the EOS era, and a post-EOS or EOS-follow-on era.

[2] Most of the information in Section 11.2 came from the 1993 edition of the *EOS Reference Handbook* (Asrar and Dokken, 1993), which is the most comprehensive and up-to-data source on future plans for Earth observation that we have found. NASA publishes a new edition of the *EOS Reference Handbook* approximately annually. We strongly suggest that readers obtain a copy of the latest edition. (See the bibliography for the address.)

11.2.1 The Pre-EOS Era

The pre-EOS era satellites are operational and experimental satellites that were planned before the advent of EOS and are scheduled for launch before the first EOS satellite in 1998. Table 11.5 lists currently planned instruments and satellites that are part of the pre-EOS MTPE program. In keeping with the MTPE philosophy, most of the satellites are interdisciplinary—they make measurements of several components of the Earth system. We briefly discuss two yet-to-be-launched satellites that will make extensive atmospheric measurements.

11.2.1.1 TRMM

The Tropical Rainfall Measuring Mission (TRMM) is a joint NASA/NASDA mission designed to measure tropical rainfall and its diurnal variability on a monthly time scale and in areas ("boxes") of 10^5 km^2. TRMM will be launched on a Japanese H-2 rocket and will fly in a 350-km orbit with a 35° inclination angle, which causes it to sample all local times each 23 days. Five instruments, which measure various related aspects of precipitation, are planned for TRMM:

- **Precipitation Radar** (PR)—An electronically scanning radar operating at 13.8 GHz; 4.3-km^2 instantaneous field of view at nadir; 220-km swath. (Provided by NASDA.)
- **TRMM Microwave Imager** (TMI)—A five-channel passive microwave imager making measurements from 10 to 91 GHz. (Provided by NASA.)
- **Visible Infrared Scanner** (VIRS)—A five-channel imaging radiometer (0.63, 1.6, 3.75, 10.7, and 12.0 μm) with nominal 2-km resolution at nadir and a 1500-km swath width; similar in design to AVHRR/3. (Provided by NASA.)
- **Lightning Imaging Sensor** (LIS)—See Section 11.2.2.10. (Provided by NASA.)
- **Clouds and Earth's Radiant Energy System** (CERES)—See Section 11.2.2.4. (Provided by NASA.)

The basic idea of TRMM is to mount all of the instruments that can be used to measure precipitation from satellite platforms on one satellite, where the measurements can be compared and combined. For example, the 220-km swath width of the PR samples only a small fraction of the area sensed by the other instruments, but it will help calibrate the visible, infrared, and microwave sensors to yield improved precipitation estimates. Similarly, the LIS, which will sense lightning flashes, will help determine the location of the more active thunderstorm cells.

11.2.1.2 ADEOS

The Advanced Earth Observing Satellite (ADEOS) is a joint Japanese/U.S./French mission to observe the atmosphere, land, and oceans. An 800-km, 98.6°-inclination sunsynchronous orbit with a 10:30 am equator crossing time is

TABLE 11.5. Mission to Planet Earth Pre-EOS Era Satellites[a]

Satellite (Estimated Launch Year)	*Mission Objectives*
	NASA Satellites
ERBS (operating) Earth Radiation Budget Satellite	Radiation budget, aerosol, and ozone data from 57° inclination orbit
TOMS/Meteor (operating) Total Ozone Mapping Spectrometer	Ozone mapping and monitoring (joint with Russia)
UARS (operating) Upper Atmosphere Research Satellite	Stratospheric and mesospheric chemistry and related processes
NASA Spacelab Series (1992 on)	A series of Shuttle-based experiments to measure atmospheric and solar dynamics (ATLAS), atmospheric aerosols (LITE), and surface radar backscatter polarization, and phase information [SIR—C and X—SAR (joint with Germany)]
TOPEX/Poseidon (operating) Ocean Topography Experiment	Ocean circulation (joint with France)
LAGEOS 2 (operating) Laser Geodynamics Satellite	Satellite laser-ranging target for monitoring crustal motions and Earth rotation variations (joint with Italy)
SeaWiFS Sea-Viewing Wide Field Study	Purchase of ocean color data to monitor ocean productivity
TOMS/Earth Probe Total Ozone Mapping Spectrometer	Ozone mapping and monitoring
NSCAT/ADEOS (1996) NASA Scatterometer/Advanced Earth Observing Satellite	Ocean surface wind vectors (joint with Japan)
TOMS/ADEOS (1996) Total Ozone Mapping Spectrometer/ Advanced Earth Observing Satellite	Ozone mapping and monitoring (joint with Japan)
TRMM (1998) Tropical Rainfall Measuring Mission	Precipitation, clouds, and radiation processes in lower latitudes (joint with Japan)
Landsat 7 (1997) Land Remote-Sensing Satellite	High-spatial-resolution visible and infrared radiance/ reflectance to monitor land surface (joint with DoD)
	Non-NASA Satellites
NOAA 9 trough J (U.S.—operational)	Visible and infrared radiance/reflectance, infrared atmospheric sounding, and ozone measurements
Landsat 4/5/6 (U.S.—operational) Land Remote-Sensing Satellite	High spatial resolution visible and infrared radiance/ reflectance
DMSP (U.S.—operational) Defense Meteorological Satellite Program	Visible, infrared, and passive microwave atmospheric and surface measurements
ERS 1 (ESA—pre-operational) European Remote-Sensing Satellite	C-band SAR, microwave altimeter, scatterometer, and sea surface temperature
JERS 1 (ESA—preoperational) Japan's Remote-Sensing Satellite	L-band SAR backscatter and high spatial resolution visible and infrared radiance—reflectance
ERS 2 (ESA—1994) European Remote-Sensing Satellite	Same as ERS 1, plus ozone mapping and monitoring
Radarsat (Canada—1995) Radar Satellite	C-band SAR measurements of Earth's surface (joint U.S./Canadian mission)
NOAA K through N (U.S.—1995 on)	Visible, infrared, and microwave radiance—reflectance infrared atmospheric sounding, and ozone measurements
ADEOS (Japan—1996) Advanced Earth Observing Satellite	Visible and near-infrared radiance—reflectance, scatterometery, and tropospheric and stratospheric chemistry (joint USA and France)

[a] See Asrar and Dokken (1993).

planned. The instrument complement includes the following:

- **Advanced Visible and Near-Infrared Radiometer** (AVNIR)—Five visible/near-infrared bands (0.420–0.8990 μm); 16- or 8-m resolution; 80-km swath; $\pm 40°$ cross-track pointing; stereo capability. (Developed by NASDA.)
- **Improved Limb Atmospheric Spectrometer** (ILAS)—Infrared occultation device; one visible band at 0.753–0.781 μm, and two infrared bands at 6.21–11.77 and 5.99–6.78 μm (16-cm^{-1} resolution); 2×10 km IFOV; observations from 10 to 60 km, approximately 4-km vertical resolution expected; retrieval of stratospheric ozone and related species at high latitudes. (Provided by Japan's Environmental Agency.)
- **Interferometric Monitor for Greenhouse Gases** (IMG)—Nadir-looking Fourier-transform infrared spectrometer; range 3.3–14.0 μm, with a spectral resolution of 0.1 cm^{-1}; $\sim$10 km^2 footprint; vertical resolution $\sim$2–6 km depending on species; observation of carbon dioxide, methane, and other greenhouse gases. (Provided by MITI.)
- **NASA Scatterometer** (NSCAT)—14 GHz (Ku band) operation; resolution of 25 km; two 600-km swaths; used for retrieval of wind speed and direction over the oceans. (Provided by NASA.)
- **Ocean Color and Temperature Scanner** (OCTS)—Six visible bands centered on 0.412, 0.443, 0.490, 0.520, 0.565, and 0.679 μm, 20-nm bandwidth, S/N = 450–500; two near-infrared bands centered on 0.765 and 0.865 μm, 40-nm bandwidth, S/N = 450–500; four thermal infrared bands centered on 3.7, 8.5, 11.0, and 12.0 μm, 330–1000-nm bandwidth, NEΔT 0.15 to 0.20 K at 300 K; 700-m resolution; 1400-km swath; $\pm 20°$ along-track tilting; measurement of ocean color and sea surface temperature. (Developed by NASDA.)
- **Polarization and Directionality of Earth's Reflectances** (POLDER)—Views $\pm 55°$ (cross- and along-track); 5-km^2 nadir footprint; eight bands in the visible and near-infrared, 0.443–0.910 μm with 10–20-nm bandwidth; all polarization measurements in three of the eight bands. (Provided by CNES.)
- **Retroreflector in Space** (RIS)—0.5-m-diameter corner-cube retroreflector to derive column density of ozone and trace species from laser absorption measurements. A ground station in the Kanto area will transmit and receive laser pulses in the wavelength region from 0.4 to 14 μm. (Provided by Japan's Environmental Agency.)
- **Total Ozone Mapping Spectrometer** (TOMS)—Six wavelength bands in the region from 0.309 to 0.360 μm, with 1-nm bandwidth; IFOV 50 km at nadir; cross-track scan 105° (35 3° steps); retrieves daily global ozone. (Provided by NASA.)

A follow-on satellite, ADEOS II, is in the planning stages.

11.2.2 The EOS Era

The Earth Observing System is designed to make observations of the processes which govern the Earth system. These include (Asrar and Dokken, 1993):

- Hydrologic processes, which govern the interactions of land and ocean surfaces with the atmosphere through the transport of heat, mass, and momentum
- Biogeochemical processes, which contribute to the formation, dissipation, and transport of trace gases and aerosols, and their global distributions
- Climatological processes, which control the formation and dissipation of clouds and their interactions with solar radiation
- Ecological processes, which are affected by and/or will affect global change, and their response to such changes through adaptation
- Geophysical processes, which have shaped or continue to modify the surface of the Earth through tectonics, volcanism, and the melting of glaciers and sea ice.

To study these processes, EOS will provide systematic, continuous observations from low Earth orbit for a minimum of 15 years. Mission objectives in support of this goal are (Asrar and Dokken, 1993):

- Create an integrated scientific observing system that will enable multidisciplinary study of the Earth's critical, life-enabling, interrelated processes involving the atmosphere, oceans, land surfaces, polar regions, and solid Earth, and the dynamic and energetic interactions among them
- Develop a comprehensive data and information system (EOSDIS), including a data retrieval and processing system, to serve the needs of scientists performing an integrated multidisciplinary study of planet Earth
- Support the overall U.S. Global Change Research Program (USGCRP) by acquiring and assembling a global database of remote-sensing measurements from space; priorities for acquiring these data will conform to the seven issues identified by USGCRP and the Intergovernmental Panel on Climate Change (IPCC) as key to understanding global climate change, including:
    - The role of clouds, radiation, water vapor, and precipitation
    - The productivity of the oceans, their circulation, and air–sea exchange
    - The sources and sinks of greenhouse gases, and their atmospheric transformations
    - Changes in land use, land cover, primary productivity, and the water cycle
    - The role of polar ice sheets and sea level
    - The coupling of ozone chemistry with climate and the biosphere
    - The role of volcanoes in climate change

To achieve these objectives, an extensive system of satellites and instruments is being planned. We are able to only very briefly describe this vast and evolving system. (Readers are again referred to the *EOS Reference Handbooks.*) Table 11.6 lists the currently planned EOS-era satellites.

In the remainder of Section 11.2.2, we cannot resist listing the instruments[3] that are currently planned for the EOS-era satellites and noting what, in our view, are the important aspects for satellite meteorology and related fields. The descriptions are sketchy and will probably change, but we think it provides readers

TABLE 11.6. EOS Era Satellites[a]

Satellite (Launch Status)	*Mission Objectives*
EOS–AM Series (1998) EOS-Morning Crossing	Clouds, aerosols and radiation balance, characterization of the terrestrial ecosystem; land use, soils, terrestrial energy and moisture, tropospheric chemical composition; contribution of volcanoes to climate, and ocean primary productivity (includes Canadian and Japanese instruments)
EOS–COLOR (1998) EOS Ocean Color Mission	Ocean primary productivity
POEM–ENVISAT (ESA—1998) Polar-Orbit Earth Observation Mission Environmental Satellite	Environmental studies in atmospheric chemistry and marine biology, and continuation of ERS mission objectives
ADEOS IIa and IIb (Japan—1999) Advanced Earth Observing Satellite IIa and IIb	Visible, near-infrared, and microwave radiance–reflectance, scatterometry, infrared and laser atmospheric sounding, tropospheric and stratospheric chemistry, and altimetry (may include French and U.S. instruments)
EOS–PM Series (2000) EOS-Afternoon Crossing	Cloud formation, precipitation, and radiative properties; air–sea fluxes of energy and moisture; and sea-ice extent (includes European instruments)
EOS–AERO (2000) EOS Aerosol Mission	Distribution of aerosols and greenhouse gases in the lower stratosphere (spacecraft to be provided through international cooperation)
POEM–METOP Series (ESA—2000) Polar-Orbit Earth Observation Mission Meteorological Operational Satellite	Operation meteorology and climate monitoring, with the future objective of operational climatology (joint with EUMETSAT and NOAA)
TRMM 2 (Japan/U.S.—proposed for 2000) Tropical Rainfall Measuring Mission	Precipitation and related variables and Earth radiation budget in tropics and higher latitudes
EOS–ALT Series (2002) EOS Altimetry Mission	Ocean circulation and ice sheet mass balance (may include French instruments)
EOS–CHEM (2002) EOS Chemistry Mission	Atmospheric chemical composition; chemistry–climate interactions; air—sea exchange of chemicals and energy (to include an as yet to be determined Japanese instrument)

[a] See Asrar and Dokken (1993).

[3] Abbreviations may be found in Appendix B.

with information that they need about the future of satellite meteorology. Much more information can be found in Asrar and Dokken (1993).

11.2.2.1 ACRIM

The Active Cavity Radiometer Irradiance Monitor, selected for flight on the EOS–CHEM Series satellites, will extend the NASA/ACRIM solar luminosity data base to three decades, which is essential for understanding the variability of the sun's output and its affect on the Earth's atmosphere. [See Willson and Hudson (1991).]

11.2.2.2 AIRS, AMSU, and MHS

The Atmospheric Infrared Sounder (AIRS) is a high-resolution grating spectrometer sounder which will measure radiance in 2300 spectral channels between 0.4 and 15.4 μm. Such high spectral resolution will improve atmospheric soundings. Together with NOAA's Advanced Microwave Sounding Unit (AMSU) and EUMETSAT's Microwave Humidity Sensor (MHS), AIRS constitutes an all-weather sounding system which may fly on future operational satellites. The AIRS/AMSU/MHS system has been selected to fly on the EOS–PM Series satellites. [See AIRS Science Team (1990).]

11.2.2.3 ASTER

The Advanced Spaceborne Thermal Emission and Reflection Radiometer (ASTER) will have 14 channels between 0.5 and 12 μm and spatial resolutions between 15 and 90 m. The high-spatial-resolution images of clouds and the Earth's surface will allow atmospheric scientists to better understand the physical processes that affect climate change. In addition, ASTER will have an aft pointing channel, which will allow stereo images of clouds to be made as the satellite flies over. This should considerably increase our knowledge of cloud properties. ASTER will complement the more frequent return period but lower-spatial-resolution MODIS. ASTER is planned to fly on the EOS–AM1 satellite. [See Fujisada and Ono (1991).]

11.2.2.4 CERES

The Clouds and Earth's Radiant Energy System (CERES) instrument is a follow-on of the ERBE Scanner. It will have a cross-track scanner and a biaxial scanner, each with three channels: shortwave, longwave, and total. CERES will fly on the EOS–AM and EOS–PM satellite series as well as TRMM 1. [See Barkstrom (1990).]

11.2.2.5 DORIS, SSALT, and TMR

The Solid-State Altimeter (SSALT) is a radar altimeter that measures ocean wave height, wind speed, and information on the ocean surface current velocity. It also maps the topography of the sea surface and polar ice sheets. DORIS (Doppler Orbitography and Radiopositioning Integrated by Satellite) precisely locates the satellite in orbit to correct the SSALT measurements. TMR (TOPEX Microwave Radiometer) provides atmospheric water vapor measurements for

SSALT corrections. The DORIS/SSALT/TMR trio is slated to fly on the EOS–ALT Series satellites.

11.2.2.6 EOS–COLOR

The EOS Ocean Color Instrument (EOS–COLOR) is a follow-on to the Nimbus 7 Coastal Zone Color Scanner and SeaWiFS, and it is designed primarily to measure marine phytoplankton. However, it will yield information on the global carbon cycle, and it may also be useful for marine aerosol studies. EOS–COLOR is the only instrument slated for the EOS–COLOR satellite.

11.2.2.7 EOSP

The Earth Observing Scanning Polarimeter (EOSP) is derived from planetary instruments on the Pioneer and Galileo satellites. It measures radiance and linear polarization of reflected sunlight in 12 spectral bands from 0.41 to 2.25 μm. EOSP data will provide information on the global aerosol and cloud distribution and on such properties as optical depth, phase, particle size, and cloud-top pressure. EOSP is scheduled for flight on the EOS–AM2 and –AM3 satellites.

11.2.2.8 GLAS

The Geoscience Laser Altimeter System (GLAS) will measure cloud heights, planetary boundary-layer heights, and vertical aerosol structure, as well as ice sheet, land, and water topography. GLAS is planned for flight on the EOS–ALT satellite. [See Cohen *et al.* (1987) and Curran *et al.* (1987a).]

11.2.2.9 HIRDLS

The High-Resolution Dynamics Limb Sounder (HIRDLS) is the latest in a long line of limb sounders (LRIR, LIMS, SAMS, ISAMS, CLAES). It measures gas and aerosol concentrations in the upper troposphere, stratosphere, and mesosphere using the spectral range 6–18 μm. HIRDLS has been selected for the EOS–CHEM Series satellites.

11.2.2.10 LIS

The Lightning Imaging Sensor (LIS) is a staring imager (no moving parts) designed to detect, locate, and quantize the radiant energy of lightning flashes. It utilizes a two-dimensional (128 $\times$ 128) CCD array to image a 600 $\times$ 600-km area with 5-km resolution. The array is read out each 2 ms and processed to yield the location of lightning flashes since the last readout. To be flown on TRMM 1, LIS will complement the other precipitation-sensing instruments by yielding an indication of storm intensity—the lightning flash rate. LIS will detect both cloud-to-cloud and cloud-to-ground lightning, whereas ground-based lightning sensors detect only cloud-to-ground strokes. LIS will detect lightning both day and night over both land and ocean. On TRMM, of course, each location will be sensed only twice per day for a few minutes as the satellite passes over. Someday, a successor instrument may be flown on a geostationary satellite (see

Section 11.3.3), where a hemisphere could be monitored constantly, and all of the lightning strokes could be observed. [See Christian *et al.* (1989, 1992).]

11.2.2.11 MIMR

The Multifrequency Imaging Microwave Radiometer (MIMR) is derived from the SSM/I on the DMSP satellites and from the SMMR on the Nimbus 7 and Seasat satellites. It will sense horizontally and vertically polarized radiation at six frequencies between 6.8 and 90 GHz in a 50° conical scan with ground resolution of from 5 to 60 km, depending on frequency. MIMR will extend previous measurements of a wide variety of geophysical parameters by including more frequencies and by having a somewhat greater swath width. MIMR is scheduled for flight on the EOS–PM Series.

11.2.2.12 MISR

The Multi-Angle Imaging SpectoRadiometer (MISR) will be the only EOS instrument that will routinely provide multiangle observations. Using nine separate charge-coupled device (CCD) pushbroom cameras, MISR will make images at nine fixed angles in the along-track direction: one at nadir, and four both fore and aft out to ±70.5°. Four spectral bands centered at 0.443, 0.555, 0.670, and 0.865 μm will be simultaneously imaged. Ground resolution will be selectable between 240 m and 1.92 km. Due to the fact that each point will be imaged nine times as the satellite passes over, stereo observations are automatic. MISR will be used for cloud, aerosol, and radiation budget studies as well as for land, biosphere, and ocean studies. MISR is scheduled for flight on the EOS–AM Series satellites.

11.2.2.13 MLS

The Microwave Limb Sounder (MLS) is derived from the UARS MLS. It will measure temperature, moisture, and a variety of gases in the stratosphere, including the chlorine compounds involved in ozone chemistry. MLS will fly on the EOS–CHEM Series.

11.2.2.14 MODIS

The Moderate-Resolution Imaging Spectrometer (MODIS) (Esias *et al.*, 1986; Salomonson and Toll, 1991; King *et al.*, 1992) is derived from the AVHRR, from the Nimbus 7 Coastal Zone Color Scanner, and from Landsat. It is designed to make 0.25–1-km resolution observations in 36 visible and infrared bands. The orbit and scan geometries are to be such that each point on Earth is sensed every day or two. A wide variety of oceanic, atmospheric, and land surface processes will be observable with MODIS. Atmospheric applications include cloud and aerosol observations. Of note are a set of channels in the Oxygen A-band near 0.76 μm that will determine cloud-top pressure (by measuring the amount of oxygen above the cloud), and a pair of polarized 0.5-μm channels that should be interesting in both cloud and aerosol studies. Because of its versatility and synergism with other EOS sensors, MODIS is slated to fly on both the EOS–AM and –PM satellites.

11.2.2.15 MOPITT

Measurements of Pollution in the Troposphere (MOPITT) is a four-channel gas-correlation spectrometer, similar to the PMR, SSU, SAMS, and ISAM, which will measure profiles of CO concentration and column-integrated CH_4 amounts. Provided by the Canadian space Agency, MOPITT is designed to fly on the EOS–AM satellites.

11.2.2.16 NSCATT II

The NASA Scatterometer II (NSCAT II) is derived from the Seasat SASS and the NSCAT to fly on ADEOS. It will utilize six fan-beam antennas to measure ocean surface wind speed and direction under all weather conditions. Having six antennas, as opposed to SASS's four antennas, should improve accuracy and reduce uncertainty in wind direction. NSCAT II is scheduled for the ADEOS II satellite.

11.2.2.17 SAGE III

The Stratospheric Aerosol and Gas Experiment III (SAGE III) is the third in the series of SAGE instruments. It will add lunar occultation and channels to improve observations of aerosols, O_3, H_2O, and NO_2 and to add retrievals of NO_3 and OClO. SAGE III has been selected for flight on the EOS–AERO and EOS–CHEM Series satellites.

11.2.2.18 SOLSTICE II

The Solar Stellar Irradiance Comparison Experiment II (SOLSTICE II) is derived from the SOLSTICE instrument on UARS. It will measure the solar irradiance at four ultraviolet wavelengths. SOLSTICE II is scheduled for the EOS–CHEM Series satellites.

11.2.2.19 TES

The Tropospheric Emission Spectrometer (TES) is a high-resolution infrared imaging Fourier-transform spectrometer with spectral coverage of 2.3–15.4 μm at a spectral resolution of 0.025 cm^{-1}. It will be capable of imaging virtually all chemical species that are active in the infrared. TES is slated for the EOS–AM2 and –AM3 satellites.

11.2.2.20 The POEM Instruments

Finally, we mention by name only the instruments that are under development for use on the ESA Polar-Orbit Earth Observation Missions (POEM) in the next century:

- AATSR—Advanced Along-Track Scanning Radiometer
- ASAR—Advanced Synthetic Aperture Radar
- GOMOS—Global Ozone Monitoring by Occultation of Stars
- IASI—Infrared Atmospheric Sounding Interferometer
- IRTS—Infrared Temperature Sounder
- MERIS—Medium-Resolution Imaging Spectrometer

- MIPAS—Michelson Interferometer for Passive Atmospheric Sounding
- MTS—Microwave Temperature Sounder
- PRAREE—Precise Range and Range Rate Equipment-Extended
- RA-2–Radar Altimeter 2
- SCARAB—Scanner for the Radiation Budget
- SCIAMACHY—Scanning Imaging Absorption Spectrometer for Atmospheric Cartography
- VIRSR—Visible Infrared Scanning Radiometer

11.2.3 The EOS Follow-on Era

Possible satellites in the future of MTPE include the Geostationary Earth Observatory (GEO) to measure processes that are too fast to be studied by the polar-orbiting EOS satellites or that require accurate observations of the diurnal cycle. Also envisioned are additional Earth Probes to make measurements which EOS satellites cannot. Plans for the EOS follow-on era will be formulated over the next decade or so.

11.3 OTHER POSSIBILITIES

Although Mission to Planet Earth is quite comprehensive, several ideas for useful instruments do not appear to be included in MTPE plans. We briefly mention these measuring techniques in the hope that they will eventually find a place in space.

11.3.1 Differential Absorption Lidar

There have been few active meteorological sensors in space. A Lidar Atmospheric Sounder and Altimeter (LASA) was proposed as part of the Earth Observing System (Curran *et al.*, 1987a). LASA was to be capable of measuring vertical profiles of ozone, water vapor, aerosol concentration, temperature, and pressure. Unfortunately, LASA is not currently scheduled for the Earth Observing System.

A lidar (light detection and ranging) is like radar but uses light instead of radio waves. A pulse of laser-generated radiation is transmitted, and the amplitude of the returned (reflected) signal is measured. The reflection can be from the surface or from atmospheric aerosols or clouds. Since aerosols are ubiquitous, a continuous return is measured. A single laser wavelength is useful for measuring distances. If the satellite orbit and attitude are accurately known, the height of the ocean or of the tops of clouds can be measured. Aerosol optical depth profiles can also be measured by a single-wavelength lidar.

Most atmospheric measurements require two laser wavelengths. To measure the concentration of a gas (ozone and water vapor, for example) the two wavelengths are chosen such that one is affected by a gaseous absorption line and the other is not. The difference between the returned signals (assuming that other things, such as scattering, are equal or can be accounted for) is proportional to the gas transmittance. This technique, called *differential absorption lidar* (DIAL), is extremely powerful. Not only can the total column depth of a gas be measured, but its vertical profile as well because the return from aerosols will give essentially continuous sampling through the depth of the atmosphere.

Further, if a well-mixed gas (like O_2) is chosen, and if measurements are made in a temperature-insensitive portion of the spectrum, the pressure at each range gate plus the surface pressure can be measured. Pressure measurements will be extremely useful; current meteorological satellites offer no means to measure atmospheric pressure.[4] If measurements are made in a portion of the spectrum which is sensitive to temperature, the temperature profile can be measured. It was planned to use the oxygen A-band, near 0.76 μm, for both pressure and temperature measurements from LASA.

11.3.2 Doppler Lidar

With the exception of winds calculated from temperature soundings, none of the techniques discussed in this chapter measure tropospheric wind profiles (wind as a function of height). Since large areas of the globe are not sampled by rawinsondes, the global tropospheric wind field remains largely unmeasured. Simulation studies indicate that global observations of wind profiles would substantially improve numerical weather predictions (Atlas *et al.*, 1985).

The Laser Atmospheric Wind Sounder (LAWS) (Curran *et al.*, 1987b), proposed as part of the Earth Observing System, would alleviate this problem. Laser pulses from a CO_2 Doppler lidar with wavelength near 10.6 μm would be transmitted. The radiation backscattered from naturally occurring aerosols would be analyzed to determine the frequency shift between the transmitted and received pulses. This Doppler shift is proportional to the relative (radial) velocity of the aerosols and the spacecraft. If two observations of each parcel of air are made from different angles (by scanning in a cone), and if it is assumed that atmospheric motions are horizontal, the motions of the spacecraft and Earth can be eliminated to yield the horizontal wind vector. Root-mean-square wind accuracies on the order of 2 m s^{-1} are thought to be possible on the basis tests using ground-based and aircraft-mounted instruments. Range gates would allow the atmosphere to be divided vertically into approximately 1-km layers. Matching samples in the layers would result in approximately 100-km horizontal resolution.

11.3.3 Lightning Mapper

Currently under development for a future geostationary satellite is a Lightning Mapper (Christian *et al.*, 1989). The Mapper would measure the location (within about 10 km) and brightness of lightning flashes both day and night. To do this, a large CCD array of about 800 × 640 pixels would be employed. The detector

[4] Except in a relative sense around tropical storms (see section 7.3.3.2).

would "stare" at the Earth and the entire array would be read out approximately every 2 ms.

Two major problems confront the Lightning Mapper: filtering out the bright background during daylight hours and the high data rate (500 frames per second). Background filtration would rely on (1) the slow time rate of change of the background versus the approximately 0.5-ms duration of a lightning stroke, (2) the nonblackbody spectrum of lightning radiation, and (3) differences in spatial scale of sunlit clouds and lightning flashes. The data-rate problem would have to be solved with some sort of on-board background subtraction routine, which might work as follows. If a lightning stroke has occurred in a 10-km pixel, that pixel should be significantly brighter than the same pixel in the last frame. If the two frames are subtracted, the only data which need to be transmitted to Earth are for those small number of pixels that are significantly different than in the last frame.

There currently exists a network of ground-based lightning detectors in the United States. The Lightning Mapper would significantly enhance this network in two ways. Of course, it would extend the range to a large fraction of the hemisphere visible from geostationary orbit. It also would be able to detect both cloud-to-cloud and cloud-to-ground lightning. Ground-based detectors only sense cloud-to-ground strokes. A wide range of storm physics, atmospheric electricity, and atmospheric chemistry applications would benefit from the Lightning Mapper data.

11.3.4 Geostationary Microwave Imager–Sounder

Though it has been discussed for years, a microwave imager–sounder for use in geostationary orbit is not in current plans. Because of its ability to measure atmospheric temperature, precipitation, and a wide variety of other important atmospheric and surface properties *through clouds*, we believe that a geostationary microwave instrument is extremely important. The large antenna needed for this instrument presents technical challenges, but we believe that the return on investment would be substantial.

11.3.5 Molniya Orbit

Lastly, we feel compelled to reiterate our suggestion (Kidder and Vonder Haar, 1990) that meteorological instruments on satellites in Molniya orbit (see Chapter 2) could significantly improve observations of the high latitudes, including Alaska, northern Europe, and the North and South Poles..

11.4 A FINAL COMMENT

It has been more than thirty years since the launch of the first meteorological satellites. In that time satellite meteorology has grown rapidly. If even a fraction of the planned satellite instruments are launched in the coming decades, we expect that growth not only to continue but to accelerate. To those of you who have just finished reading this book, we hope that it will be of use in whatever role you play in this exciting and expanding field.

Bibliography

AIRS Science Team (1990). *The Atmospheric Infrared Sounder (AIRS) Science and Measurements Requirements.* Jet Propulsion Laboratory document No. D6665, Pasadena, CA.

Asrar, G., and D. J. Dokken (eds.) (1993). *EOS Reference Handbook.* NASA, Washington, DC. [Available from the Earth Science Support Office, Document Resource Facility, 300 D Street, SW, Suite 840, Washington, DC 20024, telephone: (202) 479–0360.]

Atlas, R., E. Kalnay, W. E. Baker, J. Susskind, D. Reuter, and M. Halem (1985). Simulation studies of the impact of future observing systems on weather prediction. *Preprints: Seventh Conference on Numerical Weather Prediction*, American Meteorological Society, Boston, pp. 145–151.

Barkstrom, B. R. (1990). Long-term monitoring of the Earth's radiation budget. *Proceedings of the Society of Photo-Optical Instrumentation Engineers,* vol. 1299.

Bunting, J. T., and R. P. d'Entremont (1982). *Improved Cloud Detection Utilizing Defense Meteorological Satellite Program Near Infrared Measurements.* Air Force Geophysics Lab. Tech. Rep. AFGL-TR-82-0027, Hanscom AFB, MA.

Butler, D. M., R. E. Hartle, M. Abbott, R. Cess, R. Chase, P. Christensen, J. Dutton, L.-L. Fu, C. Gautier, J. Gille, R. Gurney, P. Hays, J. Hovermale, S. Martin, J. Melack, D. Miller, V. Mohnen, B. Moore III, J. Norman, S. Schneider, J. Sherman III, V. Suomi, B. Tapley, R. Watts, E. Wood, and P. Zinke (1986). *From Pattern to Process: The Strategy of the Earth Observing System.* Eos Science Steering Committee Report, Vol. II, NASA, Washington, DC.

Christian, H. J., R. J. Blakeslee, and S. J. Goodman (1989). The detection of lightning from geostationary orbit. *J. Geophys. Res.*, **94**, 13329–13337.

Christian, H. J., R. J. Blakeslee, and L. J. Goodman (1992). *Lightning Imaging Sensor (LIS) for the Earth Observing System.* NASA Tech. Memo. 4350, NASA/Marshall Space Flight Center, Huntsville, AL 35812.

Cohen, S. J. Degnan, J. Bufton, J. Garvin, and J. Abshire (1987). The Geoscience Laser Altimetry/Ranging System. *IEEE Trans. Geoscience Remote Sensing,* **GE-25**, 581–592.

Curran, R. J., R. R. Nelms, F. Allario, R. A. Bindschadler, E. V. Browell, J. L. Bufton, R. L. Byer, S. Cohen, J. J. Degnan, W. B. Grant, R. Greenstone, W. S. Heaps, B. M. Herman, A. Jalink, D. K. Killinger, L. Korb, J. B. Laudenslager, M. P. McCormick, S. H. Melfi, R. T. Menzies, V. Mohnen, J. Spinhirne, and H. J. Zwally (1987a). *LASA—Lidar Atmospheric Sounder and Altimeter.* Earth Observing System Instrument Panel Report, Vol. IId, NASA, Washington, DC.

Curran, R. J., D. E. Fitzjarrald, J. W. Bilbro, V. J. Abreu, R. A. Anthes, W. E. Baker, D. A. Bowdle, D. Burridge, G. D. Emmitt, S. E. D. Ferry, P. H. Flamant, R. Greenstone, R. M. Hardesty, K. R. Hardy, R. M. Huffaker, M. P. McCormick, R. T. Menzies, R. M. Schotland, J. K. Sparkman, Jr., J. M. Vaughan, and C. Werner (1987b). *LAWS—Laser Atmospheric Wind Sounder.* Earth Observing System Instrument Panel Report, Vol. IIg, NASA, Washington, DC.

Esias, W., W. Barnes, M. Abbott, S. Cox, R. Evans, R. Fraser, A. Goetz, C. Justice, E. P. McClain, M. Maxwell, R. Murphy, J. Prospero, B. Rock, S. Running, R. Smith, J. Solomon, and J. Susskind (1986). *MODIS—Moderate-Resolution Imaging Spectrometer.* Earth Observing System Instrument Panel Report, Vol. IIb, NASA, Washington, DC.

Fujisada, H. and M. Ono (1991). Overview of ASTER design concept. *Proceedings of the Society of Photo-Optical Instrumentation Engineers*, vol. 1490.

Kidder, S. Q., and T. H. Vonder Haar (1990). On the use of satellites in Molniya orbits for meteorological observation of middle and high latitudes. *J. Atmos. Ocean. Tech.*, 7, 517–522.

King, M. D., Y. J. Kaufman, W. P. Menzel, and D. Tanré (1992). Remotes sensing of cloud, aerosol, and water vapor properties from MODIS. *IEEE Trans. Geosci. Remote Sens.,* **30**, 2–27.

NASA (1988a). *Performance Specification for the NOAA–K through –M Satellites*. GSFC-S-480-25, Goddard Space Flight Center, Greenbelt, MD.
NASA (1988b). *Performance Specification for the NOAA–K, L & M Advanced Very High Resolution Radiometer/3 (AVHRR/3) Flight Models A301, A302, and A303*. GSFC-S-480-28.3, Revision B, Goddard Space Flight Center, Greenbelt, MD.
NASA (1988c). *Performance Specification for the NOAA–K, L & M High Resolution Infrared Radiation Sounder/3 (HIRS/3) Flight Models H301, H302, and H303*. GSFC-S-480-28.2, Revision B, Goddard Space Flight Center, Greenbelt, MD.
Salomonson, V. V., and D. L. Toll (1991). The Moderate Resolution Imaging Spectrometer-Nadir (MODIS–N) facility instrument. *Adv. Space Res.*, **11**, 231–236.
Willson, R. C., and H. S. Hudson (1991). The sun's luminosity over a complete solar cycle. *Nature*, **351**, 42–45.

APPENDIX

A

List of Meteorological Satellites

In this appendix, we have attempted to list all satellites that have made measurements of the Earth's atmosphere. Although the book concentrates on the troposphere and stratosphere, we have listed satellites that made measurements of any part of the atmosphere from the ionosphere down. The information comes primarily from two excellent sources: The *TRW Space Logs*,[1] whose format we have adopted, and the National Space Science Data Center's *Compendium of Meteorological Space Programs, Satellites, and Experiments*,[2] which helped determine which of the thousands of satellites launched since 1957 made atmospheric measurements. The list contains only successfully launched satellites. Manned flights are listed only if they included a specific atmospheric experiment.

The reader should be aware that the list probably includes some satellites that did not make atmospheric measurements, and it certainly omits some that did.

[1] Thompson, T. D. (ed.) (1988–1993). *TRW Space Log*, Vols. 23–29. TRW Space & Technology Group, One Space Park, Mail Station R4/2135, Redondo Beach, CA 90278.

[2] Dubach, L. L., and C. Ng (1988). *Compendium of Meteorological Space Programs, Satellites, and Experiments*. National Space Science Data Center, Goddard Space Flight Center, Greenbelt, MD.

In addition, there are undoubtedly errors. In spite of these problems, we hope that the list will be of use to readers.

A few words of explanation of some of the data are in order:

- The **Name** given a satellite is a commonly used, postlaunch name taken from one of the two sources with some modifications for consistency. One such modification is in the use of the hyphen. We use the hyphen to indicate the satellite *series*. The *number* of a satellite within a series does not receive a hyphen. Thus NOAA 6, the sixth NOAA satellite, is not hyphenated. The fourth satellite in the second generation of Meteor satellites is written as Meteor-2 4.
- The **International ID** is the one assigned by the Committee on Space Research (COSPAR) of the International Council of Scientific Unions. It consists of the year of launch, a number assigned serially to successful[3] launches for that year, and a letter that identifies the various spacecraft which resulted from that launch. For example, 1984 108B refers to the second spacecraft (B) resulting from the 108th successful launch in 1984. Before 1963 a different numbering system was used. We have attempted to renumber these early launches using the *TRW Space Log* as a guide.
- The **Date** of launch is for Coordinated Universal Time (UTC).
- For the launch **Site**, we have used a consistent name, even though names have changed through the years. For example, we have used "Cape Canav." to refer to the launch sites at Cape Canaveral, Florida; they are variously called the Kennedy Space Center, the Eastern Test Range, and the Eastern Space and Missile Center.
- The **Payload Mass** is in many cases approximate.
- The **Orbital Data** are taken from the *TRW Space Logs*. The **Period** has been adjusted slightly using Eq. 2.16 so that it agrees with the **Perigee**, **Apogee**, and **Inclination.** Note that for satellites still in orbit these are the data that were current when the *Space Logs* were compiled. They are not necessarily representative of the orbital data during the useful lifetime of the satellite. For example, the inclination angle of geostationary satellites is maintained near zero. When they are no longer operational, their inclination angles drift. The *Compendium* contains "initial" orbital data, which is often more representative of the data when the satellite was operational. For the sake of consistency, however, we have used the *Space Log* data.
- In the **Comments** section, we have tried to list the instrumentation which the satellite carried or the type of measurements which it made. In some cases this information is missing, and in others it may be incorrect.

[3] "Successful" here means that something reached orbit. Whether any useful data were collected is another question.

Name	Int'l ID	Country or Agency	LAUNCH DATA Date	Site	Vehicle	Payload Mass (kg)	ORBITAL DATA Per-iod (min)	Peri-gee (km)	Apo-gee) (km)	Incli-nation (deg)	Comments
Vanguard 2	1959 002A	NASA	Feb 17	Cape Canav.	Vanguard	9.8	123.6	559	3142	32.9	Optical Scanner; wobble degraded data
Explorer 6	1959 006A	NASA	Aug 7	Cape Canav.	Thor Able	64	764.4	245	42400	47.0	First Earth photo; first radiation budget data
Explorer 7	1959 012A	NASA	Oct 13	Cape Canav.	Juno 2	41.5	101.3	556	1088	50.2	Earth radiation budget data
TIROS 1	1960 002A	NASA	Apr 1	Cape Canav.	Thor Able	120	98.7	677	722	48.4	First dedicated meteorological satellite; 1 TV-WA, 1 TV-NA; returned 22955 photos up to 17 Jun 1960
Midas 2	1960 006A	USAF	May 24	Cape Canav.	Atlas Agena A	2300	94.3	484	511	33.0	Non-Scanning Radiometer, atmospheric density
Explorer 8	1960 014A	NASA	Nov 3	Cape Canav.	Juno 2	41	106.2	407	1699	49.9	Ionospheric research
TIROS 2	1960 016A	NASA	Nov 23	Cape Canav.	Delta	130	97.1	585	667	48.5	1 TV-WA, 1 TV-NA, Scanning Radiometer, Widefield Radiometer
Explorer 9	1961 004A	NASA	Feb 16	Wallops Is.	Scout	6.6	118.2	634	2583	38.6	3.7 m balloon to study atmospheric density
TIROS 3	1961 017A	NASA	Jul 12	Cape Canav.	Delta	129	100.0	730	801	47.9	2 TV-WA, Scanning Radiometer, Widefield Radiometer, Suomi Radiometer
Midas 3	1961 018A	USAF	Jul 12	Vandenberg	Atlas Agena B	1600	161.5	3345	3538	91.1	Scanning Radiometer
Midas 4	1961 028A	USAF	Oct 21	Vandenberg	Atlas Agena B	1800	166.0	3485	3760	91.2	Scanning Radiometer
Discoverer 35	1961 030A	USAF	Nov 15	Vandenberg	Thor Agena B	2100	89.8	238	278	81.6	Scanning Radiometer
TIROS 4	1962 002A	NASA	Feb 8	Cape Canav.	Delta	129	100.3	712	840	48.3	1 TV-WA, 1 TV-MA, Scanning Radiometer, Widefield Radiometer, Suomi Radiometer
Midas 5	1962 010A	USAF	Apr 9	Vandenberg	Atlas Agena B	1860	153.0	2782	3406	86.7	Scanning Radiometer
Kosmos 4	1962 014A	USSR	Apr 29	Tyuratam	A-1	4600	90.6	285	317	65.0	Possible met. applications
Ariel 1	1962 015A	NASA/UK	Apr 26	Cape Canav.	Delta	60	100.9	389	1214	53.9	Joint US-UK ionospheric satellite
TIROS 5	1962 025A	NASA	Jun 19	Cape Canav.	Delta	129	99.8	578	922	58.1	1 TV-WA, 1 TV-MA
Kosmos 7	1962 033A	USSR	Jul 28	Tyuratam	A-1	4600	90.1	197	356	65	Possible met. applications
TIROS 6	1962 047A	NASA	Sep 18	Cape Canav.	Delta	127	98.7	686	713	58.3	1 TV-WA, 1 TV-MA
Kosmos 9	1962 048A	USSR	Sep 27	Tyuratam	A-1	4600	90.9	292	346	65	Possible met. applications
Explorer 17	1963 009A	NASA	Apr 2	Cape Canav.	Delta	185	96.4	255	917	57.6	Atmospheric research data until 10 Jul 1963
Kosmos 14	1963 010A	USSR	Apr 13	Kapustin Yar	B-1	400	92.0	252	499	49.0	Possible metsat
Kosmos 15	1963 011A	USSR	Apr 22	Tyuratam	A-1	4730	89.7	160	358	65.0	Possible metsat
Midas 6	1963 014A	USAF	May 9	Vandenberg	Atlas Agena B	2000	166.5	3606	3677	87.3	Scanning Radiometer
TIROS 7	1963 024A	NASA	Jun 19	Cape Canav.	Delta	135	95.8	548	572	58.2	2 TV-WA, Scanning Radiometer, Suomi Radiometer
Kosmos 23	1963 050A	USSR	Dec 13	Kapustin Yar	B-1	347	93.0	240	613	49	Possible metsat
Explorer 19	1963 053A	NASA	Dec 19	Vandenberg	Scout	7	115.9	590	2395	78.8	3.7 m balloon identical to Explorer 9

Name	Int'l ID	Country or Agency	LAUNCH DATA Date	Site	Vehicle	Payload Mass (kg)	ORBITAL DATA Per-iod (min)	Peri-gee (km)	Apo-gee) (km)	Incli-nation (deg)	Comments
TIROS 8	1963 054A	NASA	Dec 21	Cape Canav.	Delta	119	98.8	682	725	58.5	First APT, 1 TV-WA
Ariel 2	1964 015A	NASA/UK	Mar 27	Wallops Is.	Scout	168	101.4	810	843	51.6	Returned data from British experiments until Nov 1964
Explorer 20	1964 051A	NASA	Aug 25	Vandenberg	Scout	44	103.8	859	1006	79.9	Returned ionospheric data until July 1966
Nimbus 1	1964 052A	NASA	Aug 28	Vandenberg	Thor Agena B	376	98.5	429	937	98.6	First sunsynchronous metsat; AVCS, APT, HRIR
Kosmos 44	1964 053A	USSR	Aug 28	Tyuratam	A-1	3800	99.0	604	813	65.1	Probable metsat
Kosmos 45	1964 055A	USSR	Sep 13	Tyuratam	A-1	4730	89.7	207	313	64.9	Weather research film capsule recovered
Explorer 22	1964 064A	NASA	Oct 10	Vandenberg	Scout	52	104.6	875	1062	79.7	Ionospheric and geodetic data satellite
Kosmos 49	1964 069A	USSR	Oct 24	Kapustin Yar	B-1	355	91.7	264	466	49	Measured Earth's IR and UV radiation flux
Explorer 24	1964 076A	NASA	Nov 21	Vandenberg	Scout	8.6	116.4	525	2498	81.4	3.7 m balloon for atmospheric density studies; part of first NASA dual-payload launch
San Marco 1	1964 084A	Italy	Dec 15	Wallops Is.	Scout	254	94.9	198	846	37.8	Measured atmospheric density
TIROS 9	1965 004A	NASA	Jan 22	Cape Canav.	Delta	138	119.1	702	2568	96.4	Cartwheel configuration; 2 TV-WA, global coverage
Kosmos 58	1965 014A	USSR	Feb 26	Tyuratam	A-1	4730	94.6	470	523	65.0	Probable metsat
Gemini 3	1965 024A	NASA	Mar 23	Cape Canav.	Titan 2	3220	88.2	160	240	32.5	Synoptic Weather Photography
Kosmos 65	1965 029A	USSR	Apr 17	Tyuratam	A-1	4730	89.8	207	319	65	Weather research film capsule recovered
Explorer 27	1965 032A	NASA	Apr 29	Wallops Is.	Scout	59.7	107.6	933	1315	41.2	Geodetic and ionospheric research satellite; battery failed 1968
Gemini 4	1965 043A	NASA	Jun 3	Cape Canav.	Titan 2	3569	88.7	162	281	32.0	Synoptic Weather Photography
TIROS 10	1965 051A	NASA	Jul 2	Cape Canav.	Delta	127	100.4	728	817	98.4	Sunsynchronous; 2 TV-WA
Gemini 5	1965 068A	NASA	Aug 21	Cape Canav.	Titan 2	3600	89.3	197	303	32.6	Synoptic Weather Photography, Cloud Top Specrometer (Oxygen A Band), Space Object Radiometry
Kosmos 92	1965 083A	USSR	Oct 16	Tyuratam	A-1	4730	89.9	202	334	65	Weather reasearch film capsule recovered
Alouette 2	1965 098A	Canada	Nov 29	Vandenberg	Thor Agena B	145	119.9	503	2837	79.8	Ionospheric research
Explorer 31	1965 098B	NASA	Nov 29	Vandenberg	Thor Agena B	99	120.6	502	2906	79.8	Ionospheric data complementing Alouette 2; launched with Alouette 2
Gemini 7	1965 100A	NASA	Dec 4	Cape Canav.	Titan 2	3658	90.1	292	298	28.9	Synoptic Weather Photography, Space Object Radiometry
Gemini 6A	1965 104A	NASA	Dec 15	Cape Canav.	Titan 2	3546	89.5	258	271	28.9	Flight delayed; Synoptic Weather Photography

Name	Int'l ID	Country or Agency	LAUNCH DATA Date	Site	Vehicle	Payload Mass (kg)	ORBITAL DATA Per-iod (min)	Peri-gee (km)	Apo-gee) (km)	Incli-nation (deg)	Comments
Kosmos 100	1965 106A	USSR	Dec 17	Tyuratam	A-1	3800	96.5	533	641	65	Probable metsat
ESSA 1	1966 008A	ESSA	Feb 3	Cape Canav.	Delta	138	100.0	689	818	97.9	2 wide-angle vidicon cameras; switched off 8 May 1967
ESSA 2	1966 016A	ESSA	Feb 28	Cape Canav.	Thor Agena D	132	113.5	1352	1412	101.2	2 APT; global operational APT
Gemini 8	1966 020A	NASA	Mar 16	Cape Canav.	Titan 2	3789	88.5	159	265	28.9	Synoptic Weather Photography, Cloud Top Spectrometer (no data)
Molniya-1 3	1966 035A	USSR	Apr 25	Tyuratam	A-2-e	1750	710.6	506	39492	65	Comsat, also transmitted cloud cover photos
Kosmos 118	1966 038A	USSR	May 11	Tyuratam	A-1	4730	93.0	415	428	65	Probable metsat
Nimbus 2	1966 040A	NASA	May 15	Vandenberg	TAT-Agena B	414	108.1	1092	1175	100.4	AVCS, HRIR, MRIR, APT/DRIR
Explorer 32	1966 044A	NASA	May 25	Cape Canav.	Delta	225	99.3	250	1194	64.5	Six aeronomy experiments; returned data until Dec 1966
Gemini 9	1966 047A	NASA	Jun 3	Cape Canav.	Titan 2	3750	89.7	270	272	28.9	Synoptic Weather Photography
Kosmos 121	1966 054A	USSR	Jun 17	Plesetsk	A-1	4730	89.9	200	333	72.9	Possible metsat
Kosmos 122	1966 057A	USSR	Jun 25	Tyuratam	A-1	4730	94.3	469	495	65	Dual Vidicon Cameras, Scanning HRIR, Actinometric Inst.
Gemini 10	1966 066A	NASA	Jul 18	Cape Canav.	Titan 2	3757	88.5	160	268	28.9	Synoptic Weather Photography
Gemini 11	1966 081A	NASA	Sep 12	Cape Canav.	Titan 2	3793	88.6	161	280	28.8	Synoptic Weather Photography
DMSP-4A F1	1966 082A	USAF	Sep 16	Vandenberg	Thor Burner 2	125	100.5	680	872	98.8	Video Camera System, C System
ESSA 3	1966 087A	ESSA	Oct 2	Vandenberg	Thor Agena D	145	114.6	1384	1484	100.9	2 AVCS, FPR; replaced ESSA 1 in TOS system; silenced 9 Oct 1968
Molniya-1 4	1966 092A	USSR	Oct 20	Tyuratam	A-2-e	1750	714.5	505	39685	64.9	Comsat, also transmitted cloud cover photos
Gemini 12	1966 104A	NASA	Nov 11	Cape Canav.	Titan 2	3655	89.8	243	310	28.8	Synoptic Weather Photography
ATS 1	1966 110A	NASA	Dec 7	Cape Canav.	Atlas-Agena D	352	1434.9	35734	35797	10.5	First geosynchronous met. data; Spin Scan Cloudcover Camera
ESSA 4	1967 006A	ESSA	Jan 26	Vandenberg	Thor Agena D	290	113.5	1324	1437	102.0	2 APT; replaced ESSA 2; camera failure caused deactivation 6 Dec 1967
DMSP-4A F2	1967 010A	USAF	Feb 8	Vandenberg	Thor Burner 2	250	101.3	778	854	98.9	Video Camera System, C System
Kosmos 144	1967 018A	USSR	Feb 28	Plesetsk	A-1	4730	97.0	574	644	81.2	Metsat similar to Kosmos 122
Kosmos 149	1967 024A	USSR	Mar 21	Kapustin Yar	B-1	375	89.7	245	285	48.4	3 ch. wide-angle radiometers, narrow-angle IR radiometer, 3 ch. telephotometer, TV camera system
ESSA 5	1967 036A	ESSA	Apr 20	Vandenberg	Thor Agena D	145	113.6	1352	1419	102.0	2 AVCS, FPR; replaced ESSA 3
San Marco 2	1967 038A	Italy	Apr 26	Indian Ocean	Scout	129	90.5	135	498	2.9	Equatorial launch; returned air density data
Kosmos 156	1967 039A	USSR	Apr 27	Plesetsk	A-1	4730	94.2	467	483	81.2	Dual Vidicon Cameras, Scanning HRIR, Actinometric Inst.
Ariel 3	1967 042A	UK	May 5	Vandenberg	Scout	90	95.8	497	608	80.2	First all-British research satellite

Name	Int'l ID	Country or Agency	LAUNCH DATA Date	Site	Vehicle	Payload Mass (kg)	ORBITAL DATA Period (min)	Perigee (km)	Apogee (km)	Inclination (deg)	Comments
Molniya-1 5	1967 052A	USSR	May 24	Tyuratam	A-2-e	1750	715.6	460	39785	64.8	Comsat, also transmitted cloud cover photos
Aurora 1	1967 065B	USA/USN	Jun 29	Vandenberg	Thor Burner 2	21	172.2	3793	3946	89.8	Investigated formation of Aurora Borealis
DODGE	1967 066F	DOD/NASA	Jul 1	Cape Canav.	Titan 3C	102	1319.0	33270	33659	6.2	DOD Gravity Experiment; Dual Vidicon Cameras; first color pics from near-geo altitude
DMSP-4A F3	1967 080A	USAF	Aug 23	Vandenberg	Thor Burner 2	195	102.0	822	878	98.8	Video Camera System, C System
Molniya-1 6	1967 095A	USSR	Oct 3	Tyuratam	A-2-e	1750	718.2	502	39868	65	Comsat, also transmitted cloud cover photos
DMSP-4A F4	1967 096A	USAF	Oct 11	Vandenberg	Thor Burner 2	195	99.6	650	822	99.2	Video Camera System, C System
Molniya-1 7	1967 101A	USSR	Oct 22	Tyuratam	A-2-e	1750	715.1	508	39710	64.7	Comsat, also transmitted cloud cover photos
Kosmos 184	1967 102A	USSR	Oct 25	Plesetsk	A-1	4730	93.7	441	459	81.2	Metsat
Kosmos 185	1967 104A	USSR	Oct 27	Tyuratam	F-1-m	4000	98.7	518	873	64.1	Probable metsat
ATS 3	1967 111A	NASA	Nov 5	Cape Canav.	Atlas Agena D	365	1436.0	35730	35844	9.2	Multicolor Spin Scan Cloudcover Camera
ESSA 6	1967 114A	ESSA	Nov 10	Vandenberg	Thor Agena D	132	114.9	1407	1483	102.1	2 APT; deactivated 4 Nov 1969
WRESAT 1	1967 118A	Australia	Nov 29	Woomera	SPARTA	45	99.4	193	1259	83.2	Returned solar radiation and upper atmospheric data
OV3 6	1967 120A	USAF	Dec 5	Vandenberg	Scout	95	93.2	412	439	90.6	Conducted ionospheric studies; transmission ended 5th day
Kosmos 206	1968 019A	USSR	Mar 14	Plesetsk	A-1	4730	93.6	449	437	81.2	Returned weather and infrared photos, thermal data
Apollo 6	1968 025A	NASA	Apr 4	Cape Canav.	Saturn 5	36806	90.2	205	392	32.5	Earth orbit; Earth photography
Molniya-1 8	1968 035A	USSR	Apr 21	Tyuratam	A-2-e	1750	713.3	391	39738	65	Comsat, also transmitted cloud cover photos
DMSP-4B F1	1968 042A	USAF	May 23	Vandenberg	Thor Burner 2	195	102.0	809	888	98.8	Video Camera System, C System
Kosmos 226	1968 049A	USSR	Jun 12	Plesetsk	A-1	4730	97.0	579	639	81.2	Returned cloud cover photos, weather data
Molniya-1 9	1968 057A	USSR	Jul 5	Tyuratam	A-2-e	1750	714.8	401	39803	65	Comsat, also transmitted cloud cover photos
OV1 15	1968 059A	USAF	Jul 11	Vandenberg	Atlas F	470	105.0	154	1818	89.8	Air density/solar radiation correlation experiment
OV1 16	1968 059B	USAF	Jul 11	Vandenberg	Atlas F	600	91.8	163	554	89.7	Ionospheric drag experiment; launched with OV1 15
Kosmos 232	1968 060A	USSR	Jul 16	Plesetsk	A-1	4730	88.9	189	248	65	Possible metsat
Explorer 39	1968 066A	NASA	Aug 8	Vandenberg	Scout	9.3	105.0	633	1341	80.6	Air Density Explorer; density & temp. variations at intermediate latitudes

Name	Int'l ID	Country or Agency	LAUNCH DATA Date	Site	Vehicle	Payload Mass (kg)	ORBITAL DATA Per-iod (min)	Peri-gee (km)	Apo-gee) (km)	Incli-nation (deg)	Comments
ESSA 7	1968 069A	ESSA	Aug 16	Vandenberg	Long-Tank Delta	145	115.0	1429	1471	101.8	2 AVCS, FPR, S-band; replaced ESSA 5; deactivated 19 Jul 1969
Kosmos 243	1968 080A	USSR	Sep 23	Tyuratam	A-2	5900	89.6	213	293	71.3	Carried first microwave radiometer, measured water vapor & liquid water
OV2 5	1968 081A	USAF	Sep 26	Cape Canav.	Titan 3C	204	-	-	-	-	Comprehensive environmental research satellite, carried 11 experiments
Aurorae	1968 084A	ESRO	Oct 3	Vandenberg	Scout	81	103.1	258	1538	93.7	ESRO 1; Investigated auroral phenomena & polar ionosphere
Molniya-1 10	1968 085A	USSR	Oct 5	Tyuratam	A-2-e	1750	712.1	436	39633	65	Comsat, also transmitted cloud cover photos
Apollo 7	1968 089A	NASA	Oct 11	Cape Canav.	Saturn 1B	14674	89.5	231	297	31.64	Earth orbit; Earth photography
DMSP-4B F2	1968 092A	USAF	Oct 23	Vandenberg	Thor Burner 2	195	101.3	792	838	98.5	Video Camera System, C System
Kosmos 258	1968 111A	USSR	Dec 10	Tyuratam	A-1	4730	89.4	202	282	65	Possible metsat
ESSA 8	1968 114A	ESSA	Dec 15	Vandenberg	Long-Tank Delta	132	114.7	1411	1461	101.2	2 APT
Kosmos 261	1968 117A	USSR	Dec 20	Plesetsk	B-1	347	93.1	207	642	71	Soviet block joint experiments in air density, polar auroras
Apollo 8	1968 118A	NASA	Dec 21	Cape Canav.	Saturn 5	28833	-	-	-	-	Lunar orbit; Earth photography
ISIS 1	1969 009A	Canada	Jan 30	Vandenberg	TAI Delta	241	128.1	575	3494	88.4	Continued ionospheric studies begun with Alouette 1
ESSA 9	1969 016A	ESSA	Feb 26	Cape Canav.	TAI Delta	145	115.3	1423	1502	102	Final ESSA launch; 2 AVCS, FPR, S-band; replaced ESSA 7
Apollo 9	1969 018A	NASA	Mar 3	Cape Canav.	Saturn 5	36511	88.6	203	229	32.6	Earth orbit; Earth photography
OV1 18	1969 025B	USAF	Mar 18	Vandenberg	Atlas F	125	95.2	466	583	98.8	Studied ionosphere measuring radio interference, electric fields, radiation
Meteor-1 1	1969 029A	USSR	Mar 26	Plesetsk	A-1	3800	97.2	600	639	81.2	Operational metsat; cloud, snow, ice cover data during daylight & dark
Nimbus 3	1969 037A	NASA	Apr 14	Vandenberg	Thor Agena D	575	107.4	1070	1130	99.6	IRIS, MRIR, HRIR, IDCS, MUSE, SIRS, IRLS
Apollo 10	1969 043A	NASA	May 18	Cape Canav.	Saturn 5	42530	-	-	-	-	Lunar orbit; Earth photography
Apollo 11	1969 059A	NASA	Jul 16	Cape Canav.	Saturn 5	43811	-	-	-	-	Lunar orbit; Earth photography
DMSP-4B F3	1969 062A	USAF	Jul 23	Vandenberg	Thor Burner 2	195	101.2	775	844	98.5	Video Camera System, C System
Boreas	1969 083A	ESRO	Oct 1	Vandenberg	Scout	80	91.5	291	389	86.0	ESRO 1B; Ionospheric and auroral experiments; early decay due to booster malfunction
Meteor-1 2	1969 084A	USSR	Oct 6	Plesetsk	A-1	3800	96.7	573	613	81.2	Monitored cloud and ice cover, IR, and thermal energy reflected and radiated from Earth's atmosphere

Name	Int'l ID	Country or Agency	LAUNCH DATA Date	Site	Vehicle	Payload Mass (kg)	ORBITAL DATA Per-iod (min)	Peri-gee (km)	Apo-gee) (km)	Incli-nation (deg)	Comments
Interkosmos 1	1969 088A	USSR	Oct 14	Kapustin Yar	B-1	320	93.3	254	626	48.4	Experiments from 3 East European countries to study sun's UV & X-ray effects on upper atmosphere
Apollo 12	1969 099A	NASA	Nov 14	Cape Canav.	Saturn 5	43848	-	-	-	-	Lunar orbit; Earth photography
Interkosmos 2	1969 110A	USSR	Dec 25	Kapustin Yar	B-1	320	98.4	200	1178	48.4	Carried East European-made ionospheric observation and survey experiments
Kosmos 320	1970 005A	USSR	Jan 16	Kapustin Yar	B-1	375	90.1	247	326	48.5	Possible metsat
ITOS	1970 008A	NASA	Jan 23	Vandenberg	TAT-Delta M	309	115.1	1432	1477	101.8	TIROS M; first of 2nd generation US metsats; 2 APT, 2 AVCS, 2 SR, FPR, 3-axis stabilization
DMSP-5A F1	1970 012A	USAF	Feb 11	Vandenberg	Thor Burner 2	195	101.1	759	850	98.8	Scanning Radiometer
Meteor-1 3	1970 019A	USSR	Mar 17	Plesetsk	A-1	3800	93.5	419	457	81.1	-
Nimbus 4	1970 025A	NASA	Apr 8	Vandenberg	Thorad-Agena D	620	107.2	1087	1097	99.5	IDCS, IRIS, BUV, SCR, FWS, THIR, SIRS, MUSE, IRLS
Apollo 13	1970 029A	NASA	Apr 11	Cape Canav.	Saturn 5	43924	-	-	-	-	Lunar orbit; Earth photography
Meteor-1 4	1970 037A	USSR	Apr 28	Plesetsk	A-1	3800	98.2	625	710	81.2	-
Meteor-1 5	1970 047A	USSR	Jun 23	Plesetsk	A-1	3300	102.0	821	879	81.2	-
Interkosmos 3	1970 057A	USSR	Aug 7	Kapustin Yar	B-1	340	99.7	200	1295	48.4	Studied ionospheric protons, electons, and alpha particles
DMSP-5A F2	1970 070A	USAF	Sep 3	Vandenberg	Thor Burner 2	-	101.4	764	874	99.1	Scanning Radiometer
Interkosmos 4	1970 084A	USSR	Oct 14	Kapustin Yar	B-1	320	93.5	255	649	43.3	Studied solar UV and X-ray effects
Meteor-1 6	1970 085A	USSR	Oct 15	Plesetsk	A-1	3800	96.2	565	574	81.2	-
Kosmos 384	1970 105A	USSR	Dec 10	Plesetsk	A-2	5900	89.5	203	292	72.8	Microwave Radiometers, Narrow-Angle IR Radiometer
NOAA 1	1970 106A	NOAA	Dec 11	Vandenberg	Long-Tank Delta	306	114.9	1422	1471	101.8	Second in the ITOS series; 2 APT, 2 AVCS, 2 SR, FPR
PEOLE 1	1970 109A	France	Dec 12	Kourou	Diamant B	70	97.0	517	747	15.0	Preliminary EOLE; gathered data from a met. balloon system
Kosmos 389	1970 113A	USSR	Dec 18	Plesetsk	A-1	3800	97.6	598	679	81.2	Possible metsat
Meteor-1 7	1971 003A	USSR	Jan 20	Plesetsk	A-1	3800	96.7	586	603	81.2	-
Apollo 14	1971 008A	NASA	Jan 31	Cape Canav.	Saturn 5	44456	-	-	-	-	Lunar orbit; Earth photography
DMSP-5A F3	1971 012A	USAF	Feb 17	Vandenberg	Thor Burner 2	195	100.7	755	817	98.3	Scanning Radiometer
ISIS 2	1971 024A	Canada	Apr 1	Vandenberg	Thor Delta	264	113.7	1354	1423	88.2	3rd in a series of international ionospheric studies satellite
Meteor-1 8	1971 031A	USSR	Apr 17	Plesetsk	A-1	3800	95.3	512	540	81.2	-
Salyut 1	1971 032A	USSR	Apr 19	Tyuratam	D-1	18500	88.5	200	210	51.5	Space station; met. obs. by crew of Soyuz 11
San Marco 3	1971 036A	Italy	Apr 24	San Marco	Scout	164	93.6	222	718	3.2	3rd Italian atmospheric satellite

Name	Int'l ID	Country or Agency	LAUNCH DATA Date	Site	Vehicle	Payload Mass (kg)	ORBITAL DATA Per-iod (min)	Peri-gee (km)	Apo-gee) (km)	Incli-nation (deg)	Comments
Meteor-1 9	1971 059A	USSR	Jul 16	Plesetsk	A-1	3800	95.7	538	554	81.2	-
Apollo 15	1971 063A	NASA	Jul 26	Cape Canav.	Saturn 5	46723	-	-	-	-	Lunar orbit; Earth photography
EOLE 1	1971 071A	France	Aug 16	Wallops Is.	Scout	84	100.1	664	870	50.1	Gathered data from a met. balloon system
DMSP-5B F1	1971 087A	USAF	Oct 14	Vandenberg	Thor Burner 2	195	101.5	782	865	99.1	Scanning Radiometer
Interkosmos 5	1971 104A	USSR	Dec 2	Kapustin Yar	B-1	340	98.5	198	1181	48.5	USSR and Czech instruments to study effects of solar activity on near-Earth radiation
Ariel 4	1971 109A	UK	Dec 11	Vandenberg	Scout	99.5	95.5	477	593	82.99	UK-4; 4 British and 1 US experiments to survey the ionosphere and radio signals
Oreol 1	1971 119A	France/USSR		Dec 27	Plesetsk	C-1	630	111.4	391	2184	74.0 Studied upper atmosphere in high latitudes and the polar lights; 1st satellite launched by USSR with French Assistance
Meteor-1 10	1971 120A	USSR	Dec 29	Plesetsk	A-1	3300	102.6	835	922	81.3	-
DMSP-5B F2	1972 018A	USAF	Mar 24	Vandenberg	Thor Burner 2	195	101.6	787	868	99.1	Scanning Radiometer, VTPR (SSE)
Meteor-1 11	1972 022A	USSR	Mar 30	Plesetsk	A-1	3300	102.5	858	883	81.2	-
Interkosmos 6	1972 027A	USSR	Apr 7	Tyuratam	A-2	6000	89.0	203	256	51.8	-
Apollo 16	1972 031A	NASA	Apr 16	Cape Canav.	Saturn 5	46733	-	-	-	-	Lunar orbit; Earth photography
Interkosmos 7	1972 047A	USSR	Jun 30	Kapustin Yar	B-1	375	92.6	260	555	48.4	-
Meteor-1 12	1972 049A	USSR	Jun 30	Plesetsk	A-1	3300	102.9	880	898	81.2	-
Landsat 1	1972 058A	NASA	Jul 23	Vandenberg	Delta	816.4	103.2	899	911	98.8	Also called ERTS 1; Multispectral Scanner (MSS), Return Beam Vidicon (RBV)
Denpa	1972 064A	Japan	Aug 19	Kagoshima	Mu 4S	75	154.3	232	6093	31.0	Investigated ionospheric phenomena
NOAA 2	1972 082A	NOAA	Oct 15	Vandenberg	Delta	344	115.0	1447	1453	101.6	First operational sounder; 2 VHRR, 2 VTPR, 2 SR, SPM
Meteor-1 13	1972 085A	USSR	Oct 26	Plesetsk	A-1	3300	102.5	858	883	81.2	-
DMSP-5B F3	1972 089A	USAF	Nov 9	Vandenberg	Thor Burner 2	195	101.5	797	855	98.8	Scanning Radiometer, VTPR (SSE)
ESRO IV	1972 092A	ESRO	Nov 22	Vandenberg	Scout	114	99.2	245	1187	91.1	Investigated polar ionosphere
Interkosmos 8	1972 094A	USSR	Nov 30	Plesetsk	B-1	340	93.6	214	679	71	-
Apollo 17	1972 096A	NASA	Dec 7	Cape Canav.	Saturn 5	46743	-	-	-	-	Lunar orbit; Earth photography
Nimbus 5	1972 097A	NASA	Dec 11	Vandenberg	Delta	772	107.2	1087	1100	99.7	First scanning microwave radiometer; THIR, SCMR, ESMR, ITPR, SCR, NEMS
Aeros 1	1972 100A	W. Germany	Dec 16	Vandenberg	Scout	127	95.7	223	867	96.9	Upper air research satellite
Meteor-1 14	1973 015A	USSR	Mar 20	Plesetsk	A-1	2000	102.6	866	883	81.2	-
Interkosmos 9	1973 022A	USSR	Apr 19	Kapustin Yar	B-1	400	102.1	199	1526	48.4	"Copurnicus-500" solar radiation monitor

Name	Int'l ID	Country or Agency	LAUNCH DATA Date	Site	Vehicle	Payload Mass (kg)	ORBITAL DATA Period (min)	Perigee (km)	Apogee (km)	Inclination (deg)	Comments
Skylab	1973 027A	NASA	May 14	Cape Canav.	Saturn 5	74783	93.1	422	442	50.0	Infrared Spectrometer, Microwave Radiometer/Scatterometer/Altimeter,L-Band Microwave Radiometer, Multispectral Scanner
Meteor-1 15	1973 034A	USSR	May 29	Plesetsk	A-1	2000	102.4	842	891	81.2	-
DMSP-5B F4	1973 054A	USAF	Aug 17	Vandenberg	Thor Burner 2	195	101.3	795	836	98.5	Scanning Radiometer
Interkosmos 10	1973 082A	USSR	Oct 30	Plesetsk	C-1	550	102.2	260	1454	74.0	Scientific research in cooperation with Czechoslovakia and East Germany
NOAA 3	1973 086A	NOAA	Nov 6	Vandenberg	Delta	345	116.2	1499	1508	101.7	2 VHRR, 2 VTPR, 2 SR, SPM
Oreol 2	1973 107A	France/USSR	Dec 26	Plesetsk	C-1	550	106.1	393	1691	74.0	-
Meteor-1 16	1974 011A	USSR	Mar 5	Plesetsk	A-1	2200	102.2	825	885	81.2	-
DMSP-5B F5	1974 015A	USAF	Mar 16	Vandenberg	Thor Burner 2A	195	101.3	767	859	99.0	Scanning Radiometer, VTPR (SSE)
Meteor-1 17	1974 025A	USSR	Apr 24	Plesetsk	A-1	2200	102.5	855	887	81.2	-
SMS 1	1974 033A	NASA	May 17	Cape Canav.	Thorad Delta	627	1340.2	32345	35439	1.9	First dedicated geosynchronous metsat; VISSR, SEM, DCP
Interkosmos 11	1974 034A	USSR	May 17	Kapustin Yar	C-1	400	95.3	483	591	50.6	-
ATS 6	1974 039A	NASA	May 30	Cape Canav.	Titan 3C	930	1411.9	35191	35438	8.0	Geosynchronous Very High Resolution Radiometer
Salut 3	1974 046A	USSR	Jun 24	Tyuratam	D-1	18500	89.3	232	257	51.1	Space station; met. obs. by Soyuz crews
Meteor-1 18	1974 052A	USSR	Jul 9	Plesetsk	A-1	2200	103.1	887	911	81.2	-
Aeros 2	1974 055A	W. Germany	Jul 16	Vandenberg	Scout	127	95.7	224	869	97.4	Upper air research satellite
DMSP-5C F1	1974 063A	USAF	Aug 9	Vandenberg	Thor Burner 2A	195	101.6	792	862	98.7	Scanning Radiometer, VTPR (SSE)
Meteor-1 19	1974 083A	USSR	Oct 28	Plesetsk	A-1	2200	102.4	834	899	81.2	-
Interkosmos 12	1974 086A	USSR	Oct 31	Plesetsk	C-1	400	94.2	243	707	74.0	Atmosphere-ionosphere research
NOAA 4	1974 089A	NOAA	Nov 15	Vandenberg	Thorad Delta	340	115.0	1443	1457	101.4	2 VHRR, 2 VTPR, 2 SR, SPM
Meteor-1 20	1974 099A	USSR	Dec 17	Plesetsk	A-1	2200	102.3	841	882	81.2	-
Salyut 4	1974 104A	USSR	Dec 26	Tyuratam	D-1	18500	91.3	337	350	51.6	Space station; met. obs. by Soyuz crews
Landsat 2	1975 004A	NASA	Jan 22	Vandenberg	Delta	953	103.3	901	914	98.9	Multispectral Scanner (MSS), Return Beam Vidicon (RBV)
SMS 2	1975 011A	NASA	Feb 6	Cape Canav.	Delta	627	1436.5	35752	35844	1.8	VISSR, SEM, DCP
Interkosmos 13	1975 022A	USSR	Mar 27	Plesetsk	C-1	400	104.5	285	1643	82.9	Upper atmosphere research satellite
Meteor-1 21	1975 023A	USSR	Apr 1	Plesetsk	A-1	2200	102.5	858	887	81.2	-
Ariabata	1975 033A	India	Apr 19	Kapustin Yar	C-1	360	95.0	512	537	50.7	Atmospheric research satellite
DMSP-5C F2	1975 043A	USAF	May 24	Vandenberg	Thor Burner 2	194	101.8	797	881	98.7	Scanning Radiometer, VTPR (SSE)
Nimbus 6	1975 052A	NASA	Jun 12	Vandenberg	Delta	829	107.5	1099	1112	99.6	LRIR, PMR, TWERLE, ERB, THIR, HIRS/1, SCAMS, ESMR, T&DRE

Name	Int'l ID	Country or Agency	LAUNCH DATA Date	Site	Vehicle	Payload Mass (kg)	ORBITAL DATA Per-iod (min)	Peri-gee (km)	Apo-gee) (km)	Incli-nation (deg)	Comments
Meteor-2 1	1975 064A	USSR	Jul 11	Plesetsk	A-1	2800	102.4	847	885	81.3	Experimental weather satellite
Soyuz 19	1975 065A	USSR	Jul 15	Tyuratam	A-2	6182	88.5	186	220	51.8	Soviet part of Apollo-Soyuz
Apollo 18	1975 066A	NASA	Jul 15	Cape Canav.	Saturn 1B	14768	88.4	170	228	51.8	US part of Apollo-Soyuz
Meteor-1 22	1975 087A	USSR	Sep 18	Plesetsk	A-1	2200	102.3	801	920	81.3	-
Explorer 54	1975 096A	NASA	Oct 6	Vandenberg	Delta	676	118.7	141	3093	90.0	Atmospheric satellite
GOES 1	1975 100A	NASA	Oct 16	Cape Canav.	Delta	627	1436.7	35797	35804	6.7	First operational geosynchronous metsat; VISSR, SEM, DCP
Explorer 55	1975 107A	NASA	Nov 20	Cape Canav.	Delta	721	93.2	447	449	19.6	Atmospheric satellite
Interkosmos 14	1975 115A	USSR	Dec 11	Plesetsk	C-1	700	96.6	300	883	74.0	-
Meteor-1 23	1975 124A	USSR	Dec 25	Plesetsk	A-1	2200	102.3	842	885	81.3	-
UME	1976 019A	Japan	Feb 29	Tanegashima	N-1	139	105.2	988	1009	69.7	Ionospheric Sounding Satellite (ISS 1)
Meteor-1 24	1976 032A	USSR	Apr 7	Plesetsk	A-1	2200	102.2	831	888	81.3	-
Meteor-1 25	1976 043A	USSR	May 15	Plesetsk	A-1	2200	102.2	832	884	81.3	-
Interkosmos 15	1976 056A	USSR	Jun 19	Plesetsk	C-1	550	94.7	483	515	74.0	-
Salyut 5	1976 057A	USSR	Jun 22	Tyuratam	D-1	19000	89.2	230	245	51.5	Space station; met. obs. by Soyuz crews
Interkosmos 16	1976 076A	USSR	Jul 27	Kapustin Yar	C-1	550	94.3	463	518	50.5	Solar X-rays
NOAA 5	1976 077A	NOAA	Jul 29	Vandenberg	Delta	340	116.3	1503	1519	101.9	2 VHRR, 2 VTPR, 2 SR, SPM; ITOS H, last of the ITOS series
DMSP-5D F1	1976 091A	USAF	Sep 11	Vandenberg	Thor Burner 2	450	101.4	806	832	98.6	Block 5D-1; OLS, SSH
Meteor-1 26	1976 102A	USSR	Oct 15	Plesetsk	A-1	2200	102.4	848	885	81.3	-
Meteor-2 2	1977 002A	USSR	Jan 6	Plesetsk	A-1	2750	102.9	881	898	81.3	-
Meteor-1 27	1977 024A	USSR	Apr 5	Plesetsk	A-1	2200	102.4	846	892	81.3	-
DMSP-5D F2	1977 044A	USAF	Jun 5	Vandenberg	Thor Burner 2	450	101.4	789	853	99.0	Block 5D-1; OLS, SSH
GOES 2	1977 048A	NOAA	Jun 16	Cape Canav.	Delta	627	1436.1	35757	35821	5.8	VISSR, SEM, DCP
Meteor-1 28	1977 057A	USSR	Jun 29	Tyuratam	A-1	2200	96.0	543	583	97.4	First sunsynchronous Meteor
GMS 1	1977 065A	Japan	Jul 14	Cape Canav.	Delta	670	1435.9	35771	35801	4.8	Also called Himawari; VISSR
Interkosmos 17	1977 096A	USSR	Sep 24	Plesetsk	C-1	550	94.5	467	511	82.9	-
Salyut 6	1977 097A	USSR	Sep 29	Tyuratam	D-1	9000	92.2	380	391	51.6	Space station; met. obs. by Soyuz crews
Meteosat 1	1977 108A	ESA	Nov 23	Cape Canav.	Delta	697	1436.1	34755	36823	6.1	Imaging Radiometer (0.5-0.9, 5.7-7.1, and 10.5-12.5 micrometers)
Meteor-2 3	1977 117A	USSR	Dec 14	Plesetsk	A-1	2700	102.4	846	886	81.2	-
Kyokko	1978 014A	Japan	Feb 4	Kagoshima	Mu 3H	103	134.2	636	3967	65.4	Scientific satellite investigating aurorae
UME 2	1978 018A	Japan	Feb 16	Tanegashima	N-1	140	107.2	974	1217	69.4	Ionospheric Sounding Satellite (ISS)
Landsat 3	1978 026A	NASA	Mar 5	Vandenberg	Delta	960	103.3	897	919	98.8	Multispectral Scanner (MSS), Return Beam Vidicon (RBV)

Name	Int'l ID	Country or Agency	LAUNCH DATA Date	Site	Vehicle	Payload Mass (kg)	ORBITAL DATA Per-iod (min)	Peri-gee (km)	Apo-gee) (km)	Incli-nation (deg)	Comments
HCMM	1978 041A	NASA	Apr 26	Vandenberg	Scout-F	134	97.3	618	627	97.6	Heat Capacity Mapping Mission (AEM 1)
DMSP-5D F3	1978 042A	USAF	May 1	Vandenberg	Thor Burner 2	513	101.2	804	817	98.6	Block 5D-1; OLS, SSH
GOES 3	1978 062A	NOAA	Jun 16	Cape Canav.	Delta	627	1436.0	35781	35795	3.7	VISSR, SEM, DCP
Seasat	1978 064A	NASA	Jun 27	Vandenberg	Atlas/Agena D	2300	100.5	778	783	108.0	Experimental ocean survey satellite; failed after 90 days
TIROS N	1978 096A	NASA	Oct 13	Vandenberg	Atlas F	1421	101.9	835	853	99.0	First of 3rd generation metsats; AVHRR/1, HIRS/2, SSU, MSU, SEM, DCS
Nimbus 7	1978 098A	NASA	Oct 24	Vandenberg	Delta	907	104.1	927	969	99.4	Last of Nimbus series; SMMR, SBUV/TOMS, CZCS, ERB, SAM II, LIMS, SAMS, THIR
Interkosmos 18	1978 099A	USSR	Oct 24	Plesetsk	C-1	1050	92.2	320	428	82.9	-
Meteor-1 29	1979 005A	USSR	Jan 25	Tyuratam	A-1	3800	97.1	581	645	97.6	Sunsynchronous orbit
SAGE	1979 013A	NASA	Feb 18	Wallops Is.	Scout-F	147	94.4	460	521	55.0	Stratospheric aerosol and gas experiment (AEM 2); solar occultation experiment
Solwind P78-1	1979 017A	US	Feb 24	Vandenberg	Atlas	1331	95.4	513	547	97.8	Ionosphere and magnetosphere research; destroyed by USAF ASAT 13 Sep 1985
Interkosmos 19	1979 020A	USSR	Feb 27	Plesetsk	C-1	1015	98.4	483	874	74.0	-
Meteor-2 4	1979 021A	USSR	Mar 1	Plesetsk	A-1	3800	102.2	845	872	81.2	-
DMSP-5D F4	1979 050A	USAF	Jun 6	Vandenberg	Thor Burner 2	513	101.3	806	828	98.7	Block 5D-1; OLS, SSH, SSM/T, SSC (1.6 micrometers)
Bhaskara 1	1979 051A	India	Jun 7	Kapustin Yar	C-1	441	92.7	407	412	50.7	Dual TV Camera, Satellite Microwave Radiometer
NOAA 6	1979 057A	NOAA	Jun 27	Vandenberg	Atlas F	1421	101.1	797	813	98.5	AVHRR/1, HIRS/2, SSU, MSU, SEM, DCS
Meteor-2 5	1979 095A	USSR	Oct 31	Plesetsk	A-1	3800	102.6	866	882	81.2	-
Interkosmos 20	1979 096A	USSR	Nov 1	Plesetsk	C-1	1100	91.7	343	363	74.0	-
SMM	1980 014A	NASA	Feb 14	Cape Canav.	Delta	2500	94.1	488	493	28.5	Solar Maximum Mission; carries Active Cavity Radiometers to measure solar constant
Meteor-1 30	1980 051A	USSR	Jun 18	Tyuratam	A-1	3475	96.2	540	600	97.7	Combined meteorological and Earth resources satellite
Meteor-2 6	1980 073A	USSR	Sep 9	Plesetsk	A-1	3300	102.4	840	889	81.2	-
GOES 4	1980 074A	NOAA	Sep 9	Cape Canav.	Delta	835	1435.9	35771	35802	3.1	First VAS; VAS, SEM, DCP
Interkosmos 21	1981 011A	USSR	Feb 6	Plesetsk	C-1	550	92.9	397	427	74.0	Oceanographic and terrestrial experiments

Name	Int'l ID	Country or Agency	LAUNCH DATA Date	Site	Vehicle	Payload Mass (kg)	ORBITAL DATA Per-iod (min)	Peri-gee (km)	Apo-gee) (km)	Incli-nation (deg)	Comments
Meteor-2 7	1981 043A	USSR	May 14	Plesetsk	A-1	2750	102.4	849	887	81.3	-
GOES 5	1981 049A	NOAA	May 22	Cape Canav.	Delta	835	1435.8	35781	35788	0.1	VAS, SEM, DCP
Meteosat 2	1981 057A	ESA	Jun 19	Kourou	Ariane 4	697	1436.2	35790	35793	0.3	Imaging Radiometer (0.5-0.9, 5.7-7.1, and 10.5-12.5 micrometers)
NOAA 7	1981 059A	NOAA	Jun 23	Vandenberg	Atlas F	1421	102.0	838	858	99.1	AVHRR/2, HIRS/2, SSU, MSU, SEM, DCS
Meteor-1 31	1981 065A	USSR	Jul 10	Plesetsk	A-1	2200	97.5	607	659	97.7	Last Meteor-1
DE 1	1981 070A	NASA	Aug 3	Vandenberg	Delta	403	410.6	464	23370	89.5	Dynamics Explorer 1
DE 2	1981 070B	NASA	Aug 3	Vandenberg	Delta	-	97.8	298	996	90.0	Dynamics Explorer 2
Interkosmos 22	1981 075A	USSR	Aug 7	Plesetsk	A-1	1500	101.9	795	890	81.2	Ionospheric and magnetospheric research
GMS 2	1981 076A	Japan	Aug 10	Tanegashima	N-2	670	1436.0	35766	35811	2.9	VISSR
Oreol 3	1981 094A	France/ USSR	Sep 21	Plesetsk	A-2	1000	108.5	400	1903	82.5	Soviet/French Magnetosphere and Ionosphere Explorer
SME	1981 100A	NASA	Oct 6	Vandenberg	Thor Delta	437	94.8	504	506	97.7	Solar Mesosphere Explorer
STS 2	1981 111A	NASA	Nov 12	Cape Canav.	STS	8516	89.5	259	259	38	Night/Day Optical Survey of Lightning, Measurement of Air Pollution from Sat., Shuttle Imaging Radar A
Bhaskara 2	1981 115A	India	Nov 20	Kapustin Yar	C-1	444	94.7	498	518	50.6	Dual TV Camera, Satellite Microwave Radiometer
Meteor-2 8	1982 025A	USSR	Mar 25	Plesetsk	F-2	2750	104.2	941	962	82.5	First Meteor-2 in this orbital regime
Salyut 7	1982 033A	USSR	Apr 19	Tyuratam	D-1-h	18900	94.0	473	474	51.6	Space station; met. obs. by Soyuz crews
STS 4	1982 065A	NASA	Jun 27	Cape Canav.	STS	11107	90.6	318	318	28.5	Night/Day Optical Survey of Lightning
Landsat 4	1982 072A	NASA	Jul 16	Vandenberg	Thor Delta	1942	98.9	699	701	98.2	Multispectral Scanner (MSS), Thematic Mapper (TM)
Meteor-2 9	1982 116A	USSR	Dec 14	Plesetsk	A-1	2750	102.0	810	890	81.2	Possible replacement for Meteor-2 6
DMSP-5D F6	1982 118A	USAF	Dec 21	Vandenberg	Atlas E	-	101.3	811	823	98.7	Block 5D-2; OLS, SSH-2, SSM/T
NOAA 8	1983 022A	NOAA	Mar 28	Vandenberg	Atlas F	1712	101.3	803	825	98.7	AVHRR/1, HIRS/2, SSU, MSU, SEM, DCS, SAR
STS 6	1983 026A	NASA	Apr 4	Cape Canav.	STS	21277	89.9	284	284	28.0	Night/Day Optical Survey of Lightning
GOES 6	1983 041A	NOAA	Apr 28	Cape Canav.	Delta 3914	835	1437.8	35782	35864	0.1	VAS, SEM, DCP
STS 7	1983 059A	NASA	Jun 18	Cape Canav.	STS	16817	90.5	314	314	28.0	Modular Optoelectric Multispectral Scanner
Insat 1B	1983 089B	India	Aug 31	STS 8	PAM-D	1152	1436.0	35636	35939	0.2	Launched from Space Shuttle; first 3-axis stabilized geosynchronous metsat
Meteor-2 10	1983 109A	USSR	Oct 28	Plesetsk	A-1	2750	101.4	752	889	81.1	Phased with Meteor-2 7 and Meteor-2 9
DMSP-5D F7	1983 113A	USAF	Nov 18	Vandenberg	Atlas F	-	101.5	815	832	98.7	Block 5D-2; OLS
STS 9	1983 116A	NASA	Nov 28	Cape Canav.	STS	15068	89.5	250	250	57.0	Spacelab 1; several met. experiments

Name	Int'l ID	Country or Agency	LAUNCH DATA Date	Site	Vehicle	Payload Mass (kg)	ORBITAL DATA Per-iod (min)	Peri-gee (km)	Apo-gee) (km)	Incli-nation (deg)	Comments
STS 10	1984 011A	NASA	Feb 3	Cape Canav.	STS	15342	90.8	326	326	28.4	STS 41-B; Modular Optoelectric Multispectral Scanner
Ohzora	1984 015A	Japan	Feb 14	Tanegashima	Mu-3S	180	92.9	317	503	74.6	EXOS-C; Middle Atmosphere Programmer; optically sensing atmosphere and ionosphere
Landsat 5	1984 021A	NASA	Mar 1	Vandenberg	Delta	1938	98.9	699	700	98.2	Multispectral Scanner (MSS), Thematic Mapper (TM)
Uosat 2	1984 021B	UK	Mar 1	Vandenberg	Delta	52	98.5	674	693	98.2	Magnetospheric studies, Earth imaging, space dust detection, amateur radio digital communications
Meteor-2 11	1984 072A	USSR	Jul 5	Plesetsk	F-2	2750	104.1	939	958	82.5	-
GMS 3	1984 080A	Japan	Aug 2	Tanegashima	N-2	303	1435.9	35783	35787	0.8	VISSR
STS 12	1984 093A	NASA	Aug 30	Cape Canav.	STS	21524	90.9	331	331	28.4	STS 41-D; CLOUDS
STS 13	1984 108A	NASA	Oct 5	Cape Canav.	STS	10629	91.6	352	352	57.0	STS 41-G; Measurement of Air Pollution from Sat., Large Format Camera, Shuttle Imaging Radar B
ERBS	1984 108B	NASA	Oct 5	STS 13	Arm	226	96.7	601	605	57.0	Carries ERBE and SAGE II
NOAA 9	1984 123A	NOAA	Dec 12	Vandenberg	Atlas	1712	102.1	841	862	99.0	AVHRR/2, HIRS/2, SSU, MSU, DCS, ERBE, SBUV/2, SAR
Meteor-2 12	1985 013A	USSR	Feb 6	Plesetsk	F-2	2200	104.1	934	960	82.5	-
Meteor-3 1	1985 100A	USSR	Oct 24	Plesetsk	F-2	-	109.5	1185	1210	82.5	First of 3rd generation metsats
Meteor-2 13	1985 119A	USSR	Dec 26	Plesetsk	F-2	2200	104.1	936	958	82.5	-
SPOT 1	1986 019A	France	Feb 22	Kourou	Ariane 1	1830	101.5	824	828	98.7	Earth resources
Viking	1986 019B	Sweden	Feb 22	Kourou	Ariane 1	538	258.9	790	13359	98.8	Studying magnetic, electric, ultraviolet properties of auroral regions
Meteor-2 14	1986 039A	USSR	May 27	Plesetsk	F-2	2000	104.2	941	960	82.5	-
NOAA 10	1986 073A	NOAA	Sep 17	Vandenberg	Atlas E	1712	101.3	808	826	98.7	AVHRR/1, HIRS/2, MSU, SEM, DCS, ERBE, SAR
Meteor-2 15	1987 001A	USSR	Jan 5	Plesetsk	F-2	2000	104.1	939	957	82.5	-
MOS 1	1987 018A	Japan	Feb 19	Tanegashima	N-2	745	103.3	909	909	99.1	Marine Observation Satellite; MESSR, VTIR, MSR
GOES 7	1987 022A	NOAA	Feb 26	Cape Canav.	Delta 3914	835	1436.1	35783	35796	0.1	VAS, SEM, DCP
DMSP-5D F8	1987 053A	USAF	Jun 20	Vandenberg	Atlas E	750	102.0	836	856	98.8	Block 5D-2; OLS, SSH-2, SSM/T, SSM/I
Meteor-2 16	1987 068A	USSR	Aug 18	Plesetsk	F-2	2000	104.1	940	957	82.6	-
Meteor-2 17	1988 005A	USSR	Jan 20	Plesetsk	F-2	2000	104.1	933	959	82.6	-
DMSP-5D F9	1988 006A	USAF	Feb 3	Vandenberg	Atlas	750	101.4	815	826	98.7	Block 5D-2; OLS
San Marco 4	1988 026A	Italy	Mar 25	San Marco	Scout	236	93.0	263	615	3.0	Upper atmosphere studies, joint Italian/FRG/U.S. mission

Name	Int'l ID	Country or Agency	LAUNCH DATA Date	Site	Vehicle	Payload Mass (kg)	ORBITAL DATA Per-iod (min)	Peri-gee (km)	Apo-gee) (km)	Incli-nation (deg)	Comments
Kosmos 1940	1988 034A	USSR	Apr 26	Tyuratum	D-1-e	2000	1430.4	35572	35784	0.2	Oceanographic, atmospheric studies; located at 12°W
Meteosat 3	1988 051A	ESA	Jun 15	Kourou	Ariane 4	696	1436.1	35786	35794	0.9	First Ariane 4 launch; stationed above 0°E
Insat 1C	1988 063A	India	Jul 21	Kourou	Ariane 3	1190	1436.0	35763	35812	0.2	Backup for Insat 1B; stationed above 93.5°East
Meteor-3 2	1988 064A	USSR	Jul 26	Plesetsk	F-2	2000	109.5	1186	1208	82.5	-
Fengyun-1 1	1988 080A	PRC	Sep 6	Taiyuan	CZ-4	750	102.8	877	899	99.2	Experimental metsat; first CZ-4 launch; new launch site
NOAA 11	1988 089A	NOAA	Sep 24	Vandenberg	Atlas E	1712	102.1	845	863	98.9	AVHRR/2, HIRS/2, MSU, SSU, SBUV/2, SEM, DCS, S&R
Exos-D	1989 016A	Japan	Feb 21	Kagoshima	M-3S-2	295	205.2	262	10021	75.1	Studying aurora borialis, aurora australis
Meteor-2 18	1989 018A	USSR	Feb 28	Plesetsk	F-2	1500	104.2	941	960	82.5	-
Meteosat 4	1989 020B	ESA	Mar 6	Kourou	Ariane 44LP	681	1436.0	35779	35798	0.7	Meteosat Operational Program 1 (MOP 1).
Resurs-F 1	1989 038A	USSR	May 25	Plesetsk	A-2	6300	89.9	254	273	82.3	Remote sensing. Deployed two satellites
Pion 1	1989 038C	USSR	Jun 9	Resurs-F 1	A-2	78	89.9	257	270	82.3	Passive atmospheric research satellite
Pion 2	1989 038D	USSR	Jun 9	Resurs-F 1	A-2	78	89.9	257	268	82.3	Passive atmospheric research satellite
Resurs-F 2	1989 049A	USSR	Jun 27	Plesetsk	A-2	6300	90.0	260	274	82.5	Remote sensing
Resurs-F 3	1989 055A	USSR	Jul 12	Plesetsk	A-2	6300	90.0	260	275	82.5	Remote sensing. Deployed two satellites
Pion 3	1989 055C	USSR	Aug 7	Resurs-F 3	A-2	78	89.9	254	272	82.5	Passive atmospheric research satellite
Pion 4	1989 055D	USSR	Aug 7	Resurs-F 3	A-2	78	89.9	255	272	82.5	Passive atmospheric research satellite
GMS 4	1989 070A	Japan	Sep 5	Tanegashima	H-1	725	1435.9	35780	35792	1.4	-
Interkosmos 24	1989 080A	USSR	Sep 28	Plesetsk	F-2	1000	116.1	505	2492	82.6	Particles and fields exp. US participation
Meteor-3 3	1989 086A	USSR	Oct 24	Plesetsk	F-2	2150	109.6	1190	1213	82.5	-
SPOT 2	1990 005A	France	Jan 22	Kourou	Ariane 4	1870	101.5	821	823	98.7	2nd in series of French remote sensing satellites
MOS 1B	1990 013A	Japan	Feb 7	Tanegashima	H-1	740	103.3	908	909	99.2	Marine observation satellite. Called Momo 1B after launch
LACE	1990 015A	US	Feb 14	Cape Canav.	Delta 2	1430	95.0	514	539	43.1	Low-power Atmospehric Compensation Experiment for Strategic Defense Initiative Office (SDIO)
Okean 2	1990 018A	USSR	Feb 28	Plesetsk	F-2	1900	97.7	632	658	82.5	Oceanographic remote sensing
Resurs F6	1990 047A	USSR	May 29	Plesetsk	A-2	6300	89.9	255	268	82.3	Remote sensing

Name	Int'l ID	Country or Agency	LAUNCH DATA Date	Site	Vehicle	Payload Mass (kg)	ORBITAL DATA Per-iod (min)	Peri-gee (km)	Apo-gee) (km)	Incli-nation (deg)	Comments
Insat 1D	1990 051A	India	Jun 12	Cape Canav.	Delta	1190	1436.0	35781	35795	0.1	Domestic comsat/metsat. Last of Insat-1 series.
Meteor-2 19	1990 057A	USSR	Jun 27	Plesetsk	F-2	1750	104.1	937	961	82.5	-
Resurs F7	1990 060A	USSR	Jul 17	Plesetsk	A-2	6300	89.3	193	276	82.3	Remote sensing
Resurs F8	1990 073A	USSR	Aug 16	Plesetsk	A-2	6300	88.7	176	229	82.3	Remote sensing
Fengyun-1 2	1990 081A	PRC	Sep 3	Taiyuan	CZ-4	881	102.9	881	896	99.0	-
PRC 31	1990 081B	PRC	Sep 3	Taiyuan	CZ-4	4	100.8	775	804	99.0	Atmospheric balloon
PRC 32	1990 081C	PRC	Sep 3	Taiyuan	CZ-4	4	102.2	833	886	99.0	Atmospheric balloon
Meteor-2 20	1990 086A	USSR	Sep 28	Plesetsk	F-2	1750	104.1	940	958	82.5	-
DMSP-5D F10	1990 105A	USAF	Dec 1	Vandenberg	Atlas E	750	100.7	729	845	98.9	Operational even though broken nozzle prevented it from reaching desired orbit
Meteosat 5	1991 015B	ESA	Mar 2	Kourou	Ariane 44LP	681	1436.1	35787	35792	0.7	Meteosat Operational Program 2 (MOP 2)
Almaz 1	1991 024A	USSR	Mar 31	Tyuratam	SL-13	18550	90.6	293	305	72.7	Remote Earth sensing; radar imaging
Meteor-3 4	1991 030A	USSR	Apr 24	Plesetsk	SL-14	2215	109.5	1184	1210	82.5	-
NOAA 12	1991 032A	NOAA	May 14	Vandenberg	Atlas E	1421	101.3	807	825	98.7	NOAA D, one of the original TIROS N type. AVHRR/1, HIRS/2, MSU, DCS
Resurs-F 10	1991 035A	USSR	May 21	Plesetsk	SL-4	6300	89.2	227	231	82.3	Earth resources; reentered 6/20/91
Okean 3	1991 039A	USSR	Jun 4	Plesetsk	SL-14	1900	97.7	627	659	82.5	Oceanography
Resurs-F 11	1991 044A	USSR	Jun 28	Plesetsk	SL-4	6300	89.8	253	268	82.3	Reentered 7/21/91
ERS 1	1991 050A	ESA	Jul 17	Kourou	Ariane 4	2384	100.5	774	775	98.5	ESA Remote Sensing satellite; microwave, IR imaging of oceans, ice, and land masses
Meteor-3 5	1991 056A	USSR	Aug 15	Plesetsk	SL-14	2215	109.4	1185	1203	82.6	Carries US-built Total Ozone Mapping Spectrometer (TOMS)
IRS 1B	1991 061A	India	Aug 29	Tyuratam	SL-3	980	103.2	890	918	99.2	Indian Remote Sensing satellite; launched commercially by USSR
UARS	1991 063B	NASA	Sep 15	STS-48	RMS	6795	96.2	573	579	57.0	Upper Atmosphere Research Satellite
DMSP-5D F11	1991 082A	USAF	Nov 28	Vandenberg	Atlas E	830	101.9	835	855	98.9	-
JERS 1	1992 007A	Japan	Feb 11	Tanegashima	H1	1340	96.1	566	570	97.7	Japanese Earth Resources Satellite
STS 45	1992 015A	NASA	Mar 24	Cape Canav.	STS	105843	90.5	292	304	57.0	Carried ATLAS science payload.
Resurs-F 14	1992 024A	CIS	Apr 29	Plesetsk	SL-4	6300	89.5	233	256	82.0	Remote sensing. Reentered 5/29/92
SROSS C	1992 028A	India	May 20	Sriharikota	ASLV	106	91.3	256	436	46.0	Streched Rohini Satellite Series. Carried gamma ray detector, ionospheric monitor. Reentered 7/14/92
Insat 2A	1992 041A	India	Jul 9	Kourou	Ariane 44L	1906	1437.2	35741	35883	0.1	Indian communications, remote sensing. Positioned above 74°E.
Topex/ Poseidon	1992 052A	US/France	Aug 10	Kourou	Ariane 42P	2402	112.4	1330	1342	66.0	Ocean sensing,mapping mission sponsored by NASA/CNES.

Name	Int'l ID	Country or Agency	LAUNCH DATA Date	Site	Vehicle	Payload Mass (kg)	ORBITAL DATA Per-iod (min)	Peri-gee (km)	Apo-gee) (km)	Incli-nation (deg)	Comments
Kitsat A	1992 052B	Korea	Aug 10	Kourou	Ariane 42P	50	112.0	1304	1325	66.0	First Korean payload. Store/forward comm., Earth imaging.
Resurs-F 16	1992 056A	CIS	Aug 19	Plesetsk	SL-4	6300	89.2	222	233	82.5	Remote sensing. Carried U.S. DoD experiment to measure beryllium atoms in low Earth orbit. Reentered 9/4/92
Pion 5	1992 056C	CIS	Aug 19	Plesetsk	SL-4	50	89.3	224	248	82.5	Passive subsatellite to determine how upper atmosphere affects spacecraft reentry projections. Reentered 9/25/92.
Pion 6	1992 056D	CIS	Aug 19	Plesetsk	SL-4	50	89.3	225	247	82.5	See Pion 5. Reentered 9/24/92.
Kosmos 2209	1992 059A	CIS	Sep 10	Tyuratam	SL-12	2200	1435.9	35765	35808	1.1	Remote sensing. Positioned above 24°E.
Freja	1992 064A	Sweden	Oct 6	Shuang Cheng Tzu	CZ-2C	259	109.0	596	1759	63.0	Swedish ionospheric, magnetospheric studies.
PRC 36	1992 064B	PRC	Oct 6	Shuang Cheng Tzu	CZ-2C	2100	89.8	211	318	63.0	Fanhul Shi Weixing remote sensing, microgravity experiments. Capsule reentered 10/13/92; instrument module, 10/31/92.
Kosmos 2224	1992 087A	CIS	Dec 17	Tyuratam	SL-12	2200	1436.0	35696	35878	2.1	Remote sensing. Located above 12°E.
Insat 2B	1993 048B	India	Jul 22	Kourou	Ariane 44L	1931	1436.0	35774	35802	0.1	Telecommunications, weather forecasting, search and rescue. Located above 93.5°E.
NOAA 13	1993 050A	NOAA	Aug 9	Vandenberg	Atlas E	1712	102.1	845	861	98.9	AVHRR/2, HIRS/2, MSU, SSU, SBUV/2, SEM, DCS, S&R. Suffered fatal power failure 8/21/93.
Meteor-2 21	1993 055A	CIS	Aug 31	Plesetsk	SL-14	2000	104.2	933	967	82.5	-
SPOT 3	1993 061A	France	Sep 26	Kourou	Ariane 40	1907	101.5	821	823	98.7	Remote sensing
Kosmos 2265	1993 067A	CIS	Oct 26	Plesetsk	SL-8	500	103.6	290	1558	82.9	Atmospheric density monitor
Meteosat 6	1993 073B	ESA	Nov 20	Kourou	Ariane 44LP	704	1432.5	35674	35757	1.2	Located above 0°
Meteor-3 6	1994 003A	CIS	Jan 25	Plesetsk	SL-14	2215	109.4	1182	1207	82.6	-
GOES 8	1994 022A	NOAA	Apr 13	Cape Canav.	Atlas	2100	1436.1	35769	35812	0.3	First of GOES I-M series.
DMSP-5D F12	1994 057A	USAF	Aug 29	Vandenberg	Atlas E	830	102.0	839	856	98.9	-
NOAA 14	1994 089A	NOAA	Dec 30	Vandenberg	Atlas E	1712	102.1	845	858	98.9	-

APPENDIX

B

Abbreviations

1D	One-dimensional
2D	Two-dimensional
3D	Three-dimensional
AATSR	Advanced Along-Track Scanning Radiometer
ACRIM	Active Cavity Radiometer Irradiance Monitor
ADEOS	Advanced Earth Observing Satellite
ADM	Angular dependence model
AEM	Applications Explorer Mission
AIRS	Atmospheric Infrared Sounder
ALT	Altimeter
AMSR	Advanced Microwave Scanning Radiometer
AMSU	Advanced Microwave Sounding Unit
API	Antecedent precipitation index
APT	Automatic picture transmission
ASAR	Advanced Synthetic Aperture Radar
ASTER	Advanced Spaceborne Thermal Emission and Reflection Radiometer
ATI	Area time integral

ATLID	Atmospheric Lidar
ATS	Applications Technology Satellite
AVCS	Advanced Vidicon Camera System
AVHRR	Advanced Very High Resolution Radiometer
AVNIR	Advanced Visible and Near-Infrared Radiometer
BIAS	Bristol/NOAA InterActive Scheme
BUV	Backscatter Ultraviolet Spectrometer
CAC	Climate Analysis Center
CCD	Charge-coupled device
CCR	Cloud Cover Radiometer
CDA	Command and data acquisition
CERES	Clouds and the Earth's Radiant Energy System
CGMS	Committee on Coordination for Geostationary Meteorlogical Satellites
CI	Current intensity
CIRA	Cooperative Institute for Research in the Atmosphere
CLAES	Cryogenic Limb Array Etalon Spectrometer
CNES	Centre National d'Etudes Spatiales
CRT	Cathode ray tube
CST	Convective—stratiform technique
CZCS	Coastal Zone Color Scanner
DAAC	Distributed Active Archive Center
DCP	Data Collection Platform
DCS	Data Collection System
DIAL	Differential absorption lidar
DMSP	Defense Meteorological Satellite Program
DORIS	Doppler Orbitography and Radiopositioning Integrated by Satellite
DRIR	Direct Readout Infrared Radiometer
DS	Dwell sounding
ECT	Equator crossing time
EOS	Earth Observing System
EOS-AERO	EOS Aerosol Mission
EOS-ALT	EOS Altimetry Mission
EOS-AM	EOS Morning Crossing (Ascending) Mission
EOS-CHEM	EOS Chemistry Mission
EOS-COLOR	EOS Ocean Color Instrument
EOS-PM	EOS Afternoon Crossing (Descending) Mission
EOSAT	Earth Observation Satellite Company
EOSDIS	EOS Data and Information System
EOSP	Earth Observing Scanning Polarimeter
EPA	Environmental Protecton Agency
ERB	Earth Radiation Budget
ERBE	Earth Radiation Budget Experiment
ERBE-NS	ERBE Nonscanner

ERBE-S	ERBE Scanner
ERBS	Earth Radiation Budget Satellite
ERTS	Earth Resources Technology Satellite
ESA	European Space Agency
ESMR	Electrically Scanning Microwave Radiometer
ESSA	Environmental Science Service Administration
EUMETSAT	European Organisation for the Exploitation of Meteorological Satellites
FACE	Florida Area Cumulus Experiment
FAR	False-alarm ratio
FOV	Field of view
FWS	Filter Wedge Spectrometer
GAC	Global area coverage
GARP	Global Atmospheric Research Program
GATE	GARP Atlantic Tropic Experiment
GDAS	Global Data Assimilation System
GEO	Geostationary Earth Observatory
GEO	Geostationary orbits
GHIS	GOES High-Resolution Interferometer Sounder
GLAS	Geoscience Laser Altimeter System
GLAS	Goddard Laboratory for Atmospheric Sciences
GLRS	Geoscience Laser Ranging System
GMS	Geostationary Meteorological Satellite
GOES	Geostationary Operational Environmental Satellite
GOMOS	Global Ozone Monitoring by Occultation of Stars
GOSTCOMP	Global Operational Sea Surface Temperature Computation
GPI	GOES Precipitation Index
GSM	Global Spectral Model
GTS	Global Telecommuncations System
GVAR	GOES Variable
HgCdTe	Mercury cadmium telluride
HIRDLS	High-Resolution Dynamics Limb Sounder
HIRIS	High-Resolution Imaging Spectrometer
HIRS	High-Resolution Infrared Radiation Sounder
HIS	High-Spectral-Resolution Interferometer Sounder
HMMR	High-Resolution Multifrequency Microwave Radiometer
HRC	Highly reflective cloud
HRIR	High-Resolution Infrared Radiometer
HRIS	High-Resolution Imaging Spectrometer
HRPT	High-resolution picture transmission
IASI	Infrared Atmospheric Sounding Interferometer
IDCS	Image Dissector Camera System
IFOV	Instantaneous field of view
ILAS	Improved Limb Atmsopheric Spectrometer
IMG	Interferometric Monitor for Greenhouse Gases

InSb	Indium antimonide
IPCC	Intergovernmental Panel on Climate Change
IR	Infrared
IRIS	Infrared Interferometer Spectrometer
IRLS	Interrogation, Recording and Location System
IRTS	Infrared Temperature Sounder
ISAMS	Improved Stratospheric and Mesospheric Sounder
ISCCP	International Satellite Cloud Climatology Program
ISRO	Indian Space Research Organization
ITCZ	Intertropical convergence zone
ITIR	Intermediate Thermal Infrared Radiometer
ITOS	Improved TIROS Operational System
ITPR	Infrared Temperature Profile Radiometer
JPL	Jet Propulsion Laboratory
LAC	Local area coverage
LASA	Lidar Atmospheric Sounder and Altimeter
LAWS	Laser Atmospheric Wind Sounder
LEO	Low Earth Orbits
LFM	Limited Fine Mesh
LIMS	Limb Infrared Monitor of the Stratosphere
LIS	Lightning Imaging Sensor
LRIR	Limb Radiance Inversion Radiometer
LW	Longwave
MCSST	Mutlichannel Sea Surface Temperatures
MERIS	Medium-Resolution Imaging Spectrometer
METSAT	Meteorlogical satellite
MFOV	Medium field of view
MHS	Microwave Humidity Sensor
MIMR	Multifrequency Imaging Microwave Radiometer
MIPAS	Michelson Interferometer for Passive Atmospheric Sounding
MISR	Multi-Angle Imaging SpectoRadiometer
MITI	Ministry of International Trade and Industry (of Japan)
MLCE	Maximum-likelihood cloud estimation
MLS	Microwave Limb Sounder
MODIS	Moderate-Resolution Imaging Spectrometer
MOPITT	Measurements of Pollution in the Troposphere
MRF	Medium-Range Forecast
MRIR	Medium-Resolution Infrared Radiometer
MSC	Meteorlogical Satellite Center
MSI	Multispectral imaging
MSSCC	Multicolor Spin Scan Cloud Camera
MSU	Microwave Sounding Unit
MTPE	Mission to Planet Earth
MTS	Microwave Temperature Sounder

MUSE	Monitor of Ultraviolet Solar Energy
MVS	Minimum variance simultaneous
NASA	National Aeronautics and Space Administration
NASDA	National Space Development Agency (of Japan)
NCDC	National Climatic Data Center
NCO_2	Narrow CO_2 channel
NEΔL	Noise-equivalent radiance difference
NEΔT	Noise-equivalent temperature difference
NEMS	Nimbus E Microwave Spectrometer
NESDIS	National Environmental Satellite, Data, and Information Service
NEXRAD	Next-Generation Radar
NFOV	Narrow field Of view
NGM	Nested Grid Model
NMC	U.S. National Meteorlogical Center
NOAA	National Oceanic and Atmospheric Administration
NOSAT	No satellite
NSCAT	NASA Scatterometer
NSSDC	National Space Science Data Center
OCTS	Ocean Color and Temperature Scanner
OLR	Outgoing longwave radiation
OLS	Operational Linescan System
PCT	Polarization-corrected brightness temperature
PE	Primitive equation
PMR	Pressure Modulator Radiometer
PMT	Photo multiplier tube
POD	Probability of detection
POEM	Polar-Orbit Earth Observation Missions
POES	Polar-orbiting Operational Environmental Satellite
POLDER	Polarization and Directionality of Earth's Reflectances
PPI	Plan position indicator
PR	Precipitation Radar
PRAREE	Precise Range and Range Rate Equipment—Extended
PRF	Pulse repetition frequency
RA-2	Radar Altimeter 2
RAFS	Regional Analysis and Forecast System
RF	Radio frequency
RIS	Retroreflector in Space
RMS	Root mean square
RPM	Revolutions per minute
RTTS	Real-Time Transmission Systems
S&R	Search and Rescue
SAB	Sorting into angular bins
SAGE	Stratosphere Aerosol and Gas Experiment
SAM	Stratospheric Aerosol Measurement

SAMS	Stratospheric and Mesospheric Sounder
SAR	Synthetic aperture radar
SAS	Solar Aspect Sensor
SASS	Seasat-A Satellite Scatterometer
SAT	Satellite
SBUV	Solar Backscatter Ultraviolet radiometer
SCAMS	Scanning Microwave Spectrometer
SCARAB	Scanner for the Radiation Budget
SCIAMACHY	Scanning Imaging Absorption Spectrometer for Atmospheric Cartography
SCMR	Surface Composition Mapping Radiometer
SCR	Selective Chopper Radiometer
SeaWiFS	Sea-Viewing Wide Field Sensor
SEM	Space Environment Monitor
SIR	Shuttle Imaging Radar
SIRS	Satellite Infrared Spectrometer
SMM	Solar Maximum Mission
SMMR	Scanning Mutlichannel Microwave Radiometer
SMS	Synchronous Meteorological Satellite
SOCC	Satellite Operations Control Center
SOLSTICE	Solar Stellar Irradiance Comparison Experiment
SPOT	Système Probatoire d'Observation de la Terre
SR	Scanning Radiometer
SSALT	Solid-State Altimeter
SSM/I	Special Sensor Microwave Imager
SSM/T	Special Sensor Microwave Temperature
SST	Sea surface temperature
SSU	Stratospheric Sounding Unit
STS	Space Transportation System
SW	Shortwave
T&DRE	Tracking and Data Relay Experiment
TES	Tropospheric Emission Spectrometer
THIR	Temperature Humidity Infrared Radiometer
TIP	TIROS Information Processor
TIROS	Television and Infrared Observational Satellite
TMI	TRMM Microwave Imager
TMR	TOPEX Microwave Radiometer
TOMS	Total Ozone Mapping Spectrometer
TOPEX	Ocean Topography Experiment
TOVS	TIROS Operational Vertical Sounder
TRMM	Tropical Rainfall Measuring Mission
TTC	Telemetry, tracking, and command
TWERLE	Tropical Wind Energy Conversion and Reference Level Experiment
UARS	Upper Atmosphere Research Satellite

UHF	Ultrahigh frequency
USGCRP	U.S. Global Change Research Program
UTC	Coordinated Universal Time
VAS	VISSR Atmospheric Sounder
VDUC	VAS Data Utilization Center
VHRR	Very High Resolution Radiometer
VIRR	Visible and Infrared Radiometer
VIRS	Visible Infrared Scanner
VIRSR	Visible Infrared Scanning Radiometer
VIS	Visible
VISSR	Visible and Infrared Spin Scan Radiometer
VTPR	Vertical Temperature Profile Radiometer
WCO_2	Wide CO_2 channel
WEFAX	Weather Facsimile
WFOV	Wide field of view
WMO	World Meteorlogical Organization
WSR	Weather Service Radar

APPENDIX

C

Principal Symbols

A	Albedo, area
a	Semimajor axis
$B_\lambda(T)$	Planck function
$\mathbf{C}$	Matrix used to retrieve temperatures from radiances
c	Speed of light
c_1	First radiation constant
c_2	Second radiation constant
d_{sun}	Earth—sun distance
E	Irradiance
e	Eccentric anomaly
ECT	Equator crossing time
f	Coriolis parameter
G	Newtonian (or universal) gravitation constant
H	Radiant exposure
h	Planck's constant, generalized height coordinate
h_s	Satellite height (above surface)
h_{TOA}	Height of top of atmosphere
i	Inclination

J_2	Coefficient of the quadrupole gravitational potential of Earth
k	Boltzmann's constant
L	Radiance
L_λ, L_κ, L_ν	Monochromatic radiance (spectral radiance)
LT	Local time
M	Radiant exitance, mean anomaly
m	Complex index of refraction, mass
m_e	Mass of the Earth
N	Cloud amount, number of pixels
N'	Effective cloud amount
$N(r)$	Drop-size distribution function
N^*	Ratio of cloud amounts
N_A	Avogadro's number
n	Index of refraction, mean motion constant
$\bar{n}$	Anomalistic mean motion constant
n'	Imaginary (absorptive) part of the complex index of refraction
P	Pressure, power
p	Pressure
$p(\psi_s)$	Scattering phase function
Q	Radiant energy
Q_a	Absorption efficiency
Q_s	Scattering efficiency
q	Mixing ratio
R	Rain depth, rain rate
R^*	Universal gas constant
r	Radius, rain rate
$\vec{r}$	Radius vector
r_e	Radius of Earth
r_{ee}	Equatorial radius of Earth
S	Line strength
$\mathbf{S}_E$	Radiance covariance matrix
S_{sun}	Solar constant
$\mathbf{S}_T$	Temperature covariance matrix
s	Distance
T	Period, temperature
$\bar{T}$	Anomalistic period
$\tilde{T}$	Synodic or nodal period
T_B	Brightness temperature
T_{BB}	Equivalent blackbody temperature
T_F	Flux temperature
t	Time
t_o	Epoch time
U	Gravitational potential, column-integrated mass
$\mathbf{V}_g$	Geostrophic wind vector
W_λ	Weighting function
W_m	Molecular weight

X_B	Backscattering cross section per unit volume
Y_n^m	Spherical harmonic function
Z	Radar reflectivity factor
z_T	Tangent height
α	Absorptance, scan angle
$\tilde{\alpha}$	Absorption number
α_P	Pitch angle
α_R	Roll angle
α_Y	Yaw angle
β	Mass absorption coefficient
Γ	Argument of perigee, sensitivity factor
γ_r	Bidirectional reflectance
ΔLON	Nodal longitude increment
δ	Declination angle, vertical optical depth
δ_o	Vertical optical depth of entire atmosphere
δ_{sl}	Slant path optical depth
ε	Eccentricity, emittance
ζ	Zenith angle
Θ	Latitude
θ	True anomaly, zenith angle
κ	Wavenumber
κ_o	Line central wavenumber
$\Delta\kappa$	Line halfwidth
λ	Wavelength
λ_m	Wavelength of maximum emission
λ_n	Eigenvalue
μ	Cosine of zenith angle ($\cos\theta$)
ν	Frequency
ξ_r	Anisotropic reflectance factor
ρ	Density, reflectance
σ	Stefan–Boltzmann constant
σ_a	Absorption coefficient
σ_e	Extinction coefficient
σ_s	Scattering coefficient
σ°	Normalized radar cross section (of ocean surface)
τ	Transmittance
τ_o	Vertical transmittance of entire atmosphere
Φ	Geopotential, radiant flux
ϕ	Azimuth angle
χ	Size parameter
Ψ	Longitude
ψ_s	Scattering angle
Ω	Solid angle, right ascension of ascending node
Ω_e	Right ascension of Greenwich
ω	Argument of perigee
$\tilde{\omega}$	Single-scatter albedo

APPENDIX D

Système International Units

MOST OF THE units in this book are Système International (SI) Units, which is the internationally accepted form of the metric system. SI Units are comprised of base units, supplementary units, and units derived from the basic units.

Quantity	*Name of unit*	*Symbol*	*Definition*
	Base Units		
Length	meter	m	
Mass	kilogram	kg	
Time	second	s	
Temperature	kelvin	K	
Amount of substance	mole	mol	
Electrical current	ampere	A	
Luminous intensity	candela	cd	

(*continues*)

Quantity	Name of unit	Symbol	Definition
	Supplementary Units		
Plane angle	radian	rad	
Solid angle	steradian	sr	
	Named Derived Units		
Force	newton	N	$kg\ m\ s^{-2}$
Pressure	pascal	Pa	$N\ m^{-2} = kg\ m^{-1}\ s^{-2}$
Energy	joule	J	$N\ m = kg\ m^2\ s^{-2}$
Power	watt	W	$J\ s^{-1} = kg\ m^2\ s^{-3}$
Electrical potential difference	volt	V	$W\ A^{-1} = kg\ m^2\ s^{-3}\ A^{-1}$
Electrical charge	coulomb	C	A s
Electrical resistance	ohm	Ω	$V\ A^{-1} = kg\ m^2\ s^{-3}\ A^{-2}$
Electrical capacitance	farad	F	$C\ V^{-1} = kg^{-1}\ m^{-2}\ s^4\ A^2$
Frequency	hertz	Hz	s^{-1}
Celsius temperature	degree Celsius	°C	K − 273.15
Magnetic flux	weber	Wb	$V\ s = kg\ m^2\ s^{-2}\ A^{-1}$
Inductance	henry	H	$Wb\ A^{-1} = kg\ m^2\ s^{-2}\ A^{-2}$
Magnetic flux density	tesla	T	$Wb\ m^{-2} = kg\ s^{-2}\ A^{-1}$
Luminous flux	lumen	lm	cd sr
Illuminance	lux	lx	$lm\ m^{-2} = cd\ sr\ m^{-2}$
	Other Derived Units (Useful in Satellite Meteorology)		
Area		m^2	
Volume		m^3	
Density		$kg\ m^{-3}$	
Velocity (speed)		$m\ s^{-1}$	
Angular velocity		$rad\ s^{-1}$	
Wavelength		m	
Wavenumber		m^{-1}	
Radiance		$W\ m^{-2}\ sr^{-1}$	
Spectral radiance	L_λ	$W\ m^{-2}\ sr^{-1}\ m^{-1}$	
	L_ν	$W\ m^{-2}\ sr^{-1}\ Hz^{-1}$	
	L_κ	$W\ m^{-2}\ sr^{-1}\ m$	
Irradiance		$W\ m^{-2}$	
Spectral irradiance	E_λ	$W\ m^{-2}\ m^{-1}$	
	E_ν	$W\ m^{-2}\ Hz^{-1}$	
	E_κ	$W\ m^{-2}\ m$	
Radiant exitance		$W\ m^{-2}$	
Spectral radiant exitance	M_λ	$W\ m^{-2}\ m^{-1}$	
	M_ν	$W\ m^{-2}\ Hz^{-1}$	
	M_κ	$W\ m^{-2}\ m$	
Radiant energy		J	
Radiant exposure		$J\ m^{-2}$	
Radiant flux		W	
Radiant flux density		$W\ m^{-2}$	
Absorption coefficient		m^{-1}	
Scattering coefficient		m^{-1}	
Extinction coefficient		m^{-1}	

Quantity	*Name of unit*	*Symbol*	*Definition*
Scattering angle		rad	
Scattering phase function		sr^{-1}	
Bidirectional reflectance		sr^{-1}	
Mean anomaly		rad	
Eccentric anomaly		rad	
True anomaly		rad	
Argument of perigee		rad	
Right ascension		rad	

The basic, supplemental, and named derived units may be prefixed to indicate order of magnitude as follows

Factor by which Unit is multiplied	*Prefix*	*Symbol*
10^{18}	exa	E
10^{15}	peta	P
10^{12}	tera	T
10^{9}	giga	G
10^{6}	mega	M
10^{3}	kilo	k
10^{2}	hecto	h
10^{1}	deka	da
10^{-1}	deci	d
10^{-2}	centi	c
10^{-3}	milli	m
10^{-6}	micro	μ
10^{-9}	nano	n
10^{-12}	pico	p
10^{-15}	femto	f
10^{-18}	atto	a

Several non-SI units are in use in satellite meteorology. Below are conversion factors for some of them.

To convert from	*To*	*Multiply by*	
angstrom	meter	1×10^{-10}	exact
atmosphere	pascal	1.01325×10^{5}	exact
bar	pascal	1×10^{5}	exact
British thermal unit	joule	1.055×10^{3}	
calorie (thermochemical)	joule	4.184	exact
day	second	8.64×10^{4}	exact

(*continues*)

To convert from	To	Multiply by	
degree (angle)	radian	1.745329×10^{-2}	
dyne	newton	1×10^{-5}	exact
erg	joule	1×10^{-7}	exact
foot	meter	3.048×10^{-1}	exact
hectare	meter^2	1×10^{4}	exact
hour (mean solar time)	second	3.6×10^{3}	exact
inch	meter	2.54×10^{-2}	exact
inch of mercury	pascal	3.386388×10^{3}	
knot	meter/second	5.144444×10^{-1}	
langley	joule/meter^2	4.184×10^{4}	exact
liter	meter^3	1×10^{-3}	exact
micron	meter	1×10^{-6}	exact
mile (nautical)	meter	1.852×10^{3}	exact
mile (statute)	meter	1.609344×10^{3}	exact
millibar	pascal	1×10^{2}	exact
millimeter of mercury	pascal	1.333224×10^{2}	
minute (angle)	radian	2.908882×10^{-4}	
minute (mean solar time)	second	6×10^{1}	exact
pound force (avoirdupois)	newton	4.4482216152605	exact
pound mass (avoirdupois)	kilogram	4.5359237×10^{-1}	exact
second (angle)	radian	$4.848136811 \times 10^{-6}$	
second (sidereal time)	second	9.9726957×10^{-1}	
ton (short, 2000 pound)	kilogram	$9.0718474\ 10^{2}$	exact
tonne (metric ton)	kilogram	1×10^{3}	exact

Bibliography

Mechtly, E. A. (1969). *The International System of Units: Physical Constants and Conversion Factors, Revised.* NASA SP-7012, Washington, DC.

Standard Practice for the Use of the International System of Units (the Modernized Metric System) (1993). American Society for Testing and Materials, ASTM E380-93, Philadelphia, PA, 35 pp.

Wallace, J. M., and P. V. Hobbs (1977). *Atmospheric Science: An Introductory Survey.* Academic Press, New York.

Wertz, J. R., and W. J. Larson (ed.) (1991). *Space Mission Analysis and Design.* Kluwer Academic Publishers, Dordrecht.

APPENDIX

E

Physical Constants

Constant	*Symbol*	*Value/Units*	*Relative Uncertainty*[a] (ppm)
		Universal Constants	
Speed of light in vacuum[b]	c	2.99792458×10^{8} m s^{-1}	exact
Newtonian gravitation constant[b]	G	6.67259×10^{-11} N m^{2} kg^{-2}	128
Planck's constant[b]	h	$6.6260755 \times 10^{-34}$ J s	0.60
Boltzmann's constant[b]	k	1.380658×10^{-23} J K^{-1}	8.5
First radiation constant[b,c]	c_1	$1.1910439 \times 10^{-16}$ W m^{2} sr^{-1}	0.60
Second radiation constant[b]	c_2	1.438769×10^{-2} m K	8.4
Stefan-Boltzmann constant[b]	σ	5.67051×10^{-8} W m^{-2} K^{-4}	34
Wein's displacement law constant[b]		2.897756×10^{-3} m K	8.4
Avogadro's number[b]	N_A	6.0221367×10^{23} mol^{-1}	0.59
Molar gas constant[b]	R^*	8.314510 J mol^{-1} K^{-1}	8.4
		Terrestrial Parameters	
Standard atmosphere[b]	atm	1.01325×10^{5} Pa	exact
Mean molecular weight of dry air[d]	W_M	2.8966×10^{-2} kg mol^{-1}	
Gas constant for dry air (R^*/W_M)	R	2.8704×10^{2} J kg^{-1} K^{-1}	

(*continues*)

Constant	Symbol	Value/Units	Relative Uncertainty[a] (ppm)
Standard acceleration of gravity[b]	g_o	9.80665 m s^{-2}	exact
Orbital constant[e]	Gm_e	$3.986005 \times 10^{14} \text{ m}^3 \text{ s}^{-2}$	0.15
Mass of Earth[f]	m_e	5.97370×10^{24} kg	128
Equatorial radius of Earth[e]	r_{ee}	6.378137×10^6 m	0.31
Polar radius of Earth[e]	r_{ep}	6.356752×10^6 m	0.31
Mean radius of Earth[e]	r_e	6.371009×10^6 m	0.31
Quadrupole gravitational coefficient of Earth[e]	J_2	1.08263×10^{-3}	2.7
Tropical year[d]		$3.15569259747 \times 10^7$ s	
Angular velocity of Earth[e]	$d\Omega_e/dt$	$7.292115 \times 10^{-5} \text{ rad s}^{-1}$	0.02
	Solar Parameters		
Solar constant[g]	S_{sun}	$1.368 \times 10^3 \text{ W m}^{-2}$	2600
Radius of sun (visible disk)[d]	r_{sun}	6.9595×10^8 m	1160
Mean Earth–sun distance[d]	$\bar{d}_{sun}$	1.4956×10^{11} m	470

[a] One standard deviation uncertainty in parts per million.

[b] From Cohen, E. R., and B. N. Taylor (1986). The 1986 Adjustment of the Fundamental Physical Constants. *CODATA Bulletin No. 63*, Pergamon Press, Elmsford, NY, and Oxford, U.K.

[c] We have divided the value given by Cohen and Taylor (1986) by π so that the Planck function yields radiance rather than radiant exitance.

[d] Weast, R. C. (ed.) (1980). *CRC Handbook of Chemistry and Physics*, 61st ed. CRC Press, Inc., Boca Raton, FL.

[e] Defense Mapping Agency (1987). *Supplement to Department of Defense World Geodetic System 1984 Technical Report: Part I—Methods, Techniques, and Data Used in WGS 84 Development.* DMA Tech. Rep. 8350.2-A, Building 56, U.S. Naval Observatory, Washington, DC.

[f] Derived from the product Gm_e.

[g] See Section 10.1 of text. The uncertainty is estimated as half of the range in the mean satellite-based measurements.

Index